Selenium and Tellurium Chemistry

J. Derek Woollins • Risto S. Laitinen
Editors

Selenium and Tellurium Chemistry

From Small Molecules to Biomolecules and Materials

Springer

Editors
J. Derek Woollins
University of St Andrews, UK
School of Chemistry
North Haugh
KY16 9ST St. Andrews
United Kingdom
jdw3@st-andrews.ac.uk

Risto S. Laitinen
University of Oulu
Dept. of Chemistry
PO Box 3000
90014 Oulu
Finland
risto.laitinen@oulu.fi

ISBN 978-3-642-43698-7 ISBN 978-3-642-20699-3 (eBook)
DOI 10.1007/978-3-642-20699-3
Springer Heidelberg Dordrecht London New York

Softcover re-print of the Hardcover 1st edition 2011

Printed on acid-free paper

Springer is part of Springer Science+Business Media (www.springer.com)

Preface

Both selenium and tellurium are very rare. Selenium has an abundance at 0.05–0.09 parts per million and is among the 25 least common elements in the Earth's crust. Tellurium is even rarer with an abundance of about one part per billion being rarer than gold, silver, or platinum and ranks about 75th in abundance of the elements in the earth. Selenium is used in glass-making and in electronics. One of the most common uses is in plain-paper photocopiers and laser printers. Selenium is also used to make photovoltaic ("solar") cells. Most of the tellurium produced today is used in alloys such as tellurium-steel alloy (approx. 0.04% tellurium), which has better machinability than steel without tellurium. Tellurium has potential for a variety of electrical devices such as CdHgTe IR detectors and it can be used to improve picture quality in photocopiers and printers.

Recent decades have witnessed significant progress in the chemistry of selenium and tellurium. New compounds with novel bonding arrangements, unprecedented structures, and unusual reactivities have been reported comprising in many cases systems which have been regarded as impossible. Such development extends the theories on molecular structures and bonding.

The driving force in the research of inorganic and organic chemistry of selenium and tellurium chemistry also arises from demands of materials science and from advances in biochemistry and medicine. There is an ever-growing need to gain firm understanding of the relationship of molecular and electronic structures with the properties and functionalities observed in the bulk phase. As a result of these investigations, both new and old materials are finding virtually unlimited number of applications. Semiconductors, insulators, coatings, ceramics, catalysts, nanotubes, polymers, and thin films all play a significant role in the current main group chemistry research as well as in the modern technological society. In this context, the increasing need for replacements for fossil fuels has driven forward the development of alternative energy sources in which selenium and tellurium compounds such as cadmium telluride are poised to play an important role. Many selenium and tellurium compounds have also found utility as reagents in synthetic organic and inorganic chemistry. Many selenium species can act as mild oxidants and conversely, organotellurium compounds have an ability to reduce different functional groups and cleave carbon-heteroatom bonds. Organotellurium ligands have also attracted interest in coordination chemistry, with the goal of designing suitable single source precursors for chemical vapor deposition processes.

The biological significance of selenium was recognised in 1973 when it was found to be an integral part of the enzyme glutathione peroxidase and is a very potent antioxidant protecting the body from damage due to oxidation by free radicals. The role of selenium compounds as antitumor agents is also under active investigation.

This volume illustrates some of the exciting developments in chemistry, materials and biochemistry of selenium and tellurium. The contributions are based on (but not limited to) the invited lectures in 11th International Conference on the Chemistry of Selenium and Tellurium (ICCST-11) held in Oulu, Finland on August 1–6, 2010. We are grateful to the contributing authors for their prompt delivery of the articles, which has made completing the volume in a timely fashion a pleasant experience.

March 2011

J. Derek Woollins
Risto S. Laitinen

Contents

List of Contributors

Eduardo E. Alberto Chemistry Department, Federal University of Santa Maria, Santa Maria, RS, Brazil

Jens Beckmann Fachbereich 2: Biologie/Chemie, Institut für Anorganische und Physikalische Chemie, Universität Bremen, Bremen, Germany

Antonio L. Braga Chemistry Department, Federal University of Santa Catarina, Florianópolis, SC, Brazil

Tristram Chivers Department of Chemistry, University of Calgary, Calgary, AB, Canada

Jason L. Dutton Department of Chemistry, The University of Western Ontario, London, ON, Canada

Pamela Finke Fachbereich 2: Biologie/Chemie, Institut für Anorganische und Physikalische Chemie, Universität Bremen, Bremen, Germany

Leopold Flohé Department of Chemistry, Otto-von-Guerricke-Universität Magdeburg, Magdeburg, Germany

Diana M. Freudendahl School of Chemistry, Cardiff University, Cardiff, UK

Richard S. Glass Department of Chemistry and Biochemistry, The University of Arizona, Tucson, AZ, USA

S. Hesse Laboratoire d'Ingénierie Moléculaire et Biochimie Pharmacologique, Institut Jean Barriol, Université Paul Verlaine Metz, Metz, France

Guoxiong Hua School of Chemistry, University of St Andrews, St Andrews, UK

G. Kirsch Laboratoire d'Ingénierie Moléculaire et Biochimie Pharmacologique, Institut Jean Barriol, Université Paul Verlaine Metz, Metz, France

Jari Konu Department of Chemistry, University of Calgary, Calgary, AB, Canada

Risto S. Laitinen Department of Chemistry, University of Oulu, Oulu, Finland

Chen-Wei Liu Department of Chemistry, National Dong Hwa University, Taiwan, China

Mohammad Azad Malik School of Chemistry, The University of Manchester, Oxford Road, Manchester, UK

Rüdiger Mews Institute for Inorganic and Physical Chemistry, University of Bremen, Bremen, Germany

Paul O'Brien School of Chemistry, The University of Manchester, Manchester, UK

Raija Oilunkaniemi Department of Chemistry, University of Oulu, Oulu, Finland

E. Perspicace Laboratoire d'Ingénierie Moléculaire et Biochimie Pharmacologique, Institut Jean Barriol, Université Paul Verlaine Metz, Metz, France

Paul J. Ragogna Department of Chemistry, The University of Western Ontario, London, ON, Canada

Karthik Ramasamy School of Chemistry, The University of Manchester, Manchester, UK

Thomas Wirth School of Chemistry, Cardiff University, Cardiff, UK

J. Derek Woollins School of Chemistry, University of St Andrews, St Andrews, UK

Andrey V. Zibarev Institute of Organic Chemistry, Russian Academy of Sciences, Novosibirsk, Russia; Department of Physics, National Research University – Novosibirsk State University, Novosibirsk, Russia

Chapter 1
Organic Phosphorus-Selenium Chemistry

Guoxiong Hua and J. Derek Woollins

1.1 Introduction

Organic phosphorus-selenium chemistry is a very fertile research area due to the wide variety of the compounds and their rich bioactivities, the potential applications in pharmaceutical industry, agrochemical industry and material science. The relatively high reactivity of P-Se containing compounds make them interesting intermediates or building blocks and thus leads to complicated and diverse molecules, expanding the horizon of organic phosphorus-selenium chemistry ever further.

In this review we will illustrate the reactivity and application of typical organic P-Se containing compounds, namely, selenotriphosphines $SePR_3$, secondary selenophosphines $SePHR_2$, selenophophates $SeP(OR)_3$, selenophosphonates $SeP(OR)_2R$, selenophosphinates $SeP(OR)R_2$, their derivatives and closely related compounds. It is difficult to draw a clear boundary line between these categories as there are many similarities in reactivity and common applications in the same area. For example, all of these systems have been involved in the synthesis of selenonucleotides, for the sake of fluency of the text, we arrange all the chemistry of selenonucleotides and related compounds in the section where it first appears. In recent years, 2,4-bis(phenyl)-1,3-diselenadiphosphetane-2,4-diselenide [$\{PhP(Se)(\mu\text{-}Se)\}_2$], the 'Woollins Reagent' (**WR**), has emerged as a powerful selenating reagent. Woollins Reagent has been commercialised by Sigma-Aldrich and found wide application in conversion of various organic substrates to organic phosphorus selenium compounds, in particular, phosphorus-selenium heterocycles and so we include discussion of this topic.

G. Hua • J.D. Woollins (✉)
School of Chemistry, University of St Andrews, St Andrews, Fife KY16 9ST, UK
e-mail: jdw3@st-and.ac.uke

J.D. Woollins and R.S. Laitinen (eds.), *Selenium and Tellurium Chemistry*,
DOI 10.1007/978-3-642-20699-3_1, © Springer-Verlag Berlin Heidelberg 2011

1.2 $R_3P{=}Se$ and Related Compounds

Selenotriphosphines are usually obtained from the oxidation of tertiary phosphines by elemental selenium. Contrary to their stable oxygen counterparts $R_3P{=}O$, which usually means a byproduct or the dead end of a reaction, selenotriphosphines, $R_3P{=}Se$, show modest reactivity due to the liability of the P=Se bond and good solubility in organic solvents. The large size of the selenium atom means not only that the orbital overlap between the phosphorus and selenium is less efficient and thus the P=Se bond is weak but also that the selenium atom is easy to polarise and vulnerable to attack. Therefore, selenotriphosphines can be used as reactants to deliver either the selenium atom or the triphosphine unit to a more complicated molecule. For example, they have been used as a synthetic building block for the construction of bicyclic cage compounds [1]. The direct tripod-tripod coupling of equi-molar amount of tris(2-azidobenzyl)amine and triphosphine $[P(Se)(CH_2PPh_3)_3]$ afforded the macrobicyclic tri-λ^5-phosphazene in 37% yield. The same compound could be also obtained in a much better yield (84%) from the reaction of the macrobicyclic triphosphazide, derived from tri(2-azidobenzyl)amine with equi-molar amount of selenotriphosphine $[P(Se)(CH_2PPh_3)_3]$ by means of a triphosphine exchange followed by a triple expulsion of molecular N_2 (Scheme 1.1).

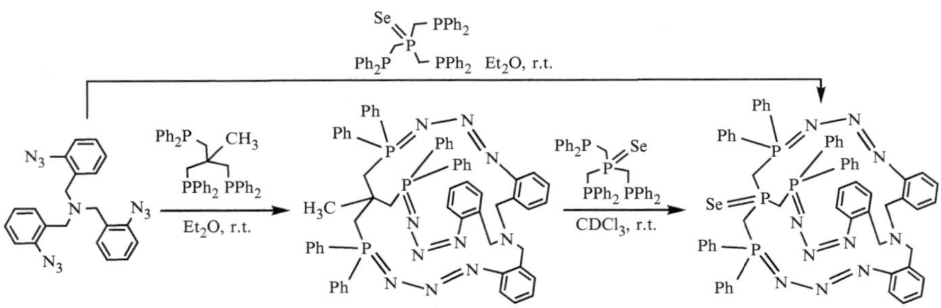

Scheme 1.1 Synthesis of the macrobicyclic tri-λ^5-phosphazene by triphosphane exchange and by tripod-tripod coupling

Selenotriphosphines are also involved in the efficient synthesis of heterocyclic λ^5-phospinines. 5-(dimethylamino)-2-(diphenylphosphoroselenoyl)-3-(pyrrolidin-1-yl)penta-2,4-dienenitrile; a selenotriphosphine generated from 3-(pyrrolidin-1-yl)but-2-enenitrile [2], was deselenated with $P(NMe_2)_3$ in benzene leading to 5-(dimethylamino)-2-(diphenylphosphino)-3-(pyrrolidin-1-yl)penta-2,4-dienenitrile, the latter continued reacting *in situ* with bromoacetophenones at room temperature giving rise to λ^5-phospinines. The cyclization reaction proceeds spontaneously, followed by elimination of dimethylamine (Scheme 1.2) [3].

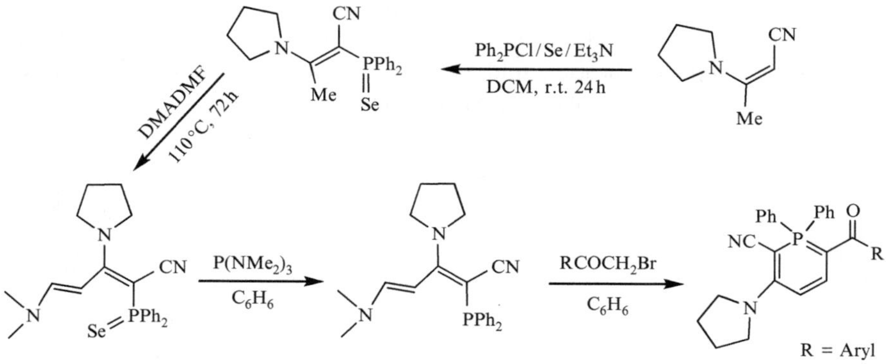

Scheme 1.2 Synthesis of Heterocyclic λ^5-phosphinines

The stereoselective hetero Diels-Alder reaction of pentavalent 3,4-dimethyl-phosphole selenides with aromatic selenoaldehydes, generated *in situ* by thermal retro Diels-Alder reaction of anthracene cycloadducts, leads to the corresponding [4 + 2] cycloadducts as a single diastereoisomer in good to excellent yield (Scheme 1.3) [4].

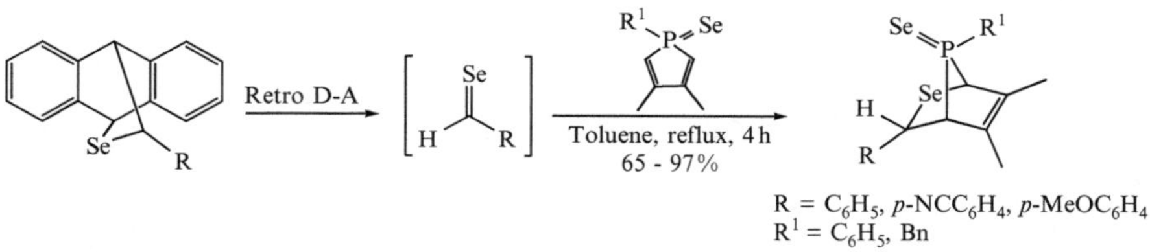

Scheme 1.3 Hetero Diels-Alder reaction of pentavalent 3,4-dimethylphosphole selenides with aromatic selenoaldehydes

1.3 $SePHR_2$

Secondary phosphine selenides are a group of excellent nucleophilic reagents. The nucleophilic addition of the selenides to aldehydes leads to α-hydroxyphosphines. For instance, secondary phosphines, easily prepared from red phosphorus and styrenes or 2-vinylnaphthalene in one step in superbase systems [5, 6], have been used for the synthesis of water soluble and hydrophilic organophosphorus ligands for use in metal catalysts in phase transfer reactions [7, 8]. The addition of this secondary phosphine selenide to 3,7-di-methyl-2,6-octadienal gave polyfunctional tertiary phosphine selenides with diene and hydroxyl moieties (Scheme 1.4) [9].

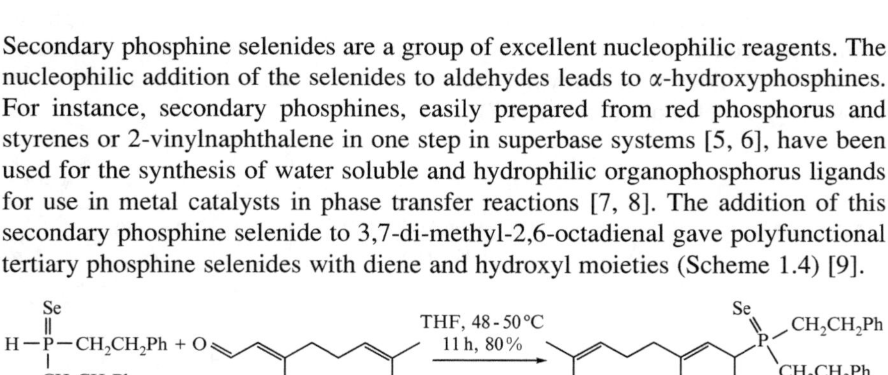

Scheme 1.4 Synthesis of polyfunctional tertiary phosphine selenide from secondary phosphine selenide and 3,7-di-methyl-2,6-octadienal

The nucleophilic addition of bis[2-(2-pyridyl)-ethyl]phosphinoselenide to 2-formyl-1-organylimidazoles and benzimidazoles generates a series of functional heterocyclic compounds bearing imidazole, benzimidazole or pyridine rings as well as hydroxyl and selenophosphoryl groups (Scheme 1.5) [10].

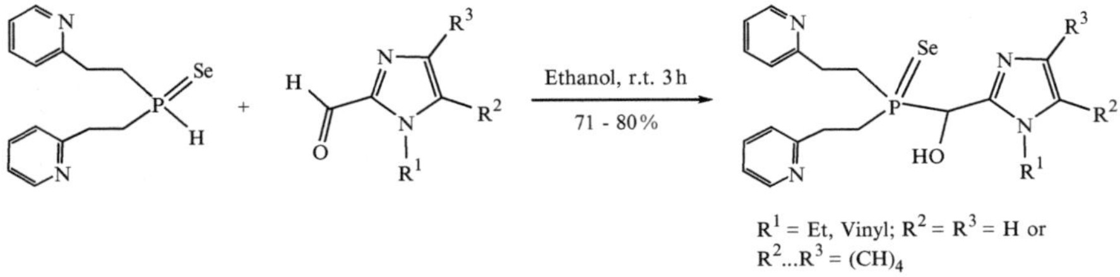

Scheme 1.5 Synthesis of polyfunctional tertiary phosphine selenides from secondary phosphine selenide and imidazoles

Secondary phosphinoselenides can also been added to vinyl ether through free radical addition. Regiospecific addition of secondary phosphinoselenides to 2-[(vinyloxy)methyl]furan, 2-[1-(vinyloxy)butan-2-yl]furan and 2-[(vinyloxy) methyl]tetrahydrofuran proceeds under UV irradiation to afford the corresponding anti-Markonikov adducts (Scheme 1.6) [11].

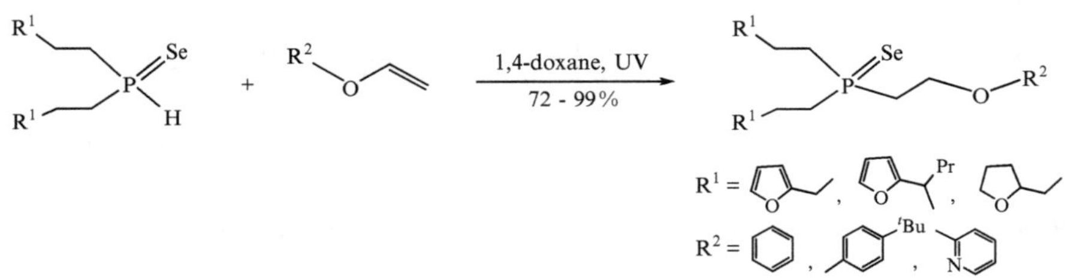

Scheme 1.6 Free radical addition of secondary phosphinoselenides and vinyl ether

The same addition has been used for the preparation of organophosphorus compounds containing two or more phosphine moieties [12]. Treating secondary phosphinoselenides with tetravinyl ether of pentaerythritol furnished a series of tetraphosphinoselenides (Scheme 1.7).

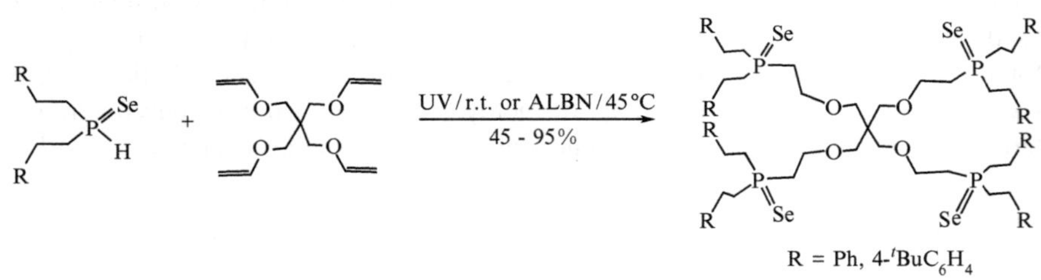

Scheme 1.7 Synthesis of tetraphosphinoselenides from secondary phosphinoselenides with tetravinyl ether of pentaerythritol

The free radical addition of secondary phosphinoselenides to aromatic and heteroaromatic acetylenes is stereoselective, leading to a series of vinylphosphine selenides in predominantly *Z*-configuration (up to 97%) as *anti*-Markovnikov adducts. On the other hand, the microwave irradiation of the reactants with the same content of AIBN reduces the reaction time from 5–7 h to 8 min, however, the stereoselectivity is lost ($Z : E = 52 : 48$) (Scheme 1.8) [13].

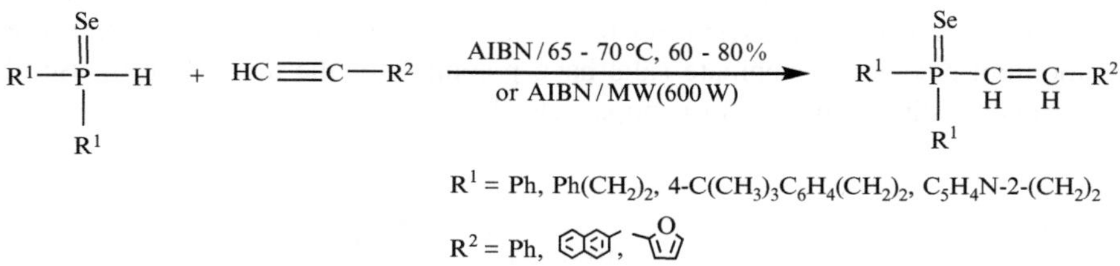

Scheme 1.8 Free radical addition of secondary phosphinoselenides to aromatic and heteroaromatic acetylenes

Under strong basic condition, the proton on the phosphorus atom in the secondary phosphine selenides can be removed, the resulting phosphorus anions then attack a suitable electrophilic substrate to give triphosphine selenides. For example, bis(2-phenethyl)phosphine selenide has been methylated by methyl iodide in THF in the presence of potassium hydroxide (Scheme 1.9) [14].

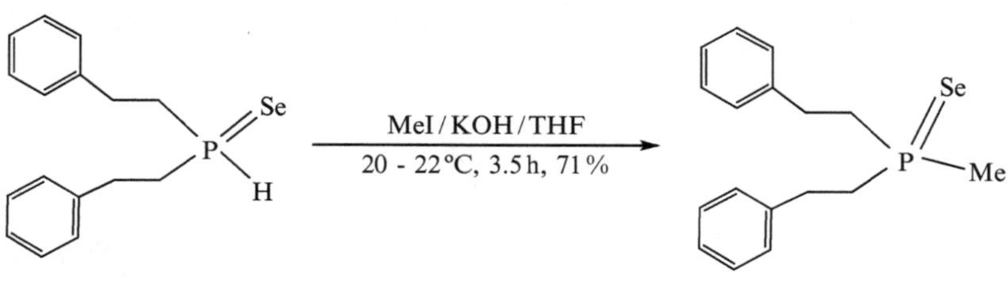

Scheme 1.9 Methylation of secondary phosphinoselenides

A more functionalized triphosphine has been achieved through the nucleophilic attack of the deprotonated diphenethylphosphine selenide to the vinyl sulfoxide in dioxane. Reduction of vinylphosphine selenide by sodium in toluene led to the formation of bis(2-phenethyl)vinylphosphine (Scheme 1.10). Both vinylphosphinoselenide and vinylphosphine are highly reactive building blocks which are potentially capable of interacting with different reagents to afford more interesting products [15].

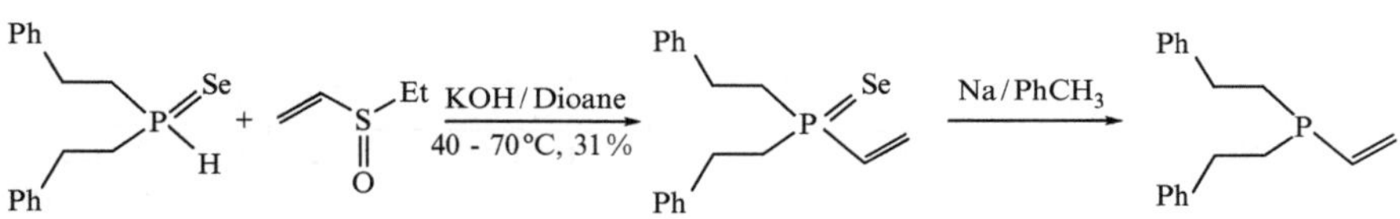

Scheme 1.10 Formation of bis(2-phenethyl)vinylphosphine from secondary phosphinoselenides and vinyl sulfoxide

Secondary phosphinoselenides can also be converted into diselenophosphonates. For instance, a series of mono-, *di-* and triorganoammonium salts of diorganodiselenophosphinates has been prepared by a simple three-component reaction of secondary phosphine selenides [$R_2P(Se)H$, R = $PhCH_2CH_2$, $PhCH(Me)CH_2$, 4-$^tBuC_6H_4CH_2CH_2$, (2-methyl-5-pyridyl)CH_2CH_2, Ph] with elemental selenium and primary, secondary, tertiary amines or diamines (Schemes 1.11 and 1.12) [16]. The synthesis of ammonium salts of diselenophosphinates of lupinine or anabasine has also been carried out in the same way (Scheme 1.13) [17]. These ammonium diselenophosphinate salts have potential for the deposition of useful materials as either thin films or nanoparticles [18].

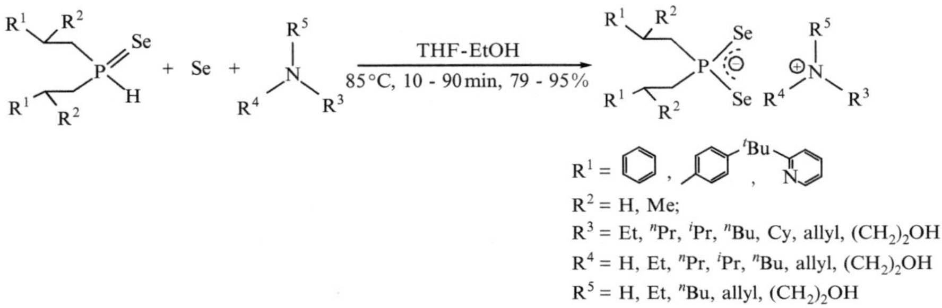

Scheme 1.11 Synthesis of mono-, *di-* and tri-ammonium salts of diselenophosphinates from secondary phosphinoselenides

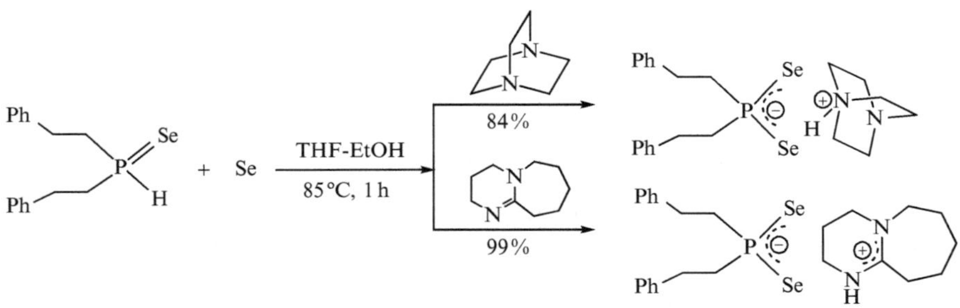

Scheme 1.12 Synthesis of ammonium salts of diselenophosphinates from secondary phosphinoselenides and tertiary diamine

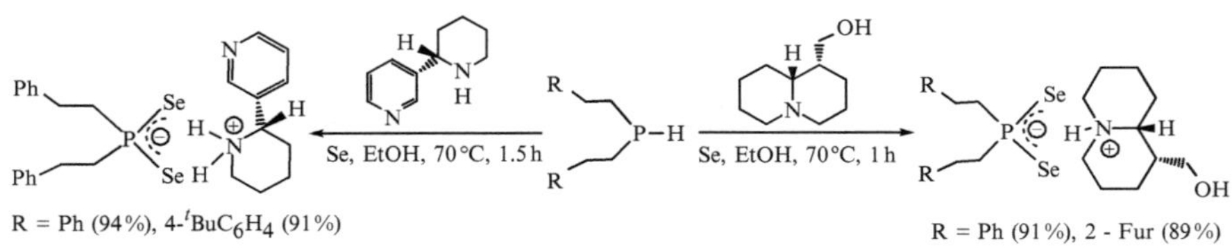

Scheme 1.13 Synthesis of ammonium salts of diselenophosphinates of lupinine or anabasine from secondary phosphinoselenides

H-phosphonoselenate, a structurally similar compound to the above secondary diphenethylphosphine selenides, has also been successfully used for oxidative phosphorylation of *3′-O*-TBDMS-thymidine to produce dinucleoside phosphoroselenoate with a modified *3′–5′* internucleotide linkage (Scheme 1.14) [19, 20] in attempts to get nucleotide analogues bearing single or multiple modification at the phosphorus center (such as phosphoroselenoates) [21].

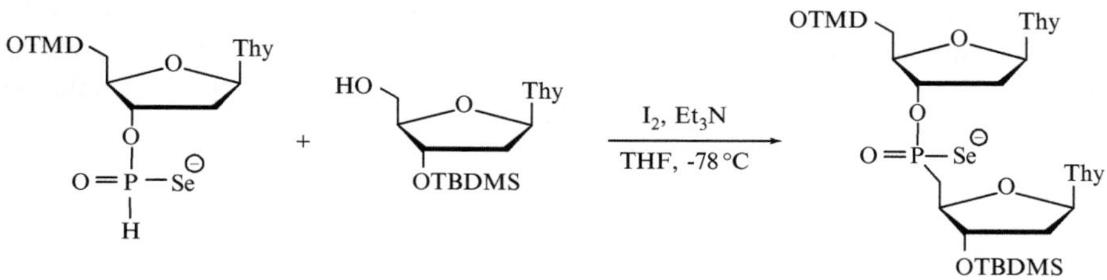

Scheme 1.14 Synthesis of dinucleoside phosphoroselenoate from *H*-phosphonoselenate

Similarly modified dinucleosides such as 3′-*O*-thymidylyl(5′-deoxy-5′-selene-thymidylyl)-*Se*-phosphoroselenolate, its *O*-methyl ester and methanephosphonate derivatives were obtained from alkylation of 5′-*O*-protected nucleoside 3′-*O*-(*O*-alkylphosphoroselenoates) and 5′-*O*-protected nucleoside 3′-(methanephosphonoselenoates). After deprotection, 5′-deoxy-5′-selene dinucleoside *Se*-phosphates and *Se*-phosphonates were formed in good yields (Scheme 1.15) [22]. The incorporation of phosphoroselenoate in DNA and RNA strands (Scheme 1.16) [23] has also been studied and the modification showed very high selectivity against point mutations. The selenium autoligation might prove useful in diagnostic strategies for direct analysis of DNAs and RNAs.

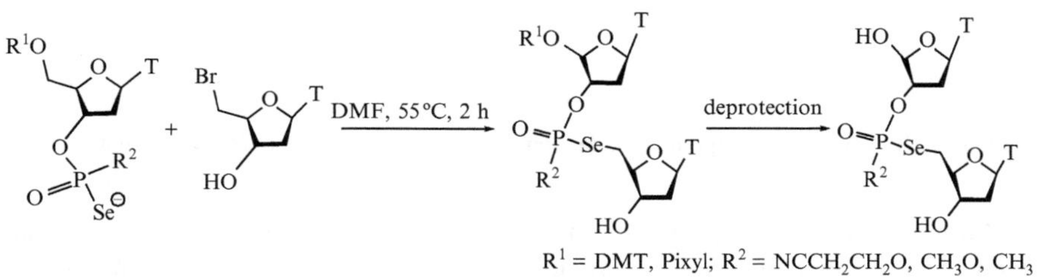

Scheme 1.15 Alkylation and deprotection of nucleoside 3′-*O*-(*O*-alkylphosphoroselenoates) and nucleoside 3′-(methanephosphonoselenoates)

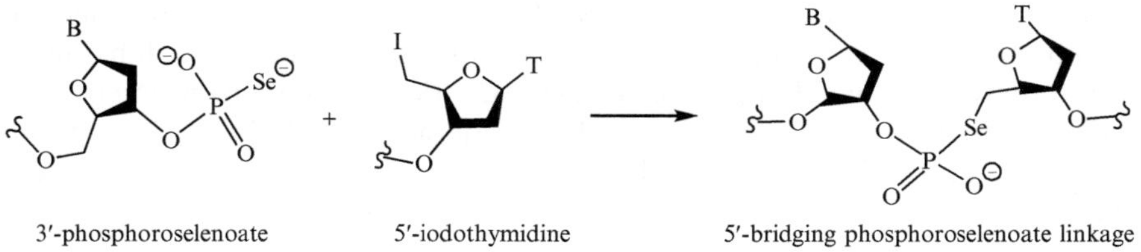

Scheme 1.16 Chemistry of phosphoroselenoate autoligation of DNA strands

Due to the scattering ability of the selenium atom, selenium derivatives have been used for multi-wavelength anomalous dispersion (MAD) phasing in X-ray crystallography [19, 24, 25] of biomacromolecules such as ribozymes, viral RNA and RNA-proton and DNA-proton complexes via replacement of sulfur with selenium in methionine residues [26–28], or replacement of a non-bridging oxygen atom of the internucleotide linkages with selenium [29–32]. Typical is the synthesis of oligonucleotide selenophosphate diesters, which involve the oxidation transfer of selenium atom to P^{III} centers, such as phosphite triesters or *H*-phosphonate diesters by using the selenium transfer reagent triphenylphosphine selenide $Ph_3P(Se)$ and its polystyryl diphenylphosphine analogue (Schemes 1.17 and 1.18) [29–32].

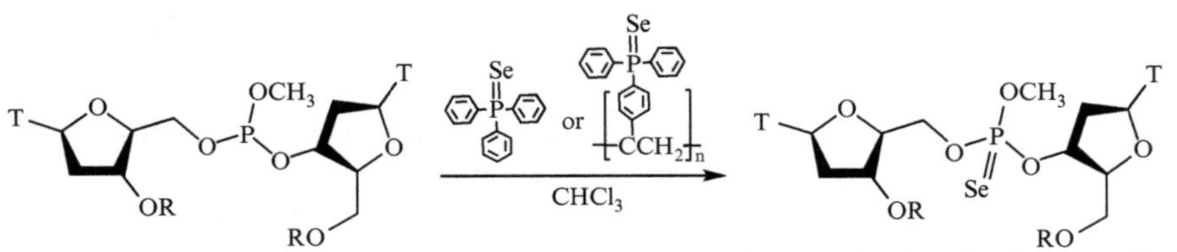

R = 4, 4′-dimethoxtrity, T = thymin-1-yl

Scheme 1.17 Synthesis of oligonucleotide selenophosphate diesters from phosphite triesters

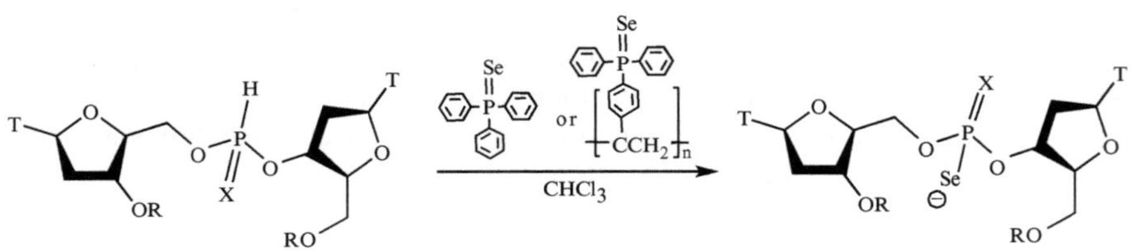

X = O, R = 4, 4′-dimethoxtrity, T = thymin-1-yl
X = S, R = 4, 4′-dimethoxtrity, T = thymin-1-yl

Scheme 1.18 Synthesis of oligonucleotide selenophosphate diesters from *H*-phosphonate diesters

Another modified approach to oligonucleotide selenophosphonates has been carried out using *Se*-(2-cyanoethyl)phthalimide as the selenium transfer reagent [33]. *Se*-(2-cyanoethyl)phthalimide can be easily obtained from the selenation reaction of potassium phthalimide with *di*-(2-cyanoethyl) diselenide. The coupling of 5’-*O*-DMTr-thymidine *H*-phosphonate with 3’-*O*-phenoxyacethymidine in the presence of pivaloyl chloride provides dinucleotide *H*-phosphonate ester, which reacts *in situ* with *Se*-(2-cyanoethyl)phthalimide to afford *Se*-(2-cyanoethyl)phosphoroselenoate triester. Subsequent removal of the protecting group DMTr with dichloroacetic acid and $CH_2CH_2C{\equiv}N$ group with DBU gave oligonucleotide phosphoroselenoate diesters in excellent yield (Scheme 1.19).

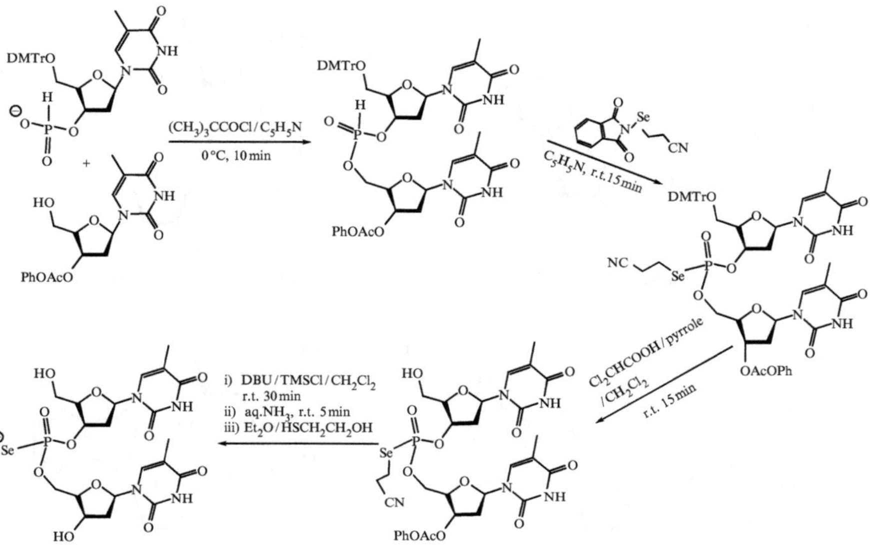

Scheme 1.19 Synthesis of oligonucleotide phosphoroselenoate diesters by using *Se*-(2-cyanoethyl)phthalimide

The introduction of a selenium atom into nucleic acids can also be accomplished by solid phase synthesis using potassium selenocyanide. The approach has been successfully used to assist crystallographic analysis of DNA hexamers containing a single phosphoroselenoate group linkage (Scheme 1.20) [19].

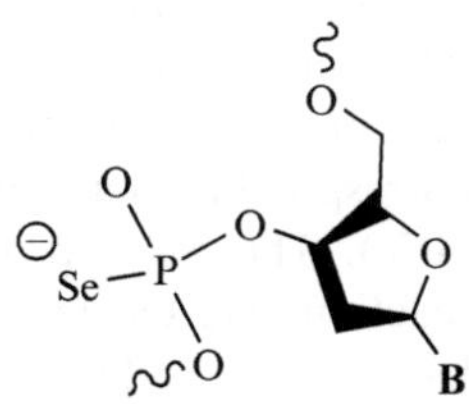

Scheme 1.20 Single phosphoroselenoate group linkage

The enzymatic synthesis of phosphoroselenoate DNA has been reported using thymidine 5′-(α-*P*-seleno)triphosphate and DNA polymerase for X-ray crystallography *via* MAD [34]. Also published was the enzymatic synthesis of phosphoroselenoate nucleic acids (phosphoroselenoate RNA) containing a selenium atom that replaced one of the bridging oxygen atoms on the phosphate group by *in vitro* transcription with T7 RNA polymerase and adenosine 5′-(α-*P*-seleno)triphosphate (ATPαSe) for X-ray structure studies (Scheme 1.21) [35].

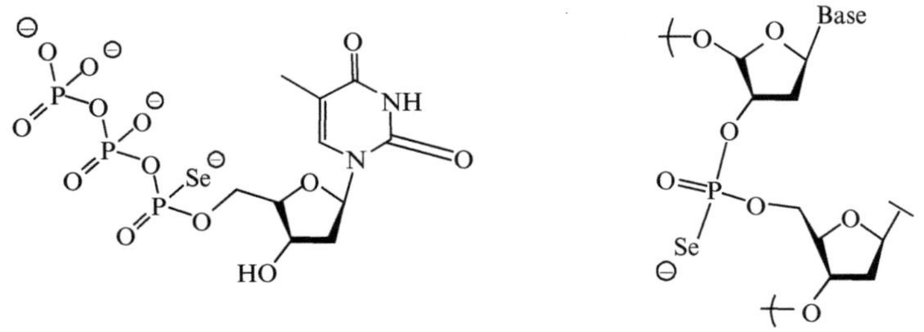

***a*-Se-TTP** **Phosphoroselenoate DNA**

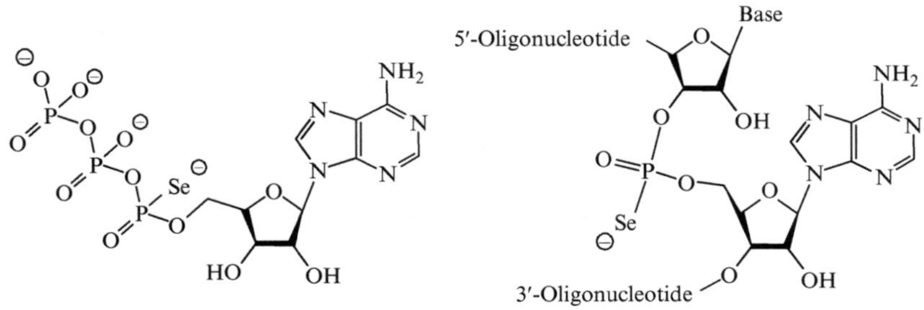

ATP*a*Se **Phosphoroselenoate RNA**

Scheme 1.21 Chemical structures of *a*-Se-TTP, phosphoroselenoate DNA, ATP*a*Se and phosphoroselenoate RNA

1.4 $SeP(OR)R_2$ and Related Compounds

Selenophosphinic chloride is the primary starting material for selenophosphinate and analogues. Diphenylselenophosphinic chloride reacts with *trans*-2,5-disubstituted pyrrolidine or *trans*-2,4-disubstituted azetidine in acetonitrile in the presence of potassium carbonate at room temperature leading to the corresponding carbon dioxide inserted products in moderate yield (60% and 40%). The same products can be obtained in 80% and 70% yields when the reactions are carried out under CO_2 atmosphere using potassium hydroxide as a base with the addition of 0.1 equiv of 18-crown-6-ether (Scheme 1.22) [36].

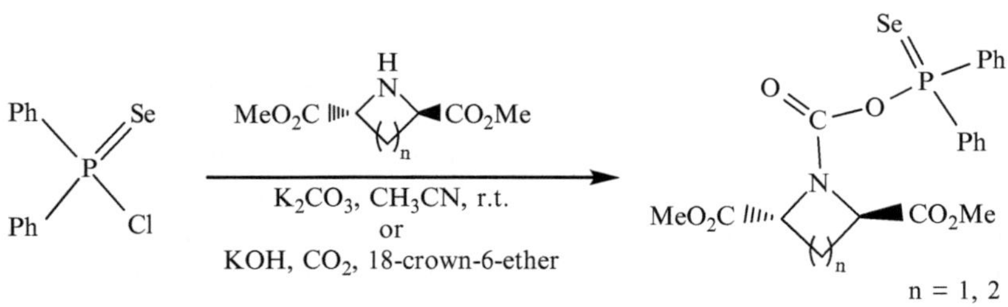

Scheme 1.22 Formation of selenophosphinates from selenophosphinic chloride and *trans*-2,5-disubstituted pyrrolidine or *trans*-2,4-disubstituted azetidine

Racemic and optically active *P*-chiral phosphinoselenoic chlorides [37, 38] can be synthesized by reacting $PhPCl_2$ with one equivalent of Grignard reagents and elemental selenium or PCl_3 with two equivalent of different Grignard reagents successively and one equivalent of elemental selenium (46–94% yields) [39]. The reaction of *P*-chiral phosphinoselenoic chlorides with alkali metal alkoxide and chalcogenolates led to a series of *P*-chiral phosphinochalcogenoselenoic acids esters bearing both a P=Se double bond and a P-E single bond. Similar products could be obtained from *P*-chiral phosphinoselenoic chlorides and alkyl iodides in the presence of sodium hydroxide, sulphide and selenide (Scheme 1.23).

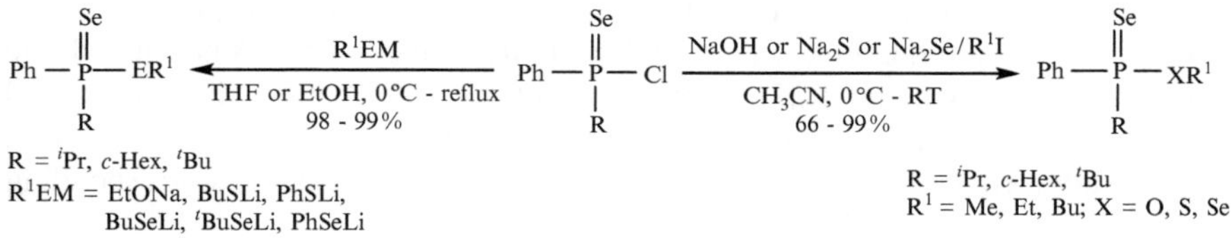

Scheme 1.23 Synthesis of *P*-chiral phosphinochalcogenoselenoic acids esters

Interaction of racemic *P*-chiral phosphinoselenoic chlorides with optically active lithium amides resulted in a two diastereomeric mixture of phosphinoselenoic amides in ratios of 35:65–58:42 (Scheme 1.24) [40]. Two optically active *P*-chiral phosphinoselenoic amides (R_p, *S*) and (S_p, *S*) have been obtained in 34–56% isolated yields.

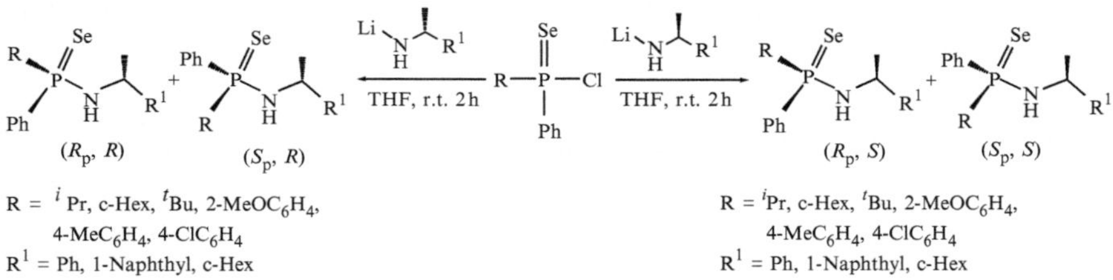

Scheme 1.24 Synthesis of optically active *P*-chiral phosphinoselenoic amides (R_p, *S*) and (S_p, *S*) from racemic *P*-chiral phosphinoselenoic chlorides and optically active lithium amides

The reaction of racemic *P*-chiral phosphinoselenoic chlorides with elemental selenium and various organolithium at 0°C or organomagnesium reagents at room temperature afforded the corresponding phosphinoselenoic acid *Se*-esters and phosphine selenides (Scheme 1.25) [41].

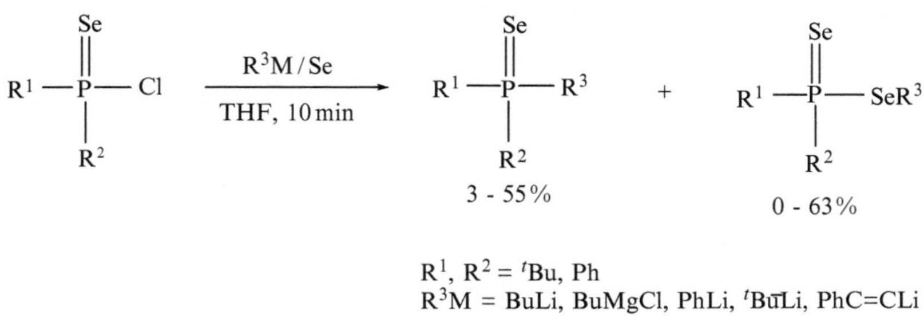

Scheme 1.25 Synthesis of phosphinoselenoic acid *Se*-esters and phosphine selenides from racemic *P*-chiral phosphinoselenoic chlorides

More studies have concentrated on phosphinoselenothioic acids and their salts with electrophiles [42]. Phosphinoselenothioic acid tetrabutylammonium salts were prepared in excellent yields by reacting phosphinoselenothioic acid *S*-[2-(trimethylsil)ethyl] esters with ammonium fluorides Bu_4NF or Me_4NF (Scheme 1.26). Phosphinoselenothioic acid alkali metal salts were obtained as 18-crown-6-ether complexes from the esters with alkali metal fluorides and 18-crown-6-ether (Scheme 1.26). Alkylation of these phosphinoselenothioic acid ammonium salts with alkyl or allyl halides selectively generates phosphinoselenothioic acid *Se*-esters (only minor *S*-esters are found), whereas acylation of these phosphinoselenothioic acid ammonium salts preferentially gives phosphinoselenothioic anhydrosulfides (Scheme 1.27). Protonation of the salts with hydrochloric acid affords selectively phosphinoselenothioic acids, the latters react *in situ* with α, *β*-unsaturated carbonyl compounds and epoxide to yield the corresponding *S*-esters and *Se* esters (Scheme 1.28).

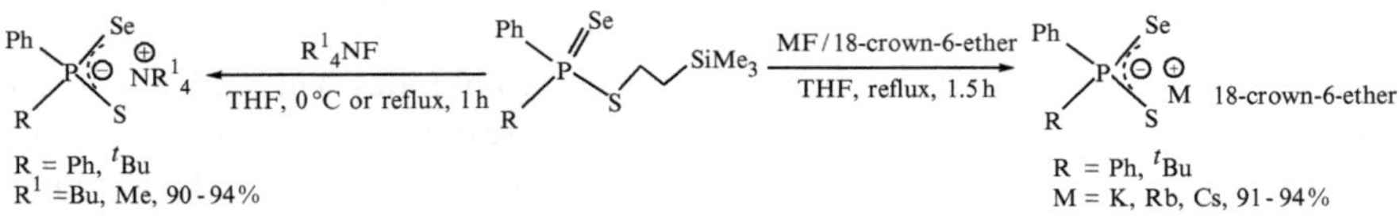

Scheme 1.26 Synthesis of phosphinoselenothioic acid tetrabutylammonium salts from phosphinoselenothioic acid *S*-[2-(trimethylsil)ethyl] esters and ammonium fluorides Bu_4NF or Me_4NF or alkali metal fluorides and 18-crown-6-ether

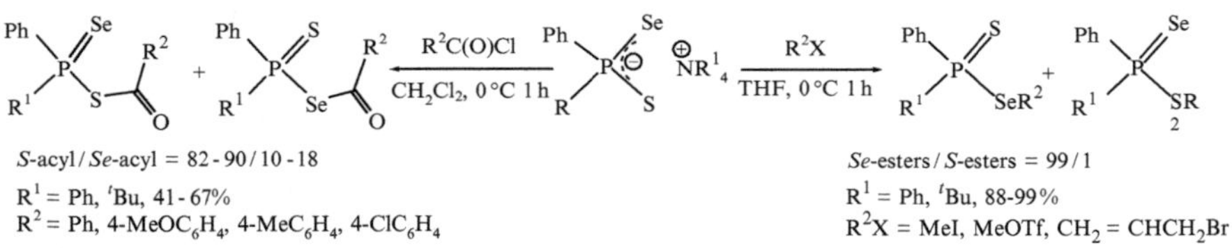

Scheme 1.27 Alkylation and acylation of phosphinoselenothioic acid ammonium salts

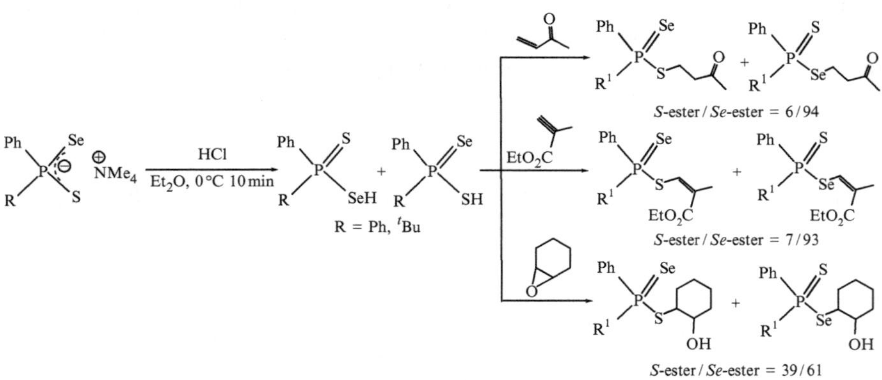

Scheme 1.28 Synthesis of *S*-esters and *Se* esters from phosphinoselenothioic acid ammonium salts

Selenophosphinate has also been incorporated into fluorescein as a fluorescent probe for mercury. In this design, fluorescein was used as a fluorophore source and the *p*-phenylphosphinoselenoic group as a reactive warhead targeting the mercury ions. Organoselenium fluorescence probe (FSe-1) was synthesized by the reaction of fluorescein with diphenylchlorophosphine followed by oxidation with elemental selenium (Scheme 1.29). The deselenation of FSe-1 by Hg^{2+} in aqueous solution lead to the formation of HgSe and the starting fluorescent agent fluorescein, the latter can be reused in next cycle [43].

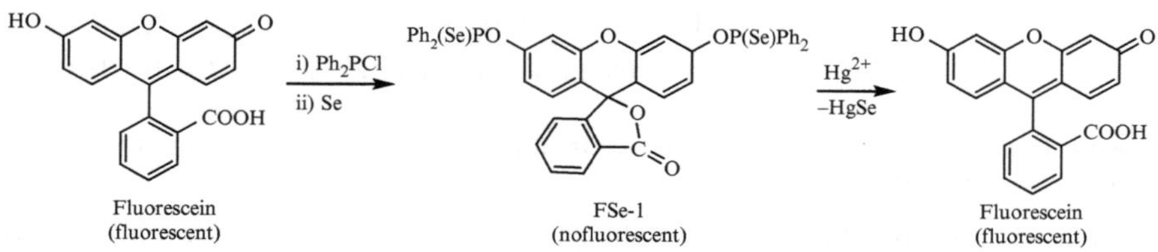

Scheme 1.29 Synthesis of organoselenium fluorescence probe (FSe-1) from fluorescein and diphenylchlorophosphine and elemental selenium

1.5 [SeP(XR)$_3$, X = O, S, Se] and Related Compounds

Selenophosphate accounts for a big slice in the pie of organic phosphorus selenium chemistry. These compounds usually start from phosphoroselenoyl chloride, which can be achieved through the reaction of PCl_3, elemental selenium and corresponding alcohol or phenol. For example, 1′-binaphthyl-2,2′-dyl phosphoroselenoyl chloride, an active compound derived from axial chiral 1,1′-binaphthyl-2,2′-diol, has been widely studied for the syntheses of chiral organics such as amines or alcohols or diselenides *via* corresponding phosphoroselenoyl amides, esters or ammonium salts [44, 45].

Reaction of 1′,1′-binaphthyl-2,2′-dyl phosphoroselenoyl chloride with primary and cyclic secondary amines affords the corresponding phosphoroselenyl amides in 3 h, whereas, with acyclic secondary amines more than 18 h is needed to give the corresponding amides due to the steric congestion around the nucleophilic N atom (Scheme 1.30). All amides products comprise of two diastereoisomers in different ratios [46]. Further extrusion reaction of selenium from these amides by use of

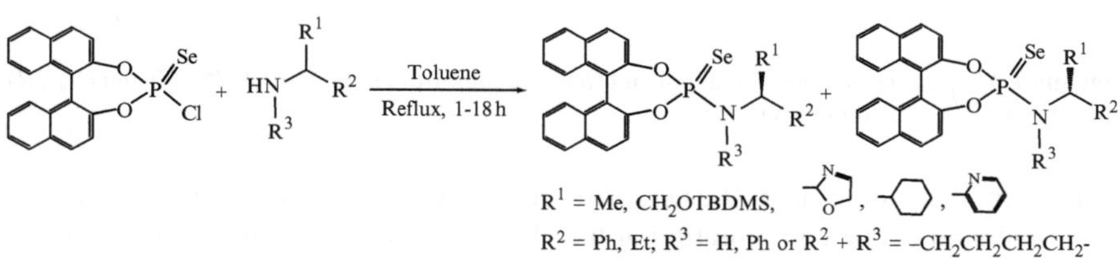

Scheme 1.30 Synthesis of phosphoroselenyl amides from 1′,1′-binaphthyl-2,2′-dyl phosphoroselenoyl chloride with primary and cyclic secondary amines

nucleophilic trivalent phosphorus PBu_3 led to the corresponding phosphoramidites (Scheme 1.31) [46].

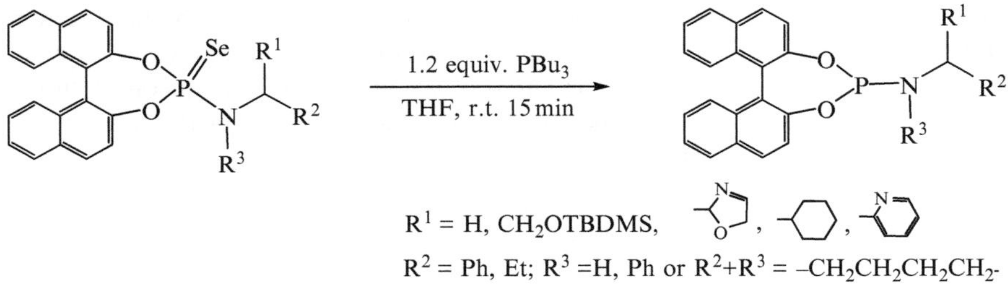

Scheme 1.31 Formation of phosphoramidites from phosphoroselenyl amides and nucleophilic trivalent phosphorus PBu_3

Treating 1,1′-binaphthyl-2,2′-dyl phosphoroselenoyl chloride with primary and secondary alcohols in the presence of Et_3N in refluxing toluene gave a series of phosphoroselenoic acid *O*-esters with in almost equal ratios of two diastereomers in modest to excellent yield (Scheme 1.32) [44, 46]. Selective reduction at phosphorus atom of some esters by butyllithium led to the corresponding enantiomerically pure products, e.g. reducing (R_{ax}, *S*) and (R_{ax}, *R*) phosphoroselenoic acid *O*-octylesters afforded (*S*)-3-octanol and (*R*)-3-octanol (Scheme 1.33).

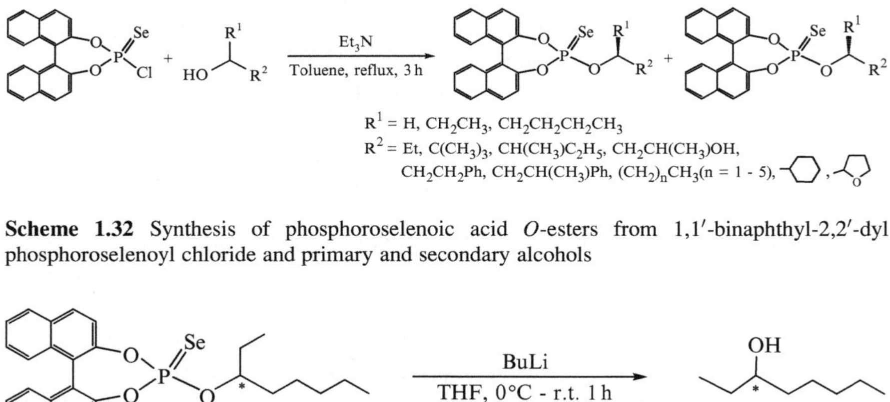

Scheme 1.32 Synthesis of phosphoroselenoic acid *O*-esters from 1,1′-binaphthyl-2,2′-dyl phosphoroselenoyl chloride and primary and secondary alcohols

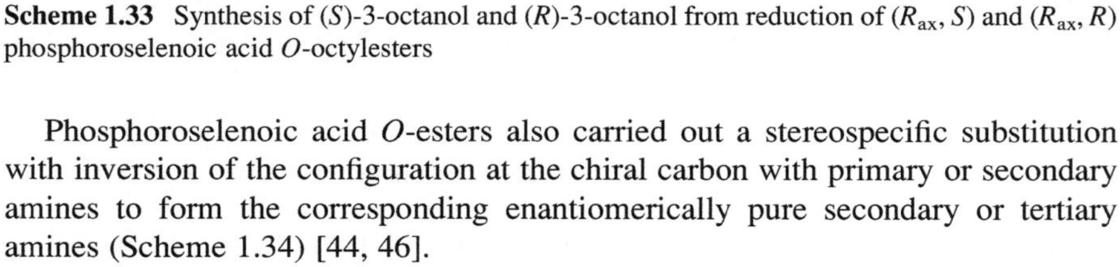

Scheme 1.33 Synthesis of (*S*)-3-octanol and (*R*)-3-octanol from reduction of (R_{ax}, *S*) and (R_{ax}, *R*) phosphoroselenoic acid *O*-octylesters

Phosphoroselenoic acid *O*-esters also carried out a stereospecific substitution with inversion of the configuration at the chiral carbon with primary or secondary amines to form the corresponding enantiomerically pure secondary or tertiary amines (Scheme 1.34) [44, 46].

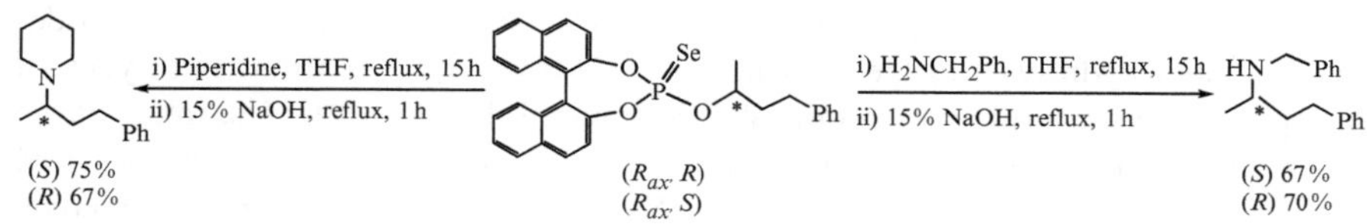

Scheme 1.34 Substitution of phosphoroselenoic acid *O*-esters

Apart from forming amide and ester, 1′,1′-binaphthyl-2,2′-dyl phosphoroselenoyl chloride can be hydrolysed into their ammonium salts in the presence of a tertiary amine in excellent yield. Alkylation of the ammonium salt with racemic alkyl halides in common organic solvents like THF, DCM or toluene generated the 1′,1′-binaphthyl-2,2′-dyl phosphoroselenoic acid *Se*-ester exclusively as diastereomers (Scheme 1.35). Cleavage of the 1,1′-binaphthyl-2,2′-dyl phosphoroselenoic acid *Se*-esters with BuN_4F provides optically active dialkyl diselenides (Scheme 1.36) [47].

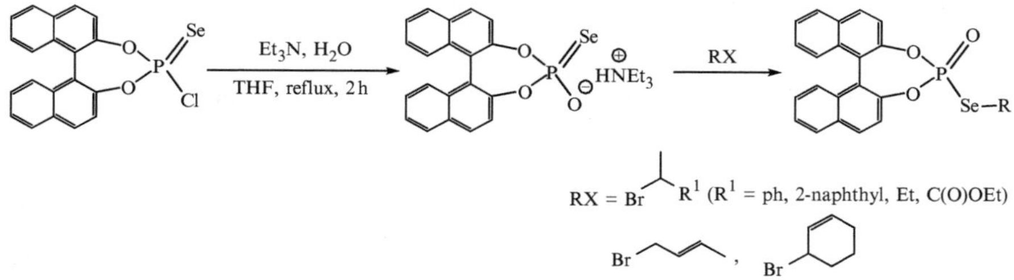

Scheme 1.35 Synthesis of 1′,1′-binaphthyl-2,2′-dyl phosphoroselenoic acid *Se*-ester from ammonium hydrolysis and alkylation of 1′,1′-binaphthyl-2,2′-dyl phosphoroselenoyl chloride

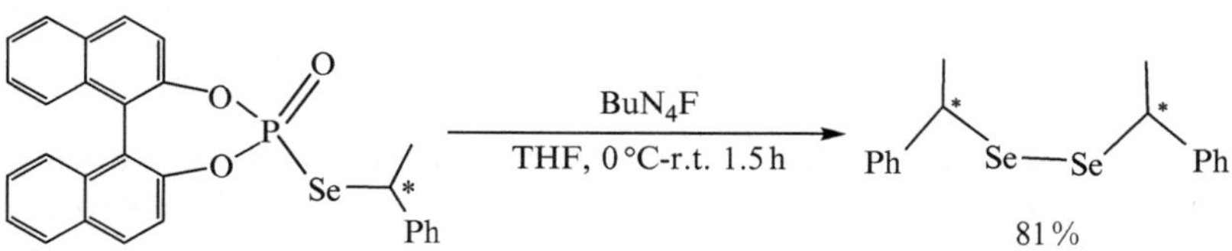

Scheme 1.36 Synthesis of optically active dialkyl diselenides from 1,1′-binaphthyl-2,2′-dyl phosphoroselenoic acid *Se*-esters

Phosphoroselenyl amides and esters may be converted into other derivatives. For instance, primary selenophosphoramides could be acylated at the nitrogen in modest yield in the presence of sodium hydride in THF (Scheme 1.37) [48].

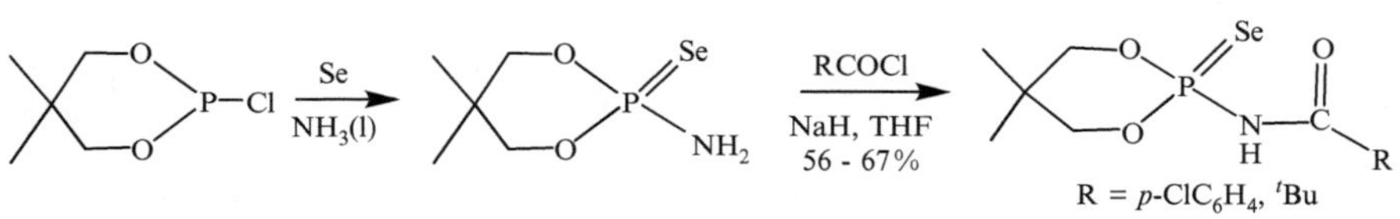

Scheme 1.37 Acylation of phosphoroselenyl amides

In another example, phosphoroselenyl amides and esters proceeded, via a fluoride-ion-mediated hydrolysis in the presence of TBAF in THF, to afford two types of phosphoroselenoic acid ammonium salts (Schemes 1.38 and 1.39). The ratios of the two ammonium salts formed are dependent on the substitutents on the oxygen atom of the ester or the nitrogen atom of the amides. These derivatives have been confirmed by the synthesis of the corresponding methyl esters by further reacting with MeI [49].

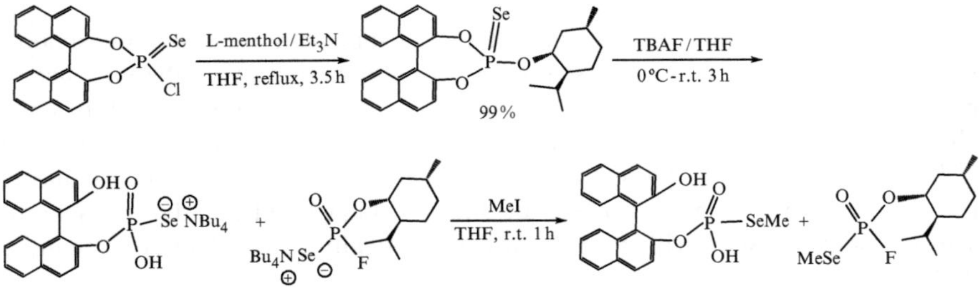

Scheme 1.38 Fluoride-ion-mediated hydrolysis of phosphoroselenyl amides

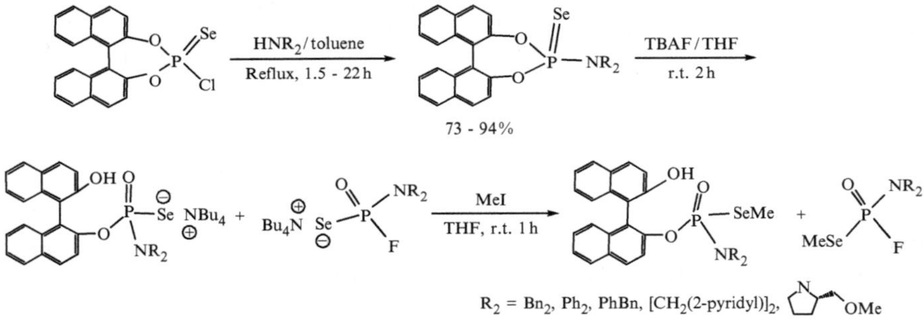

Scheme 1.39 Fluoride-ion-mediated hydrolysis of phosphoroselenyl esters

Phosphoroselenoic acid ammonium salts have also been used in carbohydrate chemistry for selenium glycosylation. Nucleophilic attack of phosphoroselenoic acid ammonium salts on glycosyl bromide/glycosyl anhydride afforded *Se*-glycoside. The P(O)-Se-C unit usually formed from the isomerization of a P(Se)-O-C moiety (Scheme 1.40) [50]. The isomerization product is thermodynamically more stable than the intermediate.

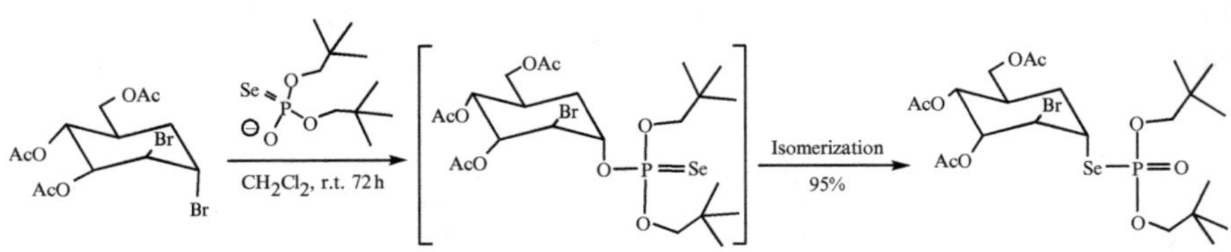

Scheme 1.40 Synthesis of *Se*-glycoside from phosphoroselenoic acid ammonium salts and glycosyl bromide or glycosyl anhydride

Other selenophosphonium salts could also form from trialkylselenophosphates and elemental halogens at low temperature. The halogenoselenophosphonium salts could be used as electrophilic reagents for the addition reaction to cyclohexene [51]. Two types of products were found indicating the ambident electrophilic characteristics of selenophosphates (Scheme 1.41). The products with *trans* configuration are dominant from the addition reaction. Low temperatures favour reaction pathway *a,* and high temperatures favour reaction pathway *b.*

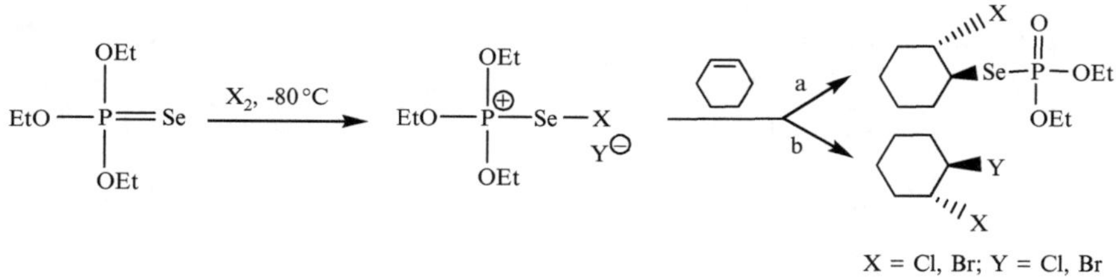

Scheme 1.41 Halogenoselenophosphonium salts as electrophilic reagents for the addition reaction to cyclohexene

When selenophosphates bear other active functional groups, the reaction becomes even more versatile. For example, tris(3-hydroxypropyl)selenophosphate $Se{=}P[O(CH_2)_3OH]_3$, obtained from the reaction of propanediol monoacetate and PCl_3 in the presence of triethylamine followed by treatment with methanolic ammonia, has been used as a dendrimer building unit. Four generations of selenophosphate dendrimers have been synthesised [first generation of dendrimer $3P_4Se_4$; second generation of dendrimer $3P_{10}Se_{10}$ + $4P_{10}Se_{10}$; third generation of dendrimer $3P_{22}Se_{22}$ + $4P_{22}Se_{22}$ and fourth generation of dendrimer $3P_{46}Se_{46}$ and $4P_{46}Se_{46}$] (Scheme 1.42) [52]. The backbone of a selenophosphate dendrimer was flexible enough to allow its chemical modification by means of the oxygenation with bulky peroxide. Treatment of the hydrophobic 3rd generation dendrimer $3P_{22}Se_{22}$ with *tert*-butylperoxytrimethylsilane resulted in water-soluble phosphate $3P_{22}O_{22}$ *via* regioselective oxygenation of phosphorus atoms at peripheral branching and extrusion of red selenium (Scheme 1.43) [52, 53].

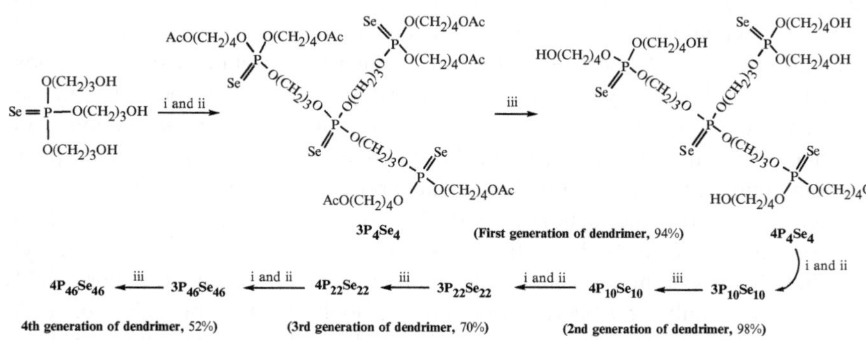

Scheme 1.42 Synthesis of four generations of selenophosphate dendrimers. Reagents and conditions: (i) $Et_2NP[O(CH_2)_4OAc]_2$, tetrazole, dichloromethane; (ii) selenium, pyridine; (iii) K_2CO_3, metnanol, water

By using the same methodology, heteroorganic dendrimers possessing three different (P=Se, P=S, P=O) branching units within the same molecule have been also prepared from tris(5-hydroxypentyl)selenophosphate $Se{=}P[O(CH_2)_5OH]_3$ (Scheme 1.44) [54].

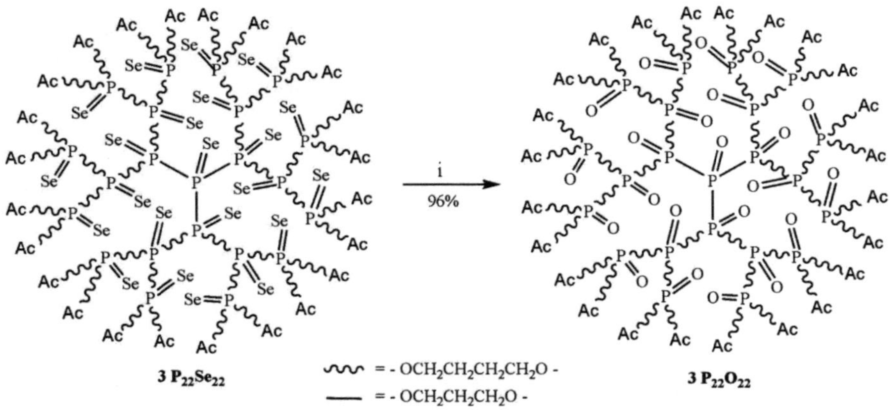

Scheme 1.43 Formation of water-soluble phosphate $3P_{22}O_{22}$ from Hydrophobic 3rd generation dendrimer $3P_{22}Se_{22}$ and *tert*-butylperoxytrimethylsilane. Reagents and conditions: (i) $^tBuOOSiMe_3$, dichloromethane, 75 h

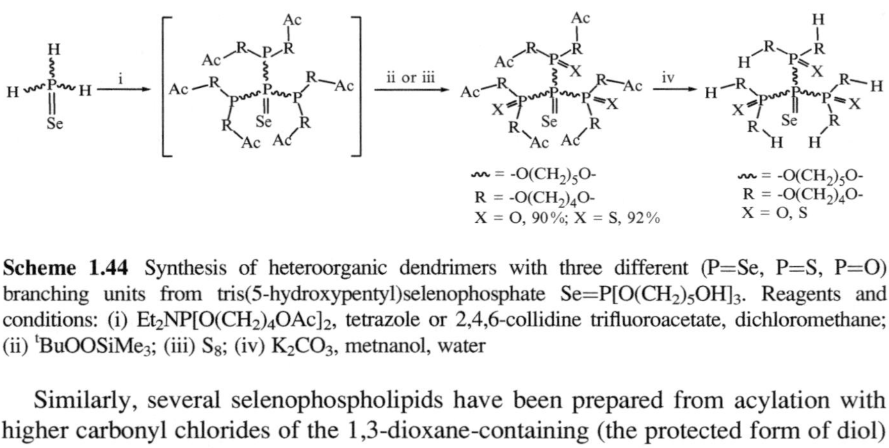

Scheme 1.44 Synthesis of heteroorganic dendrimers with three different (P=Se, P=S, P=O) branching units from tris(5-hydroxypentyl)selenophosphate $Se{=}P[O(CH_2)_5OH]_3$. Reagents and conditions: (i) $Et_2NP[O(CH_2)_4OAc]_2$, tetrazole or 2,4,6-collidine trifluoroacetate, dichloromethane; (ii) $^tBuOOSiMe_3$; (iii) S_8; (iv) K_2CO_3, metnanol, water

Similarly, several selenophospholipids have been prepared from acylation with higher carbonyl chlorides of the 1,3-dioxane-containing (the protected form of diol) selenophosphate or selenodiamidophosphate, which were obtained from the phosphorylation of 5-hydroxymethyl-2,2-dimethyl-5-propyl-1,3-dioxane with hexaethylphosphorus triamides or cyclic phosphorate, followed by selenation with elemental selenium (Scheme 1.45) [55]. Treatment of the distearoyl selenophosphate with potassium stearate led to a higher acylated potassium selenophosphorate salt.

Other selenophosphate analogues with thio- or selenium atoms in place of the oxygen or nitrogen have also been studied. As shown in Scheme 1.46, *O*-thiopivaloyl-*N*-*tert*-butylhydroxylamine reacted with two equivalents of selenophosphoric acids leading to not only *tert*-butylammonium selenophosphates, but also *bis*-(5,5-dimethyl-1,3,2-dioxoaphosphorinan-2-yl) diselenides and pivaloyl disulfide [56].

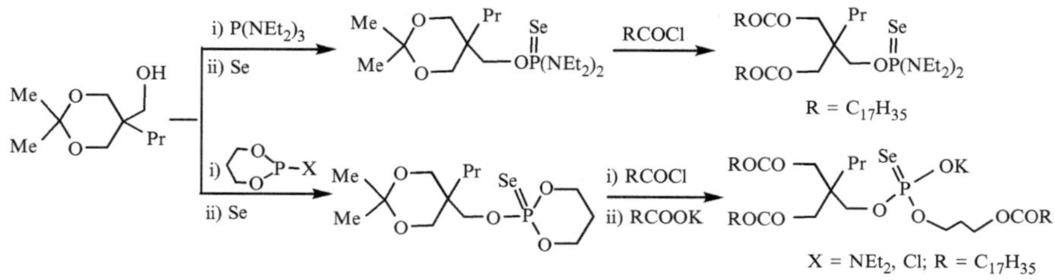

Scheme 1.45 Synthesis of selenophospholipids and potassium selenophosphorate salt

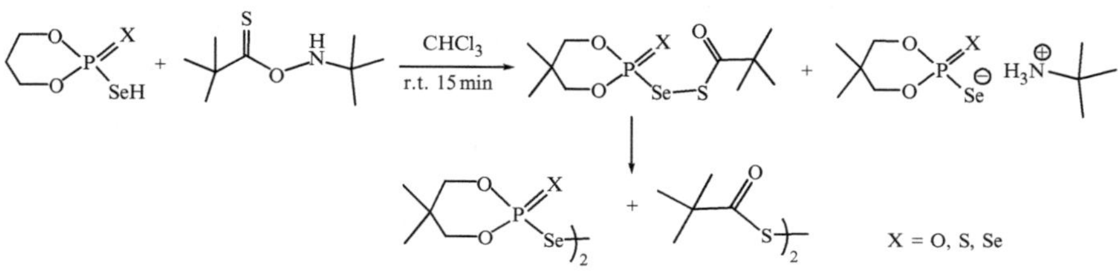

Scheme 1.46 Synthesis of *tert*-butylammonium selenophosphates and *bis*-(5,5-dimethyl-1,3,2-dioxoaphosphorinan-2-yl) diselenides from selenophosphoric acids and *O*-thiopivaloyl-*N*-*tert*-butylhydroxylamine

Meanwhile, selenyl sulfide and diselenide bridging spacer are also present in wheel-and-axle host (WAAH) molecules. Reaction of appropriate selenyl bromide and triethylamine dithiophosphoryl salt gave the following WAAH compounds, bis[*6-O*, *6-O′*-(1,2:3,4-diisopropylidene-α-D-galactopyranosyl)thiophosphoryl] disulfide, bis[*6-O*, *6-O′*-(1,2:3,4-diisopropylidene-α-D-galactopyranosyl)thiophosphoryl] diselenide and bis[*6-O*, *6-O′*-(1,2:3,4-diisopropylidene-α-D-galactopyranosyl)thiophosphoryl] selenenyl-sulfide, in 1 : 1 :1 ratio (Scheme 1.47) [57]. Bis[6-*O*, 6-*O′*-(1,2:3,4-diisopropylidene-α-D-galactopyranosyl)thiophosphoryl] selenenyl-sulfide could also be synthesized in the solid state by grinding and gentle heating of the disulfide and diselenide [58].

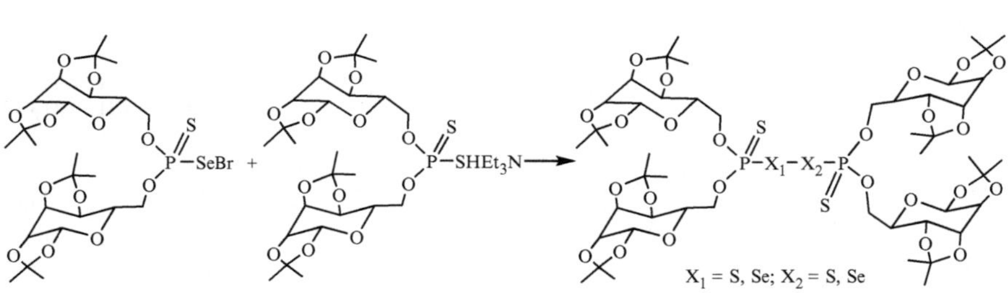

Scheme 1.47 Synthesis of WAAH compounds from the corresponding selenyl bromide and triethylamine dithiophosphoryl salt

A macrocyclic selenium-bridged hexamer $[(Se)P(\mu\text{-}N^tBu)_2P(\mu\text{-}Se)]_6$ was generated from the dimer $[(Cl)(Se)P(\mu\text{-}N^tBu)]_2$ with sodium metal in refluxing toluene *via* a formal head to tail cyclization of the intermediate anion (Scheme 1.48) [59].

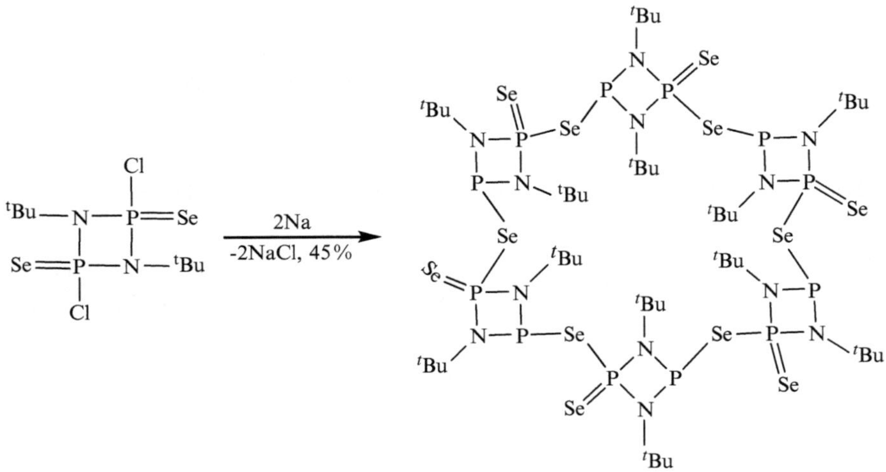

Scheme 1.48 Synthesis of a macrocyclic selenium-bridged hexamer $[(Se)P(\mu\text{-}N^tBu)_2P(\mu\text{-}Se)]_6$ from the dimer $[(Cl)(Se)P(\mu\text{-}N^tBu)]_2$

The X-ray crystal structure revealed that the P-Se bond lengths are almost equal in the ammonium salts of *O,O′*-dialkylphosphorodiselenoic acid, $NH_4Se_2P(OR)_2$. The solution state ^{31}P NMR spectra shows only one set of Se satellites flanked around the singlet peak in the ^{31}P nmr [60, 61] suggesting that the negative charge in the ammonium salts is delocalized over the entire PSe_2 unit. Thus, either of the Se atoms could act as a nucleophilic center able to carry on a series of reactions including Michael addition, epoxide ring opening addition, acylation and α-alkynation (Scheme 1.49) [62].

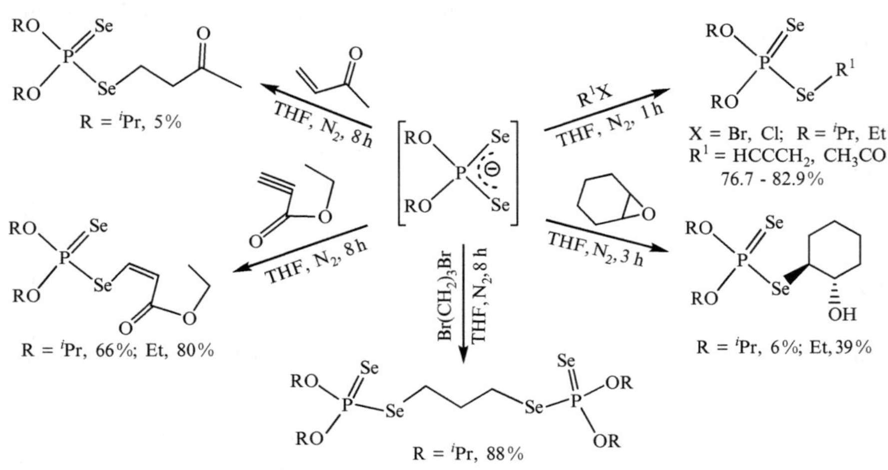

Scheme 1.49 Reactivity of ammonium salts of *O,O′*-dialkylphosphorodiselenoic acid

Other hybrid selenophosphoric acid derivatives like acylphosphoramido(thio)(seleno)ate were also synthesised from carboxamide and 2-chloro-1,3,2-oxathiaphospholane followed by selenation. The resultant phosphoramido(thio)(seleno)ate reacted with alcohol like methanol in the presence of DBU to provide the corresponding *O*-alkyl-*N*-acylphosphoroselenoamidates, which could be further oxidized *in situ* to *N*-acylphosphoramidate by treatment with $^tBuOOSiMe_3$ (Scheme 1.50). The approach has been successfully applied to prepare prolylamido-AMP containing an acylphosphoramidate linkage from *N*-tri-prolinamide by four steps (Scheme 1.51) [63].

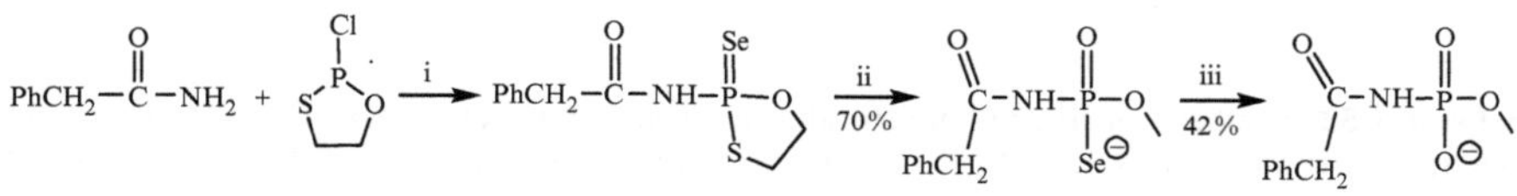

Scheme 1.50 Synthesis of *N*-acylphosphoramidate from acylphosphoramido(thio)(seleno)ate. Reagents and conditions: (i) py or DIPEA/CH_2Cl_2 /Se; (ii) MeOH/DBU/CH_3CN, 3 h; (iii) $^tBuOOSiMe_3$, 12 h

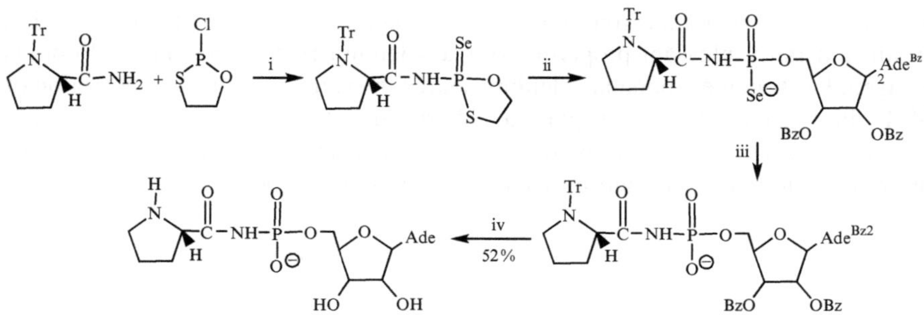

Scheme 1.51 Synthesis of prolylamido-AMP from *N*-tri-prolinamide. Reagents and conditions: (i) Se; (ii) $N^6, N^6, O^{2'}, O^{3'}$ – tetrabenzoyladenosine/DBU/CH_3CN, 12 h; (iii) $^tBuOOSiMe_3$/CH_2Cl_2, 12 h; (iv) NH_4OH/50% CF_3CO_2H, 1 h

Selenophosphates have been used for the synthesis of tri-and tetrasubstituted alkenes. Interaction of selenophosphates with sodium borohydride in alcohol or with potassium cyanide in the presence of 18-crown-6 in dimethoxyethane, after quenching with water, afforded the corresponding alkenes (Scheme 1.52) [64]. The reaction involves the addition of $NaBH_4$ or KCN to selenophosphates to give two diasrereosomeric oxyanions, which underwent a rearrangement by migration of a phosphoryl group from selenium to oxygen affording selenolate anions, and followed by cyclization with elimination of phosphate anion to give episelenides of (*E*) and (*Z*) configuration, the latter easily lost a selenium atom spontaneously to yield a series of alkenes.

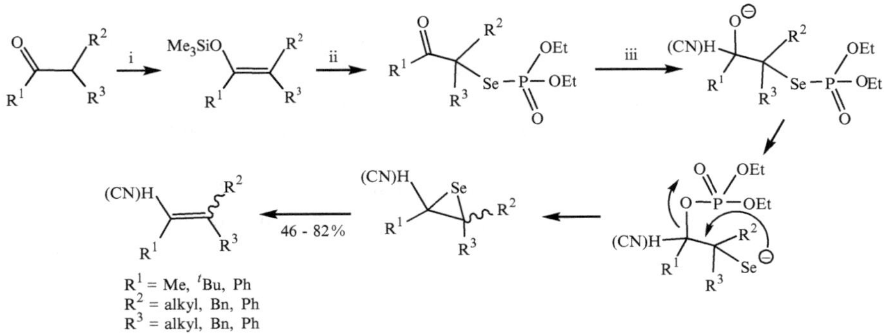

Scheme 1.52 Synthesis of tri-and tetra-substituted alkenes from selenophosphates. Reagents and conditions: (i) Me_3SiCl, NaI, Et_3N, r.t. – 60°C; (ii) $(EtO)_3PSeClSO_2Cl$, –78–0°C ; (iii) $NaBH_4$/MeOH, r.t. or KCN/18-crown-6/DME, r.t.

1.6 Woollins Reagent in Organic Synthesis

Woollins Reagent (**WR**) is available from Sigma-Aldrich. Alternatively, it can be prepared in the laboratory from $PhPCl_2$, Na_2Se and Se [65]. Due to its easy handling in air and fairly pleasant properties (for a selenium rich compound) , Woollins Reagent has become a popular selenium source in synthetic chemistry. Reaction of **WR** with organic substrates displays a wide spectrum ranging from simple oxygen-selenium exchange to the formation of complex phosphorus-selenium heterocycles as well as surprising phosphorus-selenium-free products [66].

1.6.1 Simple Exchange of Oxygen to Selenium by Woollins Reagent

Reaction of **WR** with amides, ureas, formamide and some aldehydes in common organic solvents including toluene, benzene or DCM at room temperature or at reflux led to the corresponding selenoamides, selenoureas and selenoaldehydes by simple exchange of oxygen with selenium (Scheme 1.53) [67–71].

Alternatively, primary arylselenoamides can be obtained from the reaction of arylnitriles with **WR** in refluxing toluene followed by quenching with water (Scheme 1.54) [72, 73].

When organic substrates bearing two carbonyl groups such as 1,4-diaromatic ketones or 1,4-diacylhadrazines, were treated with **WR** the reaction did not stop at the simple two oxygen-selenium exchange, but proceeded further to give 2,5-diarylselenophenes (Scheme 1.55) [74] or selenadiazoles by loss of one selenium atom (Scheme 1.56) [75]. 2,5-Diarylselenophenes can also be obtained from the reaction of arylacetylene with an equivalent of *O*-methyl *Se*-hydrogen

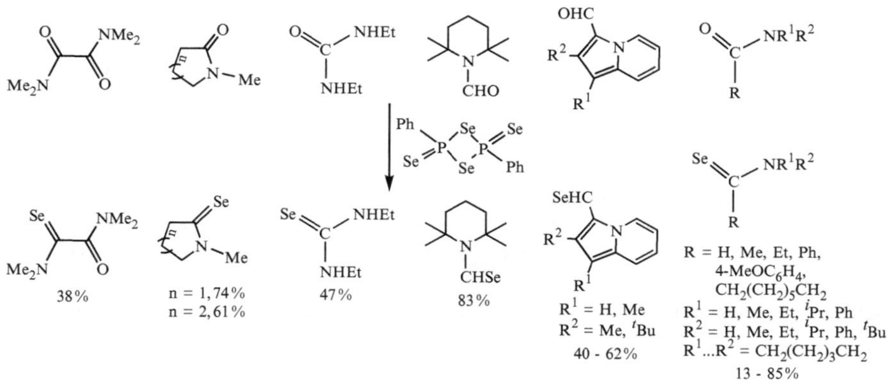

Scheme 1.53 Synthesis of selenoamides, selenoureas and selenoaldehydes from selenation of the corresponding amides, ureas, formamide and some aldehydes

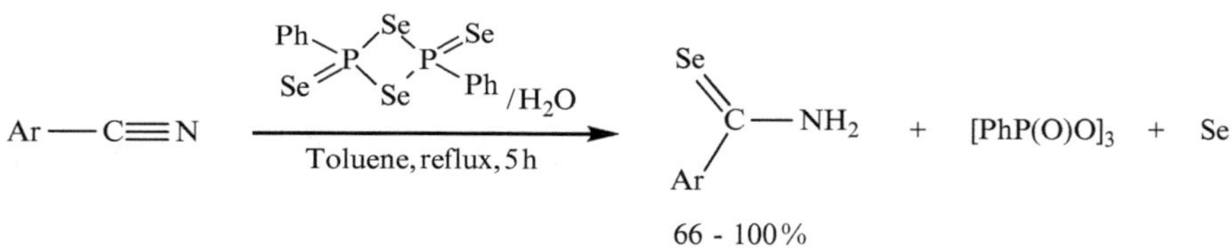

Scheme 1.54 Formation of primary arylselenoamides from arylnitriles and **WR** and water

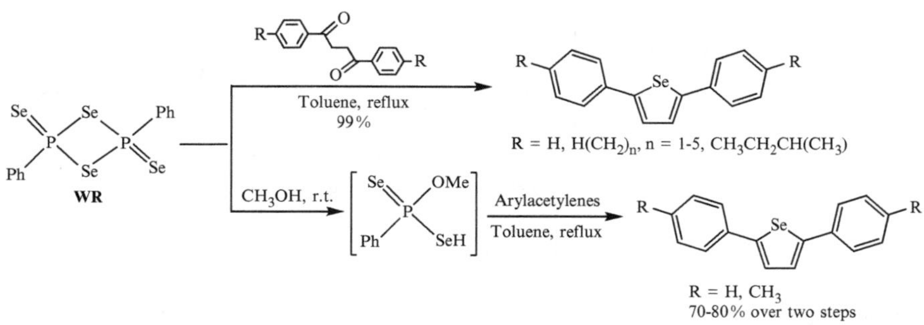

Scheme 1.55 Synthesis of 2,5-diarylselenophenes from selenation of 1,4-diaromatic ketones or from arylacetylene and *O*-methyl *Se*-hydrogen phenylphosphonodiselenoate

phenylphosphonodiselenoate; the latter was easily derived from **WR** and methanol at room temperature (Scheme 1.55) [74].

Similarly, **WR** interacted with 1,4-keto-alcohols (Scheme 1.57) or benzofurans (Scheme 1.58), which were obtained from the ring opening of the lactones with aryl/hetero-arylmagnesium bromides, to give a series of symmetrical/unsymmetrical 1,3-diarylbenzo[*c*]selenophenes [76, 77]. Bis-benz-annelated benzo[c]selenophenes were also prepared in this way (Scheme 1.59) [77].

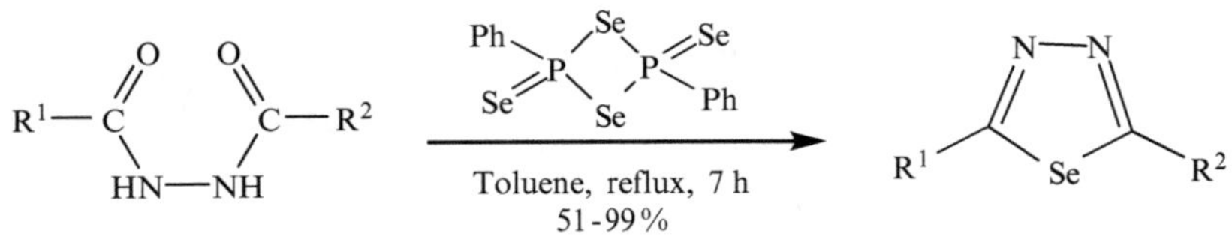

R^1 = Ph, 4-MeC_6H_4, 4-BrC_6H_4, 4-MeOC_6H_4
R^2 = Ph, 4-MeC_6H_4, 4-BrC_6H_4, 4-ClC_6H_4, 4-FC_6H_4, C_2H_5O, 4-MeOC_6H_4, pyridin-3-yl, thiophenyl-2-yl, furan-2-yl

Scheme 1.56 Synthesis of 2,5-diarylselenadiazoles from selenation of 1,4-diacylhadrazines

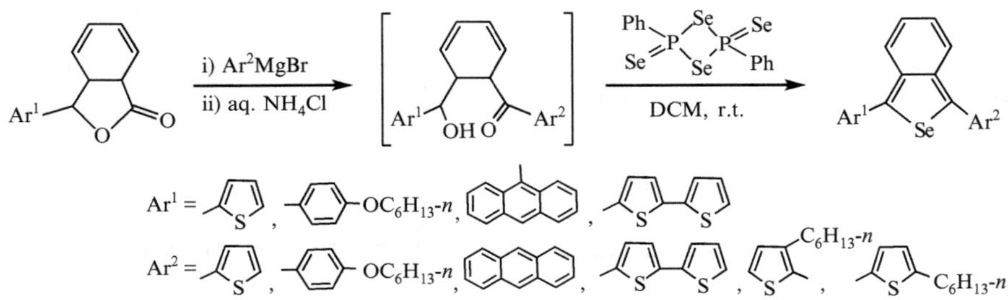

Scheme 1.57 Synthesis of symmetrical/unsymmetrical 1,3-diarylbenzo[*c*]selenophenes from selenation of 1,4-keto-alcohols

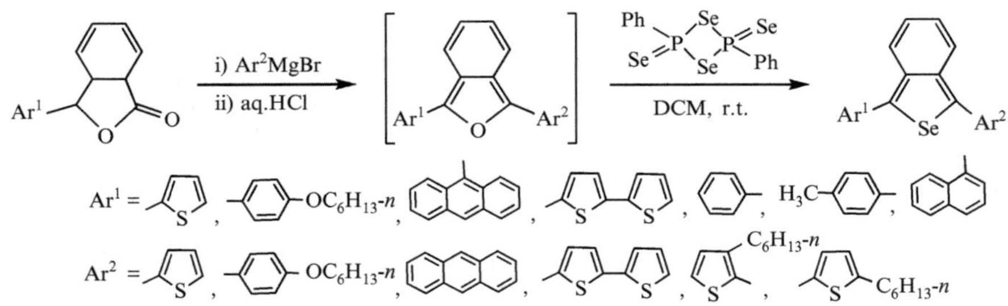

Scheme 1.58 Synthesis of symmetrical/unsymmetrical 1,3-diarylbenzo[*c*]selenophenes from selenation of benzofurans

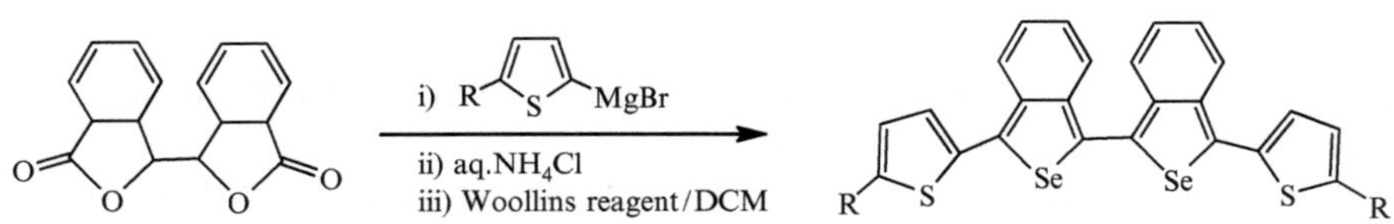

R = H, C_6H_{13}-*n*

Scheme 1.59 Synthesis of bis-benz-annelated benzo[c]selenophenes from selenation of bibenzofurane

Treatment of $CH_3C(O)OH$ or $CD_3C(O)OD$ with an excess of **WR** in a small pyrex vessel at 75°C under moisture- and air-free conditions gave the first isolated pure selenoacetic acids, $CH_3C(O)SeH$ and $CD_3C(O)SeD$ (Scheme 1.60) [78].

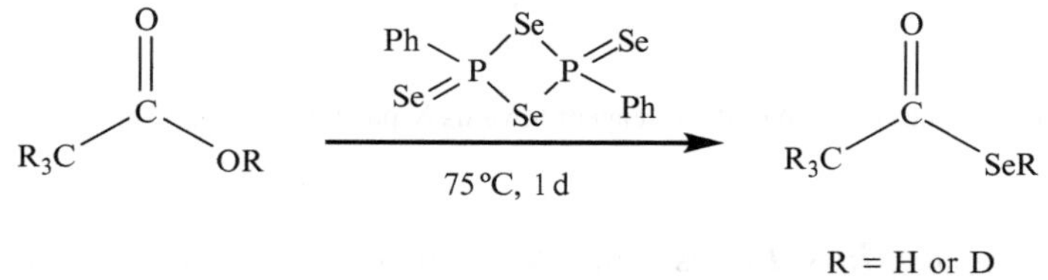

Scheme 1.60 Formation of $CH_3C(O)SeH$ and $CD_3C(O)SeD$ from selenation of $CH_3C(O)OH$ or $CD_3C(O)OD$

Other selenocarboxylic acids were generated by treating **WR** with RCO_2H. Subsequent reaction of this selenospecies *in situ* with derivatives of glucose or pseudoglucose led to selenoglucosides or pseudoselenoglucoside under very mild conditions (Scheme 1.61) [79]. In the presence of an organic base, selenocarboxylic acid interacted with organic azide to afford the corresponding amide.

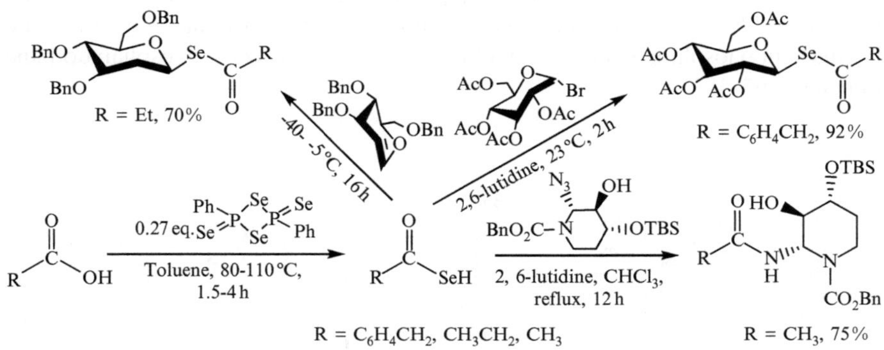

Scheme 1.61 Reactivity of selenocarboxylic acids towards various organic substrates

However, reaction of **WR** with α-amino acid *N*-phenylglycine, a secondary amine, afforded 1,4-diphenylpiperazine-2,5-diselenone [80]. The mechanism involves the seleanation of C=O by **WR** first to generate the corresponding seleno-acid intermediate; subsequent cyclization of the unstable α-amino selena-acid with loss of two molecules of H_2O to give 1,4-diphenylpiperazine-2,5-diselenone. In the case of 2-phenylglycine, in which the amino group is primary, only 2,5-diphenylpyrazine instead of 2,5-diselenopiperazine was obtained (Scheme 1.62). The formation of 2,5-diphenylpyrazine would follow the similar selenation and the intermolecular cyclocondensation, in the second step, not only two molecules of H_2O but also two selenium atoms were eliminated.

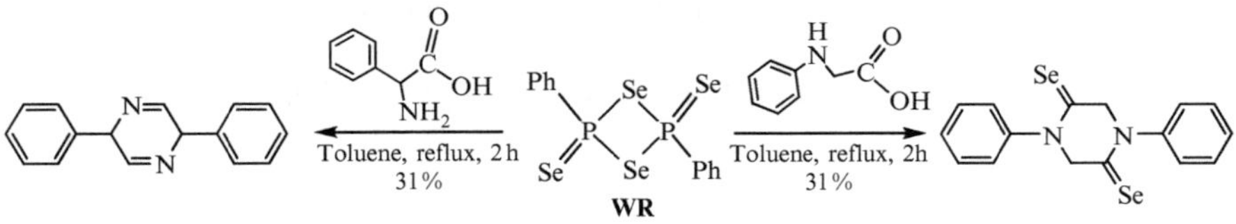

Scheme 1.62 Reactivity of Woollins' reagent towards *N*-phenylglycine and 2-phenylglycine

1.6.2 Organophosphoroselenium Heterocycles from Woollins Reagent

Woollins reagent is able not only to deliver one or two selenium atoms to a substrate, but also to incorporate a fragment of itself into the products. As shown in Scheme 1.63, **WR** reacted with phosphaindolizine to afford two products, one with two selenium atoms added, (the pale yellow crystalline air and moisture sensitive pyridinium diselenophosphinate) the other with the two phosphorus centers bridging two molecular equivalents of phosphaindolizine – the selenoanhydride consisting of nearly 1:1 mixture of two diastereomers. Using appropriate stoichiometry and in the presence of NEt_3, the diselenophosphinate could be isolated as the only product in 87% yield (Scheme 1.64). A similar reaction was utilized to prepare the diselenophosphinate in 42% yield derived from 1-aza-2-phosphaindolizine (Scheme 1.65) [81].

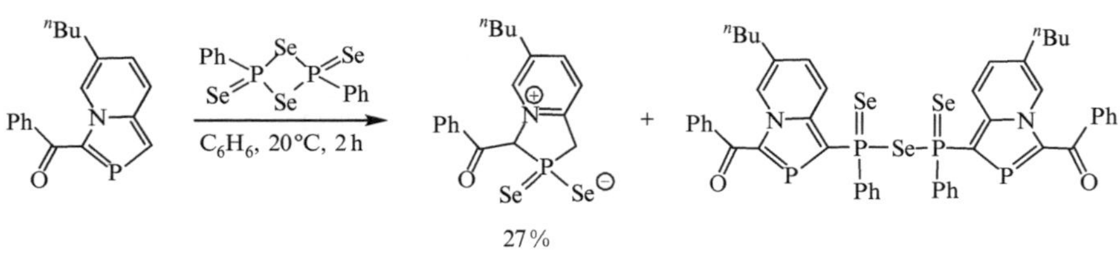

Scheme 1.63 Reaction of Woollins' reagent with phosphaindolizine

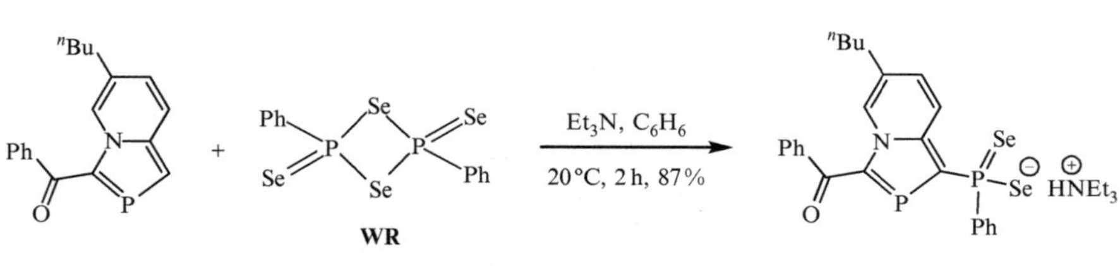

Scheme 1.64 Reaction of Woollins' reagent with phosphaindolizine in the presence of NEt_3

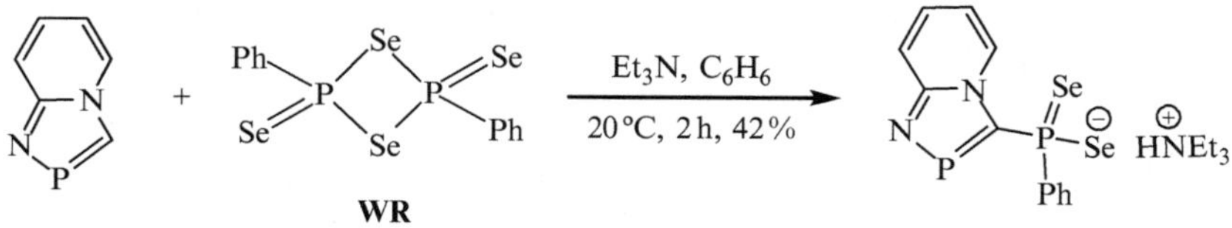

Scheme 1.65 Reaction of Woollins' reagent with 1-aza-2-phosphaindolizine in the presence of NEt_3

One of the most attractive features of **WR** lies in its capacity to form phosphorus-selenium heterocycles. The earliest example is the reaction of **WR** with small, unsaturated molecules such as acetone or CS_2 [82, 83]. Treatment of **WR** with acetone led to colourless crystals of 4,4-dimethyl-5-oxa-1,3-diphenyl-1,3-diseleno-2-selena-1,3-diphospholane, arising from insertion of acetone in a P_2Se_2 ring. Meanwhile, reacting **WR** with CS_2 slowly resulted in the formation of a dimeric motif consisting of 2 five-membered CP_3Se rings being bridged *via* their trigonal carbon atoms bonding to the phosphorus atoms of a P_2Se_2 ring (Scheme 1.66).

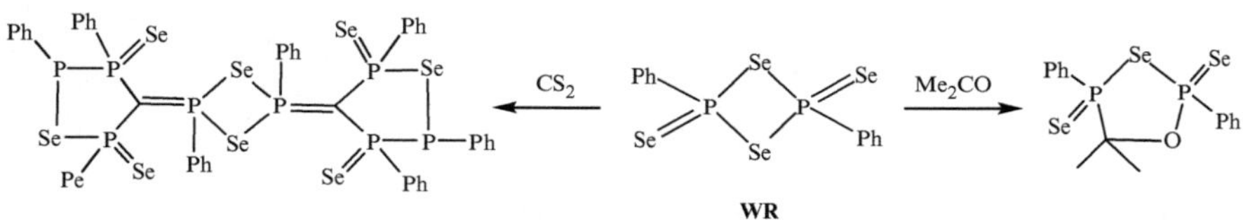

Scheme 1.66 Reaction of Woollins' reagent with acetone or CS_2

The preparation of a far more extensive series of P-Se heterocycles have been performed based on the reaction of **WR** with a wide range of organic substrates containing reactive unsaturated C=O, C=C double and C≡C triple bonds. An unusual phosphorus-selenium spirocyclic heterocycle with a four-membered P_2SeC ring (5% yield) was achieved together with an expected selenocarbonyl compound (27% yield) from **WR** and diphenylcyclopropenone (Scheme 1.67) [65].

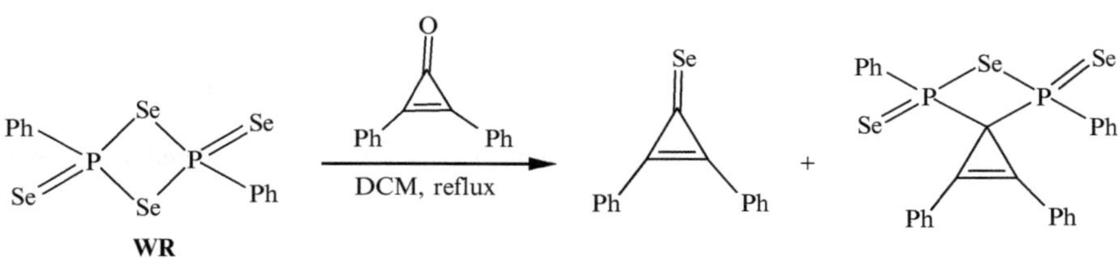

Scheme 1.67 Reaction of Woollins' reagent with diphenylcyclopropenone

The reaction of **WR** with azobenzene was carried out with cleavage of the N=N bond and substitution of a bridging selenium atom in Woollins reagent by an NPh unit, giving the first crystallographically characterised selenaazadiphosphetane (Scheme 1.68) [84].

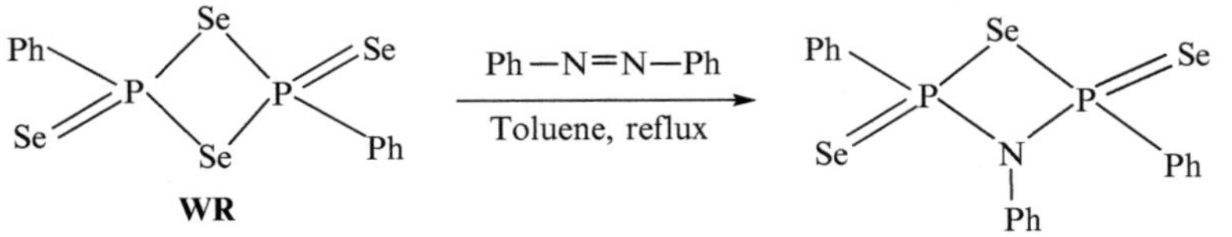

Scheme 1.68 Reaction of Woollins' reagent with azobenzene

Reaction of **WR** with methyl phenylpropiolate in refluxing toluene solution provided 4 five-membered phosphorus-selenium containing or selenium containing heterocyclic products, which were found to be moderately stable in air, degrading over a period of several weeks with the obvious expulsion of red selenium (Scheme 1.69) [65, 85].

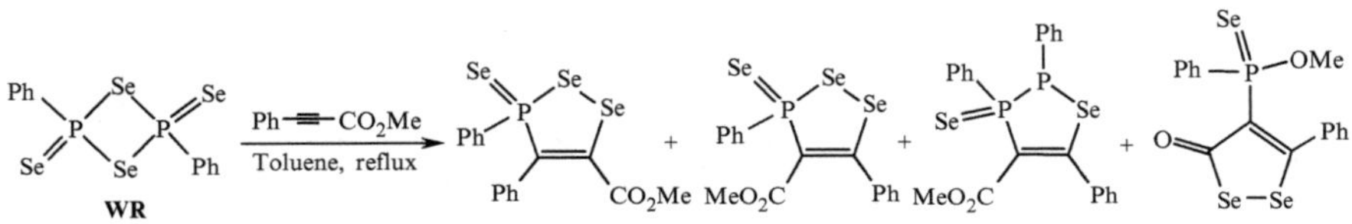

Scheme 1.69 Reaction of Woollins' reagent with methyl phenylpropiolate

A series of similarly unexpected phosphorus-selenium containing heterocycles were also obtained from the reaction of **WR** with dimethyl but-2-ynedioate, alkynes and dialkyl cyanamides. **WR** was heated with dimethyl but-2-ynedioate in toluene leading to two five-membered P-Se heterocycles and one six-membered P-Se ring in 5–19% yields (Scheme 1.70) [85]. However, only one addition product was harvested when Woollins reagent reacted with ethynylbenzene or dialkyl cyanamides in the identical conditions (Scheme 1.71) [85, 86].

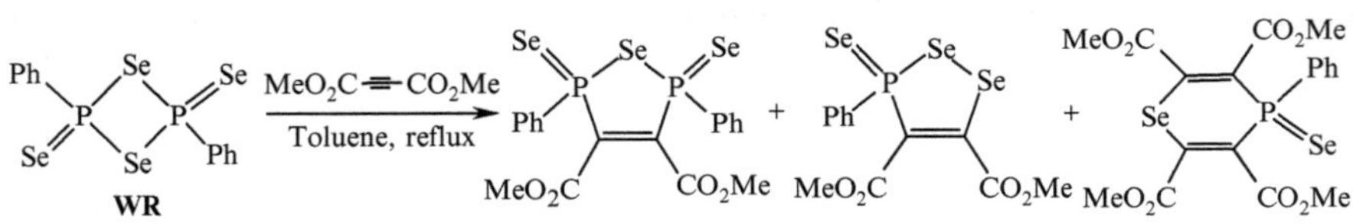

Scheme 1.70 Reaction of Woollins' reagent with dimethyl but-2-ynedioate, alkynes and dialkyl cyanamides

Several five-membered PSe_2C_2 heterocycles were synthesized from **WR** and dialkynes by formal addition of a $Ph(Se)PSe_2$ fragment to the alkyne triple bond (Scheme 1.72). An unusual diselenide was generated by an intramolecular cycloaddition/rearrangement along with a double five-membered PSe_2C_2 heterocycle formed when a sterically constrained naphthalene dialkyne was used (Scheme 1.73) [85, 86].

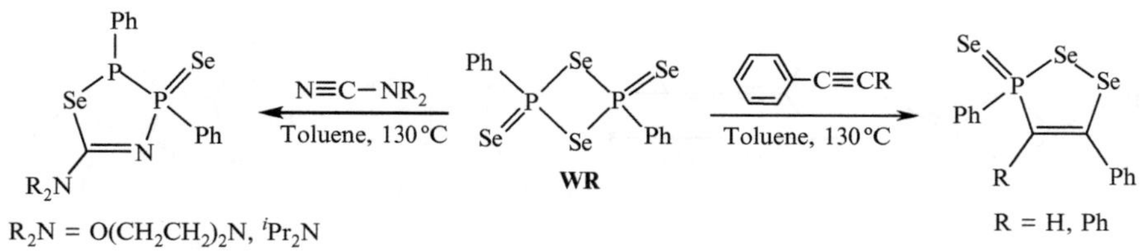

Scheme 1.71 Reaction of Woollins' reagent with ethynylbenzene or dialkyl cyanamides

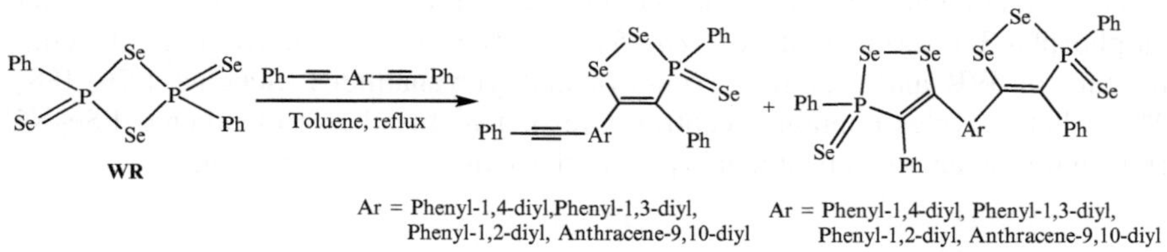

Scheme 1.72 Reaction of Woollins' reagent with dialkynes

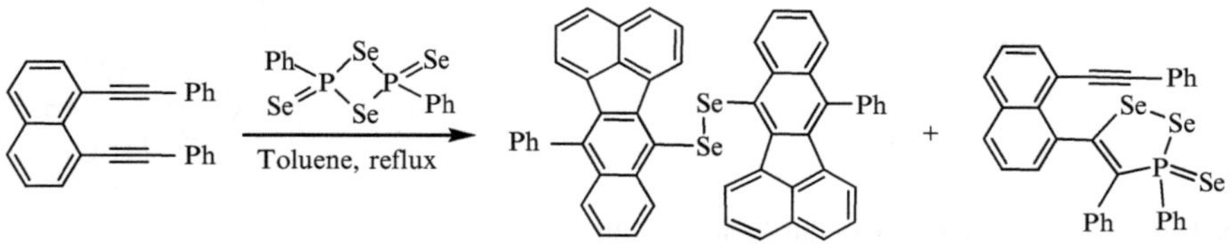

Scheme 1.73 Reaction of Woollins' reagent with naphthalene dialkyne

Similarly, treating **WR** with symmetrically disubstituted diynes (RC≡C-C≡CR, R=Ph, $Si(CH_3)_3$) in 2:1 molar ratio in refluxing toluene led to two types of phosphorus-selenium compounds: five-membered $P(Se)Se_2C_2$ heterocycles with one unreacted triple bond 'dangling' and *bis*-heterocycles with two five-membered $P(Se)Se_2C_2$ rings connected through a C-C single bond. However, Woollins reagent reacted with ^{t}Bu-C≡C-C≡C-^{t}Bu differently to afford the four-membered $P(Se)SeC_2$ heterocycle with one unreacted triple bond and the heteropentalene with two P(Se) SeC_3 rings fused at the central two carbons of the diyne to give a heteropentalene analogue of pentalene, [3.3.0]octa-1-6-diene (Schemes 1.74 and 1.75) [87].

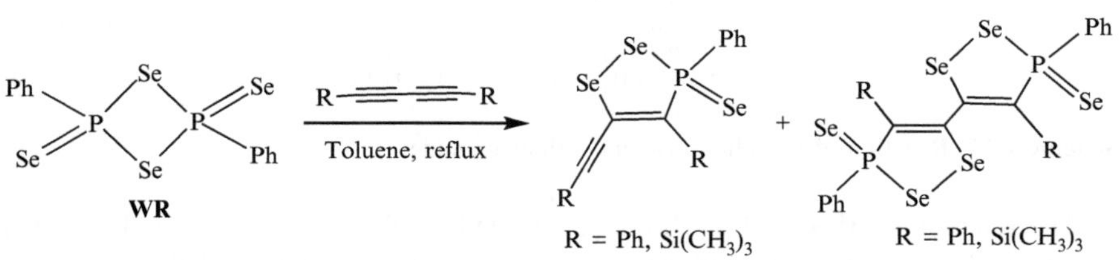

Scheme 1.74 Reaction of Woollins' reagent with symmetrically disubstituted diynes

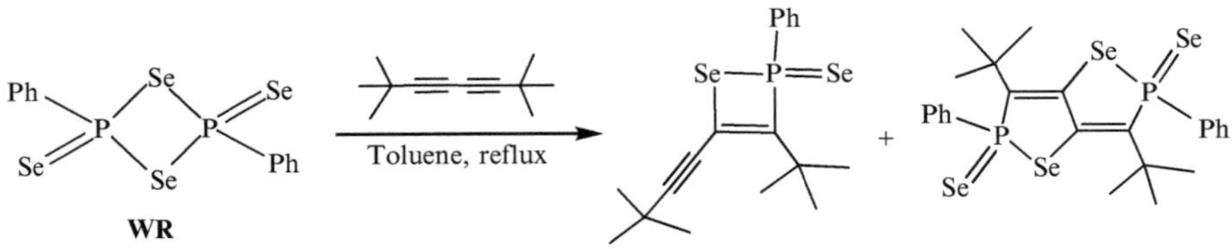

Scheme 1.75 Reaction of Woollins' reagent with 2,2,7,7-tetramethylocta-3,5-diyne

A small amount of unusual phosphorus-selenium heterocycles, 1,6,6λ^4-triselena-3a-phospha-3,4-diazapentalenes (*ca.* 5% yield) was isolated from the reaction mixture of **WR** and a tenfold excess of dialkylcyanamides (Scheme 1.76) [88]. These heterocycles contain a central core of two fused five-membered PSe_2NC rings with the amine substituents lying in the same approximate plane.

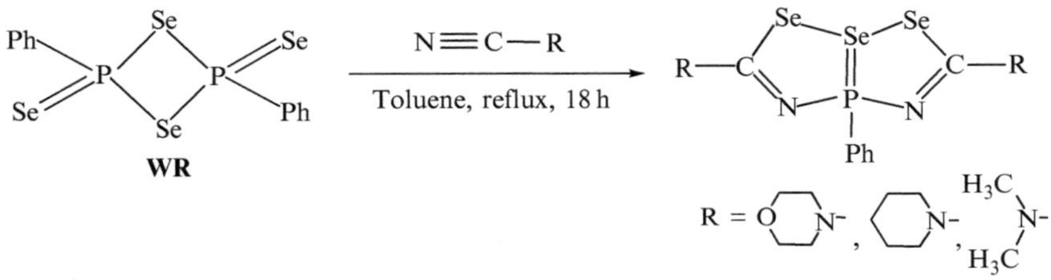

Scheme 1.76 Reaction of Woollins' reagent with dialkylcyanamides

Other cyanamides, available from the interaction of cyanogen bromide with primary or secondary amines, reacted with an equivalent of **WR** in refluxing toluene to afford a series of novel selenazadiphospholaminediselenides in 28–71% yields. Post-treatment of the reaction mixture with water led to the formation of carbamidoyl(phenyl)phosphinodiselenoic acids and selenoureas in moderate to excellent yields (20–90%) instead (Scheme 1.77) [89].

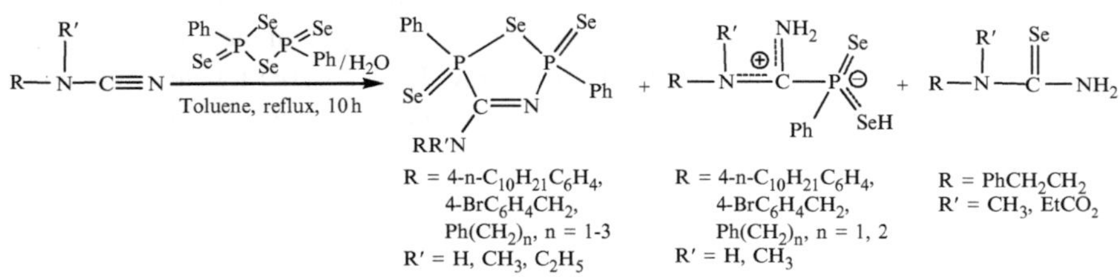

Scheme 1.77 Reaction of Woollins' reagent with cyanamides

The reaction of **WR** with norbornene proceeded with cleavage of the $P_2(\mu\text{-}Se)_2$ ring to give a 1,2-diselena-3-phospholane with a five-membered C_2PSe_2 ring *via* a [2 + 3] cycloaddition between the alkene and a Se-P=Se unit (Scheme 1.78) [84].

The formation of this compound is different from the product 1,2-thiaphosphetane (C_2PS ring) of the reaction between norbornene and $[RP(S)(\mu\text{-}S)]_2$ (R = ferrocenyl). However, **WR** and norbornadiene did give a four-membered C_2PSe heterocycle as a counterpart of the 1,2-thiaphosphetane (C_2PS) together with the five-membered (C_2PSe_2) ring product (Scheme 1.78) [90, 91].

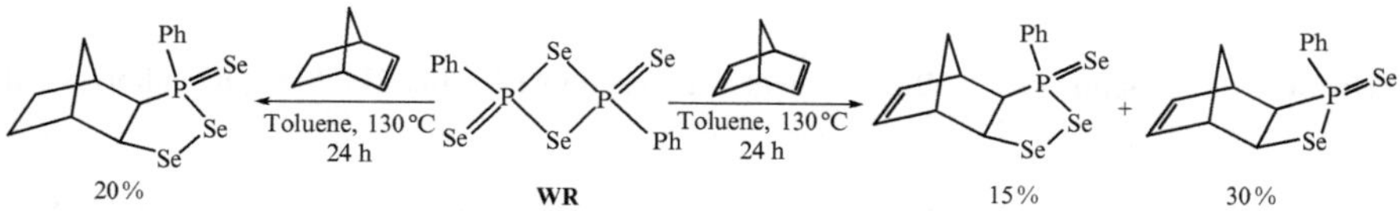

Scheme 1.78 Reaction of Woollins' reagent with norbornene and norbornadiene

Reaction of **WR** with two equivalents of difunctional organic substrates, such as diamine, dialchohol, diphenol, dithiophenol proceeded with the bridge cleavage of the central $P_2(\mu\text{-}Se)_2$ core in **WR** to afford the common structure containing a PhP (Se) motif in modest to excellent yields (Schemes 1.79–1.81) [92, 93].

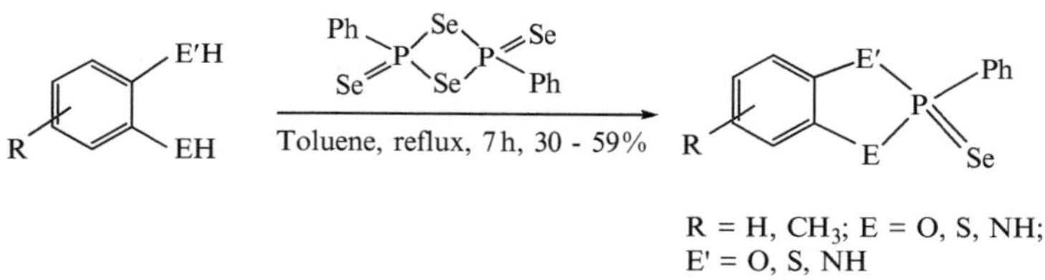

Scheme 1.79 Reaction of Woollins' reagent with organic substrates bearing difunctional groups

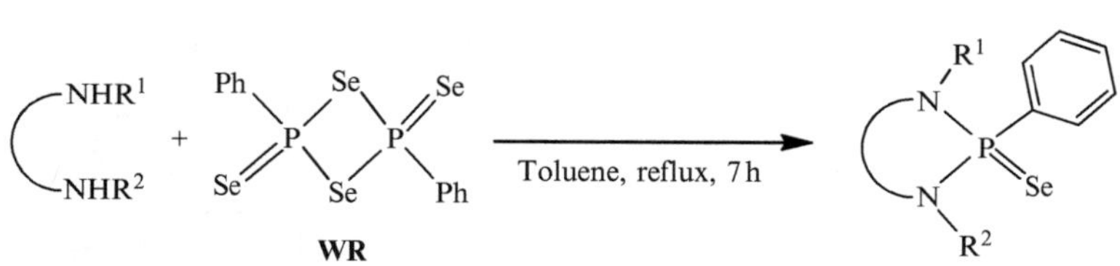

(= Phenyl-1,2-diyl, $R^1 = C_6H_5$, R^2 = H, 51 %
(= Phenyl-1,2-diyl, $R^1 = R^2 = C_6H_5CH_2$, 65%
(= Naphthalene-1,8-diyl, $R^1 = R^2$ = H, 68 %
(= 1-Phenyl-1′-naphthyl-2,2′-diyl, $R^1 = R^2$ = H, 96 %
(= 1, 1′-Dinaphthyl-2,2′-diyl, $R^1 = R^2$ = H, 98 %

Scheme 1.80 Reaction of Woollins' reagent with aryldiamine

However, as far as diols are concerned, either aliphatic or aromatic, the reaction becomes more complicated depending upon substrates. Apart from the common product bearing the PhP(Se) unit, other larger heterocycles containing PhP(Se)SeP

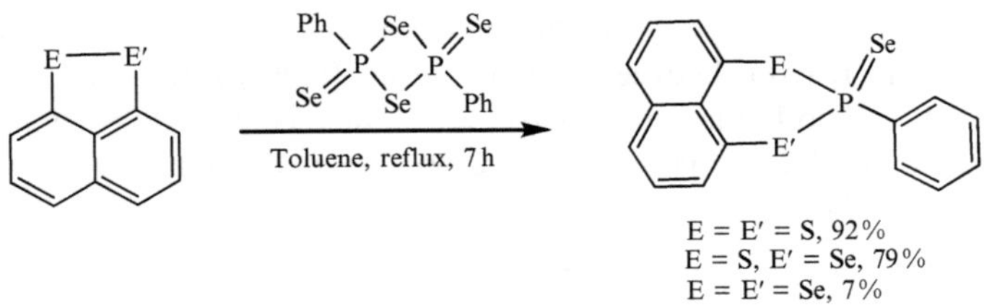

Scheme 1.81 Reaction of Woollins' reagent with naphtho[1,8-cd][1,2]dithiole, naphtho[1,8-cd][1,2]thiaselenole and naphtho[1,8-cd][1,2]diselenole

(Se)Ph and/or PhP(Se)SeSeP(Se)Ph are possible (Schemes 1.82–1.85) [94–96]. The complexity usually comes with reactions involving only one equivalent of diol.

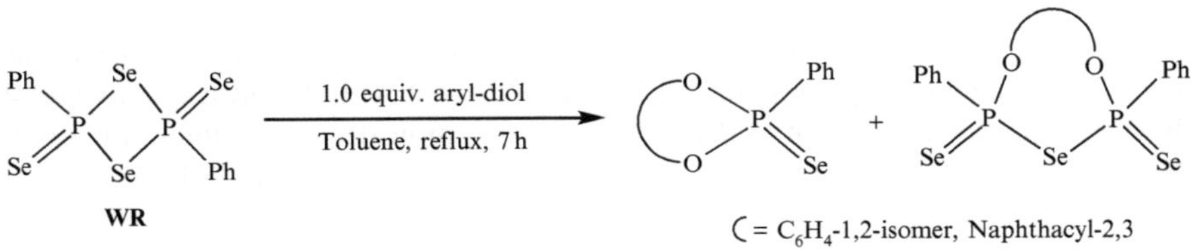

Scheme 1.82 Reaction of Woollins' reagent with one equivalent of aryl-diols

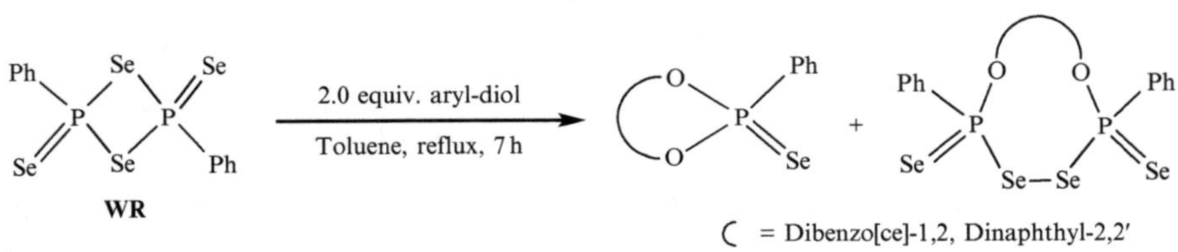

Scheme 1.83 Reaction of Woollins' reagent with two equivalents of aryl-diols

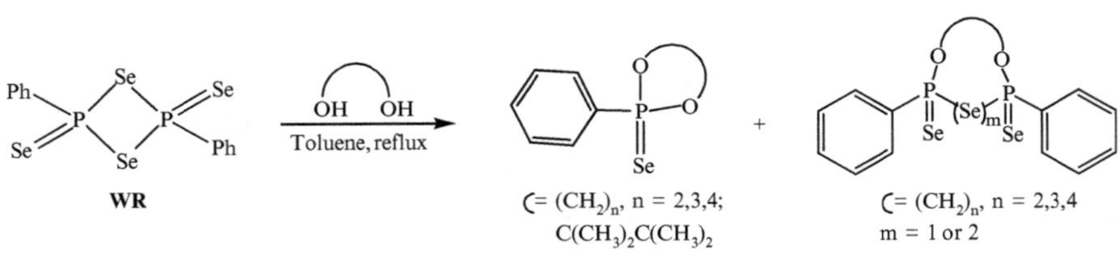

Scheme 1.84 Reaction of Woollins' reagent with aliphatic diols

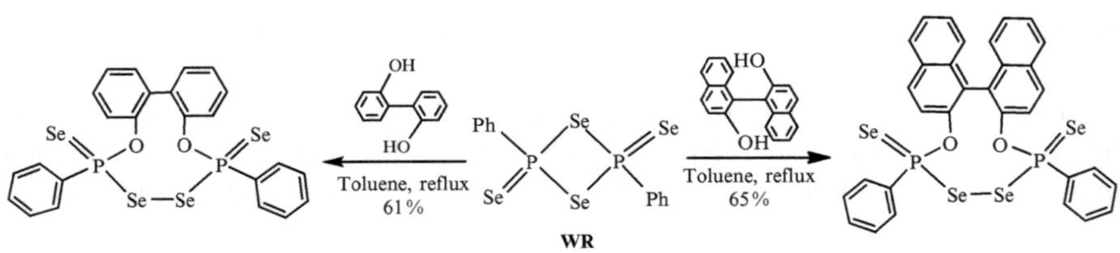

Scheme 1.85 Reaction of Woollins' reagent with [1,1′-biphenyl]-2,2′-diol and [1,1′-binaphthalene]-2,2′-diol

What is more, the larger heterocyclic diphosphorus species can be converted into the corresponding small monophosphorus rings by loss of one or two molecules of H_2Se *via* further reaction with another mole of diol in almost quantitative yields (Scheme 1.86) [94].

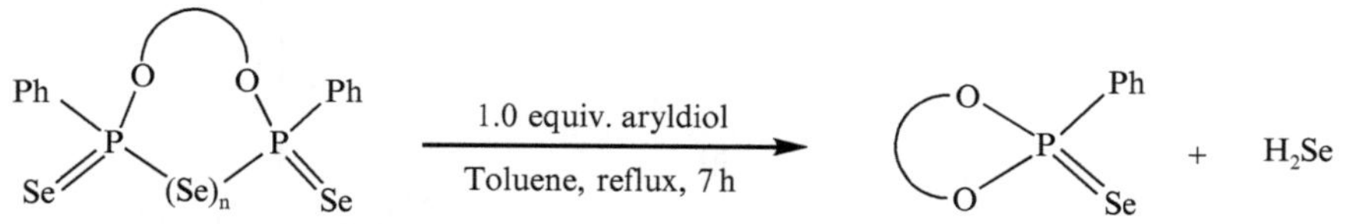

n = 1, (= C_6H_4-1,2-isomer, Naphthacyl-2,3
n = 2, (= Dibenzo[ce]-1,2, Dinaphthyl-2,2′

Scheme 1.86 Ring reduction of 10-membered heterocycles into seven-membered heterocycles

When 1,1′-methylenedi-2-naphthol was reacted with **WR**, a third six-membered Se-O heterocycle was formed in addition to the heterocycle containing the common PhP(Se) unit and the larger PhP(Se)SeP(Se)Ph heterocycle (Scheme 1.87) [94].

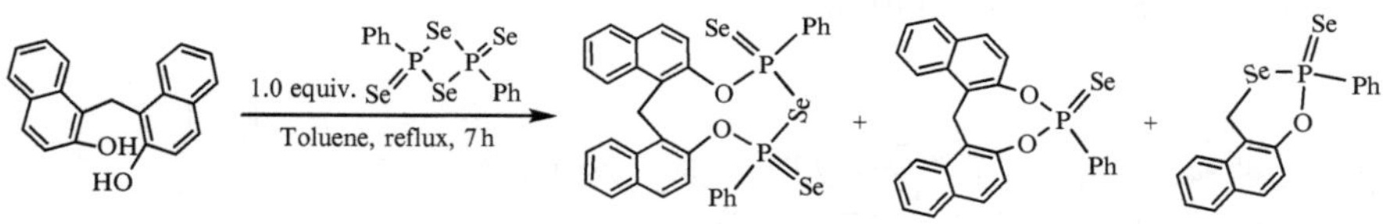

Scheme 1.87 Reaction of Woollins' reagent with 1,1′-methylenedi-2-naphthol

In the case of 2,2′-methylenebis(4-chlorophenol), partial oxidation product of P (Se)SeP(Se) heterocycle (24%) was also observed during the workup in air (Scheme 1.88) [94].

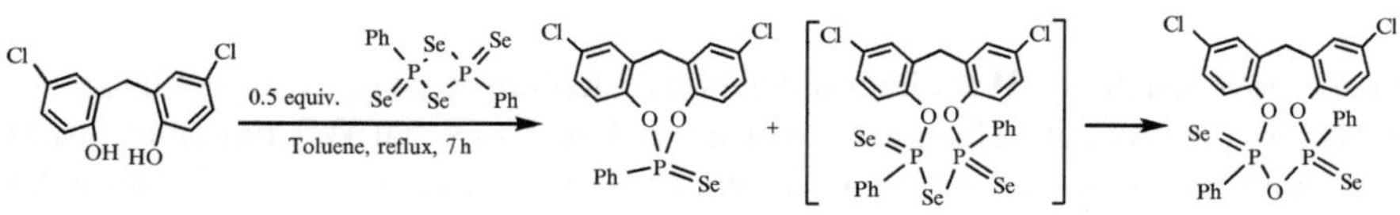

Scheme 1.88 Reaction of Woollins' reagent with 2,2′-methylenebis(4-chlorophenol)

Alternatively, large heterocycles bearing PhP(Se)SeSeP(Se)Ph moiety could also be achieved through iodine oxidation of the ammonium salts of bisdiseleno-phosphonic acids, derived from diols [$HO(CH_2)_nOH$, n = 2–6, 8 and 10] and **WR** in dry dichloromethane followed by treatment with butylamine in tetrahydrofuran. Methylation of the ammonium salts with methyl iodide afforded dimethyl esters of bisdiselenophosphonic acids (Scheme 1.89) [95, 96].

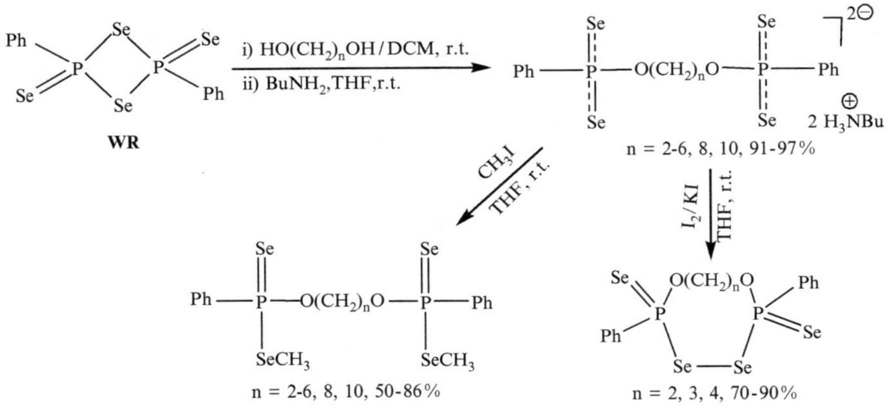

Scheme 1.89 Reaction of Woollins' reagent with diols [HO(CH$_2$)$_n$OH, n = 2 – 6, 8 and 10]

WR reacted with dibenzoylideneacetone to furnish a heterocycle with a planar C$_3$P(μ-Se)Se ring in [97]. Under similar conditions, an equivalent of 1,5-diphenylpentane-1,5-dione and **WR** gave rise to a cage [1–3] heterocycle in modest yield (Scheme 1.90) [98].

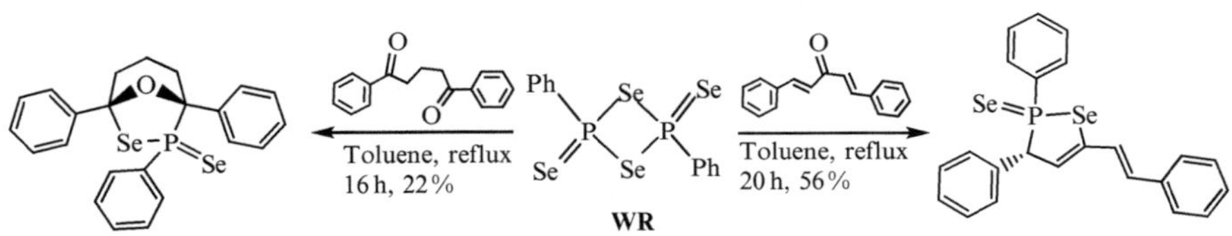

Scheme 1.90 Reaction of Woollins' reagent with dibenzoylideneacetone and 1,5-diphenylpentane-1,5-dione

1.6.3 *Woollins Reagent as Deoxygenation and Coupling Reagent*

In all above reactions, **WR** either delivers its selenium atom or incorporates part of itself into the products. However, there are striking cases that **WR** functions just as a reactant but left no selenium or phosphorus in the major products. This includes the deoxygenation of a series of sulfoxides with **WR**, leading to the corresponding sulfides under relatively mild conditions (Scheme 1.91) [99].

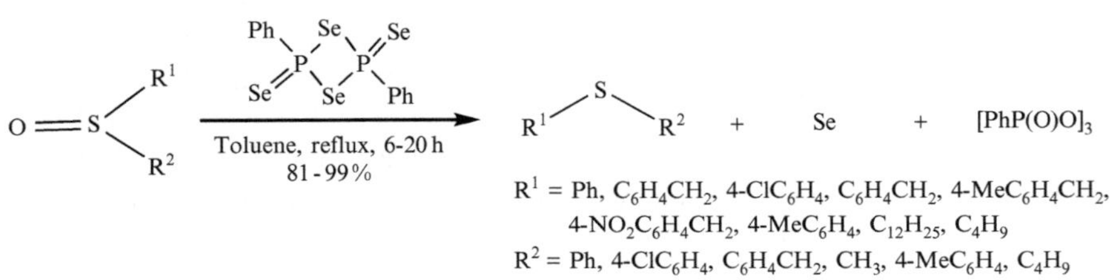

Scheme 1.91 Woollins' reagent as deoxygenation reagent for reduction of sulfoxides to the corresponding sulfides

The other application of **WR** is promoting coupling of a wide range of aromatic ketones and aldehydes to yield symmetrical and unsymmetrical (*E*)-olefins (Scheme 1.92) [100]. The method has potential for a variety of systems and might be beneficial where base sensitive substituents are present. Its speculative mechanism involves a Wittig-like reaction intermediate.

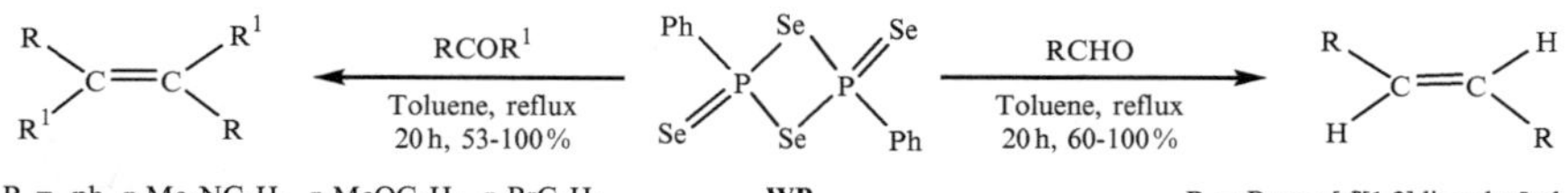

R = ph, p-$Me_2NC_6H_4$, p-$MeOC_6H_4$, p-BrC_6H_4, p-$C_lC_6H_4$, p-FC_6H_4, Biphenyl-2,2′-diyl, Oxydibenzene-2,2′-diyl, $C_{10}H_7$

R^1 = C_6H_5, p-$Me_2NC_6H_4$, p-$MeOC_6H_4$, p-BrC_6H_4, p-ClC_6H_4, p-FC_6H_4, Biphenyl-2,2′-diyl, Oxydibenzene-2,2′-diyl, Biphenyl-4-yl, CH_3

R = Benzo [*d*][1,3]dioxole-5-yl, 2,4-$(MeO)_2C_6H_4$, 4-$MeOC_6H_4$, 3-$MeOC_6H_4$, 3,5-$(MeO)_2C_6H_4$

Scheme 1.92 Woollins' reagent as coupling reagent for preparation of symmetrical and unsymmetrical (*E*)-olefins from aromatic ketones and aldehydes

Similarly, the reaction of **WR** with anthracene-9,10-dione generated 9.9'-bianthracene-10,10′-dione (Scheme 1.93) [100]. However, refluxing a toluene solution of **WR** with 9-cyanoanthracene, followed by treatment of water furnished the self-coupling product 9,9'-dicyanodianthracene as the only isolable compound, in which the C≡N group is retained (Scheme 1.93).

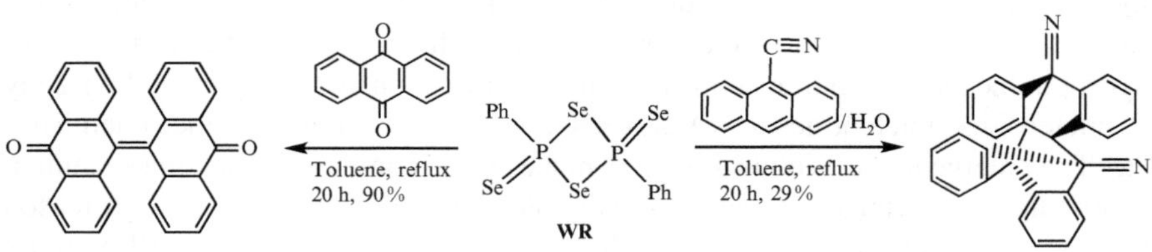

Scheme 1.93 Reaction of Woollins' reagent with anthracene-9,10-dione and 9-cyanoanthracene

A particularly interesting aromatic compound rather than an olefin has been achieved when diphenyl-1,3-diketone was utilized as starting material (Scheme 1.94). It can be anticipated that this reaction could be extended to a wide variety of β–diketones and perhaps even to heterocyclic systems to provide entry to a range of new heterocycles by using **WR** and ketones [100].

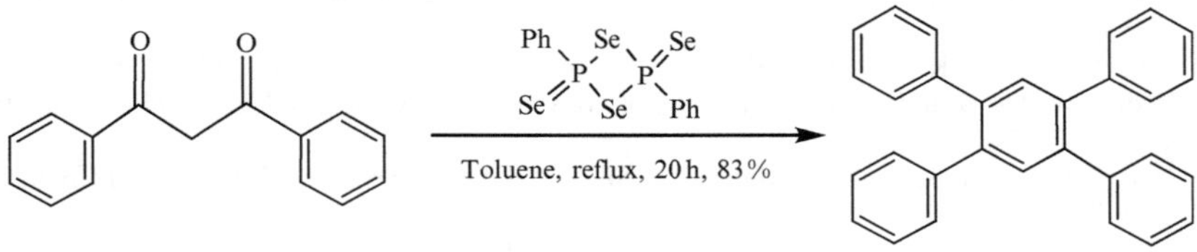

Scheme 1.94 Reaction of Woollins' reagent with diphenyl-1,3-diketone

In an attempt to synthesize the coupling product from the reaction of tetraphenylcyclopentadienone with **WR**, an unexpected product was obtained with a stable five-membered heterocycle. Steric hindrance is thought in this case to prevent the intermediate from being attacked further by another tetraphenylcyclopentadienone. Instead a hydrogen migration occurred to give an unusually organic phosphorus-selenium heterocycle (Scheme 1.95) [100].

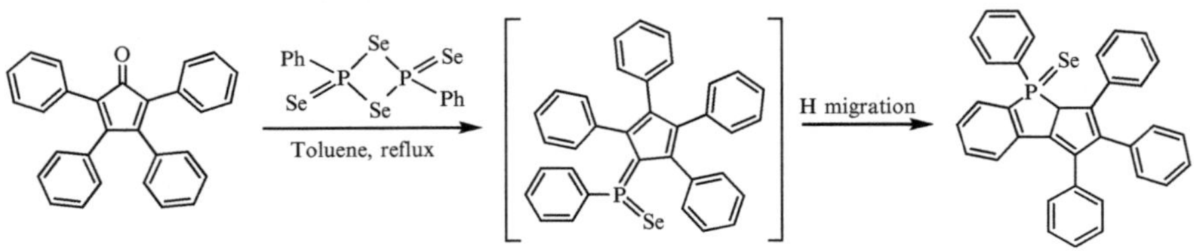

Scheme 1.95 Reaction of Woollins' reagent with tetraphenylcyclopentadienone

1.7 Conclusions

Phosphorus-selenium compounds have been involved in many aspects of organic chemistry, including oxidation/reduction, nucleophilic/electrophilic substitution, nucleophilic addition, free radical addition, Diel-Alder reaction, various cycloadditions, and so on. The resulting products have or potentially have applications in many fields such as synthetic chemistry, biochemistry and materials science. Indeed, many exemplary developments have been achieved through phosphorus-selenium chemistry, such as the construction of chiral organics, the formation of hydrophobic and hyrophilic dentrimers, the modification and X-ray crystallography of nucleosides, nucleotides, DNA and RNA, application in fluorescence chemistry, etc. The emergence of Woollins reagent makes more complex phosphorus-selenium compounds, especially heterocycles available. Therefore, we have enough reason to believe that the future of organic phosphorus-selenium chemistry will be more exciting, challenging and fruitful.

References

1. Alajarin M, Berna J, Leonardo CL, Steed JW (2008) Chem Commun 2337–2339
2. Kostyuk AN, Svyaschenko YV, Volochnyuk DM (2005) Tetrahedron 61:9263–9272
3. Svyaschenko YV, Kostyuk AN, Barnych BB, Volochnyuk DM (2007) Tetrahedron 63: 5656–5664
4. Segi M, Kawaai K, Honda M, Fujinami S (2007) Tetrahedron Lett 48:3349–3354
5. Trofimov BA, Brandsma L, Arbuzova SN, Maysheva SF, Gusarova NK (1994) Tetrahedron Lett 35:7647–7650
6. Arten'ev AV, Gusarova NK, Malysheva SF, Ushakov IA, Trofimov BA (2010) Tetrahedron Lett 51:2141–2143
7. Mann T, Riegler J (2002) Chem Eur J 8:4791–4795

8. Pavithran R, Reddy MLP, Junior SA, Freire RO, Rocha GB, Lima PP (2005) Eur J Inorg Chem 4129–4137
9. Gusarova NK, Ivanova NI, Konovalova NA, Albanov AI, Sinegovskaya LM, Avseenko ND, Sukhov BG, Mikhaleva AI, Gusarov AV, Trofimov BA (2007) Russ J Gen Chem 77:409–414
10. Ivanova NI, Volkov PA, Baikalova LV, Gusarova NK, Trofimov BA (2008) Chem Heterocyclic Comp 44:1359–1364
11. Oparina LA, Malysheva SF, Gusarova NK, Belogorlova NA, Vysotskaya OV, Stepanov AV, Albanov AI, Trofimov BA (2009) Synthesis 3427–3433
12. Trofimov BA, Malysheva SF, Belogorlova NA, Kuimov VA, Albanov AI, Gusarova NK (2009) Eur J Org Chem 3427–3431
13. Trofimov BA, Gusarova NK, Arbuzova SN, Ivanova NI, Artem'ev AV, Volkov PA, Ushakov IA, Malysheva SF, Kuimov VA (2009) J Organomet Chem 694:677–682
14. Gusarova NK, Ivanova NI, Maysheva SF, Artem'ev AV, Timokhin BV, Trofimov BA (2008) Russ J Gen Chem 78:1628–1630
15. Fedorov SV, Krivdin LB, Rusakov YY, Ushakov IA, Istomina NV, Belogorlova NA, Malysheva SF, Gusarova NK, Trofimov BA (2009) Magn Reson Chem 47:288–299
16. Trofimov BA, Artem'ev AV, Gusarova NK, Malysheva SF, Fedorov SV, Kazheva ON, Alexandrov GG, Dyachenko OA (2009) Synthesis 3332–3338
17. Gusarova NK, Artem'ev AV, Malysheva SF, Fedorov SV, Kazheva ON, Alexandrov GG, Dyachenko OA, Trofimov BA (2010) Tetrahedron Lett 51:1840–1843
18. Nguyen CQ, Afzaal M, Malik MA, Helliwell M, Raftery J, O'Brien P (2007) J Organomet Chem 692:2669–2677
19. Wilds CJ, Pattanayek R, Pan CL, Wawrzak Z, Egli M (2002) J Am Chem Soc 124: 14910–14916
20. Nawrot B, Widera K, Wojcik M, Rebowska B, Nowak G, Stec WJ (2007) FEBS J 274: 1062–1072
21. Bartoszewicz A, Kalek M, Stawinski J (2008) J Org Chem 73:5029–5038
22. Wozniak LA, Sochacki M, Mitsuya H, Kageyama S, Stec WJ (1994) Bioorg Med Chem Lett 4:1033–1036
23. Xu Y, Kool ET (2000) J Am Chem Soc 122:9040–9041
24. Sheng J, Jiang J, Salon J, Huang Z (2007) Org Lett 9:749–752
25. Du Q, Carrasco N, Teplova M, Wilds CJ, Egli M, Huang Z (2002) J Am Chem Soc 124:24–25
26. Hendrickson WA, Pahler A, Smith JL, Satow Y, Merritt EA, Phizackerley RP (1989) Proc Natl Acad Sci U S A 86:2190–2194
27. Yang W, Hendrickson WA, Crouch RJ (1990) Science 249:1398–1405
28. Hendrichson WA (2000) Trends Biochem Sci 25:637–643
29. Bollmark M, Stawinski J (2001) Chem Commun 771–772
30. Baraniak J, Korczynski D, Kaczmarek R, Stec WJ (1999) Nucleosides Nucleotides 18: 2147–2154
31. Lindh I, Stawinski J (1989) J Org Chem 54:1338
32. Stawinski J, Thelin M (1994) Org Chem 59:130–136
33. Tram K, Wang X, Yan H (2007) Org Lett 9:5103–5106
34. Carrasco N, Huang Z (2004) J Am Chem Soc 126:448–449
35. Carrasco N, Williams JC, Brandt G, Wang S, Huang Z (2006) Angew Chem Int Ed 45:94–97
36. Shi M, Jiang JK, Shen YM, Feng YS, Lei GX (2000) J Org Chem 65:3443–3448
37. Kimura T, Murai T (2004) Chem Lett 33:878–879
38. Kimura T, Murai T (2005) Chem Commun 4077–4079
39. Kimura T, Murai T (2005) Org Chem 70:952–959
40. Automu T, Murai T (2005) Tetrahedron: Asymmetry 16:3703–3710
41. Kimura T, Murai T, Mizuhata N (2005) Heteroat Chem 16:185–191
42. Kimura T, Muràì T, Miwa A, Kurachi D, Yoshikwaw H, Kato S (2005) J Org Chem 70: 5611–5617
43. Tang B, Ding B, Xu K, Tong L (2009) Chem Eur J 15:3147–3151

44. Murai T, Matsuoka D, Morishita K (2006) J Am Chem Soc 128:4584–4585
45. Murai T, Inaji S, Morishita K, Shibahara F, Tokunaga M, Obora Y, Tsuji Y (2006) Chem Lett 35:1424–1425
46. Murai T (2008) Phosphorus Sulfur Silicon Relat Elem 183:889–896
47. Murai T, Masaki Monzaki M, Shibahara F (2007) Chem Lett 852–853
48. Cholewinski G, Chojnacki J, Pikies J, Rachon J (2009) Org Biomol Chem 7:4095–4100
49. Murai T, Inaji S, Takenaka T (2009) Heteroat Chem 20:255–261
50. Borowiecka J (2000) Heteroat Chem 11:292–298
51. Krawczyk E, Skowronska A, Michalski J (2002) Dalton Trans 4471–4478
52. Salamonczyk GM, Kuznikowski M, Paniatowska E (2001) Chem Commun 2202–2203
53. Salamoczyk GM (2005) Phosphorus Sulfur Silicon Relat Elem 180:1051–1056
54. Salamonczyk GM, Kuznikowski M, Paniatowska E (2002) Tetrahedron Lett 43:1747–1749
55. Savin GA, Kammneva EA (2005) Russ J Org Chem 41:962–966
56. Cholewinski G, Witt D, Majewski R, Ossowski T, Rachon J (2007) Heteroat Chem 18:767–773
57. Potrzebowski MJ, Helinski J, Ciesielski W (2002) Chem Commun 1582–1583
58. Potrzebowski MJ, Potrzebowski WM, Jeziorna A, Ciesielski W, Gajda J, Bujacz GD, Chruszcz M, Minor W (2008) J Org Chem 73:4388–4397
59. Calera SG, Eisler DJ, Morey JV, McPartlin M, Singh S, Wright DS (2008) Angew Chem Int Ed 47:1111–1114
60. Liu CW, Shang IJ, Hung CM, Wang JC, Keng TC (2002) J Chem Soc Dalton Trans 1974–1979
61. Liu CW, Lobana TS, Xiao JL, Liu HY, Liaw BJ, Hung CM, Lin Z (2005) Organometallics 24:4072–4078
62. Sarkar B, Fang CS, You LY, Wang JC, Liu CW (2009) New J Chem 33:626–633
63. Baraniak J, Kaczmarek R, Wasilewska E, Korczynski D, Stec WJ (2004) Tetrahedron Lett 45:4269–4272
64. Maciagiewicz I, Dybowski P, Skowronska A (2003) Tetrahedron 59:6057–6066
65. Gray IP, Bhattacharyya P, Slawin AMZ, Woollins JD (2005) Chem Eur J 11:6221–6227
66. Hua G, Woollins JD (2009) Angew Chem Int Ed 48:1368–1377
67. Bhattacharyya P, Woollins JD (2001) Tetrahedron Lett 42:5949–5951
68. Bhattacharyya P, Slawin AMZ, Woollins JD (2004) Inorg Chem Commun 7:1171–1173
69. Zakrzewski J, Krawczyk M (2008) Heteroat Chem 19:549–556
70. Bethke J, Karaghiosoff K, Wessjohann LA (2003) Tetrahedron Lett 44:6911–6913
71. Hussaini SR, Hammond GB (2008) ARKIVOC 129–136
72. Hua G, Li Y, Slawin AMZ, Woollins JD (2006) Org Lett 8:5251–5254
73. Li Y, Hua G, Slawin AMZ, Woollins JD (2009) Molecules 14:884–892
74. Hua G, Henry JB, Li Y, Mount AR, Slawin AMZ, Woollins JD (2010) Org Biomol Chem 8:1655–1660
75. Hua G, Li Y, Fuller AL, Slawin AMZ, Woollins JD (2009) Eur J Org Chem 1612–1618
76. Mohanakrishann AK, Amaladass P (2005) Tetrahedron Lett 46:7201–7204
77. Amaladass P, Kumar NS, Mohanakrishnan AK (2008) Tetrahedron 64:7992–7998
78. Gomez Castano JA, Romano RM, Beckers H, Willner H, Boese R, Della Vedova CO (2008) Angew Chem Int Ed 47:10114–10118
79. Knapp S, Darout E (2005) Org Lett 7:203–206
80. Hua G, Fuller AL, Bühl M, Slawin AMZ, Woollins JD (2011) Eur J Org Chem 3067–3073
81. Bansal RK, Gupta N, Bharatiya N, Gupta G, Surana A, Hackenbracht G, Karaghiosoff K (1998) Heteroat Chem 9:445–452
82. Fitzmaurice JC, Williams DJ, Wood PT, Woollins JD (1988) J Chem Soc Chem Commun 741–743
83. Wood PT, Woollins JD (1998) J Chem Soc Chem Commun 1190–1191
84. Bhattacharyya P, Slawin AMZ, Woollins JD (2001) J Chem Soc Dalton Trans 300–303
85. Bhattacharyya P, Slawin AMZ, Woollins JD (2002) Chem Eur J 8:2705–2711

86. Hua G, Li Y, Slawin AMZ, Woollins JD (2007) Eur J Inorg Chem 891–897
87. Hua G, Li Y, Slawin AMZ, Woollins JD (2007) Chem Commun 1465–1467
88. Bhattacharyya P, Slawin AMZ, Woollins JD (2000) Angew Chem Int Ed 39:1973–1975
89. Hua G, Zhang Q, Li Y, Slawin AMZ, Woollins JD (2009) Tetrahedron 65:6074–6082
90. Foreman MStJ, Slawin AMZ, Woollins JD (1999) J Chem Soc Dalton Trans 1175–1184
91. Slawin AMZ, Woollins JD (1997) J Chem Soc Chem Commun 855–856
92. Bhattacharyya P, Slawin AMZ, Woollins JD (2001) J Organomet Chem 623:116–119
93. Hua G, Fuller AL, Li Y, Slawin AMZ, Woollins JD (2010) New J Chem 34: 1565–1571
94. Hua G, Fuller AL, Slawin AMZ, Woollins JD (2010) Eur J Org Chem 2707–2615
95. Hua G, Li Y, Slawin AMZ, Woollins JD (2008) Tetrahedron 64:5442–5448
96. Hua G, Li Y, Slawin AMZ, Woollins JD (2008) Angew Chem Int Ed 47:2857–2859
97. Li Y, Hua G, Slawin AMZ, Woollins JD (2008) Acta Crystallogr E E64:o4
98. Hua G, Li Y, Slawin AMZ, Woollins JD (2008) Acta Crystallogr E E64:o184
99. Hua G, Woollins JD (2007) Tetrahedron Lett 48:3677–3679
100. Hua G, Li Y, Slawin AMZ, Woollins JD (2007) Dalton Trans 1477–1480

Chapter 2
New Selenium Electrophiles and Their Reactivity

Diana M. Freudendahl and Thomas Wirth

2.1 Introduction

The functionalisation of activated carbon-carbon double bonds with electrophilic organoselenium compounds has been successfully applied in various cases [1–3]. Several research groups including ours have investigated stereoselective reactions of alkenes with chiral selenium electrophiles [4–22]. A range of optically active diselenides has been synthesised and the selenium electrophiles generated from these diselenides can add to alkenes with high selectivities. A variety of nucleophiles have been used to open the seleniranium intermediates **1** and the addition products **2** have been used in different subsequent reactions (Scheme 2.1).

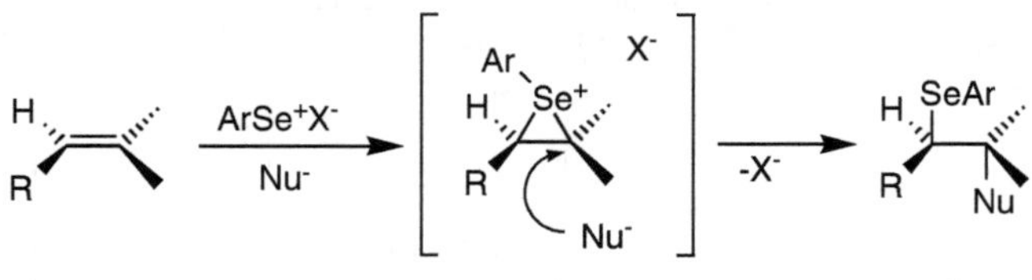

1 **2**

Scheme 2.1 Selenenylation of alkenes

2.2 Generation and Reactivity of Selenium Electrophiles

Phenylselenenyl chloride and phenylselenenyl bromide are commercially available and can be easily produced from diphenyl diselenide by treatment with sulfuryl chloride or chlorine in hexane and with bromine in tetrahydrofuran, respectively.

D.M. Freudendahl • T. Wirth (✉)
School of Chemistry, Cardiff University, Main Building, Park Place, Cardiff CF10 3AT, UK
e-mail: wirth@cf.ac.uk

J.D. Woollins and R.S. Laitinen (eds.), *Selenium and Tellurium Chemistry*,
DOI 10.1007/978-3-642-20699-3_2,

Other selenium electrophiles can be easily prepared from the corresponding diselenides employing the same strategy. In order to avoid the incorporation of the halide anions as nucleophiles during the selenenylation reaction which could, due to the reversibility, lead to a potential decrease in stereoselectivity it is possible to exchange the halide ion *in situ*. Silver salts such as triflate [23–25], hexafluorophosphate [26], hexafluoroantimonate [26], or toluenesulfonate [27, 28] can be employed for this purpose. The selenium electrophiles can also be generated by oxidation with ammonium peroxodisulfate [29, 30]. Phenylselenyl sulfate is a very efficient reagent and can be generated via this route. Other reagents like KNO_3 [29–31], $CuSO_4$ [29, 30], $Ce(NH_4)_2(NO_2)_6$ [29, 30, 32], $Mn(OAc)_3$ [33], [bis (trifluoroacetoxy)iodo]benzene [34] and (diacetoxyiodo)benzene [35] were also successfully employed to generate selenium electrophiles by oxidation from the corresponding diselenides. A third possibility is the generation of the phenylselenyl cation via photosensitised single electron transfer from the diselenide to 1,4-dicyanonaphthalene [36, 37]. However, the synthesis of the selenium electrophiles is strongly depending on the reaction requirements. Very efficient and broadly applicable are the triflate and the sulfate counterions, which generally produce clean reactions. The preparation of the sulfate is easier than the generation of the triflate, but is has the drawback that it cannot be employed at temperatures below −30°C.

The most important use of electrophilic phenylselenyl reagents is the functionalisation of carbon-carbon double and triple bonds. The selenium moiety in the addition products **2** can be used for further useful transformations as shown in Scheme 2.2.

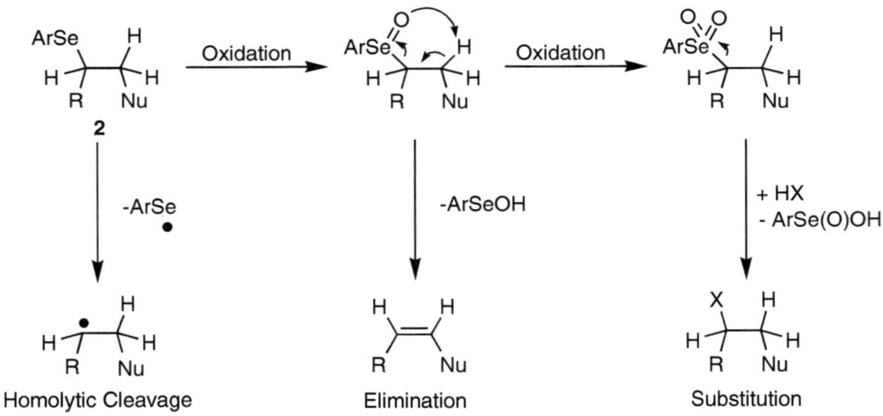

Scheme 2.2 Deselenenylation strategies

The arylseleno group can be functionalised in several ways. Treatment with tin hydrides leads to the homolytic cleavage of the carbon-selenium bond, and the generated carbon radicals can be used for subsequent radical reactions. The oxidation of the arylseleno moiety to the selenoxides leads to elimination products via the well known selenoxide elimination mechanism. Further oxidation to the selenone results in the generation of a good leaving group, which can be substituted by a variety of nucleophiles X as shown in Scheme 2.2. Likewise, the treatment with another equivalent of the selenium electrophile generates a selenonium ion which can be substituted by a nucleophile as well.

The addition of optically active selenenylating reagents to carbon-carbon double bonds leads to a mixture of two diastereomers. In some cases, these diastereomers can be separated and the subsequent deselenenylation leads to enantiomerically pure products. Asymmetric oxyselenenylation reactions with an external nucleophile (e.g., methanol) are often employed to test the efficiency of a new chiral selenium electrophile. The stereospecific *anti*-addition of an organoseleno-group and an oxygen nucleophile are used for the preparation of simple as well as complex molecules.

Wirth and co-workers investigated the methoxyselenenylation of styrene in detail and established the stereochemical course of the reaction [38–40]. The stereochemistry determining step is the formation of seleniranium intermediates **1** during the attack of the alkene onto the selenium electrophile [41, 42].

Recent advances with selenenylation reactions include an asymmetric azido selenenylation reaction by Tiecco et al. (Scheme 2.3), which allows further transformations into aziridines and triazoles [43]. It is remarkable that his reaction occurs with a very high level of facial selectivity and with "Markovnikov" orientation. The sulfur atom in the chiral side chain in electrophile **3** seems to play an important role in this reaction.

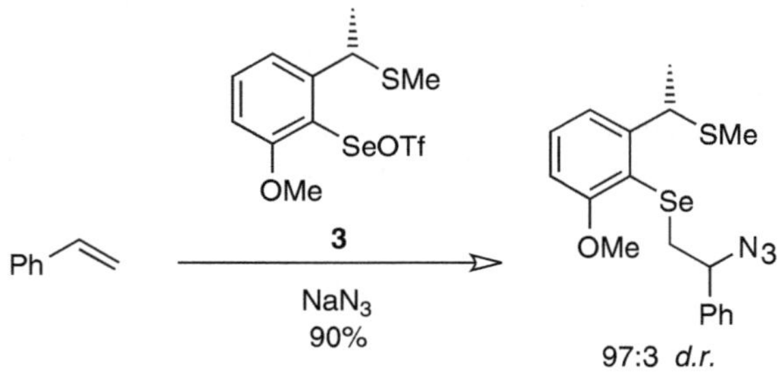

Scheme 2.3 Asymmetric azido-selenenylation by Tiecco et al.

2.3 Reactivity of Sulfoxide-Containing Selenium Electrophiles

We recently reported the synthesis of (*t*-butylsulfinyl)phenyl diselenide **4** and 2-(*t*-butylsulfinyl)-6-methoxy phenyl diselenide **5** by lithiation of the corresponding aryl sulfoxides and treatment with elemental selenium, followed by an oxidative work-up [44]. The diselenides were used for methoxyselenenylations to establish their ability to influence the stereochemical outcome of these reactions (Scheme 2.4). As already stated, there is a broad choice of methods and reagents available to generate the selenium electrophiles from the diselenides. With respect to the nature of the chiral centre of the diselenides, a sulfoxide moiety, the generation of the electrophiles with bromine was preferred over the oxidative methods.

Because of the good results which, according to literature, were obtained with triflates as counterions, silver triflate was used for the halogen exchange reaction.

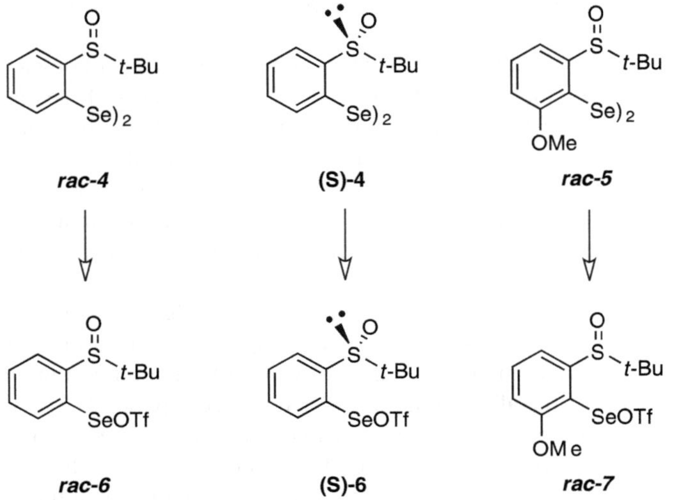

Scheme 2.4 Sulfoxide-based diselenides and their corresponding selenenyl triflates

A typical procedure proceeds by generation of the selenenyl bromide from the diselenide with bromine, the exchange of the bromide with the less nucleophilic triflate, and the addition of the selenenyl cation to the carbon-carbon double bond in the presence of a nucleophile leading to product formation (Scheme 2.5). The diastereomeric ratio (*d.r.*) of the products was determined by NMR. Initial reactions were screened with 2-chlorostyrene as substrate in different solvents.

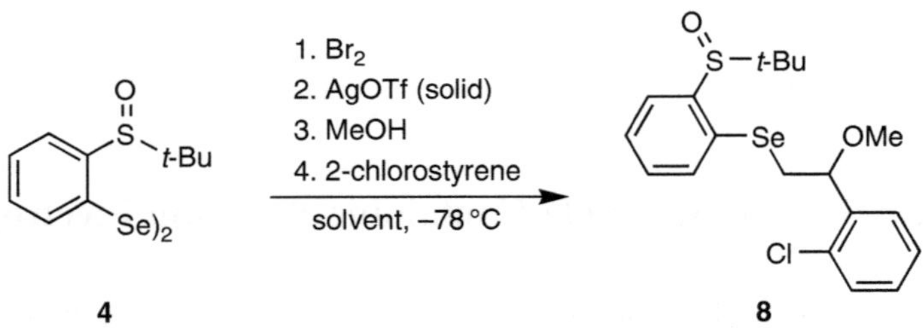

Scheme 2.5 Methoxyselenenylation of 2-chlorostyrene in different solvents

The highest diastereomeric ratios with triflate **6** were found using dichloromethane (11:1) or chloroform (7:1) (Table 2.1). These solvents also proved to give the best yields (up to 48%). When the reactions were carried out in polar ethers like tetrahydrofurane and cyclopentyl methyl ether (CPME), the selectivities dropped to 4:1. However, in diethyl ether the selectivity was slightly higher (5:1),

Table 2.1 Selectivities of the methoxyselenenylation of 2-chlorostyrene with **6** in different solvents

Entry	Solvent	Yield [%]	*d.r.*[a]
1	THF	29	4:1
2	CPME[b]	29	4:1
3	Et_2O	36	5:1
4	Et_2O:CH_2Cl_2 4:1	41	5:1
5	CH_2Cl_2	41	11:1
6	$CHCl_3$[c]	48	7:1

[a]Determined from NMR spectra of the crude products.[b]CPME: cyclopentylmethyl ether.[c]Reaction performed at −50°C

which could be caused by the difference in the solvation of the diselenide. The diastereomeric ratio in a 4:1 mixture of diethyl ether and dichloromethane was again observed as 5:1, presumably because of the high excess of the more polar solvent. The reaction was not carried out in unpolar solvents such as hexane or toluene, as the diselenide is insoluble in these solvents. It is assumed that the selectivities are a synergetic effect of the coordination of the oxygen from sulfoxide to the selenium and the bulky *t*-butyl group. The coordination of the oxygen would be stronger in less polar solvents and would force therefore the chiral centre closer to the reaction site.

Interestingly, the colour of the selenium electrophile with the triflate **6** is depending on the solvent (Fig. 2.1) and can be used as an indicator for the progress of the reaction. This is not the case with **7** which always gives yellow mixtures.

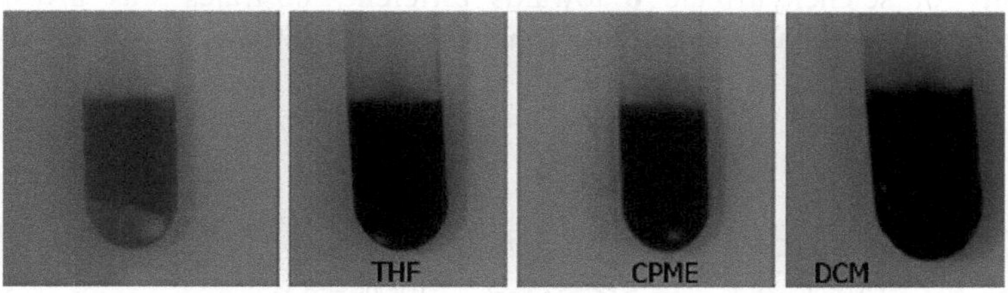

Fig. 2.1 Colour of **7** (*orange* in all solvents) and **6** in THF: *purple*, CPME: *red*, CH_2Cl_2: *green*)

Surprisingly, triflate **7** was less reactive and much less selective under the same reaction conditions (Table 2.2). The highest diastereomeric ratio was observed in tetrahydrofurane (3:1), but with low yield (18%). In chloroform and cyclopentylmethyl ether (CPME) the yields were higher but the diasteromeric ratio decreased to 2:1 and 1:1, respectively. Obviously, triflate **7** shows a slightly better selectivity in polar solvents than in unpolar solvents, compared to triflate **6**, where the reactivity is inverse. It could not yet be established why the reaction shows some selectivity in tetrahydrofuran but not in solvents of similar polarity such as diethyl ether and cyclopentylmethyl ether.

The reason for the low yields obtained with both selenenylating reagents could not be determined. Besides the products, it was generally possible to re-isolate most of the diselenide and 2-chlorostyrene with a variable amount of byproducts

Table 2.2 Selectivities of the methoxyselenenylation of 2-chlorostyrene with **7** in different solvents

Entry	Solvent	Yield [%]	*d.r.*[a]
1	THF	18	3:1
2	CPME[b]	30	1:1
3	Et_2O	10	1:1
4	CH_3CN	23	1:1
5	CH_2Cl_2	20	1:1
6	$CHCl_3$[c]	40	2:1

[a]Determined from NMR spectra of crude products.[b]CPME: cyclopentylmethyl ether.[c]Reaction performed at −50°C

(10–30%) which were not characterised. It is possible that the diselenides are not completely converted to the corresponding selenenyl bromides in the first step of the one-pot procedure. Oxidative reagents for the generation of the selenium electrophiles were ruled out due to the nature of the auxiliary and the alternative generation of the corresponding selenenyl chlorides with SO_2Cl_2 failed. The silver triflate was suspected as a possible cause for the low yields, but could be ruled out as a test reaction with commercially available selenenyl chloride showed yields of 80%. Most of the styrene derivatives were used without further purification, which could have been responsible for the low yields as well. However, when freshly distilled styrene derivatives were used it was not possible to observe significant higher yields (increase by 2–3%). To keep all experiments comparable, it was decided to continue to generate the selenenylating reagents with bromine. The reactivity of selenenyltriflate **6** towards different substrates was investigated employing methanol as standard nucleophile (Scheme 2.6).

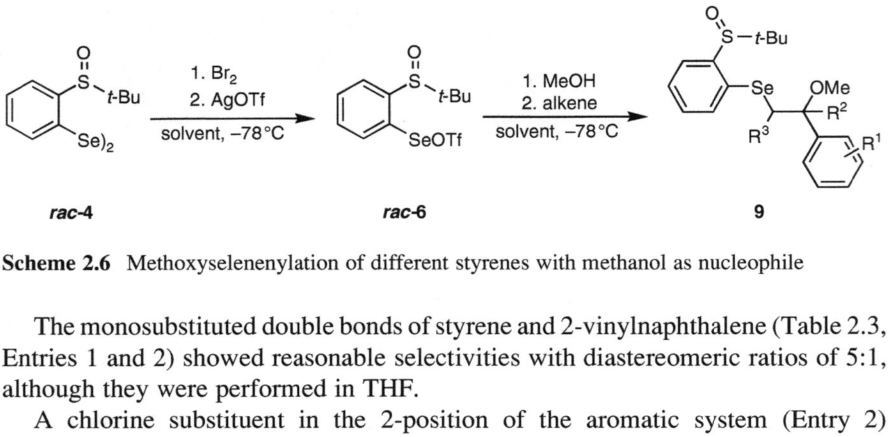

Scheme 2.6 Methoxyselenenylation of different styrenes with methanol as nucleophile

The monosubstituted double bonds of styrene and 2-vinylnaphthalene (Table 2.3, Entries 1 and 2) showed reasonable selectivities with diastereomeric ratios of 5:1, although they were performed in THF.

A chlorine substituent in the 2-position of the aromatic system (Entry 2) enhanced the selectivity considerably (11:1) and led to **9c** in 41% yield. Generally, the use of sterically more hindered alkenes led to lower yields (Entries 2–6) between 30% and 41%. However, β-substitution on the styrene enhanced the selectivity (Entries 5 and 6) to 11:1 and 9:1, respectively. The slightly electron deficient double bonds of methyl cinnamate and 3-nitrostyrene were not reactive

Table 2.3 Reaction of ***rac*-4** with different styrene derivatives

Entry	Alkene	Solvent	Yield [%]	*d.r.*[a]	Product	
1	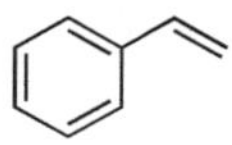	THF	52	5:1	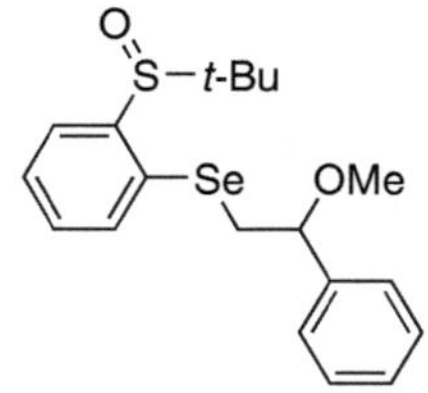	**9a**
2	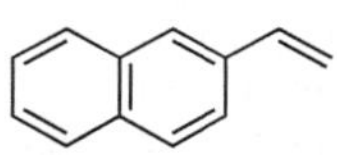	THF	32	5:1	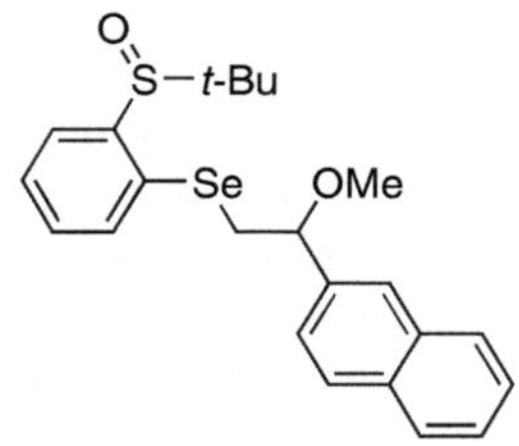	**9b**
3	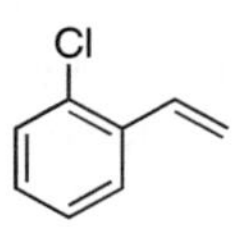	CH_2Cl_2	41	11:1	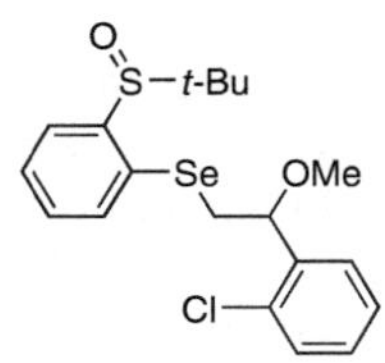	**9c**
4	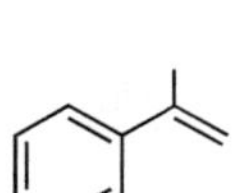	CH_2Cl_2	38	4:1	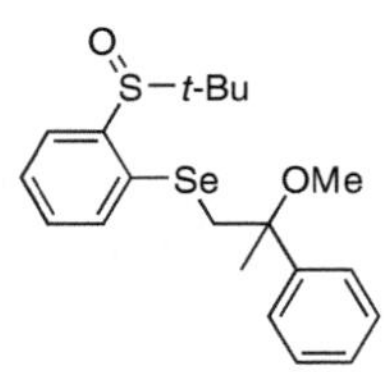	**9d**
5	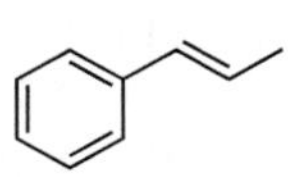	CH_2Cl_2	30	11:1	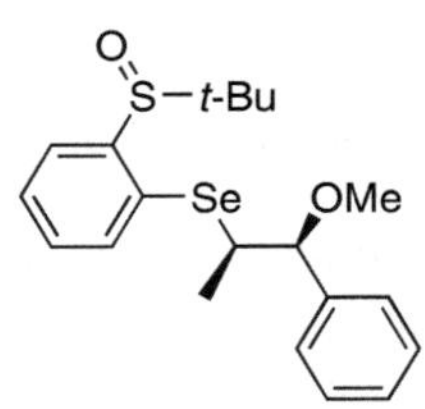	**9e**
6		CH_2Cl_2	30	9:1		

(continued)

Table 2.3 (continued)

Entry	Alkene	Solvent	Yield [%]	d.r.[a]	Product
	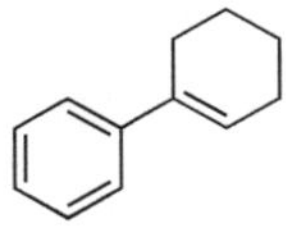				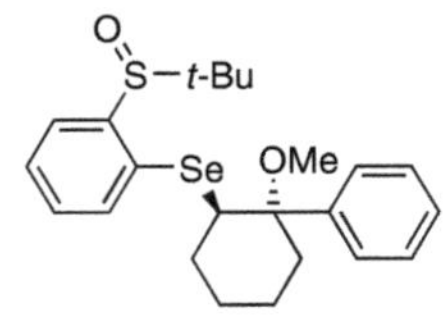 **9f**

[a]Determined from NMR spectra of crude products

enough under the conditions of the methoxyselenenylation reaction. The diastereomeric ratios obtained using the enantiomerically enriched **(*S*)-6** and the racemic selenenyltriflate ***rac*-6** are identical (5:1). Whereas the optical rotation obtained for **9a** was $[\alpha]_D^{20} = 0$ when using ***rac*-6**, the value for **9a** was $[\alpha]_D^{20} = -128$ when **(*S*)-6** was used, but the absolute stereochemistry was not determined.

The same investigation was performed with selenenyl triflate **7**. The solvent screening had already shown that **7** was much less selective than **6** using 2-chlorostyrene as substrate. As this observation could only be due to the substrate two other styrenes were tested (Scheme 2.7). Unsubstituted styrene lead to a diastereomeric ratio of 2:1 with 24% yield and *trans-ß*-methylstyrene to a product ratio of 1:1 in 22% yield. No further investigations using diselenide **5** were undertaken.

In a last series of reactions, the influence of the nature of the nucleophile on the diastereoselectivity was investigated, employing triflate **6** in the selenenylation reaction with 2-chlorostyrene as substrate.

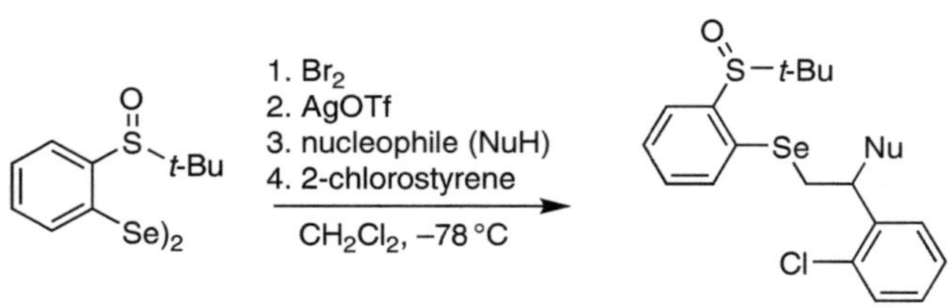

4 **9**

Scheme 2.7 Selenenylation of 2-chlorostyrene with different nucleophiles

The reaction was performed with several oxygen, nitrogen and sulfur nucleophiles. Table 2.4 presents the results obtained with different oxygen nucleophiles. The yields with small nucleophiles such as methanol, ethanol and *i*-propanol are comparable and better than that obtained with *t*-butanol, which is lower due to its steric bulkiness. The diastereomeric ratios are decreasing from methanol to ethanol and *i*-propanol to *t*-butanol. The nucleophilicity of these alcohols can be influenced by several properties [45], of which: (a) the solvation energy of the nucleophile;

Table 2.4 Selenenylations using diselenide **4** with 2-chlorostyrene and different nucleophiles

Entry	Nucleophile (NuH)	Product	Yield [%]	*d.r.*[a]
1	MeOH	**9c**	41	11:1
2	EtOH	**9g**	47	8:1
3	*i*-PrOH	**9h**	47	8:1
4	*t*-BuOH	**9i**	30	6:1
5	BnOH	**9j**	30	3.5:1
6	$PhCO_2H$	–	Traces	–
7	$TMSN_3$	**9k**	35	6:1

[a]Determined from NMR spectra of crude products

(b) the bond strength of the new Nu-C bond formed; (c) the electronegativity of the attacking atom; (d) the polarisability of the attacking atom; and (e) the steric bulk of the nucleophile are most important.

The first characteristic expresses that a strong solvation leads to an increase in the activation energy, as the solvation shell of the anionic nucleophile needs to be disrupted. The second suggests that if the newly formed bond is very stable, the transition state in a S_N2 reaction shows a higher stability as well. This leads overall to a decrease in the activation energy. The third point takes into account that a strongly electronegative nucleophilic centre is less reactive because the electrons, which are necessary for the bond formation, are more tightly bound to the nucleus. This observation is closely related to the fourth statement, the polarisability of the electron shell. Finally, a more sterically hindered nucleophile is less reactive than an unhindered centre, because it can not necessarily avoid the non-bonded repulsions in a transition state.

Taking these points into account, it can be assumed that the most important influence in the above attempted selenenylation reactions is the steric bulk which increases from methanol to ethanol and *i*-propanol to *t*-butanol. However, the selectivity obtained with benzyl alcohol (3.5:1, Table 2.4, Entry 5) can not be explained by the steric bulk involved. In this case it should be expected that the diastereoselectivity would be better than 6:1 (as for *t*-butanol). If the rate of the epimerisation of the three membered ring systems is higher than that of the nucleophilic attack of the alcohol, this would lead to a decrease in diastereoselectivity. Hence it can be assumed that another of the mentioned properties, concerning the nucleophilicity of benzyl alcohol, gains more influence. Using benzoic acid as nucleophile (Table 2.4, Entry 6) led only to traces of the product, which was not isolated.

It was also attempted to use thiophenol and several nitrogen nucleophiles such as *n*-butyl amine, benzyl amine, *N*-methylbenzyl amine, sodium azide and trimethylsilyl azide in different solvents. However, the only successful reaction occurred with trimethylsilyl azide ($TMSN_3$) as nucleophile in dichloromethane as solvent (Table 2.4, Entry 7), which resulted in 35% yield and in a diastereomeric ratio of 6:1. Employing other nitrogen or sulfur nucleophiles resulted either in the recovery of starting material or led to complex reaction mixtures.

2.4 Chiral Counteranions in Selenenylation Reactions

Most selenium electrophiles are generated *in situ* by treating the corresponding diselenides with chlorine, sulfuryl chloride or bromine. Addition reactions with these reagents can be problematic as the halide anions compete with other external nucleophiles or during cyclisation reactions, which can lead to undesired side products and to a decrease in selectivity. Halide anions can be replaced with less nucleophilic anions such as triflate, sulfate, perchlorate, tetrafluoroborane or hexafluorophosphate by treating the selenium electrophile with the appropriate silver salt as shown for silver triflate in Scheme 2.8.

Scheme 2.8 Exchange of counteranion

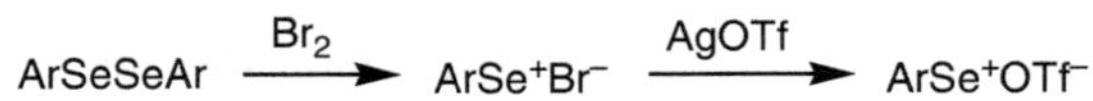

The effect of the counteranion on the course of these reactions was investigated by Tomoda et al. [16], Tiecco and co-workers [46], and Khokhar and Wirth [47]. On the basis of these results it was suggested that a decrease in the nucleophilicity of the counteranion, i.e., an increase in the electrophilicity of the selenium reagent, produces an enhancement of the diastereomeric excess in these reactions. These examples have shown that the counterion can have a strong influence on the selectivities of selenenylation reactions. Therefore it seemed to be viable to combine these observations with the concept of the "Asymmetric Counteranion Directed Catalysis" (ACDC). This is a relatively new development which emerged in the field of organocatalysis [48]. Reactions proceeding *via* anionic intermediates have been successfully influenced by chiral cationic counterions. The combination of chiral ions with enantiopure counterions can lead to two ion pairs with different stabilities and different chemical and physical properties. One ion pair can be formed preferentially if it is more stable. This concept has already been used for racemic resolutions as well as for synthesis. It is well known that high levels of asymmetric induction can be achieved using chiral cations and achiral anions. Recently, it has also been shown that this concept works also for cationic intermediates and transition states with chiral counteranions. For a long time, only cationic counterions were used and able to control the stereoselective outcome of reactions and to produce chiral products as single enantiomers. However, in recent years several independent reports in the field of enantioselective organocatalysis could show that chiral anions are also able to influence the selectivity of a reaction.

In 2003, Lacour published a review highlighting the synthesis and reactions of chiral counteranions [49]. Beside natural compounds such as chiral carboxylic acids **10** and sulfonic acids **11**, which possess a rather large number of potential conformations, tetrahedral borate **12**, a number of chiral metallo-organic complexes **13**, and phosphate anions **14** have been investigated as shown in Fig. 2.2.

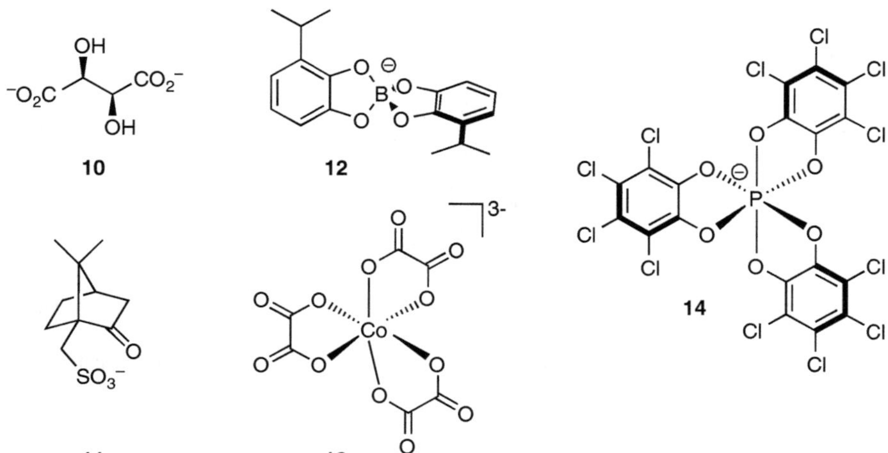

Fig. 2.2 Chiral counteranions

Toste and co-workers reported in 2007 the first highly successful application of the metal-ACDC catalysis concept in a gold(I) catalysed heteroatom cyclisations of allenes [50]. High enantioselectivities with up to 99% enantiomeric excess were observed in hydroaminations and hydroalkoxylations of allene derivatives. In 2006, Komanduri and Kirsche had already described a Rh-catalysed reductive coupling of 1,3-enynes with heterocyclic aromatic carbonyl compounds using chiral bisphosphine ligands [51]. The group of List reported the first application of the chiral counteranion strategy in the Pd-catalysed asymmetric allylic alkylation [52] and a manganese(III)-catalysed epoxidation of alkenes [53]. The same group has shown that chiral phosphoric acids together with secondary amines can modulate the enantioselective transfer hydrogenation of α,β-unsaturated aldehydes with an enantiomeric excess of up to 98% [54]. The catalytic concept of ACDC was also successfully applied for the iminium-catalysed enantioselective epoxidation of α,β-unsaturated aldehydes [55] for the enantioselective synthesis of β-alkoxy amines and for the desymmetrisation of *meso*-episulfonium ions [56].

The counteranion effects on the enantioselectivity of selenenylations and the synthesis and use of chiral counter anions using the ACDC concept seemed to be a promising tool to enhance the diastereoselectivity of selenenylation reactions. Several silver salts (Fig. 2.3) were synthesised as chiral anionic reagents for the selenenylation reactions of alkenes.

BINOL derivatives have proven to be effective in several asymmetric counter-anion mediated reactions. A tool to enhance the rigidity of the backbone is the introduction of large substituents at 3,3′-positions or 6,6′-positions on the BINOL scaffold. In this case the synthetic efforts were concentrated on the 3,3′-positions. Precursors for the silver salts of phosphoric acids **15** and **16** are commercially available which can be prepared easily [49]. Additionally, the silver salt of camphorsulfonic acid **18** [57] and the prolin derivative **19** were synthesised from camphor

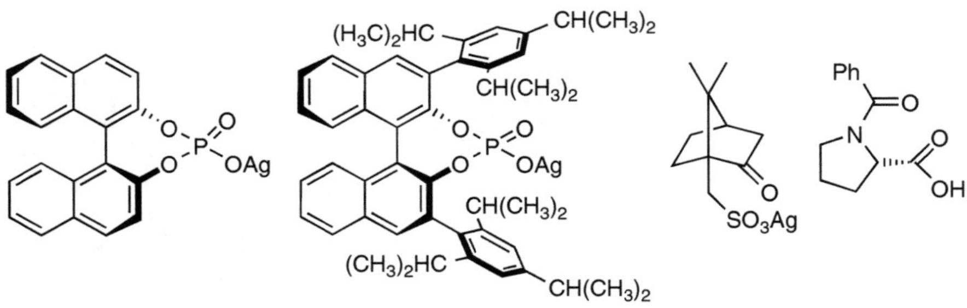

(*S*)-**15** (*S*)-**16** (*S*)-**17** (*S*)-**18**

Fig. 2.3 Chiral silver salts to introduce counteranions for electrophilic selenium reagents

and proline. Both starting materials are cheap and especially proline derivatives have already proven to be successful organocatalysts.

Ideally, the use of an unsubstituted selenenyl halide like phenylselenenyl bromide together with a chiral counteranion would form a new chiral selenenylating reagent to achieve a stereoselective reaction pathway during a selenenylation reaction. To test this hypothesis phenylselenenyl bromide **21** was generated from diselenide **20** and the chiral silver salts were added either neat or as a methanolic solution (Scheme 2.9). In principle, this approach leads to the formation of the envisioned chiral reagents and could, when reacting with styrene, lead to enantiomeric enriched products.

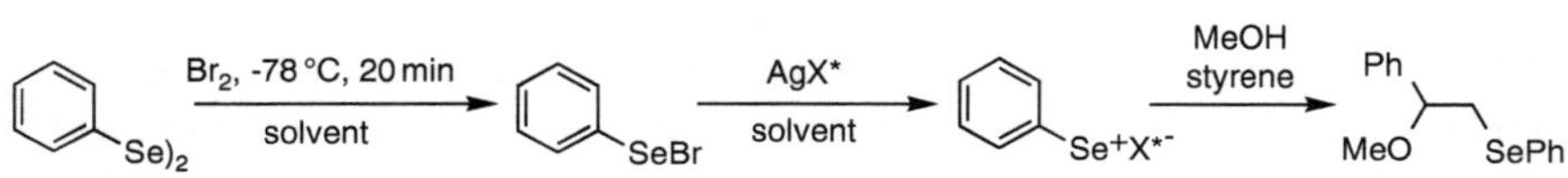

19 **20** **21** **22**

Scheme 2.9 General reaction using chiral counteranions to generate new chiral selenenylating reagents **20**

Initially, phenyl selenenyl bromide **20** was used together with the silver salt (*S*)-**15**. Three different solvents, diethylether, dichloromethane and toluene, were investigated. It was expected that the selectivity, if any could be observed, would increase from the more to the less polar solvents.

The low solubility of the silver salts in some solvents was a problem for this reaction. It was improved when the reaction was carried out in dichloromethane, but the selectivities observed by HPLC analysis were extremely small as shown in Table 2.5.

With further investigations using different silver salts it could be established if the encountered low enantiomeric excesses were generic for these reactions or if the properties of the chosen silver salt (*S*)-**15** were insufficient to positively influence the stereochemical course of the methoxyselenenylation. (*S*)-**16** was used under

Table 2.5 Methoxyselenenylation reaction with phenylselenenyl bromide, styrene and different silver salts

Silver salt	Solvent	**22** Yield [%]	*e.e.* [%][a]
(*S*)-**15**	Diethylether	21	0
(*S*)-**15**	Dichloromethane	53	3
(*S*)-**15**	Toluene	–	–
(*S*)-**16**	Diethylether	10	0
(*S*)-**16**	Dichloromethane	48	1
(*S*)-**16**	Toluene	24	1
(*S*)-**18**	Acetonitrile	18	0

[a]Determined by HPLC

identical reaction conditions as mentioned above. It was possible to obtain 1% *e.e.* using dichloromethane and toluene as solvents. However, this is again within the error limit of the HPLC system. In diethylether, the product was racemic.

To exclude the possibility that the use of styrene or methanol has major effects (*S*)-**15** was tested in a cyclisation reaction of acid **23** as shown in Scheme 2.10.

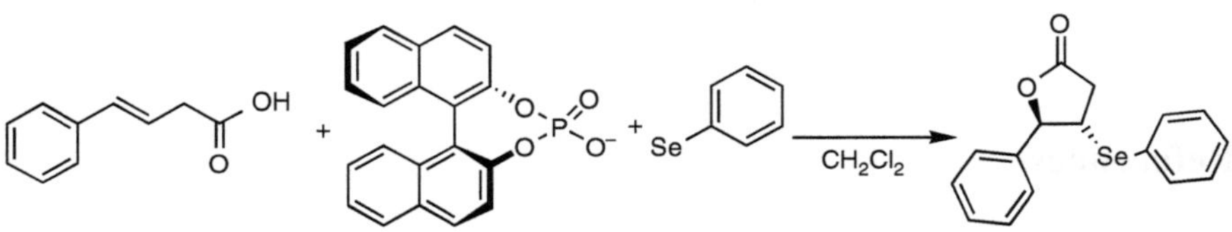

Scheme 2.10 Selenocyclisation with phenylselenenyl derivative 24

The selenenyl compound **24** was again obtained from diphenyl diselenide **19** with bromine and (*S*)-**15** in dichloromethane at −78°C. The selenocyclisation was carried out at room temperature by addition of (*E*)-4-phenylbut-3-enoic acid **23**. The product **25** was obtained in 26% yield after 4 h. According to HPLC analysis the chiral counteranion again did not show any influence in the stereochemical outcome of the reaction, the product **25** was obtained as a racemate.

As it was not possible to obtain and isolate a silver salt from acid (*S*)-**18**, a different approach was used to investigate its ability to influence the stereochemical outcome of these reactions. The chiral anion was generated *in situ* when the phenylselenyl bromide **20** was treated with silver(I) carbonate which led to the precipitation of silver bromide. The carbonate counteranion could be easily exchanged with *N*-benzoylproline (*S*)-**18** due to the higher acidity of the carboxylic acid (Scheme 2.11).

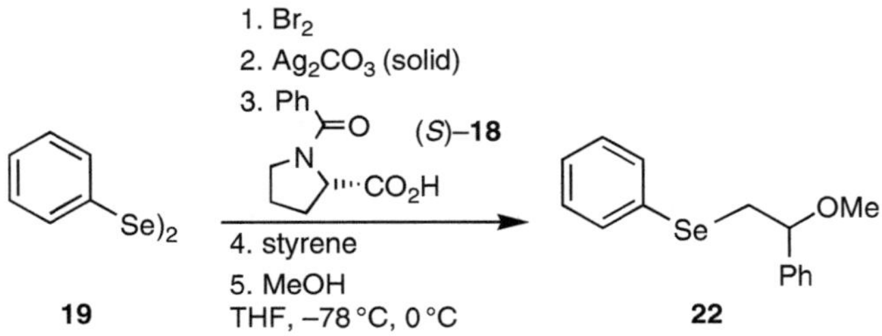

Scheme 2.11 Methoxyselenenylation with (*S*)-**18** as chiral source

This methoxyselenenylation has again been performed in different solvents (tetrahydrofurane, toluene, cyclopentylmethyl ether, toluene/cyclopentylmethyl ether 19:1, chlorobenzene). The obtained enantiomeric excesses in all reactions are only 3% or below.

In conclusion, two new diselenides were synthesised with a sulfoxide moiety as a side chain. One of these diselenides shows, depending on the solvent, good diastereoselectivities in the methoxyselenenylation of activated alkenes. Investigations towards stereoselective methoxyselenenylations using chiral counteranions have, however, been unsuccessful.

References

1. Hölzle G, Jenny W (1958) Helv Chim Acta 41:593–603
2. Back TG (ed) (1999) Organoselenium chemistry. Oxford University Press, Oxford
3. Wirth T (ed) (2000) Organoselenium chemistry: modern developments in organic synthesis, topics in current chemistry, vol 208. Springer, Berlin
4. Tiecco M (2000) Top Curr Chem 208:7–54
5. Wirth T (2000) Angew Chem 112:3890–3900; (2000) Angew Chem Int Ed 39:3740–3749
6. Paulmier C (ed) (1986) Selenium reagents and intermediates in organic synthesis. Pergamon Press, Oxford
7. Patai S, Rappoport Z (eds) (1986) The chemistry of organic selenium and tellurium compounds, vol 1. Wiley, New York
8. Patai S, Rappoport Z (eds) (1987) The chemistry of organic selenium and tellurium compounds, vol 2. Wiley, New York
9. Liotta D (ed) (1987) Organoselenium chemistry. Wiley, New York
10. Beaulieu PL, Déziel R, Back TG (eds) (1999) Organoselenium chemistry: a practical approach. Oxford University Press, Oxford, pp 35–66
11. Wirth T (1999) Tetrahedron, vol 55., pp 1–28
12. Browne DM, Wirth T (2006) Curr Org Chem 10:1893–1903
13. Freudendahl DM, Shazhad SA, Wirth T (2009) Eur J Org Chem 1649–1664
14. Déziel R, Malenfant E, Belanger G (1996) J Org Chem 61:1875–1876
15. Déziel R, Malenfant E, Thibault C (1998) Tetrahedron Lett 39:5493–5496
16. Fujita K, Murata K, Iwaoka M, Tomoda S (1997) Tetrahedron 53:2029–2048
17. Fukuzawa S, Takahashi K, Kato H, Yamazaki H (1997) J Org Chem 62:7711–7716
18. Uemura S (1998) Phosphorus Sulfur ***136–138***, 219–234
19. Back TG, Nan S (1998) J Chem Soc [Perkin 1] 3123–3124

20. Back TG, Dyck BP, Nan S (1999) Tetrahedron 55:3191–3208
21. Back TG, Moussa Z (2000) Org Lett 2:3007–3009
22. Tiecco M, Testaferri L, Santi C, Tomassini C, Marini F, Bagnoli L, Temperini A (2000) Tetrahedron Asymmetr 11:4645–4650
23. Murata S, Suzuki T (1987) Chem Lett 16:849–852
24. Murata S, Suzuki T (1987) Tetrahedron Lett 28:4297–4298
25. Murata S, Suzuki T (1987) Tetrahedron Lett 28:4415–4416
26. Jackson WP, Ley SV, Whittle AJ (1980) J Chem Soc, Chem Commun 1173–1174
27. Back TG, Muralidharan KR (1990) Tetrahedron Lett 31:1957–1960
28. Back TG, Muralidharan KR (1991) J Org Chem 56:2781–2787
29. Tiecco M, Testaferri L, Tingoli M, Bartoli D (1989) Tetrahedron Lett 30:1417–1420
30. Tiecco M, Testaferri L, Tingoli M, Bagnoli L, Marini F (1993) J Chem Soc [Perkin 1] 1989–1993
31. Tiecco M, Testaferri L, Tingoli M, Chianelli D, Bartoli D (1988) Tetrahedron 44:2273–2282
32. Bosman C, D'Annibale A, Resta S, Trogolo C (1994) Tetrahedron Lett 35:6525–6528
33. Lee DH, Kim YH (1995) Synlett 349–350
34. Roh KR, Chang HK, Kim YH (1998) Heterocycles 48:437–441
35. Tingoli M, Tiecco M, Testaferri L, Temperini A (1998) Synth Commun 28:1769–1778
36. Pandrey G, Rao VJ, Bhalerao UT (1989) J Chem Soc, Chem Commun 416–417
37. Pandrey G, Sekhar BBVS (1993) J Chem Soc, Chem Commun 780–782
38. Wirth T, Fragale G, Spichty M (1998) J Am Chem Soc 120:3376–3381
39. Wang X, Houk KN, Spichty M, Wirth T (1999) J Am Chem Soc 121:8567–8576
40. Spichty M, Fragale G, Wirth T (2000) J Am Chem Soc 122:10914–10916
41. Fujita K (1997) Rev Heteroatom Chem 16:101–117
42. Tomoda S, Iwaoka M (1988) Chem Lett 17:1895–1898
43. Tiecco M, Testaferri L, Santi C, Tomassini C, Marini F, Bagnoli L, Temperini A (2003) Angew Chem Int Ed 42:3131–3133
44. Freudendahl DM, Iwaoka M, Wirth T (2010) Eur J Org Chem 3934–3944
45. Carey FA, Sundberg RJ (2008) Advanced organic chemistry, part A: structure and mechanisms, 5th edn. Springer, New York, Chap. 4
46. Tiecco M, Testaferri L, Santi C, Marini F, Bagnoli L, Temperini A (1998) Tetrahedron Lett 39:2809–2812
47. Khokhar SS, Wirth T (2004) Angew Chem Int Ed 43:631–633
48. Adair G, Mukherjee S, List B (2008) Aldrichim Acta 41:31–39
49. Lacour J, Hebbe-Viton V (2003) Chem Soc Rev 32:373–382
50. Hamilton GL, Kang EJ, Mba M, Toste FD (2007) Science 317:496–499
51. Komanduri V, Kirsche MJ (2006) J Am Chem Soc 128:16448–16449
52. Mukherjee S, List B (2007) J Am Chem Soc 129:11336–11337
53. Liao S, List B (2010) Angew Chem Int Ed 49:628–631
54. Mayer S, List B (2006) Angew Chem Int Ed 45:4193–4195
55. Wang X, List B (2008) Angew Chem Int Ed 47:1119–1122
56. Hamilton GL, Kanai T, Toste FD (2008) J Am Chem Soc 130:14984–14986
57. Wasiak J, Michalski J (1994) Tetrahedron Lett 35:9473–9476

Chapter 3
Redox Chemistry of Sulfur, Selenium and Tellurium Compounds

Richard S. Glass

3.1 Introduction

The one- and two-electron oxidation of sulfur, selenium and tellurium compounds has attracted much interest owing to its relevance to bonding theory, electrically conducting materials and biochemistry. Although this review will focus on oxidation of chalcogen compounds generally devoid of chalcogen-chalcogen bonds, oxidation of such species provides a rich chemistry including that of the elemental chalcogens [1]. For examples, Te_4^{2+} is a square planar 6π Hückel aromatic species and the eight-membered ring E_8^{2+} species feature three weak transannular bonds, with π, $\pi^{*}\cdots\pi^{*}$ bonding as well as filled non-bonding p-lone pair to σ^{*} interactions, rather than a simple 1, 5-X, X-transannular bond [2].

One-electron oxidation of organochalcogenides (with the chalcogen formally in oxidation state −2) provides the corresponding radical cations as shown in Eq. 3.1 [3–5]. These species react with unoxidized organochalcogenide or other species with an unshared pair of electrons [6–17] to form an unusual 2c, 3e-bond [18]. A simplified view of such bonding is shown in Scheme 3.1 for two chalogen atoms using p-orbitals for σ overlap.

$$R_2X - e^- \rightarrow R_2X\cdot^+ \qquad (3.1)$$

The electronic configuration of such species $\sigma^2\,\sigma^{*1}$ connotes a formal one-electron bond due to the antibonding electron negating one of the bonding electrons, but the bonding depends on overlap [6]. Typically these species are short-lived but they have been observed by time-resolved techniques such as pulse radiolysis [19–23] and photochemistry [24, 25] as well as in the gas phase [26, 27] and by matrix isolation

R.S. Glass (✉)
Department of Chemistry and Biochemistry, The University of Arizona, Tucson, AZ 85721, USA
e-mail: rglass@email.arizona.edu

J.D. Woollins and R.S. Laitinen (eds.), *Selenium and Tellurium Chemistry*,
DOI 10.1007/978-3-642-20699-3_3, © Springer-Verlag Berlin Heidelberg 2011

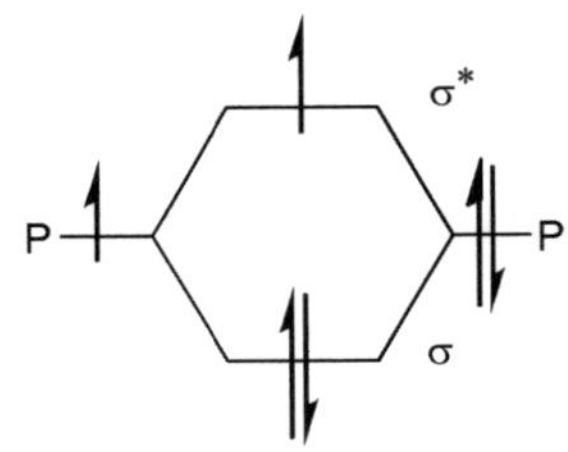

Scheme 3.1 Bonding scheme for 2c, 3e-bonds

[28, 29] at low temperatures. The 2c, 3e-bonded species can be further oxidized to dications, undergo disproportionation or react with nucleophiles [30].

The relevance of one- and two- electron oxidation of chalcogenides to materials science [31] derives from the oxidation of tetrachalcogenafulvenes, oligo- and polythiophenes and selenophenes. Tetrathiafulvalene (TTF) **1a** is electron-rich and easily oxidized. It undergoes two reversible one-electron oxidations at relatively low oxidation potentials [32]. It should be noted that one-electron oxidation provides a planar delocalized radical cation while two-electron oxidation affords two 6π aromatic1,3-dithiolium moieties which prefer a perpendicular arrangement of the two five-membered ring planes. A consequence of the electron-richness of TTF is that it forms charge-transfer complexes, for example, with 7, 7, 8, 8-tetracyano-p-quinodimethane (TCNQ), **2**, in which it is the donor species. Remarkably this complex is as electrically conducting as metals [33], thereby earning it the sobriquet of "organic metal." Not only are TTF charge-transfer complexes conducting but its stable radical cation salts, especially "Bechgaard salts" $(TTF)_2$ X, are as well.

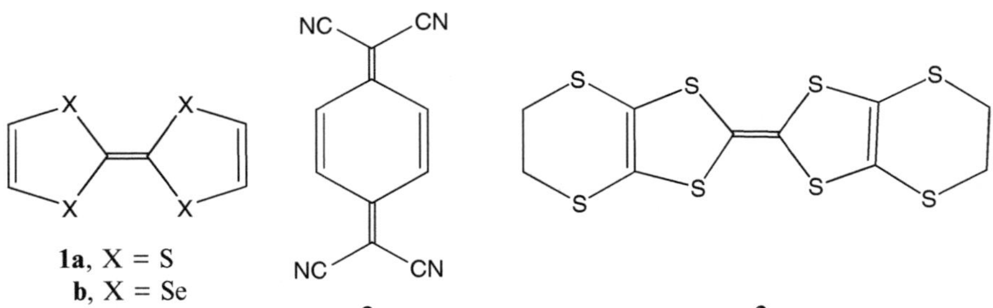

1a, X = S
b, X = Se
c, X = Te

2

3

This property has resulted in enormous interest in its derivatives, as well as selenium **1b** and tellurium **1c** analogues [34–37], of which BEDT-TTF, **3**, is the most important. The advantage of "organic metals" over metals is their lighter weight, lower density and processibility. The motivation for studying selenium and tellurium analogues of TTF is that they may be superior owing to their increased polarizability, lowered electronegativity (especially for tellurium), decreased Coulombic repulsion as a result of increased size, increased conduction bandwidth and increased intermolecular interaction. TTF-TCNQ is a low dimensional conductor and increasing the dimensionality of these materials is key to increasing their usefulness as electrical conductors. The materials have also received attention because of their superconductivity at low temperature, *e.g.* $(\mathbf{1b})_2ClO_4$ [33], and use in field effect transistors [37, 38]. In addition, polythiophene **4** [39] when "doped" with oxidants is electrically conducting owing to the presence of delocalized radical cations (polarons) or dications (bipolarons) [40, 41] which migrate through the π-system.

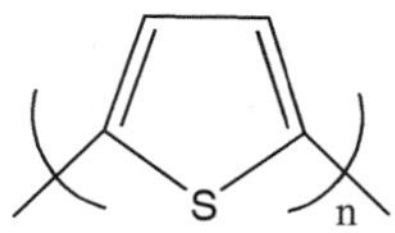

4

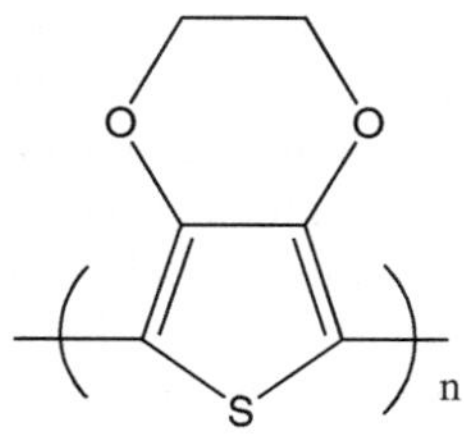

5

When paired with an n-type semiconductor (an electron acceptor in which electrons are the charge carriers), polythiophene acts as an electron donor thereby enabling it to function as a p-type semiconductor (in which positively charged holes-radical cations-are the charge carriers) on irradiation. Thus polythiophenes are important components in organic solar cells [42–45] in which sunlight is converted into electricity. Alternatively, electrons can be removed at an electrode from polythiophene and injected into an n-type semiconductor (or polythiophene). Recombination of the charges may result in an exciton (excited state) which emits light, thereby functioning as an organic light emitting diode (OLED) [46, 47]. Polythiophenes are also useful in thin film and field effect transistors [48] wherein they are hole (radical cation) transporters. The most widely used polythiophene in these applications, especially commercially, is poly-3, 4-ethylenedioxythiophene (PEDOT), **5**. Oligothiophenes are used similarly instead of polythiophenes, principally because they can be prepared chemically pure and free of defects unlike polythiophene. Polyselenophene has recently attracted interest and conducting materials have been prepared [49, 50]. Preliminary results have even been reported for a polytellurophene derivative [51].

These species are also of biological interest because of possible involvement in long-range electron-transfer in proteins. Specifically, electron-transfer in proteins can occur relatively rapidly over distances of 20 Å or greater. Such electron-transfer may occur via superexchange (which decreases logarithmically with increasing distance) or electron "hopping" (which depends on the number of sites transiently bearing the charge and their redox potentials) [52]. To determine the effect of various amino acid side chains on the rate of electron-transfer by a "hopping" mechanism, an ingenious system was designed which is depicted schematically below.

A radical cation was photochemically generated to serve as the electron-acceptor site (A) and a tyrosine side-chain served as the electron-donor site (D). The proline residues adopt a PPII helical conformation to fix the distance between the donor and acceptor (~20 Å). The $(Pro)_3$ segments are connected by an amino acid whose

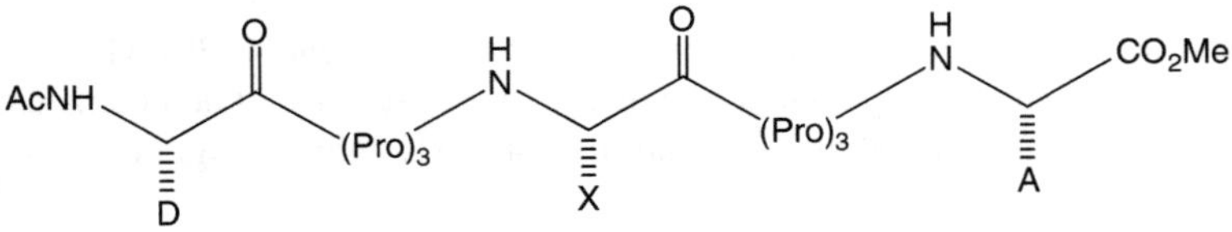

side-chain (represented by X) is probed for its effect on the rate of electron-transfer. Especially effective in promoting electron-transfer in this construct is the – SMe moiety in the methionine side-chain. This requires electron-hopping from the – SMe moiety to the acceptor site. To account for the apparent transient oxidation of – SMe to its corresponding radical cation, it was suggested [53, 54] that the oxidation potential of this moiety is lowered in a similar way to that of **6**, in which 2c, 3e-bond formation, as shown in 7, occurs on electron-transfer [55]. Analogous species have also been identified on oxidation of peptides [56, 57].

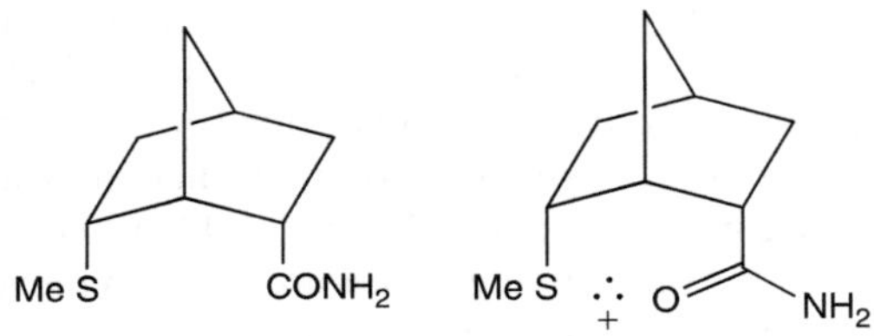

6 **7**

3.2 Cyclic Dichalcogenoethers

A system of particular interest for studying one- and two- electron oxidation of chalcogenides is medium-sized ring dichalcogenides featuring transannular chalcogens, such as **8**.

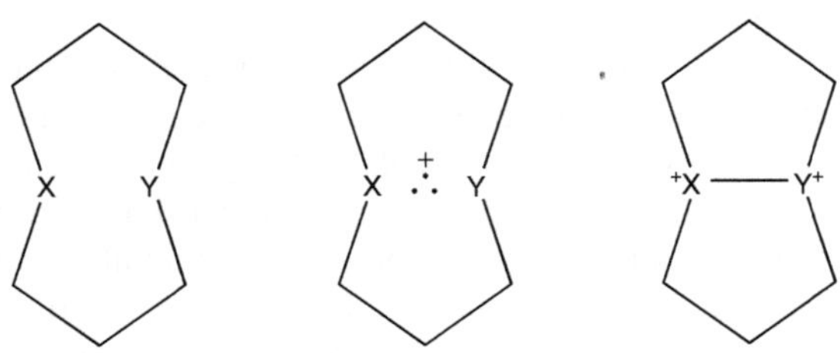

8 **9** **10**

The idea is that medium-sized rings adopt conformations amenable to transannular interactions. A classical example is the transannular 1,5-hydride shifts in cyclooctyl carbocation [58–60] and 1,5-hydride bridging in this cation under stable ion conditions [61–63]. Indeed Leonard and coworkers [64] studied transannular functional group interactions in eight-membered ring compounds comprised of two trimethylene bridges between interacting groups i.e. the functional groups are in the 1, 5-positions. They pointed out with transannular donor and acceptor groups the favorable enthalpic and entropic factors promote interaction even bond formation. The seminal results of Musker and coworkers [65] build on this foundation. These workers found that 1, 5-dithiocane, **8**, X = Y = S, could be oxidized by Cu(II) or NO^+ to radical cation **9**, X = Y = S and this species was surprisingly stable at room temperature in acetonitrile. The stability of this species is ascribed to transannular 2c, 3e-bonding [6–12, 65] between the two sulfur atoms as depicted in Scheme 3.1 in which the 3p-AOs on each of the sulfur atoms overlap [66, 67]. Further oxidation of radical cation **9**, X = Y = S with NO^+ generates dication **10**, X = Y = S, which is a stable species [30, 66, 67]. Such novel dications have not been generated by bis-alkylation of disulfides. Electrochemical studies on 1, 5-dithiocane, **8**, X = Y = S, revealed unusual features [68, 69]. In contrast to other aliphatic thioethers, it undergoes reversible oxidation and at potentials 1 V less positive than ordinary aliphatic thioethers. In addition, it undergoes oxidation with potential inversion, that is, oxidation to the dication occurs at a potential *less positive* than that for oxidation to the monocation despite expectations based on electrostatics that it should be easier to remove a negatively charged electron from neutral dithioether **8**, X = Y = S than from positively charged radical cation **9**, X = Y = S. A consequence of potential inversion is that one cannot have a pure solution of radical cation **9**, X = Y = S but rather a mixture of neutral, radical cation and dication species owing to disproportionation provided that equilibration, i.e. electron-transfer, is fast [70]. The difference in potentials for dithioether **8**, X = Y = S has been estimated to be 150 mV by electrochemical simulation of the cyclic voltammograms for this compound [71]. This difference ensures that a solution of radical cation **9**, X = Y = S will predominantly consist of dithioether **8**, X = Y = S and dication **10**, X = Y = S.

The unusual features in the electrochemistry of dithioether **8,** X = Y = S can be rationalized in terms of intramolecular 2c, 3e bonding in radical cation **9**, X = Y = S. That is, this bond stabilizes radical cation **9,** X = Y = S lowering the oxidation potential of dithioether **8,** X = Y = S and preventing its rapid decomposition (rapid decomposition is typically observed with aliphatic sulfur radical cations under these conditions) so that it persists long enough to be reduced electrochemically. It should be noted that intermolecular dialkyl sulfide radical cation dimer: $(R_2\,S\therefore SR_2)^+$, where R = Me, Et, *n*-Pr or *n*-Bu, has a bond dissociation energy for its 2c, 3e, S, S bond in the range of 90–120 kJ/mol which is somewhat less than half that of an ordinary dialkyl disulfide [72–74]. In addition, owing to the bonding scheme shown above for 2c, 3e- bonding, an electronic factor can be easily identified which accounts for the potential inversion in dithioether **8,** X = Y = S. That is, an antibonding σ^* electron is removed in radical cation **9**, X = Y = S to afford dication **10**, X = Y = S, resulting in a stronger two electron bond between the sulfur atoms (calculations reflect this increase in S, S-bond strength and show that much of the formal positive charge on sulfur is borne by the attached carbon atoms). Dication **10**, X = Y = S has also been prepared in solution and as a solid by treatment of sulfoxide **8**, X = S, Y = SO or sulfilimine **8**, X = S, Y = SNTs with sulfuric acid [75, 76] and reaction of sulfoxide **8**, X = S, Y = SO with trifluoromethanesulfonic acid anhydride [77, 78] and the structure of this bis-triflate salt was determined by X-ray crystallographic methods [79]. Related examples involving the formation of disulfonium dications have recently been reviewed [30].

Analogous results have been obtained with diselenoether **8**, X = Y = Se. That is, it undergoes two reversible one-electron oxidations at unusually low oxidation potentials [80] with potential inversion. Both radical cation **9**, X = Y = Se and dication **10**, X = Y = Se can be prepared by oxidation of diselenoether **8**, X = Y = Se with NO^+ [80–82]. The structure of dication **10**, X = Y = Se bis-tetrafluoroborate salt in the solid state was determined by X-ray crystallographic methods [83, 84]. The radical cation **9**, X = Y = Se was also formed by comproportionation of selenoether **8**, X = Y = Se and dication **10**, X = Y = Se as well as by reduction of dication **10**, X = Y = Se with ferrocene [80]. Analogous diselenonium dications have also been prepared and studied [81, 82, 85].

Ditelluroether **8**, X = Y = Te also shows two reversible one-electron oxidations at low potential [86, 87] with potential inversion [71]. Oxidation of ditelluroether **8**, X = Y = Te with two equivalents of NO^+ or 2, 3-dichloro – 5,6-dicyano-1,4-benzoquinone affords dication **10**, X = Y = Te [86]. Analogous ditelluronium dications have been made and investigated [30]. Although there is presumptive evidence for radical cation **9**, X = Y = Te [71, 86, 88] it has not been characterized as yet.

For the mixed dichalcogenoethers **11**, X = Se or Te the corresponding dications **12**, X = Se and X = Te, respectively, have been synthesized and characterized. Treatment of sulfoxide of **11**, X = Se or selenoxide **11**, X = SeO with trifluoromethanesulfonic acid anhydride or reaction of selenothioether **11**, X = Se with

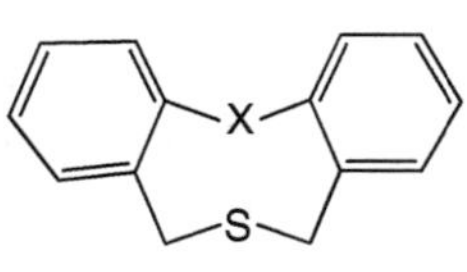

11

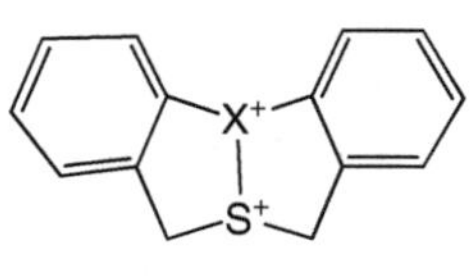

12

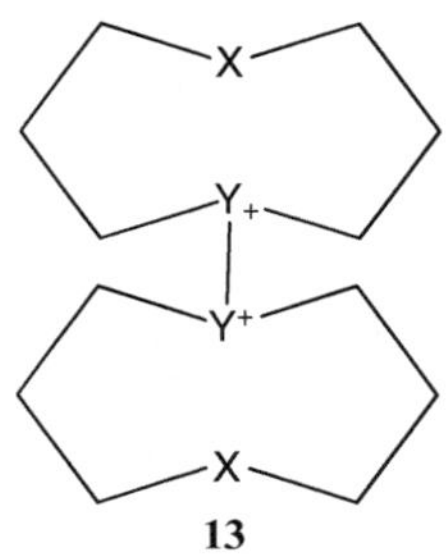

13

two equivalents of NO^+ or with sulfuric acid formed dication **12**, X = Se [89]. Reaction of tellurothioether **11**, X = Te with two equivalents of NO^+ or with sulfuric acid yielded dication **12**, X = Te [90, 91]. This dication was also produced by treating telluroxide **11**, X = TeO with trifluoromethanesulfonic acid anhydride [91]. The electrochemical oxidation potentials for selenothioether **11**, X = Se [89] and tellurothioether **11**, X = Te [90, 91] are reported and presumably correspond to a two-electron oxidation but the number of electrons transferred in these experiments is not explicitly stated. In any event, electrochemical studies of selenothioether

8, X = S, Y = Se and tellurothioether **8**, X = S, Y = Te gave anomalous results. Electrochemical oxidation of these compounds resulted in an irreversible one-electron process [71]. This suggested that oxidation to the corresponding radical cation was followed by rapid dimerization to **13**. Evidence for dimerization of the radical cation **9**, X = Y = S had been obtained previously [69, 70], but with **9**, X = S, Y = Se and **9**, X = S, Y = Te the dimerization is much more favorable with rate and equilibrium constants 10^3 times larger than those for **9**, X = Y = S. Further evidence for dimerization of **9**, X = S, Y = Se and **9**, X = S, Y = Te was the observation of a one-electron reduction peak associated with the oxidation peak. In the case of **9**, X = S, Y = Se further evidence for the reduction of the dimer as well as insight into its structure was obtained in the following way. Oxidation of **8**, X = S, Y = Se with tris (pentafluorophenyl) boron produced a crystalline dimer whose structure was unequivocally shown to be **13**, X = S, Y = Se by X-ray crystallographic methods [71]. This structure study also suggested 4c, 6e S•••$^+$Se – Se $^+$•••S bonding analogous to the 4c, 6e [92–94] S•••$^+$S–S$^+$•••S bonding suggested for **13**, X = S, Y = Se by theoretical calculations [10]. The reduction potential for **13**, X = S, Y = Se prepared chemically matched that for the reduction peak observed after electrochemical oxidation of selenothioether **8**, X = S, Y = Se. This result supports the notion that one-electron electrochemical oxidation affords dimer **13**, X = S, Y = Se in which there is a $^+$Se–Se$^+$ bond not the alternative dimers in which there is a $^+$S–S$^+$ or $^+$S–Se$^+$ bond. Based on the analogous results with electrochemical oxidation of tellurothioether **8**, X = S, Y = Te with selenothioether **8**, X = S, Y = Se (in which a one-electron reduction peak is associated with an irreversible one-electron oxidation) it is surmised that dimer **13**, X = S, Y = Te is formed on oxidation of tellurothioether **8**, X = S, Y = Te.

These results for the electrochemical oxidation the mixed chalcogenoethers **8**, X = S, Y = Se and **8**, X = S, Y = Te which diverge from those for the homochalcogenoethers **8**, X = Y = S, **8**, X = Y = Se and **8**, X = Y = Te are consistent with thermodynamic control of the product. That is, calculations [95] show that the order of bond strengths for the dicationic chalcogen-chalcogen bonds is $^+$Te – Te$^+$ > $^+$Se – Se$^+$ > $^+$S – S$^+$ and $^+$Te – Te$^+$ > $^+$Te – S$^+$ and $^+$Se – Se$^+$ > $^+$Se – S$^+$. Thus oxidation of homodichalcogenoethers **8**, X = Y = S, **8**, X = Y = Se and **8**, X = Y = Te gives the corresponding dications **10**, X = Y = S, **10**, X = Y = Se and **10**, X = Y = Te, respectively, which are favored entropically over the corresponding dimeric dications. However, oxidation of **8**, X = S, Y = Se or **8**, X = S, Y = Te yields dimeric dications **13**, X = S, Y = Se and **13**, X = S, Y = Te respectively, because enthalpic control to give $^+$Se – Se$^+$ and $^+$Te – Te$^+$, respectively, outweighs entropic favoring of dications **10**, X = S, Y = Se and **10**, X = S, Y = Te, respectively, which have weaker $^+$S – Se$^+$ and $^+$S – Te$^+$ bonds, respectively.

In addition to electrochemical oxidation, removal of an electron from a compound can be studied by photoelectron spectroscopy (PES) [96–99]. In our case, the molecule of interest in the gas phase is irradiated with monochromatic radiation (typically He I, 21.21 eV). This results in ejection of electrons from valence orbitals and their kinetic energy is measured. The difference in energy between the monochromatic radiation and electron kinetic energy is the binding energy of the electron i.e. its ionization energy. Assuming Koopmans' theorem [100], this ionization energy is the negative of the orbital energy so one gets insight into the orbital from which the electron is ejected. Another useful feature of the method is that the ionization cross-section changes with different energies of the monochromatic radiation with different atoms and orbitals. For example, the cross-section for sulfur ionizations drops substantially on going from He I (21.21 eV) to He II (40.8 eV) relative to carbon ionizations [101]. This greatly assists the assignment of orbital character – whether carbon, sulfur or a combination of both. In comparing ionization energies determined by gas-phase photoelectron spectroscopy with oxidation potentials in solution, there are important differences to consider.

Since the ionization energies are measured in the gas-phase at low pressure, solvent, supporting electrolyte effects and intermolecular interactions which may be important in the electrochemical experiment are absent in PES measurements. In addition, if the electrochemical oxidation is reversible then E^0 measures the difference in free energy between oxidized product and starting material. In the PES experiment the ionization process is a non-adiabatic, vertical "Franck-Condon" process. Alternatively stated, the time scales for the two processes are very different. In the electrochemical experiment there is enough time for the oxidized product to undergo atomic movements to obtain the lowest energy molecular geometry. However, the PES time scale is much shorter and significant molecular movements cannot occur. Thus the peak maximum corresponds to non-adiabatic ionization (however, the onset of the peak is believed to reflect adiabatic ionization). If the electrochemical oxidation is irreversible then, not only the formal potential E^0, but the heterogeneous electron-transfer rate constant k_0, the electron-transfer coefficient $\propto$, the number of electrons transferred, the diffusion coefficient, the scan rate and rate constant k_1 for the irreversible reaction following electron-transfer affect the observed potential [102, 103].

In the gas-phase PES of **8**, X = Y = S, **8**, X = Y = Se and **8**, X = Y = Te there are several overlapping ionizations of lowest energy. These multiple ionizations can be fitted to pairs of ionizations with each pair of comparable amplitude. Each pair corresponds to ionizations from one conformer. That is, there is more than one conformer present for each compound. Generally there is one predominant conformer for each compound but calculations support the notion that more than one conformer is present. The reason that each conformer shows a pair of ionizations is that the lowest energy ionizations correspond to removal of an electron from the chalcogen p-type orbital and this orbital on one of the chalcogens interacts with the p-type orbital on the other chalcogen. That is, the HOMO and HOMO-1 for the molecules consist of the symmetric and antisymmetric combinations of the p-type orbitals. Furthermore, calculations show that this interaction is predominantly through-space rather than through-bonds [104–106]. Consequently, the lone-pair splitting (difference in energy between HOMO and HOMO-1) depends on their overlap which, in turn, depends on the distance between chalcogens and the orientation of the p-orbitals relative to each other, e.g. if the p-orbitals point directly at each other then there will be more overlap than if they point somewhat askew (the geometric parameter determining this orientation is the dihedral angle between the C-X-C planes) for the same distance between chalcogens. A useful application of this geometry dependence of lone-pair splitting is that it can be used with computations to do conformational analysis of these heterocycles in the gas-phase i.e. deduce the geometry of each of the conformers in the gas phase [107, 108].

A further consequence of this transannular lone-pair interaction is that it raises the HOMO energy thereby lowering the lowest ionization energy. In essence this splitting provides a quantitative measure of lone pair-lone pair repulsion which has often been invoked qualitatively but only indirectly quantitatively. The two lowest ionization energies for the major conformer, their energy difference (a measure of lone-pair, lone-pair interaction) and the electrochemical oxidation potentials for **8**, X = Y = S, **8**, X = Y = Se, and **8**, X = Y = Te are listed in Table 3.1.

As seen in the table the lowest ionization energies for **8**, X = Y = S, **8**, X = Y = Se, and **8**, X = Y = Te track the corresponding oxidation potentials. However, the lowest ionization energies and oxidation potentials for **8**, X = S, Y = Se, and **8**, X = S, Y = Te do not fit this correlation. It is not surprising that these compounds do not fit the correlation because their electrochemical oxidations are irreversible and involve intermolecular interaction, i.e. dimerization. Calculations on these compounds also reveal another difference with **8**, X = Y = S, **8**, X = Y = Se, and **8**, X = Y = Te. The two lowest ionization energies for **8**, X = Y = S, **8**, X = Y = Se, and **8**, X = Y = Te are the ionizations from the symmetric and antisymmetric combinations of the p-lone pair orbitals that interact transannularly. However, for the lowest energy conformer of **8**, X = S, Y = Te the HOMO consists primarily of the Te 5p lone-pair orbital and HOMO-1 corresponds to predominantly the S 3p lone-pair orbital. That is, the S and Te p-type lone-pair orbitals do not mix predominantly due to their difference in energy (the lowest energy conformations are similar to those of the other 1, 5-dichalcogenoethers). Clearly the separation of the two lowest ionizations in the case of **8**, X = S, Y = Te is not a measure of transannular interaction but rather a reflection of the different ionization energies for S and Te. For **8**, X = S, Y = Se there is mixing of the S and Se p-orbitals but there is greater Se character in its HOMO and greater S character in its HOMO-1.

Table 3.1 Lowest ionization energies[a] and oxidation potentials[b] for **8**, X = Y = S, **8**, X = Y = Se, and **8**, X = Y = Te

Compound	I.E. (eV)	ΔI.E. (eV)	E^{o} [c] (V)
8, X = Y = S	8.27, 8.69[d]	0.42	0.404[e]
8, X = Y = Se	8.00, 8.35[d]	0.35	0.102[e]
8, X = Y = Te	7.59, 7.91[d]	0.32	−0.294[e]

[a]Obtained from gas phase PES [b]Obtained by cyclic voltammetric studies in acetonitrile [c]Oxidation potentials are reported versus the ferrocene/ferrocenium couple in CH_3CN [d]Refs. [109, 110] [e]Ref. [71]

3.3 Cyclic Dichalcogenoethers with Group 14 Heteroatoms

It is known that a neighboring C-Sn σ-bond [111–113] and, to a lesser extent, C-Si σ-bond [114–116] can interact with a chalcogen lone-pair orbital thereby raising its orbital energy and lowering its ionization energy. This interaction depends on orbital energies (the closer the orbitals are in energy, the greater the interaction) and geometry (which determines overlap, e.g. orthogonal orbitals will not interact but aligned orbitals will interact). A dramatic illustration of this effect is illustrated by 1, 3-dithiane **14a** and its 2, 2-distannyl derivative **14b**. The lowest ionization energy for distannyl **14b**, determined by PES, is 1 eV lower than that for 1, 3-dithiane **14a** [111].

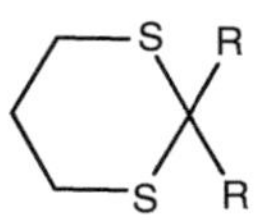

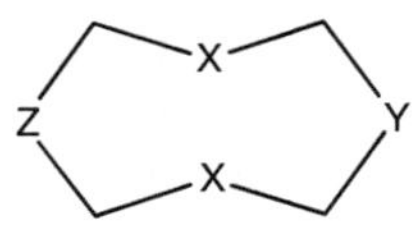

$$MeS \overset{\theta}{—} CH_2 \overset{\varphi}{—} SiMe_2SiMe_3$$

16

14a, R = H
b, R = $SnMe_3$
c, R = $SiMe_3$

15a, X = S; Y = Z = $SiMe_2$
b, X = S; Y = $SiMe_2$; Z = $SnMe_2$
c, X = S, Y = Z = $SnMe_2$
d, X = Se; Y = Z = $SiMe_2$
e, X = Se, Y = Z = $SnMe_2$
f, X = Te; Y = Z = $SiMe_2$

In addition, the irreversible oxidation potential for distannyl **14b**, determined electrochemically, is about 1 V less positive than that for 1, 3-dithiane **14a** [111]. The oxidation potential for disilyl **14c** is 0.48 V less positive than that for 1, 3-dithiane **14a** [111]. Similarly the adjacent C-Si σ-bond and, even more so, adjacent C-Sn σ-bond in dithioethers **15a–c**, diselenoethers **15d, e** and the adjacent C-Si σ-bond in ditelluroether **15f** lower the ionization energies compared with dithioether **8**, X = Y = S, diselenoether **8**, X = Y = Se and ditelluroether **8,** X = Y = Te, respectively [109, 110]. However, electrochemical oxidation of the silyl and stannyl substituted dithioethers **15a–c** and diselenoethers **15d, e** are irreversible and their oxidation potentials do not correlate with their ionization energies. The dichalogenium dications derived from these silyl and stannyl substituted compounds are not stable. Studies [88, 117] of the products obtained by chemical oxidation of the silyl compounds have revealed the interesting reaction products obtained from these dications as illustrated below:

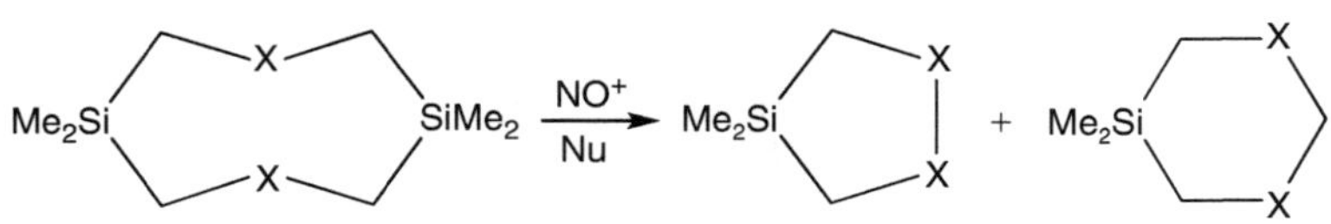

In the case of silyl substituted ditelluroether **15f** its lowest ionization energy is lowered compared with ditelluroether **8**, X = Y = Te and its oxidation is reversible. However, its oxidation potential is more positive than that of 8, X = Y = Te.

Even in this case there is no correlation between oxidation potential and lowest ionization energy.

The reason for the greater effect of a C-Sn than C-Si σ-bond in lowering the ionization energy of an adjacent chalcogenoether is that the orbital energy of a C-Sn σ-bond is closer to that of the chalcogenoether p-orbital than is a C-Si σ-bond. However a Si-Si-C-σ-bond is closer in energy to a chalcogen p-orbital than a C-Si σ-bond. To provide insight into the effect of a β-Si-Si σ-bond on the HOMO of a thioether, theoretical calculations were carried out on disilylthioether **16** in which dihedral angles θ and ϕ (rotation around the S-CH_2 bond and CH_2-Si bond respectively) were varied [118] Varying θ determines interaction between the sulfur 3p-orbital and C-Si σ-bond. When $\theta = 0°$ (S–Me and CH_2-Si bonds are eclipsed) the sulfur 3p lone-pair orbital and C–Si σ–bond are orthogonal and they do not interact. When $\theta = 90°$, the sulfur 3p lone-pair orbital and C–Si σ–bond are eclipsed and aligned for maximum interaction. Varying ϕ (when S and the Si of the $SiMe_3$ group are eclipsed $\phi = 0$) provides insight into Si–Si interaction with the sulfur 3p lone-pair orbital. The calculations reveal an important feature in such interactions which the following example illustrates. For $\theta = 0$ or 180° one might have expected greatest Si-Si through-space interaction when $\phi = 90°$ in which the Si–Si σ–bond and sulfur 3p lone-pair orbital are aligned. But this is not the case. The greatest interaction (resulting in the lowest ionization energy) occurs when $\phi = 120°$. The reason for this is that the HOMO is a combination of sulfur 3p orbital and "π-like" antisymmetric combination of C-H σ-orbitals of the CH_2 group attached to sulfur. With $\phi = 120°$ the Si–Si σ–bond eclipses a C-H σ-bond of this CH_2 and provides maximal interaction. That is, the β–Si–Si σ–bond exerts its maximal effect by through-bond not through-space interaction. Analogous through-bond effects are manifested in **15**, X = S, Y = Z = $Si(SiMe_3)_2$, in which its HOMO is strongly destabilized (the lowest ionization energy for this compound is 0.7 eV lower than the for **8**, X = Y = S). Using the calculated lowest energy conformers for **15**, X = S, Y = Z = $Si(SiMe_3)_2$ and the conformer observed in the solid state by X-ray crystallographic analysis, the destabilized HOMO shows that all of the ring atoms participate as well as the combined exocyclic Si-Si orbitals. That is, the marked destabilization is mediated by extensive through-bond interaction.

3.4 Chalcogenoethers with Arene Moieties

As reviewed above, medium-sized rings, especially eight-membered1,5-dichalcogenoethers, provide an excellent template for studying through-space interactions. This results from their conformational preferences and favorable formation of transannular bonds. Of course, this is tempered with through-bond interactions as well as intermolecular interactions and fragmentation in some cases. These studies have resulted in a deeper understanding of the redox chemistry of chalcogenoethers and their novel one- and two-electron oxidation products namely their radical cations and dications. In this section, other conformationally constrained systems are studied in which chalcogenoethers and arene π moieties are juxtaposed to elucidate their redox chemistry and through-space interaction. The motivation for such studies is that little is known about such interactions and they may be of biochemical relevance. Intermolecular stabilization of aliphatic radical cations by arenes is known, [119–121] but the basis for this stabilization is unknown. In addition, several hypotheses of biochemical relevance involving sulfur ••• π interaction in redox systems have been proposed. Such interactions have been speculated to be conduits for electrons in proteins [122] and to modulate biomolecule reduction potentials [123–125]. The causative agents in Alzheimer's disease are believed to be β–amyloid peptides. It has been conjectured that the soluble oligomers of these peptides kill nerve cells by initiating the formation of reactive oxygen species in which the only methionine residue in the β–amyloid peptides, Met-35, is invoked as a reducing agent [126–129] i.e. metals ions Fe(III) or Cu(II) are reduced and the reduced metal ions initiate a Fenton-type reaction with the H_2O_2 known [130–133] to be produced in such systems. Since methionine residues are typically not good enough reducing agents to effect this reduction on a thermodynamic basis (or, alternatively stated, Fe(III) and Cu(II) are not strong enough oxidizing agents), it has been suggested that the structure of these oligomeric peptides renders this redox reaction more favorable. Based on modeling studies, Tycko et al. [134, 135] have suggested that the aromatic side chain of Phe 19 or Phe 20 in these peptides is close in space to the methionine side-chain [136–139].[1,2] Hence the chemical question: what is the oxidation potential of thioethers juxtaposed with aromatic rings.

To address this question, aryl-sulfides **17** and **18** were stereospecifically synthesized and their redox chemistry studied [140].

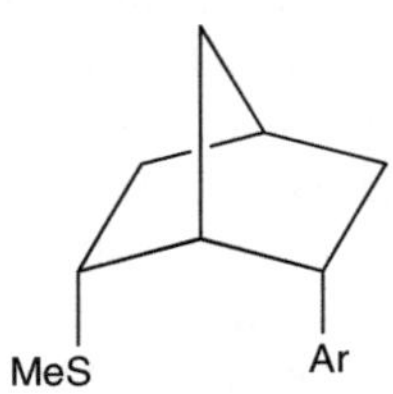

17a, Ar = 1-Nphth
b, Ar = 2-Nphth
c, Ar = Ph
d, Ar = p-MeOC$_6$H$_4$

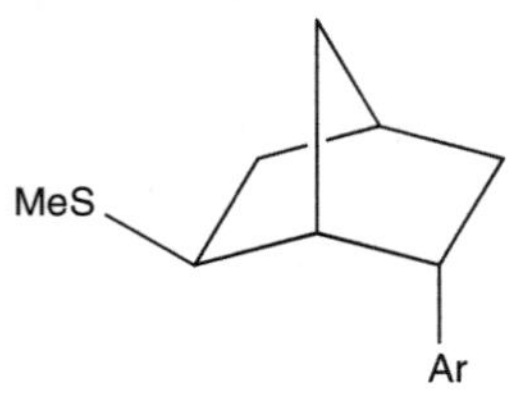

18a, Ar = 1-Nphth
b, Ar = 2-Nphth
c, Ar = Ph
d, Ar = p-MeOC$_6$H$_4$

[1]These model studies are challenging in their own right but further complicated by β-amyloid polymorphism [136].

[2]Interestingly this model has been supplanted by one in which Met-35 is directed away from the hydrophobic interior of the peptide (containing the side-chain of Phe-19) toward the exterior where it is close to a second Met residue of another peptide in the oligomer. If oxidation of Met-35 is important in the pathogenesis of Alzheimer's disease, this suggests the possibility of interaction between the two methionine sulfur atoms. As mentioned above, 1, 5-dithiocane **8**, X = Y = S reduces Cu(II).

The idea is that the thioether and arene moieties are constrained to be close to each other when in the 2-endo, 6-endo positions as in **17** rendering through-space interaction possible but the 2-exo, 6-endo substitution pattern in **18** precludes through-space interaction. Thus **18** provides a reference for **17**. Since the number of bonds separating the thioether and arene moieties are the same in **17** and **18**, inductive effects are the same between the MeS and Ar moieties in **17** and 18. However, through-bond orbital effects depend on geometry and must be evaluated computationally. The specific compounds studied by cyclic voltammetry are **17a–d** and **18a–d**. Unfortunately, their oxidations are irreversible and, therefore, depend not only on the formal potential E^0 but on the other factors enumerated in the preceding discussion of irreversible oxidation. Nevertheless for a related series of thioethers peak potentials appear to track formal potentials [141]. Thus comparison of the peak potentials listed in Table 3.2 shows that for each pair of comparably substituted compounds **17** and **18**, **17** oxidizes at a less positive potential by 0.31–0.51 V than **18**.

The gas-phase photoelectron spectra of **17a** and **18a** were measured. Both showed four ionizations at low energy with the lowest ionization for **17a** 0.20 eV lower in energy than that for **18a**. Furthermore, comparison of the He I/He II ionization intensity ratios revealed that the two lowest ionizations of **17a** corresponded to orbitals with mixed sulfur lone-pair-arene π character; whereas for **18a** the lowest ionization corresponds to an orbital of predominantly arene π character and the next lowest ionization from an orbital with predominantly sulfur lone pair character. Calculations for the lowest energy orbitals of the conformers of lowest energy for **17a** and **18a** were in accord with the experimental results. In particular, the HOMO for **17a** showed extensive sulfur p-orbital and naphthalene π-MO through-space mixing (with little through-bond interaction). Similar results were found experimentally and computationally for **17c**. Thus the lowered oxidation potentials for **17** compared with **18** appear to be due to through-space sulfur 3p lone-pair-arene π interaction. The intriguing question of the structure and bonding in the radical cation obtained after removal of an electron for **17** is currently under investigation. Stabilization of this radical cation could be achieved by charge-transfer interaction, charge-quadrupole interaction or bonding between the sulfur and arene moieties.

Table 3.2 Peak potentials[a] for oxidation of thioether, arenes **17** and **18**

Cmpd 17	E_p (V)	Cmpd 18	E_p (V)	ΔE_p (V)[b]
a	0.74	a	1.16	0.42
b	0.73	b	1.20	0.47
c	0.96	c	1.27	0.31
d	0.76	d	1.27	0.51

[a]Peak potentials of the first oxidation peaks determined at a Pt electrode, 0.1 V/s scan rate in CH_3CN, 0.1 M $NaClO_4$ *vs* Ag/0.1 M $AgNO_3$ in CH_3CN reference electrode [b]Difference in peak potentials for corresponding compounds **17** and **18**

Another conformationally constrained system in which chalcogenoether lone-pair- π interaction, and possibly extended π-lone-pair- π interaction, can be evaluated is *m*-terphenyl chalcogenoethers **19**. The idea is that these compounds adopt the geometry shown in **20** in which the chalcogen "sandwiched" between the two neighboring arene rings and π-conjugation between the chalcogen lone-pair and phenyl ring to which it is attached and between the arene substituents and the central phenyl ring is mitigated.

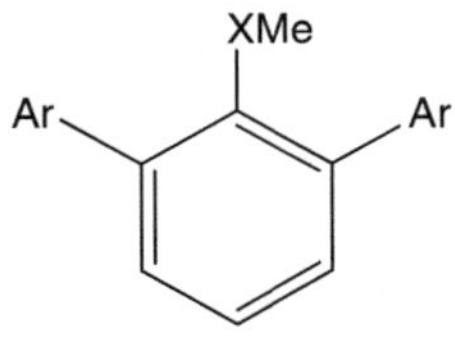

19

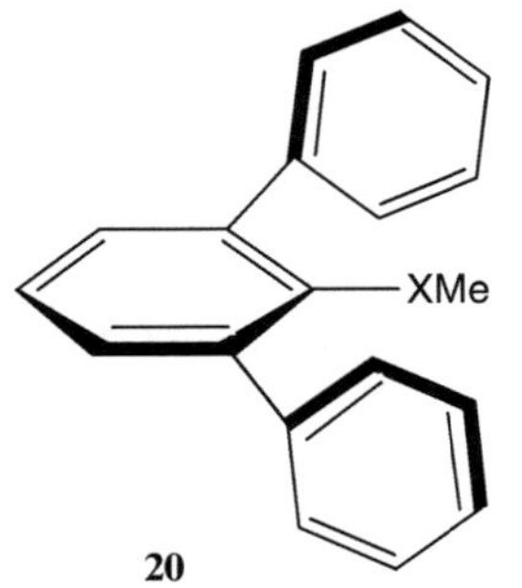

20

Evidence that indeed these compounds adopted the desired geometry for chalcogen•••π interactions was twofold [142]. First the structure of **19**, X = S, Ar = *o*-$MeOC_6H_4$ in the solid state was unequivocally established by X-ray crystallographic methods. Second, with suitably substituted Ar groups the compounds, if they adopted the geometry shown in **20**, would be capable of atropisomerism. That is, **19**, X = S, Se or Te, Ar = *o*-$MeOC_6H_4$ could adopt a syn-conformation, in which the MeO- substituents point in the same direction, or an anti-conformation, in which the substituents point in opposite directions. In principle, an additional atropisomer could be present in which the XMe group points in the same direction as the MeO groups in the syn-isomer or in the opposite direction. However, computational studies show that the barrier for rotation around the aryl C-X bond is low. Thus, these additional atropisomers are readily interconverted. Evidence for atropisomerism in **19** could conveniently be obtained by NMR spectroscopy. These compounds, **19**, X = S, Se or Te, Ar = *o*-$MeOC_6H_4$, show two singlets for the MeO and MeX groups in their ^{1}H NMR spectra. Furthermore, the barriers for interconversion of atropisomers, by rotation about the central-aryl C–C(Ar) bond, could be readily determined by variable temperature NMR spectroscopic studies [142]. Interestingly the barrier depends on the chalcogen and follows the order Te > Se > S. One might have anticipated that increasing the bond length would move the XMe further away from steric interaction with the rotating Ar groups and lower the barrier but the opposite is observed. Thus the size of the chalcogen (Te > Se > S) is the dominating factor in hindering rotation about the aryl-aryl bonds. Computational analysis provides further insight into the basis for the barriers [142]. Calculations suggest that the barrier for rotation about aryl C-XMe decreases on going from X = S to X = Se because the increase in bond length on going from X = S to X = Se places the Me group further away from the adjacent Ar group

Table 3.3 Peak potentials[a] for oxidation of *m*-terphenyl thio- and selenoethers **19**

X	Ar	E_p, V
S	*o*-$MeOC_6H_4$	1.08
S	*p*-$MeOC_6H_4$	1.02
S	2-Nphth	0.95
S	Ph	1.12
Se	*o*-$MeOC_6H_4$	0.96
Se	*p*-$MeOC_6H_4$	0.86
Se	1-Nphth	0.86

[a]Peak potentials of the first oxidation peaks determined at a Pt electrode, 0.1 V/s scan rate in CH_3CN, 0.1 M $NaClO_4$ vs Ag/ 0.1 M $AgNO_3$ in CH_3CN reference electrode

(because of the low barrier variable temperature NMR spectroscopic studies do not provide experimental evidence for this effect). However, the barrier for rotating the aryl group which interconverts syn and anti isomers is determined by steric interaction between X and the ortho-H of the rotating aryl group. Although the distance between the ortho-H and Se is greater than that for the ortho-H and S for their corresponding transition states, the increased size of Se relative to S results in increased steric interaction and an increased barrier as observed experimentally.

To evaluate through-space chalcogen-π interaction in these *m*-terphenyl chalcogenoethers their oxidation was studied electrochemically using cyclic voltammetry [143]. Their irreversible oxidation potentials are shown in Table 3.3. As with thioether-arenes **17** and **18** it is assumed that peak potentials track formal potentials in this closely related series of compounds. To ascertain whether the oxidation potentials for **19**, X = S are rendered less positive by the Ar substituents, a model compound devoid of such substituents is required. However, comparison with thioanisole, PhSMe, (E_p = 1.12 V) [144] is flawed because the MeS group in thioanisole is coplanar with the phenyl ring [145–148] in contrast to the perpendicular conformation in **19**, X = S. Removal of an electron from thioanisole generates the planar resonance stabilized radical cation whose energy has been calculated [149] to be 18.131 kcal/mol lower than that of the perpendicular radical cation in which resonance is precluded. Indeed it has been suggested that the lowest ionization energy for perpendicular thioanisole is higher than that for the planar conformer (8.55 vs 8.02 eV) [150–153]. Consequently, the peak potentials for **19**, X = S are less positive than expected pointing to through-space S•••π interaction. A similar argument pertains to the oxidation potentials of selenoether **19**, X = Se compound with selenoanisole ($E_{1/2}$ = 0.965 V) [154, 155]. Indeed the oxidation potential for 2, 6 – dimethylselenoanisole, which adopts a perpendicular conformation, is more positive than that for selenoanisole despite the electron-donating effects of the Me groups which lower its oxidation potential. The lowest ionization energy of perpendicular selenoanisole is at higher energy than that for the planar conformer [156–158]. Consequently the peak potentials for **19**, X = Se are less positive than expected pointing to through-space S•••π interaction. In contrast with the sulfur and selenium compounds there does not appear to be Te•••π interaction in the tellurium compounds studied.

Further evidence for S•••π interaction in thioether **19**, X = S, Ar = *o*-$MeOC_6H_4$ was obtained from its photoelectron spectrum and calculations. The lowest ionization energy for X = S, Ar = *o*-$MeOC_6H_4$ is 7.86 eV which is lower than that for planar (8.02 eV) or perpendicular (8.55 eV) thioanisole [150–153]. Furthermore, the He I/He II ionization intensity ratio show that the ionization corresponds to an orbital with both sulfur lone-pair and π-MO character. Computational analysis supports the sulfur lone-pair and arene π compostition of the HOMO for this compound.

Thus, the *m*-terphenyl system provides evidence for S and Se•••π interaction. The question of the structure and bonding in the radical cations obtained by one electron oxidation is under investigation. In comparison with similar studies with **17** mentioned above, there are two differences. In **19** the flanking arene groups may both interact with sulfur or selenium. In addition, there is the possibility of resonance stabilization in the radical cations of **19**, X = S, if the MeS moiety becomes coplanar with the central phenyl ring despite the increase in steric interactions.

3.5 Conclusions

The understanding of functional group chemistry has traditionally been viewed as the key for understanding organic chemistry in general. That is, the behavior of a compound with several functional groups connected only by "localized" σ-bonds is the sum of the behavior of each individual functional group. This concept was tempered with interactions between functional groups as a result of resonance. That is, functional groups connected by π-systems affect each other's reactivity (as well as that of the π-system). Cases in which interaction between remote functional groups or reactive site of intermediates, e.g. carbocations, and remote functional group occurred were considered exceptional e.g. "nonclassical carbocations." However, as the chemistry reviewed here illustrates, through-space interactions and interactions mediated through σ-bonds are important. Thus, such interactions extend our understanding of the redox chemistry of chalcogenoethers, the chemistry of new reactive intermediates such as chalcogen radical cations and dichalcogenium dications, unusual 2c, 3e-bonding and its chemical consequences, and chalcogen lone-pair -π interactions. Such understanding extends bonding theory, may have important applications in materials science and may provide additional insight into long-range electron-transfer in biological systems and redox-related diseases.

References

1. Krossing I (2007) In: Devillanova FA (ed) Handbook of chalcogen chemistry. RSC Publishing, Cambridge; chapter 7.1
2. Cameron TS, Deeth RJ, Dionne I, Du H, Jenkins HDB, Krossing I, Passmore J, Roobottom HK (2000) Inorg Chem 39:5614–5631
3. Glass RS (2002) Spec Chem Mag 22:34–36
4. Glass RS (1999) Top Curr Chem 205:1–87

5. Detty MR, Logan ME (2004) Adv Phys Org Chem 39:79–145
6. Baird NC (1977) J Chem Educ 54:291–293
7. Clark T (1990) In: Chatgilialogu C, Asmus K-D (eds) Sulfur-centered reactive intermediates in chemistry and biology, vol 197, NATO ASI Ser A. Plenum, New York, pp 13–18
8. Stowasser R, Glass RS, Hoffmann R (1999) J Chem Soc Perkin Trans 2:1559–1561
9. Asmus K-D (2001) In: Jonah CD, Rao DSM (eds) Radiation chemistry: present status and future trends. Elsevier, Amsterdam, pp 341–393
10. Nakayama N, Takahashi O, Kikuchi O, Furukawa N (2001) J Mol Struct THEOCHEM 542: 215–226
11. Maity DK (2002) J Am Chem Soc 124:8321–8328
12. Fourré I, Silvi B (2007) Heteroatom Chem 18:135–160
13. King JE, Illies AJ (2003) Int J Mass Spectrom 228:429–437
14. King JE, Illies AJ (2004) J Phys Chem A 108:3581–3585
15. Mishra B, Priyadarsini KJ, Mohan H (2006) J Phys Chem A 110:1894–1900
16. Mishra B, Sharma A, Naumov S, Priyadarsini KI (2009) J Phys Chem B 113:7709–7715
17. Naumov S, Bonifacic M, Glass RS, Asmus K-D (2009) Res Chem Intermed 35:479–496
18. Asmus K-D (1990) In: Chatgilialoglu C, Asmus K-D (eds) Sulfur-centered reactive intermediates in chemistry and biology. Plenum Press, New York, pp 155–172
19. von Sonntag C, Schuchman H-P (1980) In: Patai S (ed) The chemistry of ethers, crown ethers, hydroxyl groups, and their sulphur analogues, vol Pt 2, Suppl E. Wiley, Chichester; chapter 24
20. Asmus K-D (1984) Meth Enzymol 105:167–178
21. von Sonntag C (1987) The chemical basis of radiation biology. Taylor and Francis, London
22. Waltz WL (1988) In: Fox MA, Chanon M (eds) Photoinduced electron transfer, vol Pt B. Elsevier, Amsterdam, pp 57–109
23. Takamuku S, Yamamoto Y (1991) In: Tabata Y (ed) Pulse radiolysis. CRC Press, Boca Raton; chapter 18
24. Pienta NJ (1988) In: Fox MA, Chanon M (eds) Photoinduced electron transfer, vol Pt C. Elsevier, Amsterdam, pp 421–486
25. Kavarnos GJ (1993) Fundamentals of photoinduced electron transfer. Verlag Chemie, New York
26. van de Sande CC (1983) In: Patai S (ed) The chemistry of ethers, crown ethers, hydroxyl groups, and their sulphur analogues, vol Pt 1. Wiley, Chichester; chapter 7
27. Nibbering NMM, Ingermann S, de Kooning LJ (1996) In: Baer T, Ng CY, Powis I (eds) The structure, energetics, dynamics of organic ions. Wiley, New York, pp 283–290
28. Symons MCR (1984) Chem Soc Rev 13:393–439
29. Lund A, Shiotani M (eds) (1991) Radical ion systems properties in condensed phases. Kluwer, Dordrecht
30. Nenajedenko VG, Schevchenko NE, Balenkova ES, Alabugin IV (2007) In: Devillanova FA (ed) Handbook of chalcogen chemistry. RSC Publishing, Cambridge; chapter 7.2
31. Westgate TD, Skabara PJ (2007) In: Devillanova FA (ed) Handbook of chalcogen chemistry. RSC Publishing, Cambridge; chapter 12.2
32. Coffen DL, Chambers JQ, Williams DR, Garrett PE, Canfield ND (1971) J Am Chem Soc 93:2258–2268
33. Ferraris J, Cowan DO, Walatka V Jr, Perlstein JH (1973) J Am Chem Soc 95:948–949
34. Bechgaard K, Carneiro K, Rasmussen FG, Olsen M, Rindorf G, Jacobsen CS, Pedersen HJ, Scott JC (1981) J Am Chem Soc 103:2440–2442
35. Williams JM, Ferraro JR, Thorn RJ, Carlson KD, Geiser U, Wang HH, Kini AM, Whangbo M-H (1992) Organic superconductors. Prentice Hall, Englewood Cliffs
36. Ishiguro T, Yamaji K, Saito G (1998) Organic superconductors, 2nd edn. Springer, Berlin
37. Mori T, Kawamoto T, Bando Y, Noda B, Wada H, Matsuzawa T, Taguchi T, Katsuhara M, Aoyagi I, Kambayashi T, Ishikawa K, Takezoe H, Uji S, Takimaya K, Otsubo T (2007)

In: Saito G, Wudl F, Haddon RC, Tanigaki K, Enoki T, Katz HE, Maesato M (eds) Multifunctional conducting molecular materials. RSC Publishing, Cambridge, pp 15–22
38. Gao X, Qiu W, Liu Y, Yu G, Zhu D (2008) Pure Appl Chem 80:2405–2423
39. Perepichka IF, Perepichka DF (eds) (2009) Handbook of thiophene-based materials. Wiley-VCH, Chichester
40. Heeger AJ (2001) Angew Chem Int Ed 40:2591–2611
41. Farchioni R, Grosso G (eds) (2001) Organic electronic materials. Berlin, Springer
42. Roncali J (2005) Chem Soc Rev 34:483–495
43. Kim JY, Lee K, Coates NE, Moses D, Nguyen T-Q, Dante M, Heeger AJ (2007) Science 317:222–225
44. Günes S, Neugebauer H, Sariciftci NS (2007) Chem Rev 107:1324–1338
45. Thompson BC, Fréchet JMJ (2008) Angew Chem Int Ed 47:58–77
46. Perepichka IF, Perepichka DF, Meng H, Wudl F (2005) Adv Mater 17:2281–2305
47. Grimsdale AC, Chan KL, Martin RE, Jokisz PG, Holmes AB (2009) Chem Rev 109: 897–1091
48. Murphy AR, Fréchet JMJ (2007) Chem Rev 107:1066–1096
49. Patra A, Wijsboom YH, Zade SS, Li M, Sheynin Y, Leitus G, Bendikov M (2008) J Am Chem Soc 130:6734–6736
50. Patra A, Bendikov M (2010) J Mater Chem 20:422–433
51. Patra A, Wijsboom YH, Leitus G, Bendikov M (2009) Org Lett 11:1487–1490
52. Cordes M, Giese B (2009) Chem Soc Rev 38:892–901
53. Wang M, Gao J, Müller P, Giese B (2009) Angew Chem Int Ed 48:4232–4234
54. Giese B, Wang M, Gao J, Stotlz M, Müller P, Graber M (2009) J Org Chem 74:3621–3625
55. Glass RS, Hug GL, Schöneich C, Wilson GS, Kuznetsova L, Lee T-M, Ammam M, Lorance E, Nauser T, Nichol GS, Yamamoto T (2009) J Am Chem Soc 131:13791–13805
56. Schöneich C, Pogocki D, Wisniowski P, Hug GL, Bobrowski K (2000) J Am Chem Soc 122: 10224–10225
57. Schöneich C, Pogocki D, Hug GL, Bobrowski K (2003) J Am Chem Soc 125:13700–13713
58. Prelog V, Traynham JG (1963) In: de Mayo P (ed) Molecular rearrangements. Interscience, NewYork, pp 593–615
59. Cope AC, Martin MM, McKervey MA (1966) Quart Rev Chem Soc 20:119–152
60. Roberts AA, Anderson CB (1969) Tetrahedron Lett 10:3883–3885
61. Kirchen RP, Sorensen TS (1979) J Am Chem Soc 101:3240–3243
62. Buzek P, Schleyer PVR, Vancik H, Sunko DE (1991). Chem Commun: 1538–1540
63. Galasso V (1998) Int J Quantum Chem 170:313–320
64. Leonard NJ (1979) Acc Chem Res 12:423–429
65. Musker WK, Wolford TL (1976) J Am Chem Soc 98:3055–3056
66. Musker WK, Wolford TL, Roush PD (1978) J Am Chem Soc 100:6416–6421
67. Musker WK (1980) Acc Chem Res 13:200–206
68. Wilson GS, Swanson DD, Klug JT, Glass RS, Ryan MD, Musker WK (1979) J Am Chem Soc 101:1040–1042
69. Ryan MD, Swanson DD, Glass RS, Wilson GS (1981) J Phys Chem 85:1069–1075
70. Brown TG, Hirschon AS, Musker WK (1981) J Phys Chem 85:3767–3771
71. Evans DH, Gruhn NE, Jin J, Li B, Lorance E, Okumura N, Macias-Ruvalcaba NA, Zakai UI, Zhang S-Z, Block E, Glass RS (2010) J Org Chem 75:1997–2009
72. Deng Y, Illies AJ, James MA, McKee ML, Peschke M (1995) J Am Chem Soc 117:420–428
73. de Visser SP, de Konning LJ, Nibbering NMM (1996) Int J Mass Spectrum Ion Process 157/158:283–291
74. James MA, Illies AJ (1996) J Phys Chem 100:15794–15799
75. Furukawa N, Kawada A, Kawai T (1984). J Chem Soc Chem Commun: 1151–1152
76. Fujihara H, Akaishi R, Furukawa N (1987) J Org Chem 52:4254–4257
77. Fujihara H, Akaishi R, Furukawa N (1987) J Chem Soc Chem Commun: 930–931

78. Fujihara H, Furukawa N (1989) J Mol Struct THEOCHEM 55:261–272
79. Iwasaki F, Toyoda N, Akaishi R, Fujihara H, Furukawa N (1988) Bull Chem Soc Jpn 61: 2563–2567
80. Fujihara H, Akaishi R, Nakamura A, Furukawa N (1990) Tetrahedron Lett 31:6375–6378
81. Fujihara H, Akaishi R, Erata T, Furukawa N (1989). Chem Commum: 1789–1790
82. Fujihara H, Akaishi R, Furukawa N (1993) Tetrahedron 49:1605–1618
83. Iwasaki F, Morimoto M, Yasui M, Akaishi R, Fujihara H, Furukawa N (1991) Acta Crystallogr C47:1463–1466
84. Fujihara H, Furukawa N (1992) Phosphorus Sulfur Silicon Relat Elem 67:131–134
85. Fujihara H, Yabe M, Chiu J-J, Furukawa N (1991) Tetrahedron Lett 32:4345–4348
86. Fujihara H, Ninoi T, Akaishi R, Erata T, Furukawa N (1991) Tetrahedron Lett 32:4537–4540
87. Fujihara H, Ninoi T, Akaishi R, Erata T, Furukawa N (1991) Tetrahedron Lett 32:4537–4540; footnote 4
88. Block E, Dikarev EV, Glass RS, Jin J, Li B, Li X, Zhang S-Z (2006) J Am Chem Soc 128: 14949–14961
89. Fujihara H, Mima H, Chiu J-J, Furukawa N (1990) Tetrahedron Lett 31:2307–2310
90. Fujihara H, Takaguchi Y, Chiu J-J, Erata T, Furukawa N (1992) Chem Lett 21:151–154
91. Takaguchi Y, Fujihara H, Furukawa N (1996) Organometallics 15:1913–1919
92. Nakanishi W, Hayashi S, Toyota S (1998) J Org Chem 63:8790–8800
93. Hayashi S, Nakanishi W (1999) J Org Chem 64:6688–6696
94. Nakanishi W, Hayashi S, Morinaka S, Sasamori T, Tokitoh N (2008) New J Chem 32: 1881–1889
95. Nakayama N, Takahashi O, Kikuchi O, Furukawa N (2000) Heteroatom Chem 11:31–41
96. Brundle CR, Baker A (1977) Electron spectroscopy: theory techniques, and applications, vol 1. Academic, London
97. Rabalais JW (1977) Principles of ultraviolet photoelectron spectroscopy. Wiley, New York
98. Ghosh PK (1983) Introduction to photoelectron spectroscopy. Wiley, New York
99. Barr TL (1994) Modern ESCA. The principles and practice of x-ray photoelectron spectroscopy. CRC Press, Boca Raton
100. Koopmans T (1933) Physica 1:104–113
101. Glass RS, Broeker JL, Jatcko ME (1989) Tetrahedron 45:1263–1272
102. Nadjo L, Savéant JM (1973) Electroanal Chem Interfac Electrochem 48:113–145
103. Houmam A (2008) Chem Rev 108:2180–2237
104. Hoffmann R (1971) Acc Chem Res 4:1–9
105. Gleiter R (1974) Angew Chem Int Ed 13:696–701
106. Brunck TK, Weinhold F (1976) J Am Chem Soc 98:4392–4393
107. Setzer WN, Glass RS (1988) In: Glass RS (ed) Conformational analysis of medium-sized heterocycles. Verlag Chemie, Weinheim
108. Drouin BJ, Madden JF, Gruhn NE, Kukolich SG, Barfield M, Glass RS (1997) J Phys Chem A 101:9180–9184
109. Glass RS, Block E, Gruhn NE, Jin J, Lorance E, Zakai UI, Zhang S-Z (2007) J Org Chem 72:8290–8297
110. Block E, Glass RS, Gruhn N, Jin J, Lorance E, Zakai UI, Zhang S-Z (2008) Phosphorus Sulfur Silicon 183:856–862
111. Glass RS, Radspinner AM, Singh WP (1992) J Am Chem Soc 114:4921–4923
112. Glass RS, Guo Q, Liu Y (1997) Tetrahedron 53:12273–12286
113. Li H, Nishiwaki K, Itami K, Yoshida J (2001) Bull Chem Soc Jpn 74:1717–1725
114. Yoshida J, Maekawa T, Murata T, Matsunaga S, Isoe S (1990) J Am Chem Soc 112: 1962–1970
115. Block E, Yencha AJ, Aslam M, Eswarakrishnan V, Lao J, Sano J (1988) J Am Chem Soc 110: 4748–4753

116. Block E, Aslam M (1988) Tetrahedron 44:281–324
117. Block E, Glass RS, Dikarev EV, Gruhn NE, Jin J, Li B, Lorance E, Zakai UI, Zhang S-Z (2007) Heteroatom Chem 18:509–515
118. Glass RS, Block E, Lorance E, Zakai UI, Gruhn NE, Jin J, Zhang S-Z (2006) J Am Chem Soc 128:12685–12692
119. Werst DW, Trifunac AD (1991) J Phys Chem 95:3466–3477
120. Werst DW (1991) J Am Chem Soc 113:4345–4346
121. Werst DW (1992) J Phys Chem 96:3640–3646
122. Morgan RS, Tabsch CE, Gushard RH, McAdon JM, Warme PK (1978) Int J Peptide Protein Res 11:209–217
123. Breinlinger EC, Keenan CJ, Rotello VM (1998) J Am Chem Soc 120:8608–8609
124. Rotello VM (1998) Heteroatom Chem 9:605–606
125. Low DW, Hill MG (1998) J Am Chem Soc 120:11536–11537
126. Butterfield DA (2002) Free Radic Res 36:1307–1313
127. Butterfield DA, Kanski J (2002) Peptides 23:1299–1309
128. Butterfield DA (2003) Curr Med Chem 10:2651–2659
129. Schöneich C (2005) Biochim Biophys Acta 1703:111–119
130. Huang X, Atwood CS, Hartshorn MA, Multhap G, Goldstein LE, Scarpa RC, Cuajungco MP, Gray DN, Lim J, Moir RD, Tanzi RE, Bush AI (1999) Biochemistry 38:7609–7616
131. Huang X, Cuajungco MP, Atwood CS, Hartshorn MA, Tyndall JDA, Hanson GR, Stokes KC, Leopold M, Multhaup G, Goldstein LE, Scarpa RC, Saunders AJ, Lim J, Moir RD, Glabe C, Bowden EF, Masters CL, Fairlie DP, Tanzi RE, Bush AI (1999) J Biol Chem 274: 37111–37116
132. Opuzo C, Huang X, Cherny RA, Moir RD, Roher AE, White AR, Cappai R, Masters EL, Tanzi RE, Inestrosa NC, Bush AI (2002) J Biol Chem 277:40302–40308
133. Del Rio MJ, Velez-Pardo C (2004) Curr Med Chem Cent Nerv Syst Agents 4:279–285
134. Petkova AT, Ishii Y, Bulbach JJ, Antzutkin ON, Leapman RD, Delaglio F, Tycko R (2002) Proc Natl Acad Sci USA 99:16742–16747
135. Petkova AT, Yau W-M, Tycko R (2006) Biochemistry 45:498–512
136. Miller Y, Ma B, Nussinov R (2010) Chem Rev 110:4820–4839
137. Luhrs T, Ritter C, Adrian M, Riek-Loher D, Bohrman B, Döbeli H, Schubert D, Riek R (2005) Proc Natl Acad Sci USA 102:17342–17347
138. Ma B, Nussinov R (2006) Curr Opin Chem Biol 10:445–452
139. Horn AHC, Sticht H (2010) J Phys Chem B 114:2219–2226
140. Chung WJ, Ammam M, Gruhn NE, Nichol GS, Singh WP, Wilson GS, Glass RS (2009) Org Lett 11:397–400
141. Coleman BR, Glass RS, Setzer WN, Prabhu UDG, Wilson GS (1982) Adv Chem Ser 201: 417–441
142. Zakai UI, Bloch-Mechkour A, Jacobsen NE, Abrell L, Lin G, Nichol GS, Bally T, Glass RS (2010) J Org Chem 75:8363–8371
143. Ammam M, Zakai UI, Wilson GS, Glass RS (2010) Pure Appl Chem 82:555–563
144. Elinson MN, Simonet J, Toupet L (1993) J Electroanal Chem 350:117–132
145. Jones IW, Tebby JC (1979) J Chem Soc Perkin Trans 2:217–218
146. Emsley JW, Longeri M, Veracini CA, Catalano D, Pedulli GF (1982) J Chem Soc Perkin Trans 2:1289–1296
147. Vondrak T, Sato S, Spirko V, Kimura K (1997) J Phys Chem A 101:8631–8638
148. Frolov YuL, Vashchenko AV (2003) Russ J Org Chem 39:1412–1414
149. Baciocchi E, Gerini MF (2004) J Phys Chem A 108:2332–2338
150. Dewar PS, Ernstbrunner E, Gilmore JR, Godfrey M, Mellor JM (1974) Tetrahedron 30:2455–2459
151. Schweig A, Thon N (1976) Chem Phys Lett 38:482–485
152. Colle MD, Distefano G, Jones D, Modelli A (2000) J Phys Chem A 104:8227–8233

153. Bossa M, Morpurgo S, Stranges S (2002) J Mol Struct THEOCHEM 618:155–164
154. Jouikov V (1995) J Electroanal Chem 398:159–164
155. Jouikov V, Ivkor V (1995) Electrochim Acta 40:1617–1622
156. Tschmutowa G, Block H (1976) Z Naturforsch B 31:1611–1615
157. Baker AD, Armen GH, Guang-di Y (1981) J Org Chem 46:4127–4130
158. Bzheovskii VM, Kapustin EG (2003) Russ J Gen Chem 73:54–60

Chapter 4
Redox and Related Coordination Chemistry of PNP- and PCP-Bridged Selenium and Tellurium-Centred Ligands

Tristram Chivers and Jari Konu

4.1 Introduction

The discovery of chalcogen-centred chelating ligands of the type $[N(PR_2E)_2]^-$ (**1**, E = O, S, Se; R = alkyl, aryl) in the 1960s [1] led to extensive studies of main group, transition-metal and lanthanide complexes during the following decades [2–4]. Some early potential applications of these complexes included their use as lanthanide shift reagents, in luminescent materials, or for metal-extraction processes. At the turn of the twenty-first century renewed interest in metal complexes of **1** was stimulated, especially for E = Se, by the findings of the O'Brien group that isopropyl derivatives (R = *i*-Pr) are suitable single-source precursors for the generation of thin films or quantum dots of certain semiconducting metal selenides, e.g. MSe (M = Zn, Cd, Hg), M_2Se_3 (M = Ga, In, Bi) and PbSe [5–8]. A suitable synthesis of the tellurium-centred congener was not developed until 2002 [9]. Since that time numerous main group [10–13], transition-metal [14–16] and lanthanide complexes [17, 18] of ligands of the type **1** (E = Te) have been prepared and structurally characterized. A review of the early work on the metal complexes of **1** (E = Te) appeared in 2007 [19]. In certain cases, these metal complexes have been used to produce thin films or nanorods of binary metal tellurides, e.g. CdTe [20, 21], Sb_2Te_3 [22], In_2Te_3 [23] or PbTe [12] by using Aerosol-Assisted Chemical Vapor Deposition (AACVD) or atmospheric pressure CVD. The use of AACVD for generating these important semiconducting materials has been elaborated in a tutorial review [24].

T. Chivers (✉) • J. Konu
Department of Chemistry, University of Calgary, Calgary, AB, Canada, T2N 1N4
e-mail: chivers@ucalgary.ca

J.D. Woollins and R.S. Laitinen (eds.), *Selenium and Tellurium Chemistry*,
DOI 10.1007/978-3-642-20699-3_4, © Springer-Verlag Berlin Heidelberg 2011

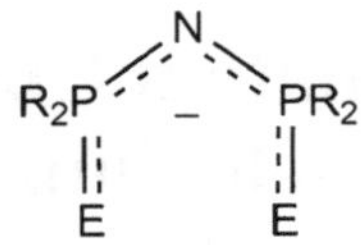

1

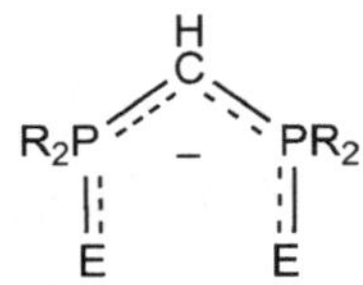

2

Despite extensive investigations of the coordination complexes of **1**, the redox chemistry of these monoanions has only been investigated in a systematic manner during the past 5 years. In 2009, synthetic routes to the *isoelectronic* carbon-bridged monoanions $[HC(PR_2E)_2]^-$ (**2**, E = Se, Te; R = Ph, *i*-Pr) were developed for the heavy chalcogens [25]. The focus of this chapter will be a comparative discussion of the redox behaviour of **1** (E = Se, Te) and mixed chalcogen systems, with that of the isoelectronic *C*-bridged systems **2**. Coordination complexes are discussed in the context of the related redox chemistry.

4.2 Synthesis

4.2.1 N-Bridged Systems

Woollins and co-workers have shown that the neutral diseleno ligands $HN(PR_2Se)_2$ (R = Ph, *i*-Pr) are easily generated by the direct reaction of the corresponding P(III)/P(III) systems with elemental selenium in hot toluene [26, 27]. Deprotonation to give the corresponding anions is effected in a straightforward manner by treatment with a base (Scheme 4.1). The relative stability of the PSe linkage in these ligands is illustrated by the observation that KO*t*-Bu in ethanol can be used as the base [26].

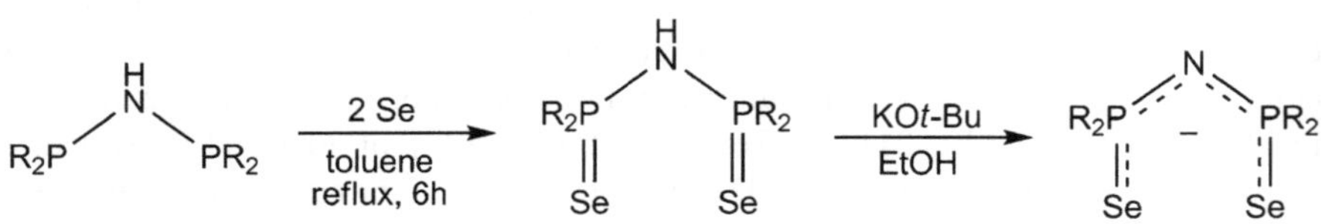

Scheme 4.1 Synthesis of PNP-bridged Se-centred anions

By contrast, tellurium does not oxidise $HN(PPh_2)_2$, even on heating, although the monotelluride is accessible for the *iso*propyl derivative TeP*i*-Pr_2NP(H)*i*-Pr_2 [28]. To circumvent the lack of reactivity of the P(III) centres towards tellurium it is necessary to generate the monoanions $[N(PR_2)_2]^-$ prior to reaction with tellurium. By using this "metallation-first" approach, the ditelluro ligands $[N(PR_2Te)_2]^-$ (**3a**, R = Ph [9]; **3b**, R = *i*-Pr [10]; **3c**, R = *t*-Bu [29]) are obtained in good yields as sodium salts (Scheme 4.2).

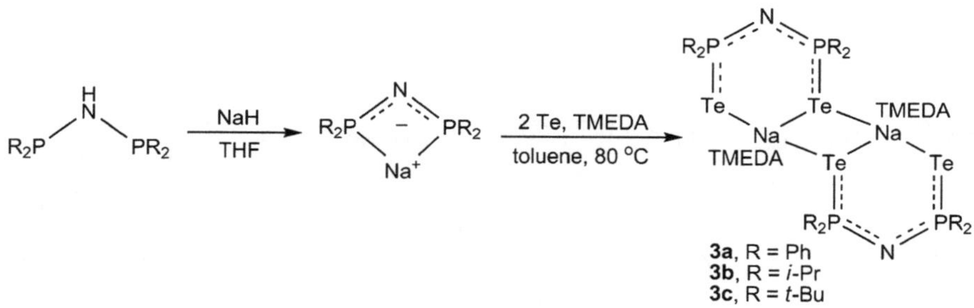

Scheme 4.2 Synthesis of PNP-bridged Te-centred anions

The monochalcogenides EPi-$Pr_2NP(H)i$-Pr_2 (**4**, E = Te; **5**, E = Se) are obtained as the P-H tautomers, which are readily metallated by n-BuLi to give the corresponding monoanions as Li derivatives [30, 31]. These reagents are indispensable for the synthesis of mixed chalcogen N-bridged ligands [**6** (S/Te) and **7** (Se/Te)] in high purity and reasonable yields, by the three-step approach depicted in Scheme 4.3 [30]. The final step formally involves insertion of the second chalcogen into a P(III)-Li bond.

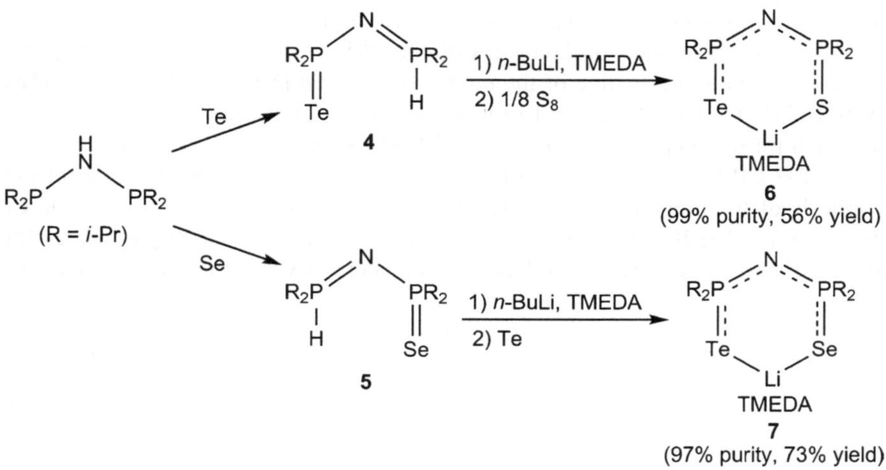

Scheme 4.3 Synthesis of PNP-bridged mixed chalcogen anions

A related approach to the mixed chalcogen N-bridged ligands entails chalcogen insertion into the Cu-P(III) bond of the homoleptic Cu(I) complex of the monotelluride ligand (**9a**) [32]. The trimeric Cu(I) complex **9a** and the silver analogue **9b** are obtained by the in situ metathesis of the thermally unstable lithium derivative of the P(III)/P(V) monotelluride Li[TePi-Pr_2NPi-Pr_2] (**8**) [31] with the appropriate coinage metal halide (Scheme 4.4) [32]. The tellurium centre in the trimeric complexes **10a-c** preferentially adopts the bridging position.

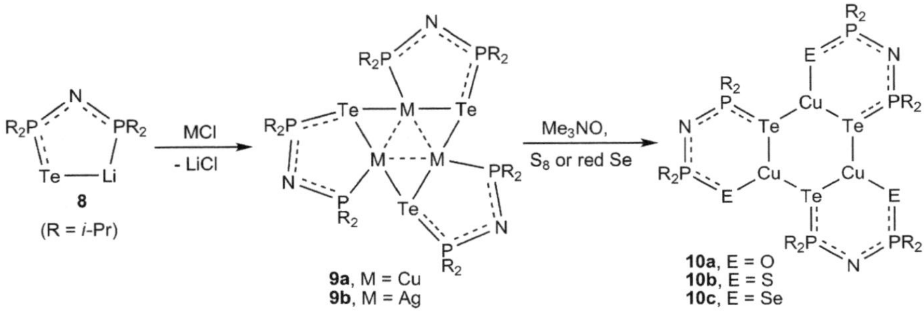

Scheme 4.4 Synthesis of Cu(I) complexes of PNP-bridged mixed chalcogen anions

4.2.2 C-Bridged Systems

In early work the *C*-bridged diseleno anion Li[HC$(PPh_2Se)_2$] (**12a**) was generated by reaction of $H_2C(PPh_2Se)_2$ with *n*-BuLi at −70°C and used as an in situ reagent to prepare homoleptic complexes of Fe, Co and Ni [33]. Transition-metal complexes of the anion $[HC(PPh_2Se)_2]^-$ have also been obtained by deprotonation of coordination complexes of the neutral ligand $H_2C(PPh_2Se)_2$ with NaH, e.g. [(η^5-C_5Me_5)Rh{η^3-$(SePPh_2)_2$CH}][ClO_4] [34, 35]. The anionic ligand in this Rh(III) complex is *C,Se,Se'* coordinated via a strong Rh–C interaction.

More recently, in view of the observation of P-Se bond cleavage in the preparation of the lithium derivative of the monoseleno anion $[SePPh_2C(H)PPh_2]^-$ (**11a**) from [$SePPh_2C(H_2)PPh_2$] and *n*-BuLi [36], we have employed the metallation-first approach for the preparations of **11a** and **12a** (Scheme 4.5) [25]. The lack of reactivity of $H_2C(PPh_2)_2$ towards tellurium necessitates the use of this method for the synthesis of the monotelluride Li[$TePPh_2C(H)PPh_2$] (**11b**) and the ditelluride Li[HC$(PPh_2Te)_2$] (**12b**). As indicated in Scheme 4.5, this methodology is successful for the preparation of TMEDA-solvated lithium derivatives of the heavy chalcogen-

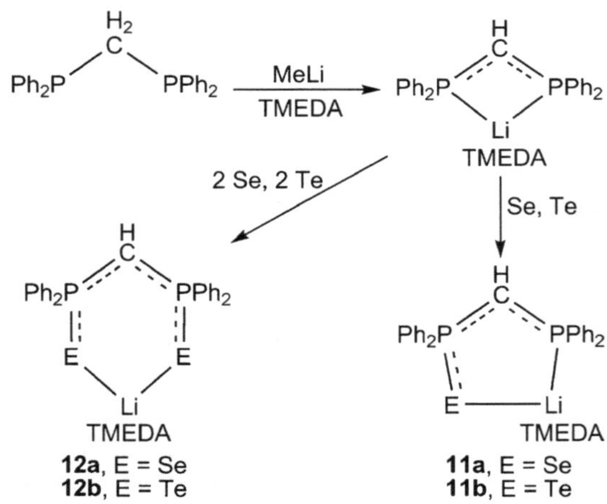

Scheme 4.5 Synthesis of PCP-bridged Se and Te-centred anions

centred *C*-bridged ligands. In the case of the telluro derivatives, however, both **11b** and **12b** are unstable in solution and the latter has been characterised only by multinuclear NMR spectra [25].

A different approach is necessary for the synthesis of *iso*propyl derivatives of the C-bridged monoanions since, unlike $H_2C(PPh_2)_2$, alkyl derivatives of the P(III)/P(III) reagent are not available from commercial sources and can only be obtained by multi-step and/or low-yield processes. As indicated in Scheme 4.6, a three-step process starting from the commercially available tetrachloro compound $H_2C(PCl_2)_2$ produces the monoanionic diseleno ligand Li[HC(P*i*-Pr_2Se$)_2$] (**13**) in excellent yield; the use of LiN*i*-Pr_2 for the deprotonation step avoids P-Se cleavage in this case [25].

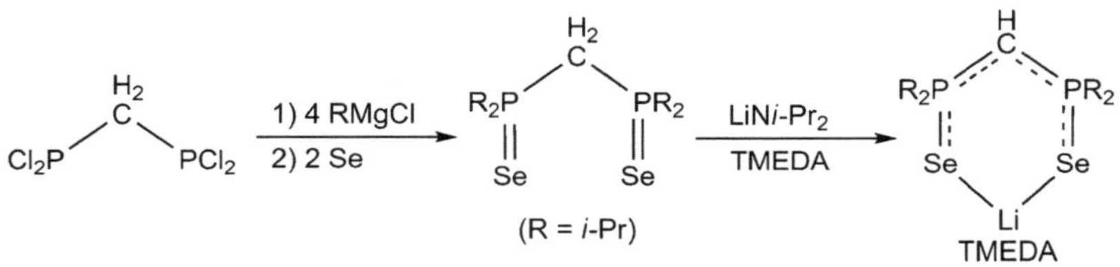

13

Scheme 4.6 Synthesis of PCP-bridged Se-centred anions with P*i*-Pr-substituents

Attempts to generate the diseleno dianion $[C(PPh_2Se)_2]^{2-}$ from $[H_2C(PPh_2Se)_2]$ and two equivalents of RLi reagents have to date been unsuccessful owing to complications arising from P-Se bond cleavage concomitant with deprotonation [25]. By contrast, Le Floch and co-workers have demonstrated that the sulfur-centred dianion $[C(PPh_2S)_2]^{2-}$ is formed as the dilithium derivative by the reaction of $[H_2C(PPh_2S)_2]$ with two equivalents of MeLi at low temperatures [37]. Metathetical reactions of this reagent with halides of a variety of main group elements [38–40], transition metals [37, 41, 42], lanthanides [43, 44] and actinides [45, 46] produce complexes that incorporate strong metal-carbon interactions. Other examples of the *C*-centred reactivity of the dianion $[C(PPh_2S)_2]^{2-}$ are discussed in Sect. 4.4.1. An alternative approach to complexes of $[C(PPh_2S)_2]^{2-}$ is amine elimination in the reaction of the neutral compound $H_2C(PPh_2S)_2$ with $M[N(SiMe_3)_2]_2$ (M = Sn, Pb) [39].

4.3 Redox Behaviour

4.3.1 One-Electron Oxidation: N-Bridged Systems

The first indication of an interesting redox behaviour for the *N*-bridged monoanions **1** (R = *i*-Pr; E = Se, Te) came from the serendipitous observation of the air oxidation of yellow solutions of the neutral monotelluride TeP*i*-Pr_2NP(H)*i*-Pr_2 (**4**)

to give a low yield of red crystals identified as the unusual ditelluride (TeP*i*-Pr_2NP*i*-Pr_2Te−$)_2$ (**14b**), which was subsequently prepared in high yield by the one-electron oxidation of the anion $[N(Pi\text{-}Pr_2Te)_2]^-$ with I_2 [28]. This oxidation was shown to be adaptable to the synthesis of analogous diselenides as either *iso*propyl or *tert*-butyl derivatives (**14a** and **15a**, Scheme 4.7); the disulfide could only be obtained as the *tert*-butyl derivative owing to the hydrogen-abstraction reactions for R = *i*-Pr [29].

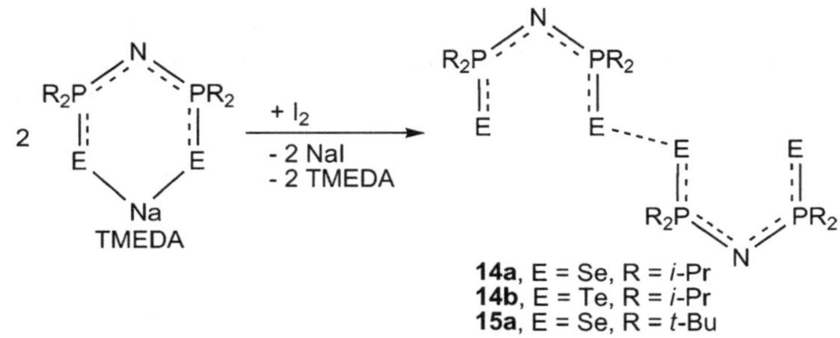

Scheme 4.7 One-electron oxidation of PNP-bridged Se- and Te-centred anions

The formation of analogous dimers by oxidation of the mixed chalcogen anions in **6** or **7** could, in principle, generate E−E, Te−Te or E−Te (E = S, Se) linkages. In practice, the I_2 oxidation does not proceed cleanly for these mixed chalcogen systems. However, the one-electron reduction of the corresponding cations **16a** and **16b** (see Sect. 4.3.3) with cobaltocene produces the neutral dimers **17a** and **17b** that both display a central Te−Te bond (Scheme 4.8) [47].

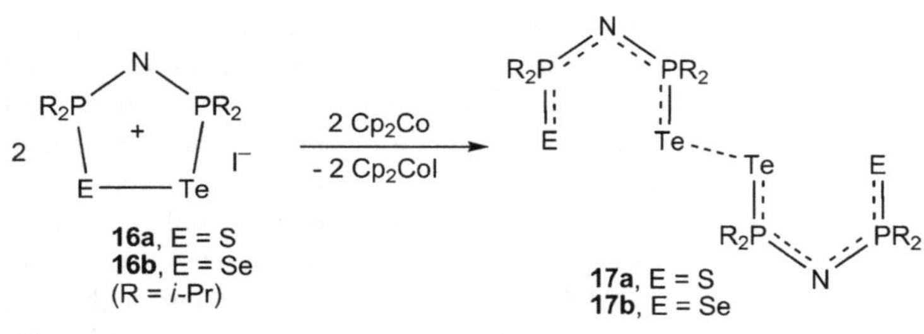

Scheme 4.8 One-electron reduction of cyclic mixed chalcogen cations

The symmetrical dichalcogenides **14a**, **14b**, **15a**, **17a** and **17b** all exhibit elongated E−E bonds, although the lengthening is smaller for the diselenides **14a** and **15a** compared to the ditelluride **14b** (ca. 4% vs. ca. 8%) [29]. In the mixed chalcogen dimers **17a** and **17b** the elongation of the central Te−Te bond is significantly less pronounced (E = Se, ca. 6%; E = S, ca. 3%) than the value of ca. 8% observed for the all-tellurium system **14b** [47]. DFT calculations for the neutral radical $[TePi\text{-}Pr_2NPi\text{-}Pr_2Te]^{\bullet}$ show that the SOMO is comprised of an almost pure linear combination of tellurium p_x and p_y orbitals (Fig. 4.1) [28]. The spatial orientation and multicentred nature of the SOMO effectively weaken the Te–Te bonding interaction, thus accounting for the elongated Te–Te bond in **14b**. The diminution of the Te–Te bond elongation for the mixed chalcogen systems **17a** and **17b** is attributed to the polarization of the SOMO of the radicals [EP*i*-Pr_2NP*i*-Pr_2Te]$^•$ (E = S, Se) towards the more electropositive tellurium atom resulting in stronger Te–Te overlap [47, 48].

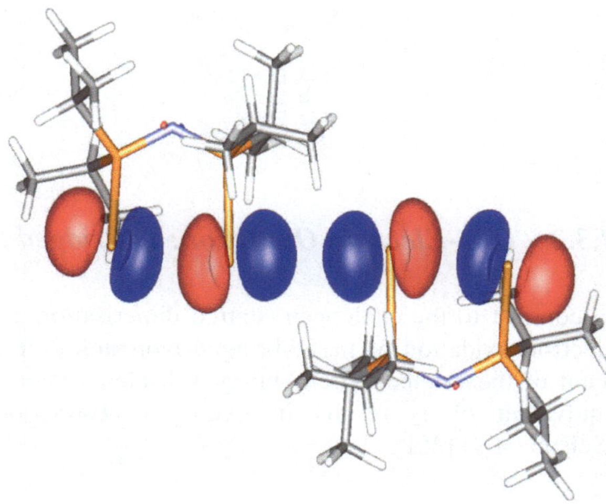

Fig. 4.1 Bonding interaction between the SOMOs of two [TeP*i*-Pr_2N*i*-Pr_2PTe]$^•$ radicals [28]

In addition to the dichalcogenides (DCs) **14a, 14b** and **15a**, a second structural isomer with a spirocyclic structure was identified for these dimers in the case of the *tert*-butyl derivative of the all-tellurium system (**18a**) [29]. DFT calculations support the description of this isomer as a contact-ion pair (CIP) in which an acyclic anion is chelated to a cyclic cation in an *Te,Te'*-bidentate fashion. The calculations also reveal that the differences in energy for the DC and CIP arrangements are insignificant for certain combinations of E and R groups, suggesting that the preferred structure could be influenced by the method of synthesis [29].

The anion-cation reactions between $[Na(TMEDA)]^+[N(PR_2E)_2]^-$ (E = Se, S) and $[N(PR_2Te)_2]^+I^-$ provide an excellent route to dichalcogenides in which the two halves of the molecule incorporate different chalcogens [48]. Interestingly, this protocol led to two different CIP structures depending on the nature of the R groups attached to phosphorus. *Tert*-butyl substituents give rise to spirocyclic structures (**18b** and **18c**) as in the case of the all-tellurium system (**18a**), whereas the corresponding *i*-Pr derivatives adopt a new CIP framework in which the acyclic anion is coordinated to the cyclic cation in an *E-monodentate* mode, **19a** and **19b**. DFT calculations for the CIP structures show that the bidentate coordination mode is more stable than the monodentate arrangement by 35–45 kJ mol^{-1} for the *tert*-butyl derivatives, but preferred by only 5–10 kJ mol^{-1} for the *iso*propyl analogues [49]. Crystal packing forces may be responsible for the isolation of monodentate structures in the latter case.

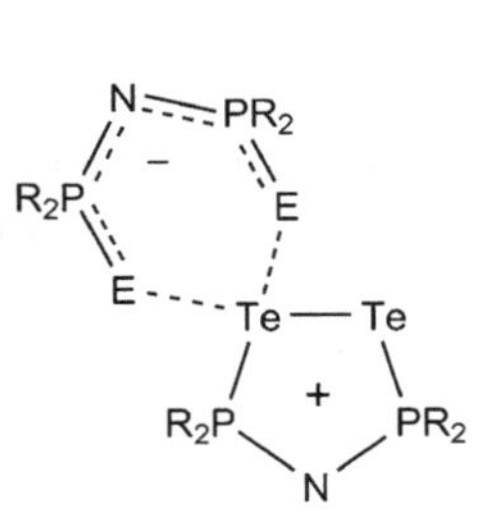

18a, E = Te
18b, E = Se
18c, E = S
(R = *t*-Bu)

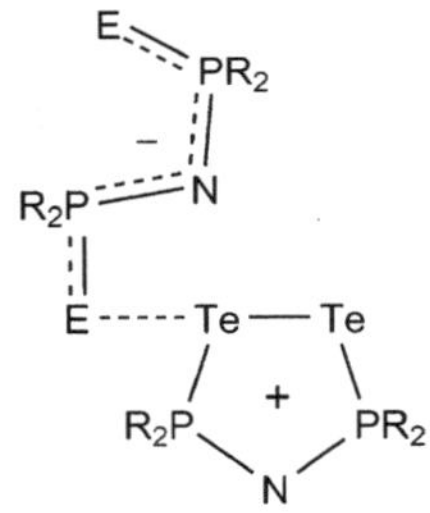

19a, E = Se
19b, E = S
(R = *i*-Pr)

4.3.2 One-Electron Oxidation: C-Bridged Systems

In contrast to the chalcogen-centred dimerisation process observed in the one-electron oxidation of the *N*-bridged monoanions **1** (R = *i*-Pr, *t*-Bu), the treatment of the isoelectronic *C*-bridged diseleno anions **12a** and **13** with one-half equivalent of I_2 results in hydrogen abstraction to give $[H_2C(PR_2Se)_2]$ (Scheme 4.9) [25].

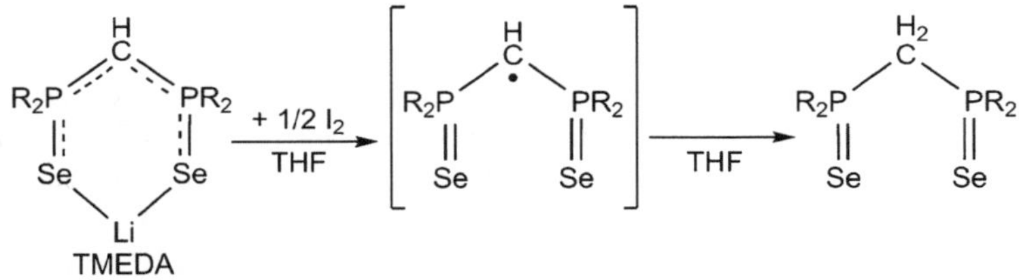

12a, R = Ph
13, R = *i*-Pr

Scheme 4.9 One-electron oxidation of PCP-bridged Se-centred anions

This carbon-centred reactivity is attributed to the different spatial morphologies of the SOMOs of the neutral radicals $[N(PR_2Te)_2]^{\bullet}$ and $[HC(PR_2Se)_2]^{\bullet}$ [25]. In the *N*-bridged radical the SOMO is almost solely localised on the chalcogen centres (Fig. 4.1), whereas the SOMOs in the *C*-bridged analogues have a large contribution from the *p* orbital on the PCP carbon (Fig. 4.2) thus predisposing them to the formation of C–H bonds. Although a red intermediate was observed in the oxidation of the diseleno *C*-bridged monoanions $[HC(PR_2Se)_2]^-$ (R = Ph, *i*-Pr) (Scheme 4.9) the associated weak EPR signals could not be definitively attributed to the corresponding radicals $[HC(PR_2Se)_2]^{\bullet}$ [25].

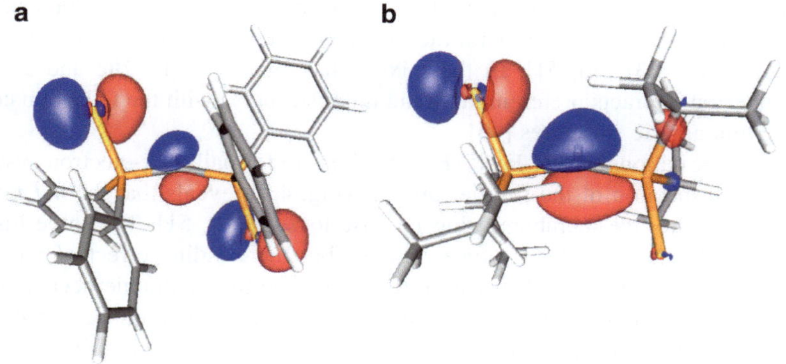

Fig. 4.2 Calculated geometries and SOMOs of the neutral radicals $[HC(PR_2Se)_2]^{\bullet}$ (**a**) R = Ph, (**b**) R = *i*-Pr [25]

4.3.3 Two-Electron Oxidation: Dichalcogeno N-Bridged Systems

The two-electron oxidation of the dichalcogeno monoanions $[N(Pi\text{-}Pr_2E)_2]^-$ (E = Se, Te) with I_2 produces the corresponding cyclic cations $[N(Pi\text{-}Pr_2E)_2]^+$ as surprisingly air-stable iodide salts (**20a**, E = Se; **20b**, E = Te) [50]. The hexafluoroantimonate salts of these cyclic cations (**21a** and **21b**) are readily prepared by metathesis of the iodide salts with $Ag[SbF_6]$ (Scheme 4.10) [51]. The mixed chalcogen cations are obtained in a similar manner as either iodide or SbF_6^- salts (**20c,d** and **21c,d**, respectively) [47].

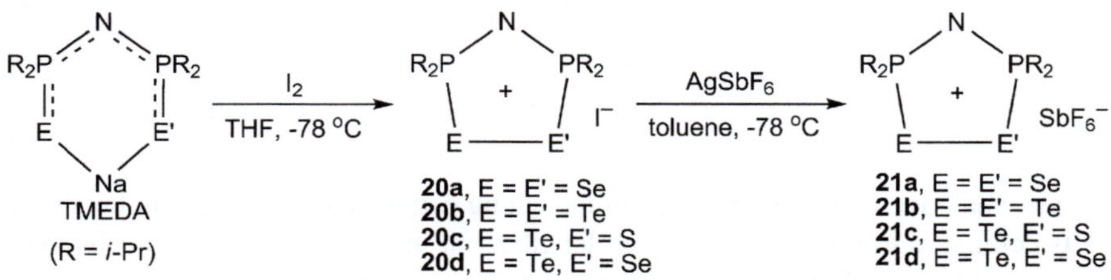

Scheme 4.10 Two-electron oxidation of PNP-bridged S-, Se- and Te-centred anions

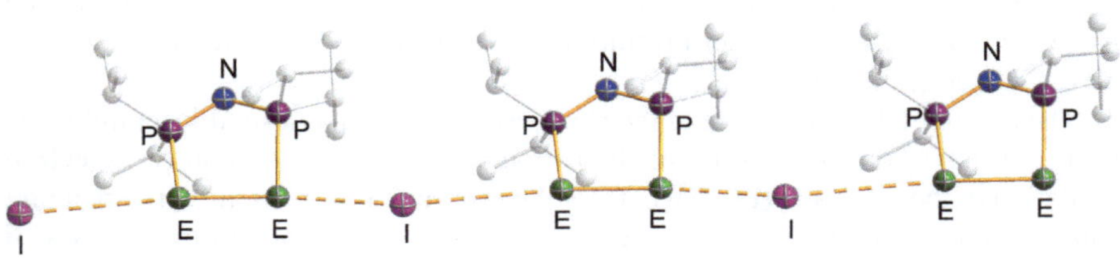

Fig. 4.3 Polymeric strand of $[N(Pi\text{-}Pr_2E)_2]^+I^-$ ion pairs (**20a**, E = Se; **20b**, E = Te) [50]

The five-membered cyclic cation in the *iso*propyl derivatives **20a** and **20b** engages in a weak interaction with the iodide counterion to form an infinite chain structure (Fig. 4.3) [50, 51]. In the mixed chalcogen systems **20c** and **20d** the I^- counterion interacts preferentially (and more strongly) with the tellurium centre to give monomeric structures [46].

The cyclic cations $[N(PR_2E)_2]^+$ (E = Se, Te) are formally 6 π-electron systems. However, inspection of the frontier orbitals (Fig. 4.4) reveals that the *net* π-bond order within the five-membered ring is close to zero [50, 51]. The three highest occupied orbitals are indeed π-type orbitals, but the bonding effect of the E–E π-bonding orbital (HOMO-2) is essentially cancelled by the double occupation of the HOMO which is the E–E π^*-antibonding orbital. The third occupied E–E π-orbital (HOMO-1) is primarily a non-bonding nitrogen-centred orbital.

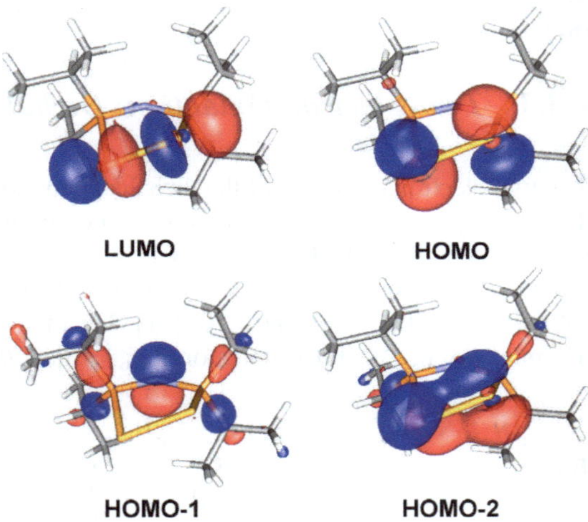

Fig. 4.4 Frontier molecular orbitals of the cations $[N(P\textit{i}\text{-}Pr_2E)_2]^+$ (E = Se, Te) [49]

The chalcogen-chalcogen bonds in the cyclic cations are elongated in the iodide salts compared to typical single-bond lengths, but this lengthening is not observed for the SbF_6^- analogues. This structural feature is attributed to the transfer of electron density from I^- to the LUMO of the cyclic cations, which is the E–E σ^* antibonding orbital (Fig. 4.4). The stronger iodide-tellurium interaction in the mixed chalcogen systems **20c** and **20d** produces a more pronounced elongation of the chalcogen-chalcogen bonds (4% for **20b**, 12% for **20c**, and 8% for **20d**) [47]. This trend is attributed to the polarisation of the σ^* orbital toward tellurium (the more electropositive chalcogen) resulting in an enhanced Te–I interaction for the mixed chalcogen cations.

The nature of the substituent of the phosphorus centres has a significant influence on the structure or composition of the cations that are formed upon two-electron oxidation of the anions $[N(PR_2E)_2]^-$ (E = Se, Te). For R = *t*-Bu the corresponding cations are obtained as five-membered rings, but the iodide salts form dimers with close E···E contacts in the solid state [29]. The phenyl derivative $[N(PPh_2Te)_2]^+I^-$

(**22**) also forms a dimeric structure (Fig. 4.5a). However, attempts to prepare the selenium congener produced the *six-membered* ring $[N(PPh_2Se)_2(\mu\text{-}Se)]^+I^-$ (**23**), which has a dimeric structure with iodide bridges (Fig. 4.5b), together with a neutral, acyclic species $[SePPh_2NP(I)Ph_2]$ (cf. **24** in Scheme 4.11) [51].

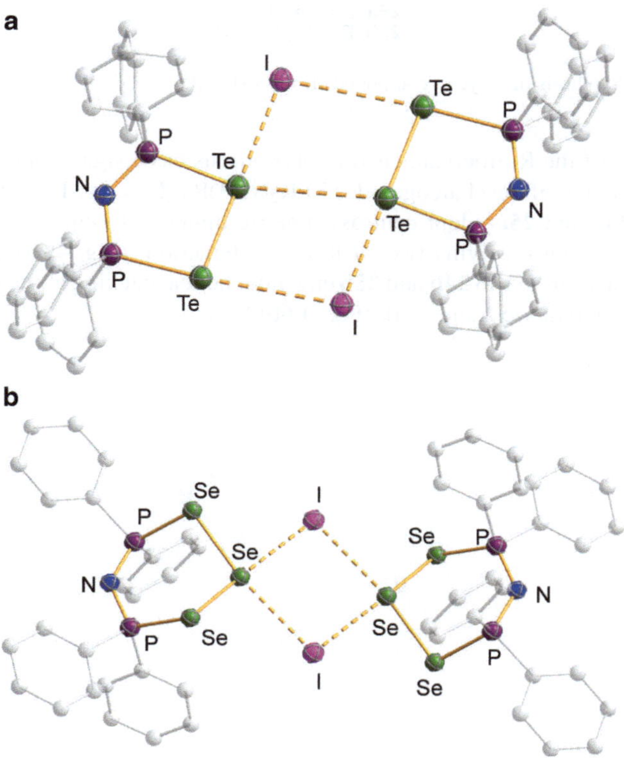

Fig. 4.5 Dimeric arrangement of (**a**) $[N(PPh_2Te)_2]^+I^-$ (**22**), and (**b**) $[N(PPh_2Se)_2(\mu\text{-}Se)]^+I^-$ (**23**) [51]

4.3.4 Two-Electron Oxidation: Monochalcogeno N-Bridged Systems

The two-electron oxidation of the monochalcogeno monoanions $[EP\textit{i}\text{-}Pr_2NP\textit{i}\text{-}Pr_2]^-$ with I_2 produces the bifunctional P(V)/P(V) compounds $EP\textit{i}\text{-}Pr_2NP(I)\textit{i}\text{-}Pr_2$ (**24a**, E = Se; **25a**, E = Te), although the yield is low in the case of the tellurium derivative **25a** owing to the concomitant formation of the cyclic cation $[N(P\textit{i}\text{-}Pr_2Te)_2]^+I^-$ (**20b**) [52]. In a different approach, the *tert*-butyl derivatives $EP\textit{t}\text{-}Bu_2NP(I)\textit{t}\text{-}Bu_2$ (**24b**, E = Se; **25b**, E = Te) are prepared in good yields by oxidation of $\textit{t}\text{-}Bu_2PNP(I)\textit{t}\text{-}Bu_2$ with elemental chalcogen (Scheme 4.11).

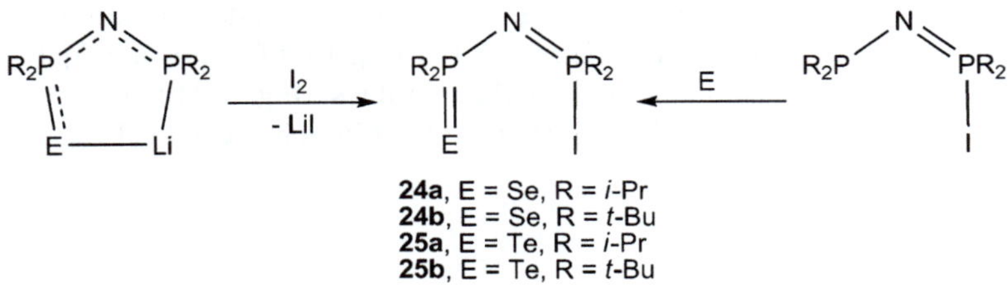

Scheme 4.11 Synthesis of acyclic compounds $E{=}PR_2NP(I)R_2$

The nature of the R group attached to phosphorus has a significant influence on the crystal packing of the chalcogenides $EPR_2NP(I)R_2$ (E = Se, Te). The *iso*propyl derivatives **24a** and **25a** adopt centrosymmetric dimeric structures with a "head-to-tail" arrangement involving two weak E···I interactions (Fig. 4.6a), whereas the bulkier *tert*-butyl groups in **24b** and **25b** engender helical packing about the 2_1 screw axis to give a polymeric framework (Fig. 4.6b) [52].

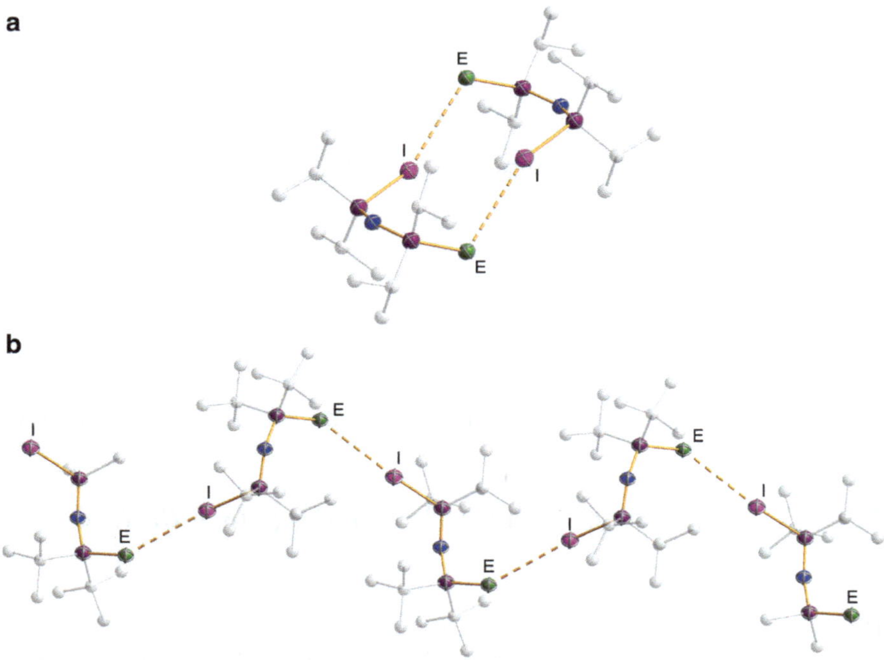

Fig. 4.6 The E···I interactions in $EPR_2NP(I)R_2$ (E = Se, Te) (**a**) R = *i*-Pr, and (**b**) R = *t*-Bu [52]

4.4 Carbon-Centred Reactivity of *C*-Bridged Systems

4.4.1 Formation of Stable Carbenoids and a $C_2P_2S_2$ Ring by Oxidation of $[C(PPh_2S)_2]^{2-}$

The behaviour of the sulfur-centred dianion $[C(PPh_2S)_2]^{2-}$ upon oxidation provides a compelling illustration of the novel carbon-centred reactivity that may be expected for *C*-bridged systems. Le Floch and co-workers demonstrated that oxidation of this dianion with C_2Cl_6 produces the remarkably stable carbenoid **26** (Scheme 4.12) [53]. Subsequently, Konu and Chivers found that the use of I_2 as a mild oxidising agent generates a dimeric form of this carbenoid that incorporates the LiI by-product (**27**) together with a novel, unsaturated six-membered $C_2P_2S_2$ ring (**28**) (Scheme 4.12) [54]. The heterocycle **28** is formally comprised of two molecules of the carbene, **:**$C(PPh_2S)_2$, one of which has undergone a P → C sulfur-transfer process [54].

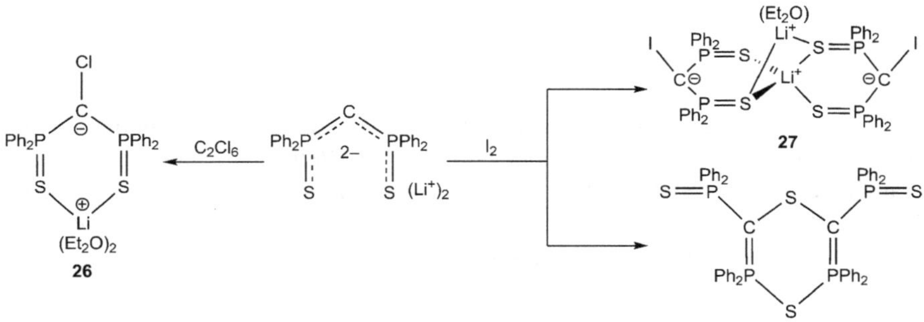

Scheme 4.12 Formation of stable carbenoids by oxidation of $[C(PPh_2S)_2]^{2-}$

4.4.2 Formation of $[SeC(PPh_2Se)_2]^{2-}$ from $[HC(PPh_2Se)_2]^-$ at Metal Centres

The diseleno *C*-bridged ligand $[HC(PPh_2Se)_2]^-$ forms homoleptic group 12 complexes with mercury or zinc [25], which exhibit distorted tetrahedral structures similar to those of the analogous complexes of the isoelectronic *N*-bridged ligands **1** (E = Se) [2–4]. In contrast to this predictable behaviour, the metathetical reactions of $Li[HC(PPh_2Se)_2]$ with MCl_2 (M = Sn, Te) unexpectedly produce the homoleptic complexes **29** and **30** in which the metal centre has been oxidized from M(II) to M(IV) and the ligand is the *triseleno dianion* $[SeC(PPh_2Se)_2]^{2-}$ formed from the diseleno mononanion $[HC(PPh_2Se)_2]^-$ by a process that formally involves proton-selenium exchange (Scheme 4.13) [55]. The tin(IV) complex **29** adopts a distorted octahedral structure in which the Sn-Se(C) bonds are ca. 0.17 Å shorter than the

Sn-Se(P) bonds, while the stereochemical activity of the lone pair on the tellurium centre in **30** imposes a see-saw geometry and one of the Se(P) atoms in each ligand is only weakly coordinated to Te.

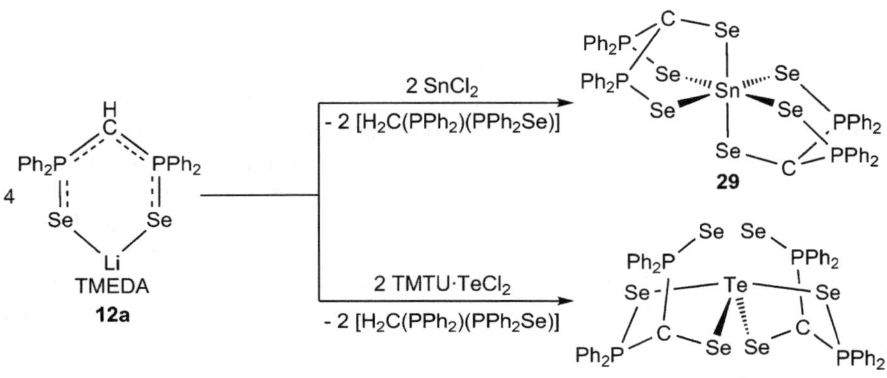

Scheme 4.13 H^{+}-Se exchange in reactions of $[HC(PPh_2Se)_2]^{-}$ with metal halides

A change in the oxidation state of the metal centre is not essential for the transformation of the diseleno monoanion of $[HC(PPh_2Se)_2]^{-}$ into the triseleno dianion $[SeC(PPh_2Se)_2]^{2-}$. Mild heating of the homoleptic mercury(II) complex $Hg[HC(PPh_2Se)_2]_2$ (**31**) in THF or toluene produces a binuclear Hg(II)/Hg(II) complex of $[SeC(PPh_2Se)_2]^{2-}$ (**32**, Scheme 4.14) [55].

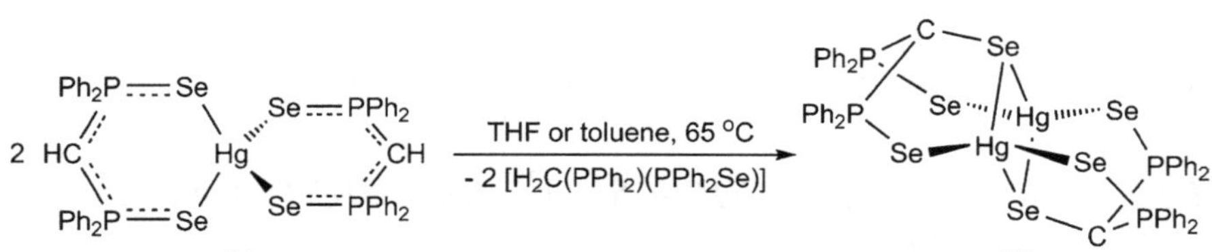

Scheme 4.14 A Hg(II)/Hg(II) complex of $[SeC(PPh_2Se)_2]^{2-}$ formed via H^{+}-Se exchange

An intriguing transformation occurs in the reaction of $Li[HC(PPh_2Se)_2]$ (**12a**) with TlOEt (Scheme 4.15) [56]. The initial product is the homoleptic thallium(I) complex $Tl[HC(PPh_2Se)_2]$ obtained as an adduct with the by-product LiOEt. This intermediate is converted upon mild heating (60°C) for 2–3 days into the mixed-oxidation state complex $Tl[Tl\{SeC(PPh_2Se)_2\}_2]$ (**33**). The crystal structure of **33** reveals a polymeric arrangement in which Tl^{+} ions are weakly hexa-coordinated to $[Tl\{SeC(PPh_2Se)_2\}_2]^{-}$ anions that are comprised of a Tl^{3+} centre octahedrally coordinated by two tripodal *Se,Se',Se'*-anions, cf. **29** [56].

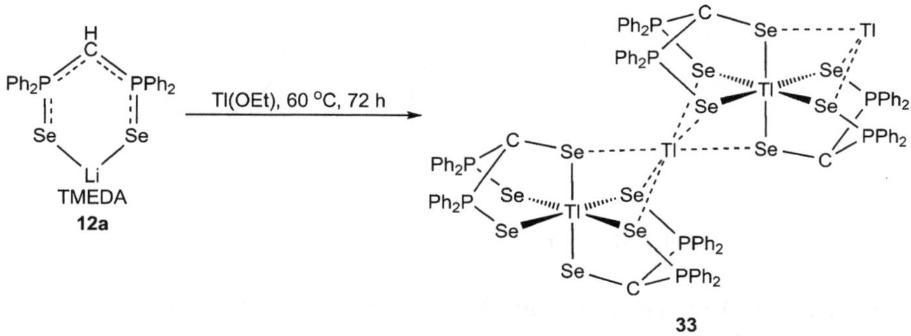

Scheme 4.15 A Tl(I)/Tl(III) complex of $[SeC(PPh_2Se)_2]^{2-}$ formed via H^+-Se exchange

4.4.3 Synthesis of Trichalcogeno PCP-Bridged Dianions $[EC(PPh_2S)_2]^{2-}$ (E = Se, S)

The mechanistically obscure formation of the triseleno dianion $[SeC(PPh_2Se)_2]^{2-}$ at metal centres (Sect. 4.4.2) prompted a search for a direct synthesis of alkali metal derivatives that could be used to explore the coordination chemistry of this tridentate ligand via metathetical reactions with halides of main group elements, transition metals and lanthanides. Since the dianionic diseleno ligand $[C(PPh_2Se)_2]^{2-}$ is not yet accessible owing to P–Se bond cleavage by RLi reagents (Sect. 4.2.2), the dithio analogue $[C(PPh_2S)_2]^{2-}$, which is known to react readily with CS_2 [57], was employed as a starting point. As a further example of the nucleophilic reactivity of $[C(PPh_2S)_2]^{2-}$ the reaction with sulfur or selenium occurs readily in toluene to give the $[Li^+(TMEDA)]$ salts of the trichalcogeno dianions $[EC(PPh_2S)_2]^{2-}$ (**34a**, E = Se; **34b**, E = S) as orange-red or red solids in excellent yields (Scheme 4.16). The solid-state structure of **34a** is monomeric with both $[Li^+(TMEDA)]$ cations *S*, *Se*-chelated by the tridentate ligand $[SeC(PPh_2S)_2]^{2-}$ [58].

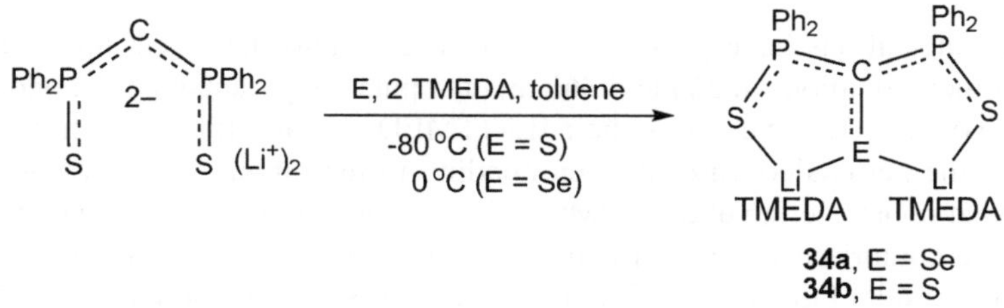

Scheme 4.16 Synthesis of PCP-bridged trichalcogeno dianions

The dianions $[EC(PPh_2S)_2]^{2-}$ (E = Se, E = S) have also been generated as the monomeric Pb(II) complexes **35a** and **35b** by chalcogen insertion into the Pb-C bonds of the dimeric Pb(II) complex of $[C(PPh_2S)_2]^{2-}$ as illustrated in Scheme 4.17 [39].

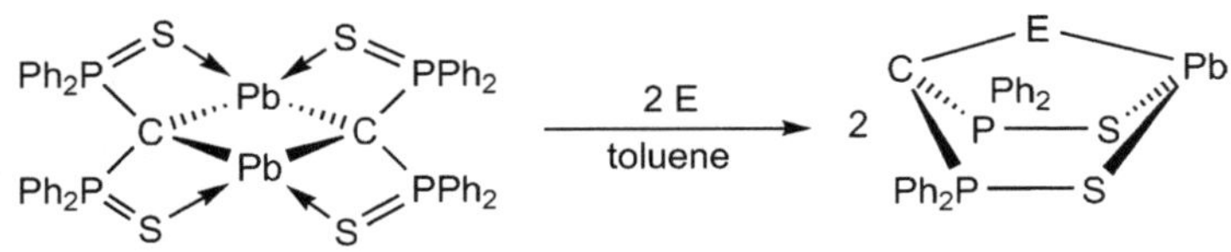

Scheme 4.17 Synthesis of Pb(II) complexes of $[EC(PPh_2S)_2]^{2-}$ (E = S, Se) by chalcogen insertion

4.5 Redox Behaviour of the Trichalcogeno Dianions $[EC(PPh_2S)_2]^{2-}$ (E = Se, S)

4.5.1 One-Electron Oxidation

The one-electron oxidation of the dilithium derivatives **34a** and **34b** with one-half equivalent of I_2 was investigated with a view to generating and assessing the stability of the corresponding radical anions $[EC(PPh_2S)_2]^{\bullet-}$. In both cases dimers of the radical anions joined by E–E bonds (E = Se, S) were formed as $[Li^+(TMEDA)]$ derivatives (**36a** and **36b**, Scheme 4.18). The central chalcogen-chalcogen bonds in both of these dimers are elongated by ca. 8% compared to typical single-bond values [58].

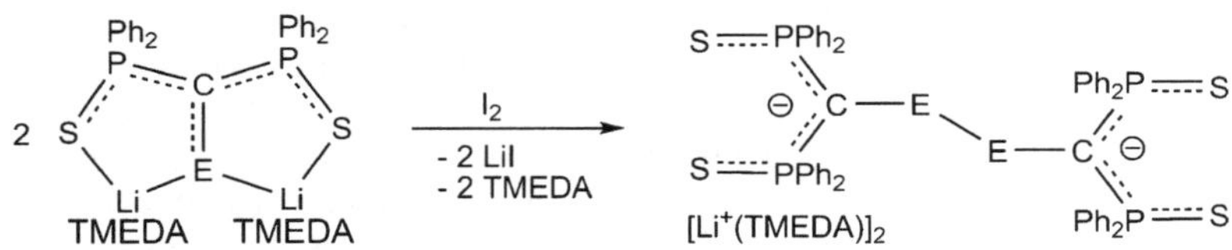

Scheme 4.18 One-electron oxidation of $[EC(PPh_2S)_2]^{2-}$ (E = S, Se)

DFT calculations have provided useful insights into the electronic structures of the dimeric dianions in **36a** and **36b** and the putative paramagnetic monoanions $[EC(PPh_2S)_2]^{\bullet-}$ (E = Se, S) as their $[Li^+(TMEDA)]$ salts [58]. As illustrated in Fig. 4.7, the net chalcogen-chalcogen bonding in the dimers is solely due to the (somewhat poor) overlap of the SOMOs of the mononanionic radicals which themselves are comprised of an anti-bonding combination of p-orbitals on the C=E bond. As a result of the weak E–E bonds, solutions of **36a** and **36b** undergo decomposition to form paramagnetic species. The EPR spectra of such solutions show that a common *persistent radical* identified by a binomial triplet is generated from both **36a** and **36b**. The EPR spectrum of this species can be simulated by using the hyperfine coupling constants $a(^{31}P) = 19.2$ and $a(^{33}S) = 5.70$ G. This observation indicates that the radical monoanions $[EC(PPh_2S)_2]^{\bullet-}$, which are presumably formed initially by dissociation of the dimers **36a** and **36b**, readily decompose with loss of the carbon-bound chalcogen to produce a persistent (as yet unidentified) radical that incorporates two equivalent phosphorus (and sulfur) nuclei [58].

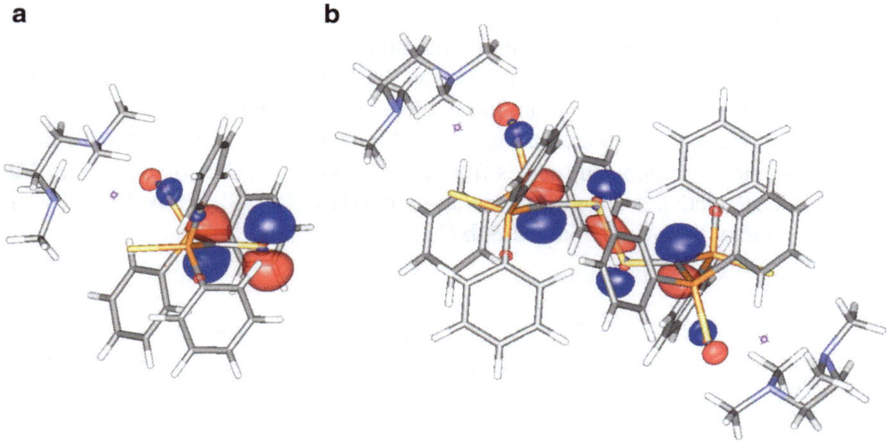

Fig. 4.7 Frontier molecular orbitals of [Li^+(TMEDA)] salts of (**a**) [$EC(PPh_2S)_2$]$^{\bullet-}$ (SOMO) and (**b**) [$(SPh_2P)CEEC(PPh_2S)_2$]$^{2-}$ (HOMO) (E = Se, S) [58]

4.5.2 Two-Electron Oxidation

The two-electron oxidation of the trichalcogeno dianions [$EC(PPh_2S)_2$]$^{2-}$ in **34a** and **34b** is a potential source of the corresponding neutral chalcogenocarbonyls [$EC(PPh_2S)_2$]. Indeed, the treatment of the Se-containing derivative **34a** with one equivalent of iodine in toluene produces a dark compound that was identified by X-ray crystallography as the LiI adduct of the expected selone [$SeC(PPh_2S)_2$] (**37**, Scheme 4.19) [58]. The Se-I contact of 2.722(1) Å in **37** represents a moderately strong interaction, cf. 2.50 Å for the sum of the covalent radii of Se and I, and values in the range 2.56–2.73 Å for charger-transfer complexes of selones with I_2 or IBr, which are described by Devillanova et al. as "strong" Se-I bonds [59]. In the current example it appears that the Se-I interaction is crucial to the stability of **37**, since attempts to remove LiI from this adduct, e.g. with 12-crown-4, resulted in decomposition. The C-Se distance of 1.815(4) Å found for **37** falls within the range of 1.77–1.84 Å reported for a wide variety of selenocarbonyl compounds [60].

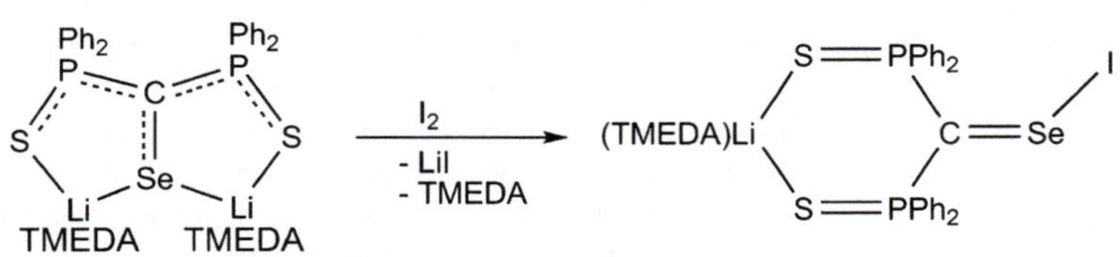

34a

37

Scheme 4.19 Two-electron oxidation of [$SeC(PPh_2S)_2$]$^{2-}$

By contrast to the behaviour of the selenium-containing dianion $[SeC(PPh_2S)_2]^{2-}$ in **34a**, the attempted two-electron oxidation of the all-sulfur system **34b** in a variety of solvents generated a protonated species **38**, which was also obtained as a part of the decomposition process of the dimeric dianion **36b** (Scheme 4.20). The diamagnetic compound **38** is formally comprised of the anion radical $[SC(PPh_2S)_2]^{\bullet-}$ and the neutral radical $[H(S)C(PPh_2S)_2]^{\bullet}$ linked by an S-S bond. In this context we note that the HOMO of the dimeric dianion **36b** exhibits a significant contribution from the *p*-orbitals of the two backbone carbon atoms (Fig. 4.7b). However, the formation of the corresponding protonated species from the Se-containing dimeric dianion in **36a** could not be detected, despite the observation that $[H_2C(PPh_2S)_2]$ is the final decomposition product from both **36a** and **36b** [58].

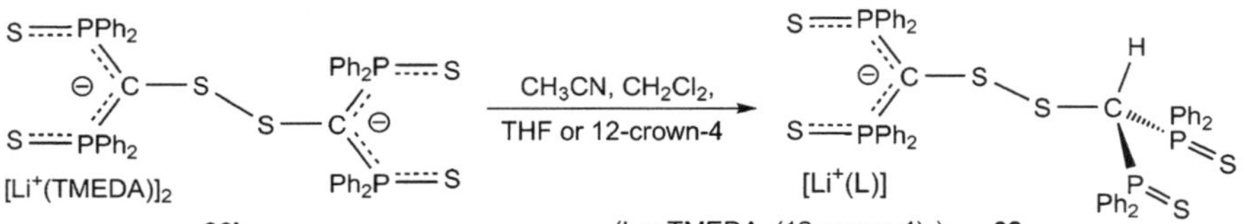

36b **38**

Scheme 4.20 Formation of the protonated species $[(SPPh_2)_2CSSC(H)(PPh_2S)_2]^-$

4.6 Copper and Silver Complexes of the Dichalcogenides $[(EPh_2P)_2SeSeC(PPh_2E)_2]^{2-}$ (E = Se, S)

An interesting consequence of the selenium-proton exchange process described in Sect. 4.4.2 was observed in the metathetical reaction between $[Li(TMEDA)][HC(PPh_2Se)_2]$ (**12a**) and $CuCl_2$ in a 2:1 M ratio, which was shown by ^{31}P NMR spectroscopy to involve the formation of the P(III)/P(V) monoselenide $[H_2C(PPh_2)(PPh_2Se)]$. The solid-state structure of the binuclear Cu(I)/Cu(I) complex $\{Cu_2\text{-}\eta^2:\eta^2\text{-}[(SePh_2P)_2CSeSeC(PPh_2Se)_2]\}$ (**39**) formed in this reaction revealed the presence of a dimeric dianionic ligand in which two $[SeC(PPh_2Se)_2]^{-\bullet}$ radical anions are connected by a Se–Se bond with a separation of 2.683(2) Å (Scheme 4.21), cf. 2.508(1) Å for the dilithium derivative of this dimeric dianion, **36a** [61]. By contrast, homoleptic M(IV) complexes of the monomeric $[SeC(PPh_2Se)_2]^{2-}$ dianion are obtained from the reactions of MCl_2 (M = Sn, Te) and **12a** (Scheme 4.13). The structure of **39** displays notably different features compared to those of the binuclear Hg(II)/Hg(II) complex **32** (Scheme 4.14) as a result of the different charges on the metal and the ligands. As a result of the Se–Se bond formation, the two metal centres in **39** are well-separated (Cu···Cu = 4.374(1) Å, cf. Hg(II)···Hg(II) = 3.105(1) Å in **32**). In summary, the reaction between **12a** and $CuCl_2$ involves not only a selenium-proton exchange, but also an internal redox process in which the dianionic ligands $[SeC(PPh_2Se)_2]^{2-}$ so formed are oxidized to the dimeric dianion $[(SePh_2P)_2CSeSeC(PPh_2Se)_2]^{2-}$ and the Cu(II) centres are reduced to Cu(I).

The η^2-Se_2 bonding mode adopted by the Cu(I) centres in the all-selenium complex **39** is the first example of this bonding arrangement for transition-metal complexes of diselenide ligands of the type RSe–SeR, for which oxidative addition (insertion of the metal into the Se–Se bond) is a more common behaviour [62]. A related binuclear Cu(I)/Cu(I) complex involving this structural motif, $\{Cu_2$-η^2:η^2-$[(SPh_2P)_2CSeSeC(PPh_2S)_2]\}$ (**40**), is obtained by metathetical reactions of (a) the dilithium salt of the monomeric dianion **34a** and $Cu(II)Cl_2$ or (b) the dilithium salt of the dimeric dianion **36a** and Cu(I)Cl (Scheme 4.21) [61]. The formation of the same Cu(I)/Cu(I) complex **40** in both reactions corroborates the occurrence of a redox process in reaction (a). The Se–Se distances of 2.610(2) and 2.688(2) Å in the two independent molecules of the structure of **40** are comparable with that in the all-selenium derivative **39**.

The corresponding all-sulfur system $\{Cu_2$-η^2:η^2-$[(SPh_2P)_2CSSC(PPh_2S)_2]\}$ (**41**) is prepared in a similar manner to **40** by metathetical reactions of **34b** or **36b** and copper halides (Scheme 4.21) [63]. Intriguingly, the S–S distances in the two molecules in the unit cell of **41** are 2.540(4) and 2.720(3) Å, i.e. elongated by 0.32 and 0.50 Å, respectively, compared to the corresponding distance in the dilithium derivative **36b**. The weak S···S contacts in **41** are in the same range as those found for transannular S···S interactions in unsaturated eight-membered S-N rings, e.g. 2.60 Å in S_4N_4 [64] and values in the range 2.43–2.55 Å for the bicyclic compounds 1,5-$R_4P_2N_4S_2$ (R = alkyl, aryl) [65] for which Breher has suggested biradical character [66]. Intriguingly, DFT calculations reveal biradical character associated with the weak S···S interactions for **41** [63].

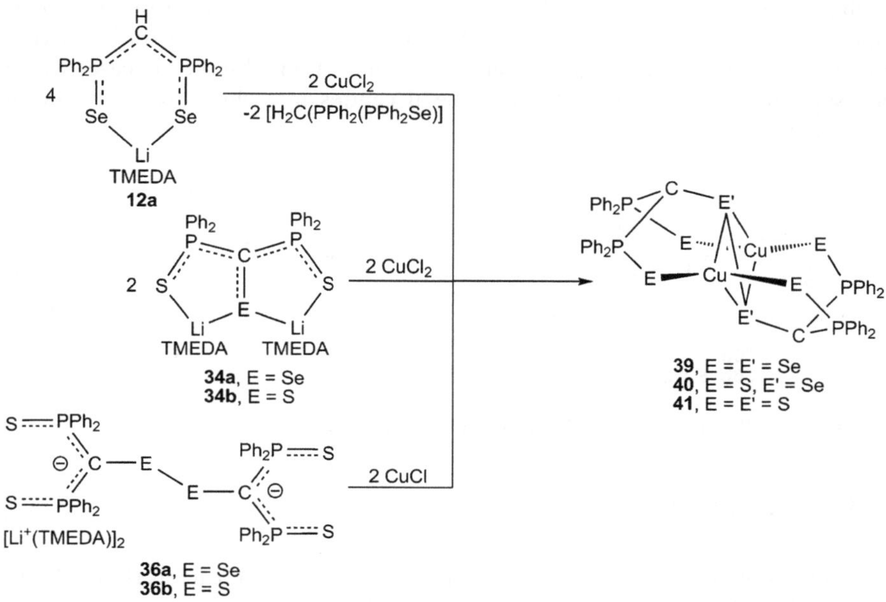

Scheme 4.21 Synthesis of Cu(I)/Cu(I) complexes of $[(EPPh_2)_2CEEC(PPh_2E)_2]^{2-}$ (E = S, Se)

The binuclear Ag(I)/Ag(I) complexes {Ag_2-η^2:η^2-[$(SPh_2P)_2CEEC(PPh_2S)_2$]} (**42**, E = Se; **43** E = S) are obtained by metathesis between Ag[OSO_2CF_3] and **36a** or **36b**, respectively [63]. The molecular structure of **42** is similar to that of the copper analogue **40** except that the silver-selenium contacts in the markedly asymmetrical Ag-η^2-Se_2 interactions are weaker than those in **40**. This structural difference may account for the fluxional behaviour that is apparent from variable temperature ^{31}P NMR studies of **42** and which is not exhibited by the Cu(I)/Cu(I) complex **40** [63].

4.7 Gold Complexes of the Trichalcogeno Dianions $[EC(PPh_2S)_2]^{2-}$ (E = Se, S)

In contrast to the metathesis observed for Cu(I) and Ag(I) reagents, the reactions of **36a** or **36b** with the Au(I) source Au(CO)Cl involve a redox process in which the dimeric dichalcogenide ligands are reduced to the corresponding monomeric dianions $[EC(PPh_2S)_2]^{2-}$ (E = Se, S) and one of the gold centres is oxidized to generate the mixed-oxidation state Au(I)/Au(III) complexes {Au[$EC(PPh_2S)_2$]}$_2$ (**44**, E = Se; **45** E = S) (Scheme 4.22) [63]. The gold centres in **44** and **45** exhibit the typical square-planar (four-coordinate) and linear (two-coordinate) geometries expected for Au(III) and Au(I), respectively [67]. However, the S–Au(I)–S bond angles in both compounds deviate noticeably from linearity (by *ca.* 25°) consistent with a relatively strong Au(I)···Au(III) interaction [63]. The Au(I)···Au(III) separations of ca. 3.12 Å lie at the low end of the normal range reported for Au(I)···Au(III) interactions [68–71], although shorter contacts have been observed in cases where the supporting ligands bring the two metal centres into close proximity owing to steric hindrance [72–74].

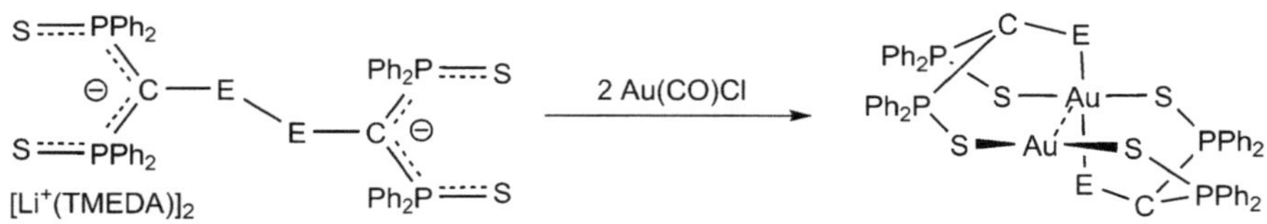

36a, E = Se
36b, E = S

44, E = Se
45, E = S

Scheme 4.22 Synthesis of Au(I)/Au(III) complexes of $[EC(PPh_2S)_2]^{2-}$ (E = S, Se)

4.8 Conclusions

A comparison of the redox chemistry of the isoelectronic monoanions $[N(PR_2E)_2]^-$ and $[HC(PR_2E)_2]^-$ has revealed markedly different behaviour. The *N*-bridged systems give rise to compounds with chalcogen-chalcogen bonds upon oxidation either in the form of a variety of neutral dimers (one-electron oxidation) or as cyclic cations (two-electron oxidation). By contrast, the analogous *C*-bridged monoanions exhibit carbon-centred reactivity, i.e. hydrogen abstraction, upon oxidation. This fundamental difference is attributed to the disparity in the composition of the SOMOs of the corresponding neutral radicals. In the case of the *N*-bridged systems, the SOMOs are almost exclusively chalcogen-based whereas there is a large contribution from the *p*-orbital on carbon for the C-bridged congeners.

A more significant aspect of this carbon-centred reactivity is the proton-selenium exchange process that results in the transformation of the diseleno monoanion $[HC(PPh_2Se)_2]^-$ into the triseleno dianion $[SeC(PPh_2Se)_2]^{2-}$ in the presence of a metal centre. The mechanism of this transformation is unclear, however a related observation has been reported by Grim et al. [75]. These authors found that deprotonation of $[HC(PPh_2S)_3]$ to give the monoanion $[C(PPh_2S)_3]^-$ occurs by direct reaction with mercuric halides in ethanol, although it cannot be achieved by treatment with a variety of bases; they suggest that coordination of one or more of the PS groups to mercury increases the acidity of the methine proton thus facilitating proton transfer to ethanol [75]. In a related observation the metal-coordinated neutral ligand $[H_2C(PPh_2Se)_2]$ is readily deprotonated to the corresponding anion $[HC(PPh_2Se)_2]^-$, which engages in a strong metal-carbon interaction [34, 35]. The possibility that chelation of the monoanionic ligand $[HC(PPh_2Se)_2]^-$ [25] to a metal centre makes the methine proton more acidic merits further investigation.

The tridentate trichalcogeno dianions $[EC(PPh_2S)_2]^{2-}$ (E = Se, S) are readily prepared by treating $[C(PPh_2S)_2]^{2-}$ with the appropriate chalcogen, but the all-selenium derivative $[SeC(PPh_2Se)_2]^{2-}$ is not yet available via this route owing to PSe cleavage in attempts to make $[C(PPh_2Se)_2]^{2-}$ from $[H_2C(PPh_2Se)_2]$ and RLi reagents (Sect. 4.2.2). The use of alternative deprotonation agents may circumvent this problem. The one-electron oxidation of the dianions $[EC(PPh_2S)_2]^{2-}$ (E = Se, S) produces the novel dichalcogenides $[(SPh_2P)_2CEEC(PPh_2S)_2]^{2-}$ with chalcogen-chalcogen bonds that are elongated by ca. 8%. Two-electron oxidation of the Se-containing dianion $[SeC(PPh_2S)_2]^{2-}$ gives the corresponding neutral selone $[SeC(PPh_2S)_2]$ as the LiI adduct.

The tridentate $[SeC(PPh_2S)_2]^{2-}$ and the multidentate $[(SPh_2P)_2CSeSeC(PPh_2S)_2]^{2-}$ dianions are potentially versatile ligands for the production of a wide variety of metal complexes as illustrated by the preliminary investigations of metathetical reactions with coinage metal reagents. The redox behaviour involved in some of these reactions, notably for the gold systems, suggests the intriguing possibility that, in some cases, metathesis of these dianions with metal halides could lead to complexes of the monoanion radical $[SeC(PPh_2S)_2]^{\bullet-}$ either via one-electron oxidation of $[SeC(PPh_2S)_2]^{2-}$ or as a result of Se−Se bond cleavage in the dimer $[(SPh_2P)_2CSeSeC(PPh_2S)_2]^{2-}$. In this context the pronounced S–S bond-stretching and resulting biradicaloid character induced by metal coordination in $\{Cu_2\text{-}\eta^2\text{:}\eta^2\text{-}[(SPh_2P)_2CSSC(PPh_2S)_2]\}$ may confer unusual reactivity on this Cu(I)/Cu(I) system.

Acknowledgements We thank former members of the Chivers group at the University of Calgary, Jamie S. Ritch, Drs S. R. Robertson and D. J. Eisler, for their contributions to the synthetic and structural work on the PNP-bridged systems, Dr M. Risto for her investigations of Group 13 and copper complexes of PCP-bridged chalcogen-centred ligands, and Dr Heikki Tuononen (University of Jyväskylä, Finland), who has provided helpful insights to our understanding of the chemistry described herein through DFT calculations. The financial support of NSERC (Canada) is also gratefully acknowledged.

References

1. Schmidpeter A, Böhm R (1964) Angew Chem Int Ed 3:704
2. Ly TQ, Woollins JD (1998) Coord Chem Rev 176:451–481
3. Silvestru C, Drake JE (2001) Coord Chem Rev 223:117–216
4. Haiduc I (2004) Comprehensive coordination chemistry II. In: Lever ABP (ed) Elsevier Pergamon, Oxford, vol 1, Amsterdam, pp 323–347
5. Afzaal M, Crouch D, Malik MA, Motevalli M, O'Brien P, Park JH (2003) J Mater Chem 13:639–640
6. Afzaal M, Ellwood K, Pickett NL, O'Brien P, Raftery J, Waters J (2004) J Mater Chem 14:1310–1315
7. Waters J, Crouch DJ, Raftery J, O'Brien P (2004) Chem Mater 16:3289–3298
8. Crouch DJ, O'Brien P, Malik MA, Skabara PJ, Wright SP (2003) Chem Commun 1454–1455
9. Chivers T, Parvez M, Briand GG (2002) Angew Chem Int Ed 41:3468–3470
10. Chivers T, Eisler DJ, Ritch JS (2005) Dalton Trans 2675–2677
11. Copsey M, Chivers T (2005) Chem Commun 4938–4940
12. Ritch JS, Chivers T, Ahmad K, Afzaal M, O'Brien P (2010) Inorg Chem 48:1198–1205
13. Ritch JS, Chivers T (2010) Dalton Trans 39:1745–1750
14. Copsey MC, Panneerselvam A, Afzaal M, Chivers T, O'Brien P (2007) Dalton Trans 1528–1538
15. Levesanos N, Robertson SD, Maganas D, Raptopoulou CP, Terzis A, Kyritsis P, Chivers T (2008) Inorg Chem 47:2949–2951
16. Eisler DJ, Robertson SD, Chivers T (2009) Can J Chem 87:39–46
17. Gaunt AJ, Scott BL, Neu MP (2006) Angew Chem Int Ed 45(10):1638–1641
18. Gaunt AJ, Reilly SD, Enriquez AE, Scott BL, Ibers JA, Sekar P, Ingram KIM, Kaltsoyannis N, Neu MP (2008) Inorg Chem 47:29–41
19. Chivers T, Konu J, Ritch JS, Eisler DJ, Copsey MC, Tuononen HM (2007) J Organomet Chem 693:2658–2668
20. Garje SS, Ritch JS, Eisler DJ, Afzaal M, O'Brien P, Chivers T (2006) J Mater Chem 16: 966–969
21. Ahmad K, Afzaal M, Ritch JS, Chivers T, O'Brien P (2010) J Am Chem Soc 132:5964–5965
22. Garje SS, Eisler DJ, Ritch JS, Afzaal M, O'Brien P, Chivers T (2006) J Am Chem Soc 128: 3120–3121
23. Garje SS, Copsey MC, Afzaal M, O'Brien P, Chivers T (2006) J Mater Chem 16:4542–4547
24. Ritch JS, Afzaal M, Chivers T, O'Brien P (2007) Chem Soc Rev 36:1622–1631
25. Konu J, Tuononen HM, Chivers T (2009) Inorg Chem 48:11788–11798
26. Bhattacharyya P, Novosad J, Phillips J, Slawin AMZ, Williams DJ, Woollins JD (1995) J Chem Soc Dalton Trans 1607–1613
27. Cupertino D, Birdsall DJ, Slawin AMZ, Woollins JD (1999) Inorg Chim Acta 290:1–7
28. Chivers T, Eisler D, Ritch JS, Tuononen HM (2005) Angew Chem Int Ed 44:4953–4956
29. Ritch JS, Chivers T, Eisler DJ, Tuononen HM (2007) Chem Eur J 13:4643–4653
30. Robertson SD, Chivers T (2008) Dalton Trans 1765–1772

31. Ritch JS, Chivers T (2009) Inorg Chem 48:3857–3865
32. Ritch JS, Chivers T (2008) Dalton Trans 957–962
33. Davison A, Reger DL (1971) Inorg Chem 10:1967–1970
34. Valderrama M, Contreras R, Bascuñan M (1994) Polyhedron 13:1101–1103
35. Contreras R, Valderrama M, Yañez S (1993) Transit Met Chem 18:73–76
36. Lusser M, Peringer P (1987) Inorg Chim Acta 127:151–152
37. Cantat T, Mézailles N, Ricard L, Jean Y, Le Floch PA (2004) Angew Chem Int Ed 43: 6382–6385
38. Foo C, Lau K-C, Yang Y-F, So C-W (2009) Chem Commun 6816–6818
39. Leung W-P, Wan C-L, Kan K-W, Mak TCW (2010) Organometallics 29:814–820
40. Leung W-P, Wan C-L, Mak TCW (2010) Organometallics 29:1622–1628
41. Cantat T, Demange M, Mézailles N, Ricard L, Jean Y, Le Floch PA (2005) Organometallics 24:4838–4841
42. Cantat T, Ricard L, Mézailles N, Le Floch P (2006) Organometallics 25:6030–6038
43. Cantat T, Jaroschik F, Nief F, Ricard L, Mézailles N, Le Floch P (2005) Chem Commun 5178–5180
44. Cantat T, Jaroschik F, Ricard L, Le Floch P, Nief F, Mézailles N (2006) Organometallics 25:1329–1332
45. Cantat T, Arliguie T, Noël A, Thuéry P, Ephritikine M, Le Floch P, Mézailles N (2009) J Am Chem Soc 131:963–972
46. Tourneux J-C, Berthet J-C, Thuéry P, Mézailles N, Le Floch P, Ephritikine M (2010) Dalton Trans 39:2494–2496
47. Robertson SD, Chivers T, Tuononen HM (2008) Inorg Chem 47:10634–10643
48. Chivers T, Ritch JS, Robertson SD, Konu J, Tuononen HM (2010) Acc Chem Res 43: 1053–1062
49. Robertson SD, Chivers T, Tuononen HM (2009) Inorg Chem 48:6755–6762
50. Konu J, Chivers T, Tuononen HM, (2006) Chem Commun 1634–1636
51. Konu J, Chivers T, Tuononen HM (2006) Inorg Chem 45:10678–10687
52. Ritch JS, Robertson SD, Risto M, Chivers T (2010) Inorg Chem 49:4681–4686
53. Cantat T, Jacques XF, Ricard L, Le Goff XF, Mézailles N, Le Floch P (2007) Angew Chem Int Ed 46:5947–5950
54. Konu J, Chivers T (2008) Chem Commun 4995–4997
55. Konu J, Chivers T (2010) Chem Commun 46:1431–1433
56. Risto M, Konu J, Chivers T (2011) Dalton Trans DOI: 10.1039/C1DT10646E
57. Cantat T, Nief F, Ricard L, Le Floch P, Mézailles N (2006) Organometallics 25:4965–4976
58. Konu J, Chivers T, Tuononen HM (2010) Chem Eur J 16:12977–12987
59. Cristiani F, Demartin F, Devillanova FA, Isaia F, Lippolis V, Verani G (1994) Inorg Chem 33: 6315–6324
60. Murai T, Kato S (2000) Top Curr Chem 208:177–199
61. Risto M, Konu J, Chivers T (2010) Inorg Chem 50:406–408
62. Oilunkaniemi R, Laitinen RS, Ahlgren M (2001) J Organomet Chem 623:168–175
63. Konu J, Tuononen HM, Chivers T (2011) Chem Eur J (in press)
64. Delucia ML, Coppens P (1978) Inorg Chem 17:2336–2338
65. Chivers T, Hilts RW, Jin P, Chen Z, Lu X (2010) Inorg Chem 49:3810–3815
66. Breher F (2007) Coord Chem Rev 251:1007–1043
67. Schmidbaur H, Dash KC (1982) Adv Inorg Chem Radiochem 25:239–266
68. Calhorda MJ, Canales F, Gimeno MC, Jimenez J, Jones PG, Laguna A, Veiros F (1997) Organometallics 16:3837–3844
69. Canales S, Crespo O, Gimeno MC, Jones PG, Laguna A, Mendizabal F (2001) Organometallics 20:4812–4818
70. Gimeno MC, Laguna A (2008) Chem Soc Rev 37:1952–1966

71. Schmidbaur H, Schier A (2008) Chem Soc Rev 37:1931–1951
72. Raptis RG, Porter LC, Emrich RJ, Murray HH, Fackler JP Jr (1990) Inorg Chem 29:4408–4412
73. Schmidbaur H, Hartmann C, Reber G, Muller G (1987) Angew Chem Int Ed 26:1146–1148
74. Ortner K, Hilditch L, Zheng Y, Dilworth JR, Abram U (2000) Inorg Chem 39:2801–2806
75. Grim SO, Smith PH, Nittolo S, Ammon HL, Satek LC, Sangokoya SA, Khanna RK, Colquhoun IJ, McFarlane W, Holden JR (1985) Inorg Chem 24:2889–2895

Chapter 5
Synthesis, Structures, Bonding, and Reactions of Imido-Selenium and -Tellurium Compounds

Risto S. Laitinen, Raija Oilunkaniemi, and Tristram Chivers

5.1 Introduction

Though progress in the chemistry of Se-N and Te-N compounds has been slower compared to that of S-N compounds due to lack of suitable reagents, impressive developments have been made in recent decades [1–4]. The heavier chalcogen derivatives show significant differences in their structures, reactivities, and properties compared to the sulfur analogues. In addition, the lability of Se-N and Te-N bonds has led to applications of these reactive functionalities in organic synthesis. Selenium and tellurium imide and amide derivatives have played a major role in this development. Their synthesis, structural features, bonding properties, some reactions, and metal complexes are reviewed in this chapter. Comparisons with the corresponding sulfur species will be made, where appropriate, in order to illustrate the group trends that are observed for these intriguing chalcogen-nitrogen compounds.

5.2 Selenium and Tellurium Diimides

5.2.1 Synthesis

The first selenium(IV) diimide derivative, $Se(N^tBu)_2$, was prepared by the reaction of *tert*-butylamine with $SeCl_4$ modifying the approach that was used to synthesize the analogous sulfur(IV) diimide $S(N^tBu)_2$ [5, 6]. By contrast to the sulfur(IV)

R.S. Laitinen (✉) • R. Oilunkaniemi
Department of Chemistry, University of Oulu, P.O. Box 3000, FI-90014 Oulu, Finland
e-mail: risto.laitinen@oulu.fi

T. Chivers
Department of Chemistry, University of Calgary, 2500 University Drive NW, Calgary, AB, Canada, T2N 1N4

J.D. Woollins and R.S. Laitinen (eds.), *Selenium and Tellurium Chemistry*,
DOI 10.1007/978-3-642-20699-3_5,

diimide, which is a stable monomeric species under ambient conditions, monomeric Se(N^{t}Bu)$_2$ decomposes at room temperature to give a mixture of cyclic selenium imides and ^{t}BuN = N^{t}Bu [7] (see Sect. 5.5.1). The thermal stability of selenium diimides is enhanced by the introduction of supermesityl substituents in Se(NMes*)$_2$ (Mes* = 2,4,6-^{t}Bu$_3$C$_6$H$_2$), which is prepared by the reaction of $SeCl_4$ with Mes*NHLi [8]. The tellurium diimide ^{t}BuNTe(μ-N^{t}Bu)$_2$TeN^{t}Bu (see Sect. 5.2.2) is obtained in good yields as a thermally stable, orange solid from the reaction of lithium *tert*-butylamide with $TeCl_4$ in THF [9]. This synthesis was originally conducted in toluene for which the six-membered tellurium(II) imide (TeN^{t}Bu)$_3$ is a minor product (see Sect. 5.5.1) [10].

5.2.2 Structures

Three different conformations are possible for monomeric chalcogen diimides (Scheme 5.1).

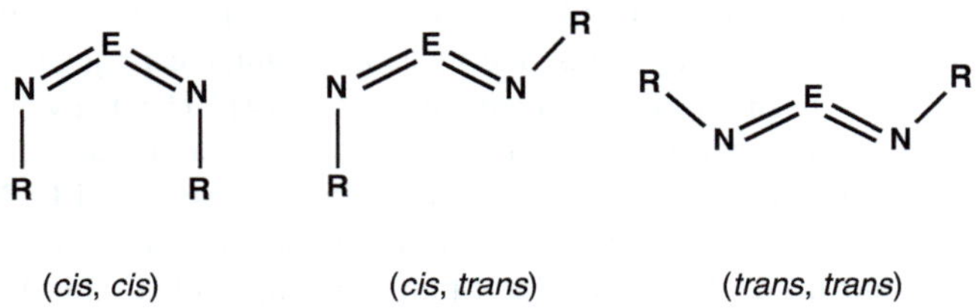

Scheme 5.1 Conformational isomers of chalcogen diimides

Ab initio and DFT molecular orbital computations for E(NR)$_2$ (E = S, Se; R = H, Me, ^{t}Bu, and SiMe$_3$) (see Fig. 5.1) have predicted that, with the exception of the parent imides E(NH)$_2$ [11] and some aryl derivatives [8], the *cis,trans* conformation is the most stable conformation for the majority of chalcogen

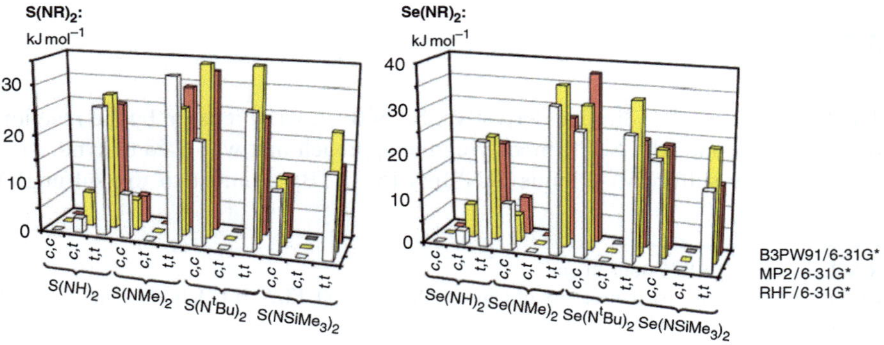

Fig. 5.1 RHF/6-31G*, MP2/6-31G*, and B3PW91/6-31G* relative energies of the different conformations of S(NR)$_2$ and Se(NR)$_2$ (R = H, Me, ^{t}Bu, SiMe$_3$) [11]

diimides. This is indeed found to be the case for sulfur diimides. The *cis,trans* isomer is most common [12, 13], but the *cis,cis* conformation is found for R = H [14], as well as for C_6F_5, 2,6-$Me_2C_6H_3$, and 2,4,6-$Br_3C_6H_2$ [12, 13]. In the gas phase $(Me_3SiN)_2S$ also adopts a *cis,cis* arrangement [15].

While structural information on selenium diimides is rather sparse, the crystal structure determination of adamantyl selenium diimide shows that this species also adopts the *cis,trans*-conformation [16]. The ^1H and ^{14}N NMR spectrum of $Se(N^tBu)_2$ has also been interpreted in terms of the *cis,trans* conformation [6]. On the other hand, both DFT calculations and an approximate crystal structure determination of the supermesityl derivative $Se(NMes^*)_2$ indicate an unprecedented *trans,trans* conformation [8].

In contrast to their sulfur and selenium analogues, tellurium diimides adopt dimeric structures [10]. Dimerization is attributed to the increasing reluctance of the heavier chalcogens to engage in pπ –pπ multiple bonds, as clearly manifested by the trends in the structures of the chalcogen dioxides EO_2 (E = S, Se, Te). Sulfur dioxide is monomeric both in the gas phase and in the solid state [17]. By contrast, selenium dioxide is a two-dimensional polymer in the solid state with both single and double SeO bonds [18–20], although a dimeric species has been identified in the gas phase [21]. Tellurium dioxide is a three-dimensional polymer with only single TeO bonds [22, 23].

The two most common conformations for tellurium diimide dimers are shown in Scheme 5.2. The *tert*-butyl derivative $^tBuNTe(\mu\text{-}N^tBu)_2TeN^tBu$ has a *cis,endo, endo* arrangement of terminal ^tBu groups with respect to the Te_2N_2 ring [10], whereas a *trans,exo,exo* arrangement of the exocyclic groups is observed for the unsymmetrical derivatives $RNTe(\mu\text{-}NR')_2TeNR$ (R = PPh_2NSiMe_3; R′ = ^tBu, ^tOct) in the solid state [24].

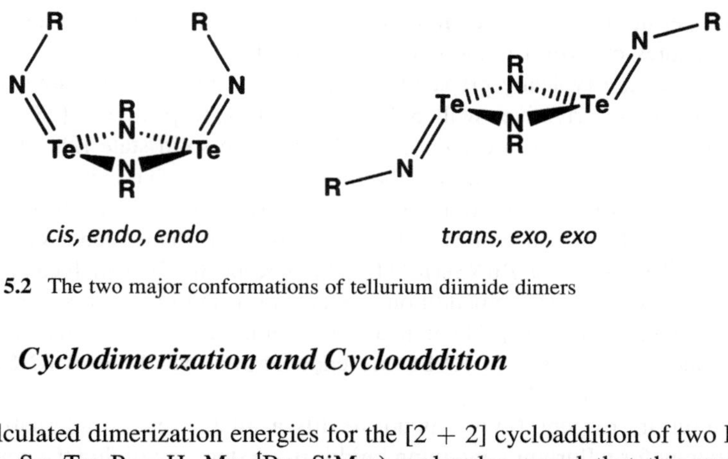

Scheme 5.2 The two major conformations of tellurium diimide dimers

5.2.3 Cyclodimerization and Cycloaddition

The calculated dimerization energies for the [2 + 2] cycloaddition of two $E(NR)_2$ (E = S, Se, Te; R = H, Me, ^tBu, $SiMe_3$) molecules reveal that this process is strongly endothermic for sulfur diimides, approximately thermoneutral for selenium diimides and strongly exothermic for tellurium diimides [25, 26], consistent with experimental observations, as exemplified in Fig. 5.2 for the N-methyl derivatives $E(NMe)_2$. Whereas the cycloaddition of $S(NR)_2$ and $Se(NR)_2$ is not energetically favourable, that of $S(NR)_2$ and $Te(NR)_2$ is energetically neutral, and that of $Se(NR)_2$ and $Te(NR)_2$ is highly favourable [26]. The mixed chalcogen system $RNSe(\mu\text{-}NR)_2TeNR$ has not, however, been prepared.

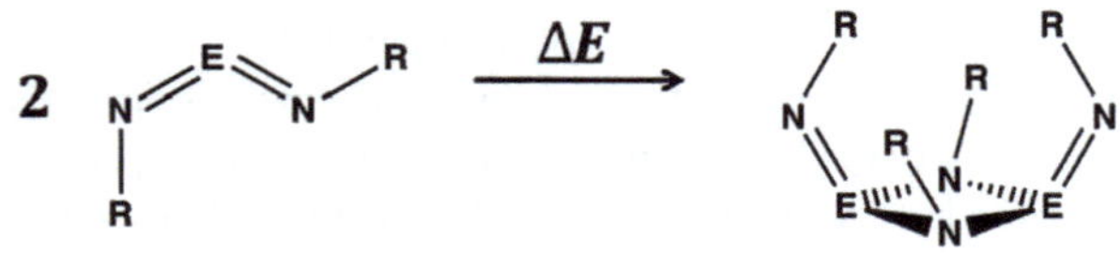

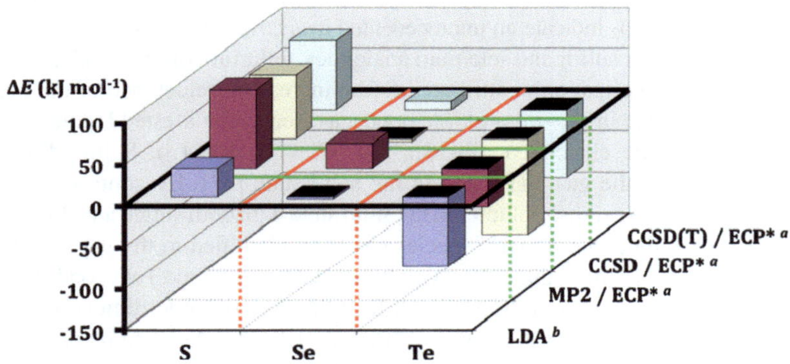

Fig. 5.2 Cyclodimerization energies of $E(NMe)_2$ (E = S, Se, Te). [a]LDA calculations, see Ref. [25][b] MP2 and CCSD calculations, see Ref. [26]. (Adapted from Ref. [27]; reproduced with permission by Taylor & Francis)

The propensity for Se and Te imides to undergo cyclodimerization is also evident in the structures of hybrid imido-oxo systems of the type RNEO. For E = Se or S these species are typically prepared by the reaction of a primary amine, or the trimethylsilylated derivative, with $SeOCl_2$ or $SOCl_2$, respectively [28–30]. The sulfur derivatives (E = S) are monomeric both in the solid state [28, 29] and in the gas phase [31, 32]. By contrast, the only structurally characterized selenium analogue $OSe(\mu\text{-}N^tBu)_2SeO$ is dimeric in the solid state [16]. The reaction of $SeCl_4$ with $^tBuNH_2$ in the presence of SO_2Cl_2 or $SeOCl_2$ yields the related cyclic species $^tBuNSe(\mu\text{-}N^tBu)_2SO_2$ and $^tBuNSe(\mu\text{-}N^tBu)_2SeO$, respectively, which can formally be considered as [2 + 2] cycloaddition products of $^tBuNEO_x$ (E = S, $x = 2$; E = Se, $x = 1$) and $Se(N^tBu)_2$ [7]. Theoretical calculations have shown consistently that the cycloaddition reactions of RNSO are endothermic, but those of RNSeO and RNTeO are exothermic [26].

The tellurium reagent $TeOCl_2$ is not readily available, but an alternative synthetic strategy for tellurium-containing imido-oxo systems involving the controlled hydrolysis of the tellurium diimide dimer $^tBuNTe(\mu\text{-}N^tBu)_2TeN^tBu$ by use of $(C_6F_5)_3B{\cdot}H_2O$ as a stoichiometric reagent has been developed [33]. This approach allows the successive replacement of terminal N^tBu groups by oxo ligands to give the imidotelluroxanes $OTe(\mu\text{-}N^tBu)_2TeN^tBu$ and $[OTe(\mu\text{-}N^tBu)_2Te(\mu\text{-}O)]_2$ (a tetramer of ^tBuNTeO) as Lewis acid adducts with $(C_6F_5)_3B$. Adduct formation prevents the energetically favoured production of the polymer $({}^tBuNTeO)_\infty$ [33]. The structural trend monomer → dimer → polymer observed for the hybrid systems ^tBuNEO (E = S, Se, Te) provides a cogent illustration of the increasing reluctance of the heavier chalcogens to form multiple bonds with oxygen or nitrogen.

The presence of two Te = N^{t}Bu groups in the tellurium diimide dimer $^tBuNTe(\mu\text{-}N^tBu)_2TeN^tBu$ provides an opportunity to study the outcome of double cycloadditions. Indeed, the reaction of the dimer with four equivalents of ^tBuNCO proceeds via an initial cycloaddition to generate an *N,N′*-ureato tellurium imide, which is converted to the corresponding telluroxide by reaction with ^tBuNCO (Scheme 5.3). The final product is the dimeric ureatotelluroxide, $[OC(\mu\text{-}N^tBu)_2TeO]_2$, which forms an extended helical network in the solid state as a result of weak C = O···Te interactions [34].

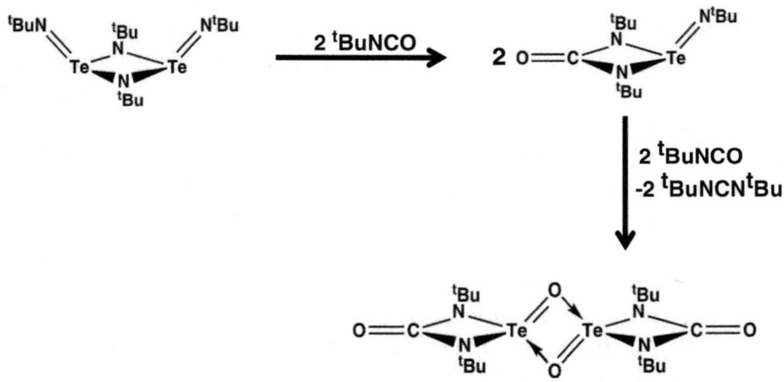

Scheme 5.3 Cycloaddition reaction of $^tBuNTe(\mu\text{-}N^tBu)_2TeN^tBu$ and ^tBuNCO [34]

5.2.4 *Reactions*

Selenium diimides are reactive species that are efficient in situ reagents for allylic amination of olefins and 1,2-diamination of 1,3-dienes [5, 35]. The thermal decomposition of selenium(IV) diimides provides a fruitful source of a medley of cyclic selenium imides (see Sect. 5.5.1). Recently, the thermally unstable selenium(IV) monoimide $Ph_2Se = NH$ has been generated in situ by treatment of the cation $[Ph_2SeNH_2]^+$ with LDA. In the presence of *N*-bromosuccinimide the cation $[Ph_2SeNSePh_2]^+$ is produced and can be isolated as the $[BPh_4]^-$salt [36].

The chalcogen diimides are also readily susceptible to attack by nucleophiles. For example, $Se(N^tBu)_2$ has been used to generate the pyramidal dianion $[Se(N^tBu)_3]^{2-}$, isoelectronic with SeO_3^{2-}, by reaction with 2 equivalents of $LiNH^tBu$ [37]. In a similar manner *tert*-butyl tellurium diimide dimer reacts with $LiNH^tBu$ or KO^tBu to afford the pyramidal anions $[Te(N^tBu)_3]^{2-}$ and $[Te(N^tBu)_2(O^tBu)]^-$, respectively [38, 39].

The dilithium triimidochalcogenites $[Li_2\{E(N^tBu)_3\}]_2$ form dimeric structures in which two pyramidal $[E(N^tBu)_3]^{2-}$ dianions are bridged by four lithium cations to form distorted, hexagonal prisms (see Scheme 5.4). Lithium halides disrupt the dimeric hexameric structures to give distorted cubes, in which a molecule of the lithium halide is entrapped by a $Li_2[E(N^tBu)_3]$ monomer [40, 41].

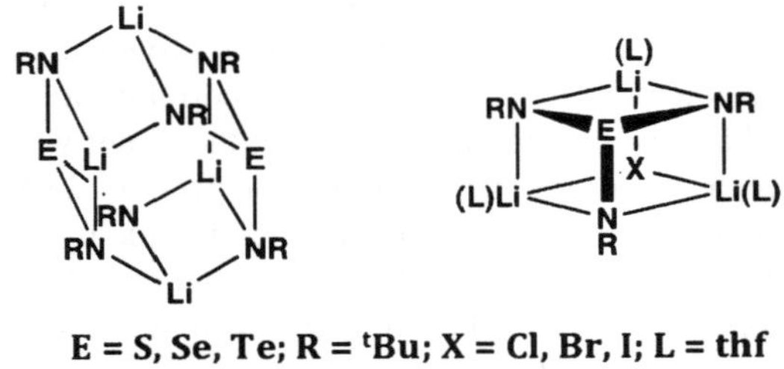

Scheme 5.4 Dilithium triimidochalcogenites and their LiX adducts

The attempted oxidation of $^tBuNTe(\mu\text{-}N^tBu)_2TeN^tBu$ with iodine affords $[(^tBuIN)Te(\mu\text{-}N^tBu)_2Te(\mu\text{-}O)]_2(I_3)_2$ [42], a complex that contains a cation with an N-I bond and a triiodide counter-ion. Concomitantly, one of the terminal TeN^tBu groups is converted to a TeO linkage emphasizing the extreme hydrolytic sensitivity of Te imides.

The dimer $^tBuNTe(\mu\text{-}N^tBu)_2TeN^tBu$ is readily protonated at one of the terminal N^tBu groups by Brønsted acids. For example, reaction with HCF_3SO_3 produces the monoprotonated derivative $[(^tBuNH)Te(\mu\text{-}N^tBu)_2Te(N^tBu)][CF_3SO_3]$ in quantitative yields [42] (see Scheme 5.5). An analogous complex with a chloride counterion $[(^tBuNH)Te(\mu\text{-}N^tBu)_2Te(N^tBu)]Cl$ is formed in the reaction of $TeCl_4$ with ^tBuNHLi [10]. The *cis-endo,endo* arrangement of exocyclic N^tBu groups with respect to the Te_2N_2 ring in the dimer $^tBuNTe(\mu\text{-}N^tBu)_2TeN^tBu$ (see Scheme 5.2) becomes *cis-endo,exo* in the protonated complexes [10, 42]. Mono- or di-methylation of the exocyclic N^tBu groups in the dimer $^tBuNTe(\mu\text{-}N^tBu)_2TeN^tBu$ can be achieved by addition of the appropriate amounts of CF_3SO_3Me [43].

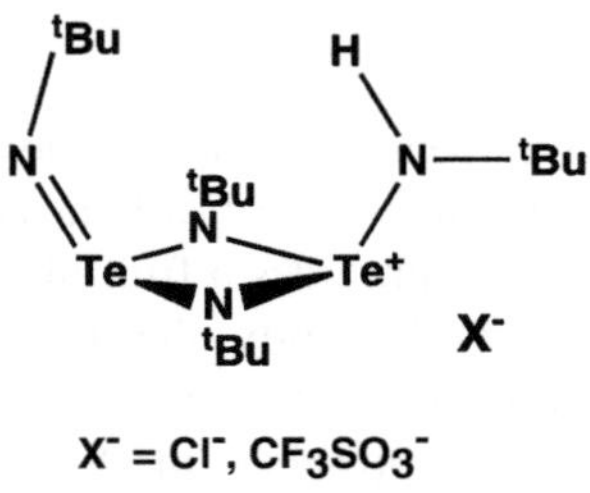

Scheme 5.5 $[(^tBuNH)Te(\mu\text{-}N^tBu)_2Te(N^tBu)]X$

5.2.5 *Metal Complexes*

All chalcogen diimides form metal complexes with both main group and transition metal centres. The coordination chemistry of sulfur diimides is particularly rich due to the availability of three potential donor sites and two formal π-bonds [44]. As shown in Scheme 5.6, mononuclear *N,N′*-chelated complexes of the selenium diimide $Se(N^tBu)_2$ are also known both with main group metals, e.g., $[SnCl_4\{Se(N^tBu)_2\}]$ [45], and with transition metals, such as $[MCl_2\{N,N'\text{-}Se(N^tBu)_2\}]$ (M = Pd [46], Pt [47]) and $[Co_2(\mu\text{-}Cl)_3\{N,N'\text{-}Se(N^tBu)_2\}(C_4H_8O)_2][CoCl_3(N^tBu)]\cdot$thf [48]. In these complexes the ligand is constrained to adopt a *trans,trans* conformation, though it is energetically the least favorable conformer [11] (see Fig. 5.1). Dialkyl selenium diimides $Se(NR)_2$ undergo a redox process with bis(amino) stannylenes to produce a spirocyclic Sn(IV)NSeN ring system [49].

The dimeric tellurium diimide $^tBuNTe(\mu\text{-}N^tBu)_2TeN^tBu$ acts as a chelating ligand with a larger bite than the chalcogen diimide monomers in the formation of metal complexes via the exocyclic nitrogen donors. The first example of this behaviour involved the complex $\{Li(thf)_2[Te_2(N^tBu)_4]\}(\mu_3\text{-}I)\{LiI[Te_2(N^tBu)_4]\}$, which is also be formed in the oxidation of $[Te(N^tBu)_3]^{2-}$ with iodine (see Scheme 5.6) [41]. In this complex the tellurium diimide dimer acts as a chelating ligand towards both a $[Li(thf)_2]^+$ cation and a molecule of LiI and adopts an extended structure via weak Te···I interactions [41]. Simple 1:1 adducts are exemplified by $[MCl_2\{^tBuNTe(\mu\text{-}N^tBu)_2TeN^tBu\}]$ (M = Co [48], Hg [50]).

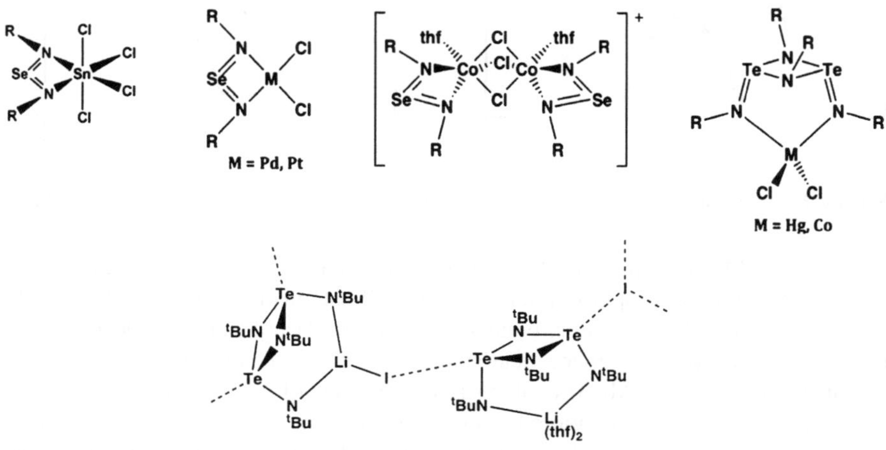

Scheme 5.6 Metal complexes of selenium and tellurium dimides

Both selenium diimide monomers and tellurium diimide dimers react with Ag(CF_3SO_3) to form binuclear complexes with bridging chalcogen diimide ligands that coordinate through the nitrogen atoms and link the metal centers into metallacycles [51, 52]. Interestingly, the macrocyclic dinuclear cations $[Ag_2\{\mu\text{-}N,N'\text{-}Se(NR)_2\}_2]^{2+}$ (R = ^{t}Bu, Ad) [52] and $[Ag_2\{\mu\text{-}N,N'\text{-}Te_2(N^tBu)_4\}_2]^{2+}$ [51] show Ag···Ag close contacts of 2.7384(9) and 2.751(2) Å, and 2.888(2) Å, respectively (see Scheme 5.7). The analogous copper complexes $[Cu_2\{\mu\text{-}N,N'\text{-}Se(NR)_2\}_2]^{2+}$ (R = ^{t}Bu, Ad) show a similar metallacycle with Cu···Cu close contacts of 2.531(1)–2.569(2) Å [52], whereas the related tellurium diimide complex $[Cu_2\{\mu\text{-}N,N'\text{-}Te_2(N^tBu)_4\}\{N,N'\text{-}Te_2(N^tBu)_4\}_2](CF_3SO_3)_2$ exhibits an open-chain dinuclear structure in which one tellurium diimide dimer acts as a bridging ligand between the metal centers [51] (Scheme 5.7).

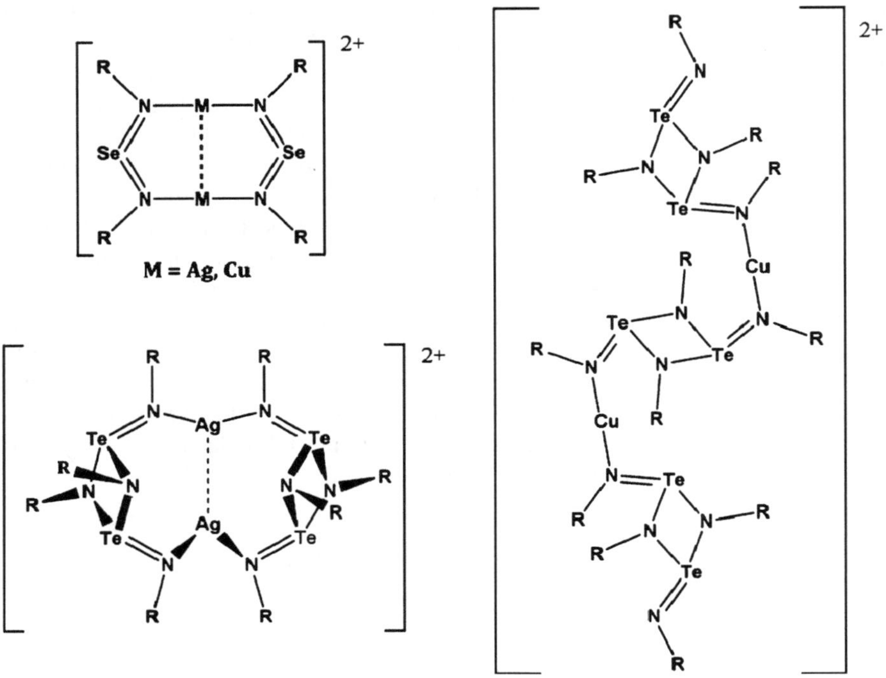

Scheme 5.7 Dinuclear coinage metal complexes of selenium diimide and tellurium diimide dimer

The close M···M contact has been attributed to a d^{10}-d^{10} closed shell interaction [51, 52]. Such a metallophilic interaction is well-established for a variety of gold complexes [53–56], but is rather controversial for Ag(I) and Cu(I) [see Ref. 52, and references therein]. Early theoretical studies [57, 58] indicated no net interaction and attributed the close M···M contacts to the bite size of the bridging ligand. With improvements in the computational sophistication, the presence of an attractive interaction was deduced and has been ascribed to dispersion forces [59] and correlation [60, 61], relativistic, and excitation effects [60]. Recent DFT calculations [52] of the series of complexes $[M_2\{\mu\text{-}N,N'\text{-Se(NR)}_2\}_2]^{2+}$ (M = Ag, Cu; R = H, Me, Bu, Ad) have predicted that the Ag···Ag and Cu···Cu distances become shorter as the organic group becomes bulkier. Concurrently, the metallacyclic framework deviates more significantly from planarity and the $<$ N-M-N bond angles deviate from linearity. While the bite size of the bidentate ligand bridging the two metal centers may correlate with the M···M distance, especially when the bite size is short, the deviation of the $D_2M\cdots MD_2$ fragment (D is the donor atom of the bidentate ligand) from planarity has a more significant effect on the M···M distance. AIM calculations for $[M_2\{\mu\text{-}N,N'\text{-Se(NR)}_2\}_2]^{2+}$ show the presence of a bond critical point between the metal atoms as well as two ring critical points in the centers of the two pseudo-five-membered MMNSeN rings [52]. These results are suggestive of metallophilic interactions.

5.3 Imido Selenium and Tellurium Halides

5.3.1 Synthesis

In contrast to their sulfur analogues [62, 63], imido selenium dihalides of the type $RNSeCl_2$ are either unknown (R = alkyl, aryl) or thermally unstable when R = CF_3 or C_2F_5 [64]. These perfluoroalkyl derivatives are obtained as yellow liquids by reaction of the corresponding dichloroamines $RNCl_2$ with Se_2Cl_2 in CCl_3F [64]. By contrast, *tert*-butylimidotellurium dihalides $(^tBuNTeX_2)_n$ (X = Cl, Br) are thermally stable in the solid state [65]. They are prepared by the redistribution reaction between TeX_4 and the tellurium diimide dimer (Eq. 5.1) in a 2:1 molar ratio. When this reaction is carried out in a 3:2 molar ratio, the unsymmetric imidotellurium chloride $^tBuNTe(\mu\text{-}N^tBu)_2TeCl_2$ is obtained. Halogen exchange between $(^tBuNTeCl_2)_n$ and Me_3SiBr provides a cleaner preparation of the corresponding dibromide [65].

$$^{t}BuNTe(\mu{-}N^{t}Bu)_2TeN^{t}Bu + 2TeX_4 \rightarrow 4/n(^{t}BuNTeX_2)_n(X = Cl,\ Br) \quad (5.1)$$

Acyclic imidoselenium(II) dihalides $ClSe[N(^tBu)Se]_nCl$ ($n = 1, 2$) are obtained from the reaction of $SeCl_2$ with *tert*-butylamine in a 2:3 molar ratio in thf (see Scheme 5.8) [66]. There are no sulfur or tellurium analogues of this class of imido chalcogen halides. The cyclic selenium(II) imide 1,3,5-$(SeN^tBu)_3$ is also formed in this reaction. The synthesis of cyclic selenium imides from these bifunctional imidoselenium(II) dihalides via cyclocondensation with primary amines or reduction with $Me_3SiSiMe_3$ is discussed in Sect. 5.5.1.

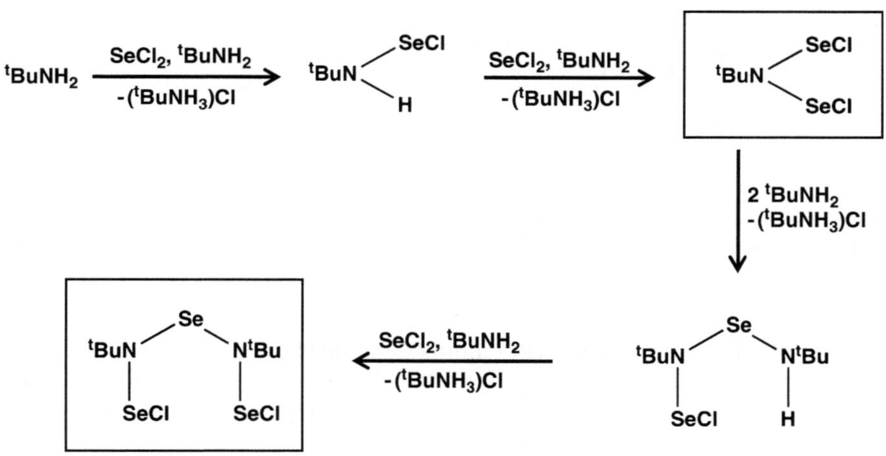

Scheme 5.8 Reaction of $SeCl_2$ with $^tBuNH_2$ in thf at $-80^\circ C$ [66]

5.3.2 Structures

The structures and physical properties of imido chalcogen halides of the type $RNECl_2$ (E = S, Se, Te) demonstrate the decreasing propensity of the heavier chalcogens to form –N = E< double bonds. The sulfur derivatives $RNSX_2$ (X = F, Cl) are monomers [62, 63] and the selenium analogues RN = $SeCl_2$ (R = CF_3 or C_2F_5) are unstable liquids [64]. By contrast, crystalline $(^tBuNTeCl_2)_n$ has a highly associated structure in which the fundamental building block is the dimer $(^tBuNTeCl_2)_2$ formed by cycloaddition of two Te = N^tBu units. In the solid state a layered arrangement of hexameric units is formed by linking three of these dimers by chloride bridges [65]. The reaction of $(^tBuNTeCl_2)_2$ with potassium *tert*-butoxide produces $(^tBuO)_2Te(\mu\text{-}N^tBu)_2Te(O^tBu)_2$, which is a weakly associated dimer (Te-N ~1.94 and 2.22 Å) [65], suggesting that extremely bulky alkoxy or aryloxy groups might preempt dimerization. A monomeric tellurium imide is stabilized in the boraamidinate complex $PhB(\mu\text{-}N^tBu)_2Te$ = N^tBu [38].

The crystal structure of $ClSe[N(^tBu)Se]_2Cl$ is shown in Fig. 5.3a. The close ClSe···SeCl contact of 2.891(1) Å and the alternations in the Se-N and Se-Cl bond lengths can be explained by two l.p.(Se)-σ* interactions as shown in Fig. 5.3b [66].

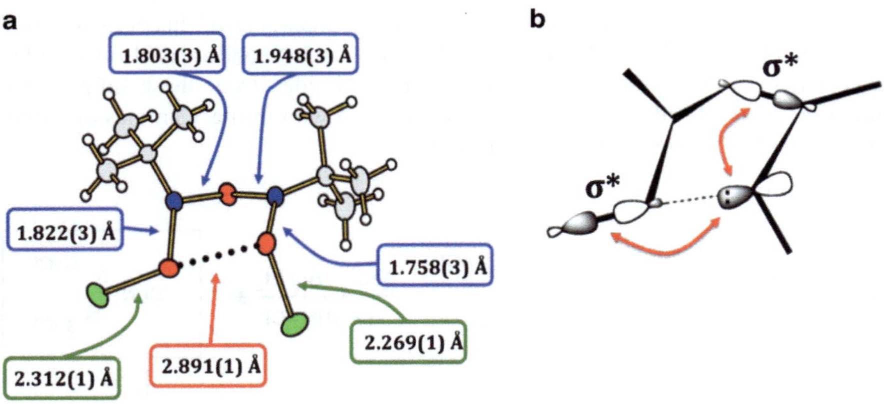

Fig. 5.3 (**a**) Crystal structure of $ClSe[N(^tBu)Se]_2Cl$ (redrawn from data in Ref. [66]). (**b**) Hyperconjugation in $ClSe[N(^tBu)Se]_2Cl$ rationalizing the observed Se···Se interaction and bond length alternation

The reaction of the sulfur(IV) diimide $S(NSiMe_3)_2$ with $TeCl_4$ produces compounds in which the –N = S = N- functionality bridges tellurium chloride moieties. The identities of the reaction products are dependent on the molar ratios of the reactants and the solvent, as indicated in Scheme 5.9.

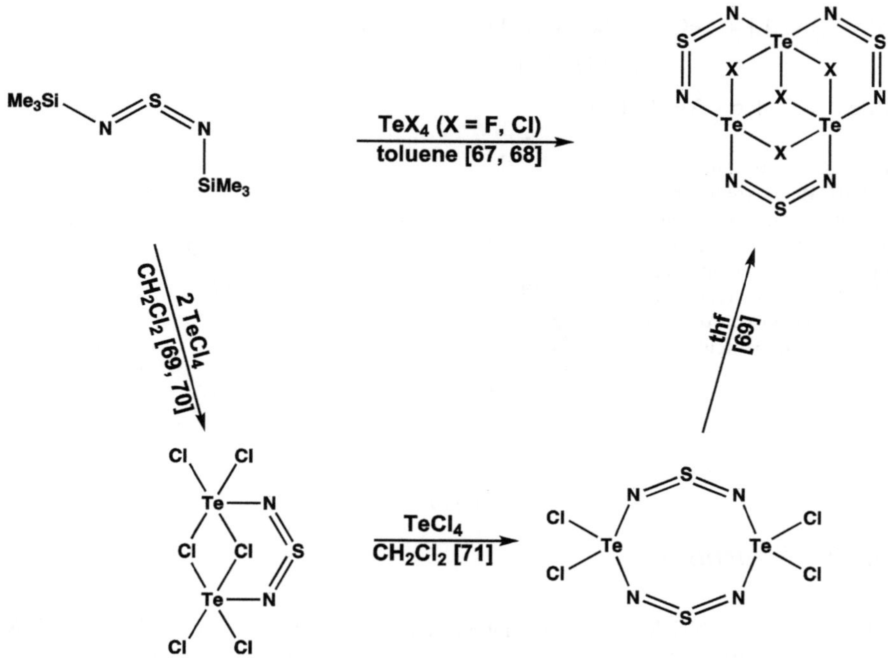

Scheme 5.9 The reaction of $S(NSiMe_3)_2$ with TeX_4

5.4 Selenium and Tellurium Diamides

5.4.1 Synthesis

Sulfur(II) and selenium(II) diamides are obtained by the reaction of sulfur or selenium chlorides with an aliphatic secondary amine [1]. Dialkylaminopolyselanes $Se_x(NR_2)_2$ ($x = 2$–4, NR_2 = morpholinyl; x = 4, NR_2 = piperidinyl) are formed in the reaction of elemental selenium with the boiling amine in the presence of PbO_2 [72]. The acyclic tellurium(II) diamide $[Te(NMe_2)_2]_\infty$ is prepared by the reaction of $TeCl_4$ with $LiNMe_2$ [73]. A similar reduction of the chalcogen centre occurs in the treatment of $LiN(SiMe_3)_2$ with $TeCl_4$ that affords $Te[N(SiMe_3)_2]_2$ [74]. The selenium analogue is obtained (together with Se_8) from $LiN(SiMe_3)_2$ and Se_2Cl_2 [74]. The addition of elemental selenium or sulfur to the initial reaction mixture results in the formation of a mixture of polyselanes or polysulfanes $E_x[N(SiMe_3)_2]_2$ (E = Se, S; $x = 2$–4), which have been identified by mass spectrometry, as well as by ^{77}Se NMR spectroscopy [75].

The reaction of mesitylaminolithium with $SeCl_4$ also results in the reduction of the selenium tetrahalide to give $Se[NH(Mes)]_2$, a selenium(II) diamide containing two Se-N single bonds (Eq. 5.2), cf. the formation of the expected selenium(IV) diimide $Se(NMes^*)_2$ from treatment of $SeCl_4$ with Mes*NHLi [8] (see Sects. 5.2.1 and 5.2.2).

$$4(\text{Mes})\text{NHLi} + \text{SeCl}_4 \rightarrow \text{Se}\{\text{NH(Mes)}\}_2 + {}^1\!/\!_2(\text{Mes})\text{N} = \text{N(Mes)} + (\text{Mes})\text{NH}_2 + 4\ \text{LiCl} \quad (5.2)$$

$(\text{Mes} = 2,4,6\text{ - Me}_3\text{C}_6\text{H}_2)$

Fig. 5.4 Two isomers of $Se[NH(Mes)]_2$ (redrawn from data in ref. 8). Carbon atoms are indicated in gray, nitrogen in blue, and selenium atoms in red. Only the hydrogen atoms bound to nitrogen is displayed in the figure

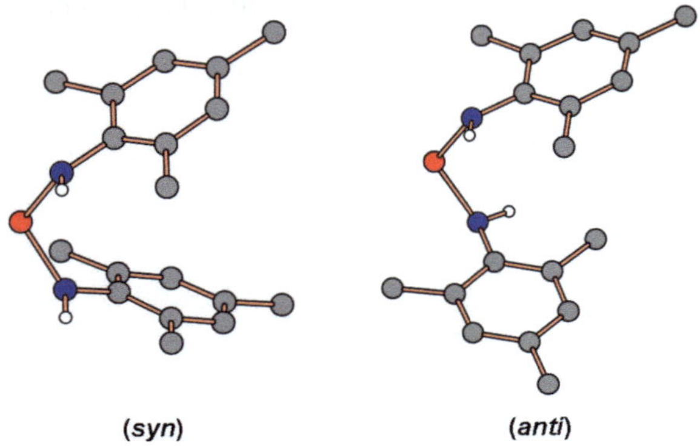

5.4.2 *Structures*

The diamide $Se\{NH(Mes)\}_2$ was characterized by X-ray crystallography and NMR spectroscopy [8]. In solution two isomers (*syn* and *anti*) that exhibit two close-lying ^{77}Se NMR resonances at 1,077 and 1,076 ppm co-exist (see Fig. 5.4). This assignment is supported by DFT calculations of the chemical shifts. In the solid state the *anti* conformation is stabilized as an adduct that incorporates two molecules of $MesNH_2$ [8].

The series $E[N(SiMe_3)_2]_2$ (E = S, Se, Te) are all monomers with single E-N bond lengths and a steady decrease in the NEN bond angles that reflects the increasing *s* character of the lone pair on the chalcogen [74]. By contrast, the dimethylamino derivative $[Te(NMe_2)_2]_\infty$ has a polymeric structure as a result of intermolecular Te···N contacts that give rise to trapezoidal Te_2N_2 rings [73].

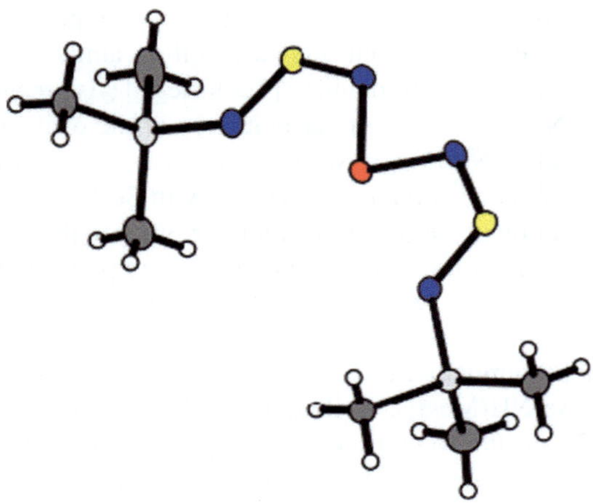

Se(1)-N(1):	1.844(3) Å
Se(1)-N(2):	1.844(3) Å
S(1)-N(1):	1.541(2) Å
S(1)-N(3):	1.521(10) Å
S(2)-N(2):	1.548(4) Å
S(2)-N(4):	1.523(8) Å

Fig. 5.5 The molecular structure of $Se(NSNSiMe_3)_2$ [76]

The SiNSNSeNSNSi chain in $Se(NSNSiMe_3)_2$ is approximately planar with the two diimide fragment showing *cis,trans*-conformations for the two -N = S = N- groups [76] (see Fig. 5.5).

5.4.3 Reactions

Silylated amino derivatives of the type $E[N(SiMe_3)_2]_2$ (E = S, Se, Te) are useful for the synthesis of a variety of chalcogen–nitrogen ring systems via reactions with chalcogen halides and elimination of Me_3SiCl (Scheme 5.10).

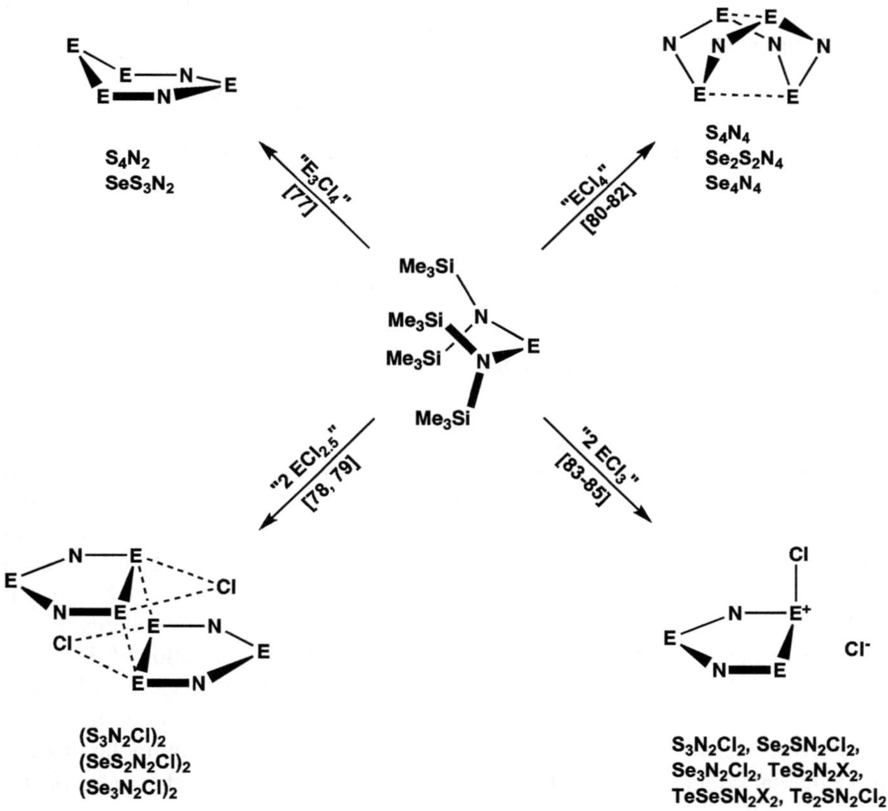

Scheme 5.10 Synthesis of chalcogen-nitrogen rings from $E[N(SiMe_3)_2]_2$ (E = S, Se, Te)

The polar Te-N bond in $[Te(NMe_2)_2]_\infty$ is readily susceptible to protolysis by weakly acidic reagents, e.g., Ph_3CSH produces the monomeric thiolato derivative $Te(SCPh_3)_2$ [73]. However, such reactions are often accompanied by formation of elemental tellurium and the reaction of $[Te(NMe_2)_2]_\infty$ with primary amines RNH_2 was not successful for the synthesis of cyclic tellurium(II) imides [73] (see Sect. 5.5.1).

The acylic compounds $E(NSNSiMe_3)_2$ (E = S, Se) are potential sources of chalcogen-nitrogen rings or longer chains via reactions of the Si-N linkages with chalcogen halides. For example, the all-sulfur system, $S(NSNSiMe_3)_2$, has been utilized in a good yield preparation of S_4N_4 by the reaction with SCl_2 (Eq. 5.3) [86]. In a similar fashion, the reaction of $Se(NSNSiMe_3)_2$ and $SeCl_2$ affords the hybrid chalcogen nitride 1,5-$Se_2S_2N_4$ (Eq. 5.4), which is also obtained from the reactions of (a) $S[N(SiMe_3)_2]_2$ and $SeCl_4$ or (b) $Se[N(SiMe_3)_2]_2$ and a mixture of SCl_2 and SO_2Cl_2 [82]. The reactions of (a) $Se(NSNSiMe_3)_2$ and SCl_2 and (b) $S(NSNSiMe_3)_2$ and $SeCl_2$ produced an equimolar mixture of S_4N_4 and 1,5-$Se_2S_2N_4$ rather than SeS_3N_4 [76]. A reaction pathway that invokes the intermolecular elimination of the four-membered rings S_2N_2 and $SeSN_2$ followed by dimerization has been invoked to explain the production of this mixture [76].

$$S(NSNSiMe_3)_2 + SCl_2 \rightarrow S_4N_4 + 2\ Me_3SiCl \quad (5.3)$$

$$Se(NSNSiMe_3)_2 + SeCl_2 \rightarrow 1,5\text{-}Se_2S_2N_4 + 2\ Me_3SiCl \quad (5.4)$$

5.5 Cyclic Selenium and Tellurium Imides

5.5.1 Synthesis and Structures

Cyclic sulfur imides are predominantly eight-membered rings, in which one or more of the sulfur atoms in *cyclo*-S_8 are replaced by an imido (NH) group [1, 87]. Ring systems involving 6, 7, 9 or 10 atoms have also been synthesized, but are much less common [88, 89]. Cyclic imides of the heavier chalcogens are also known, especially for selenium, for which several derivatives with no sulfur analogues have been characterized. For example, the 15-membered ring 1,3,6,8,11,13-$Se_9(N^tBu)_6$, in addition to the major product 1,5-$Se_6(N^tBu)_2$, was obtained in low yields by the reaction of $LiN(^tBu)SiMe_3$ with Se_2Cl_2 or $SeOCl_2$ [90]; the formation of the lower homologue 1,3-$Se_3(N^tBu)_2$ was reported in the reaction of this reagent with $SeCl_4$ [30]. The best yields of 1,5-$Se_6(N^tBu)_2$ and 1,3,6,8,11,13-$Se_9(N^tBu)_6$ (11 and 64%, respectively) have been obtained by treatment of $^tBuNH_2$ with an equimolar mixture of elemental selenium and $SeCl_4$ in THF [91, 92], which has been shown to contain 70 mol-% of $SeCl_2$ and 30 mol-% of Se_2Cl_2 [7].

Pure $SeCl_2$, prepared by oxidative addition of SO_2Cl_2 to elemental selenium in THF [93], reacts with $^tBuNH_2$ in THF in a 3:1 molar ratio to yield the five-membered ring 1,3-$Se_3(N^tBu)_2$ and the six-membered ring 1,3,5-$Se_3(N^tBu)_3$ in addition to 1,5-$Se_6(N^tBu)_2$ and $Se_9(N^tBu)_6$ [7, 66]. A better preparation of 1,3-$Se_3(N^tBu)_2$ involves the decomposition of the selenium(IV) diimide $Se(N^tBu)_2$ in toluene at 20°C, which generates a mixture of products from which 1,3-$Se_3(N^tBu)_2$ and 1,5-$Se_6(N^tBu)_2$ can be isolated in good yields [7]. The decomposition of $Se(NAd)_2$ (Ad = 1-adamantyl) affords crystalline 1,3-$Se_3(NAd)_2$ in good yields and no 1,3,5-$Se_3(NAd)_3$ is isolated [26]. The only known cyclic tellurium imide 1,3,5-$Te_3(N^tBu)_3$ is obtained as a minor product from the reaction of lithium *tert*-butylamide with $TeCl_4$ in toluene (see Sect. 5.2.1) [10].

The known crystal structures of heavy cyclic chalcogen imide derivatives are shown in Fig. 5.6. In general, the Se–N and Se–Se bond lengths fall within the typical single-bond ranges. However, those distances are significantly elongated in the puckered five-membered ring 1,3-$Se_3(NAd)_2$ as a result of ring strain [26]. In addition,

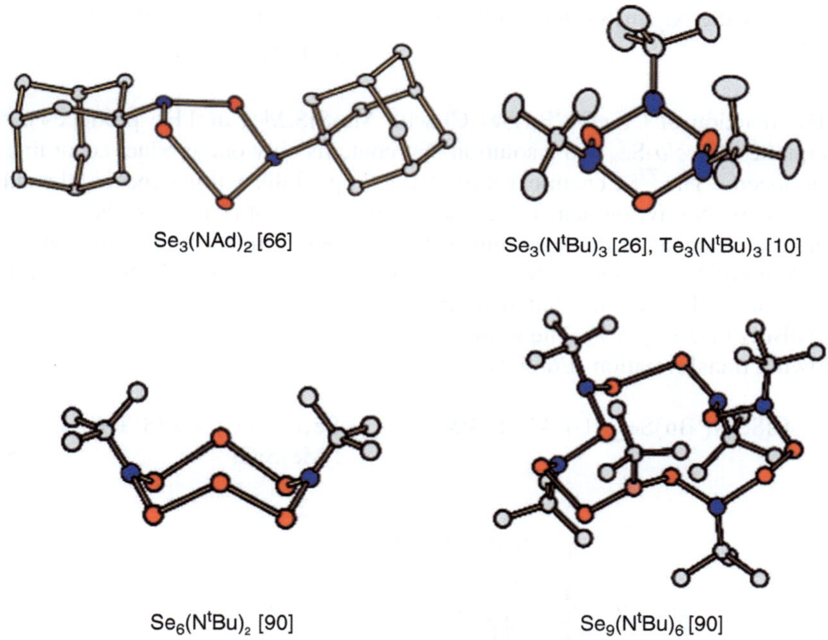

Fig. 5.6 Crystal structures of $Se_3(NAd)_2$, $Se_3(N^tBu)_3$, $Se_6(N^tBu)_2$, and $Se_9(N^tBu)_6$. (Redrawn from data in Refs. [10, 26, 66, 90])

the geometry at nitrogen in 1,3-$Se_3(NAd)_2$ is distinctly pyramidal in contrast to the almost planar configurations observed for 1,3,5-$Se_3(N^tBu)_3$ and 1,5-$Se_6(N^tBu)_2$.

The ^{77}Se NMR spectra provide an incisive analysis of the components of a mixture of cyclic selenium imides. Rings with both –Se–Se– and –Se– bridges, 1,3-$Se_3(NAd)_2$ and 1,3,6,8,11,13-$Se_9(N^tBu)_6$, exhibit two resonances with relative intensities 2:1. The eight-membered ring 1,5-$Se_6(N^tBu)_2$ also shows two resonances attributable to the two different selenium environments. The ^{77}Se NMR chemical shifts are strongly influenced by the electronegativity of nearest neighbours as indicated by the following trend: δ 1400–1625 (N*Se*N), 1100–1200 (N*Se*Se), 500–600 (Se*Se*Se) [7].

The acyclic imidoselenium halide $ClSe[N(^tBu)Se]_2Cl$ (see Sects. 5.3.1 and 5.3.2) is a potentially promising building block for the preparation of cyclic selenium imides. For example, it can be observed by ^{77}Se NMR spectroscopy that the cyclocondensation reaction of this bifunctional reagent with $^tBuNH_2$ in a molar ratio of 1:3 affords an approximately equimolar mixture of 1,3-$Se_3(N^tBu)_2$ and 1,3,5-$Se_3(N^tBu)_3$ [47], the formation of which can be rationalized by the following competing reactions (Eqs. 5.5 and 5.6). The presence of $^tBuN = N^tBu$ has also been detected by NMR spectroscopy [47].

$$ClSe[N(^tBu)Se]_2Cl + 3^tBuNH_2 \rightarrow Se_3(N^tBu)_3 + 2^tBuNH_3Cl \quad (5.5)$$

$$ClSe[N(^tBu)Se]_2Cl + 3^tBuNH_2 \rightarrow Se_3(N^tBu)_2 + 2^tBuNH_3Cl + {}^1\!/_2{}^tBuN = N^tBu \quad (5.6)$$

The reaction of $ClSe[N(^tBu)Se]_2Cl$ with $Me_3SiSiMe_3$ in THF produces a red precipitate of *cyclo*-Se_8 and a solution that contains only one product according to NMR spectra. The ^{77}Se chemical shift of 1,487 ppm falls within expected the range for an N–*Se*–N environment [47]. The resonance at 1,487 ppm is also observed as the major signal in the spectrum of the products formed by decomposition of the selenium(IV) diimide $Se(N^tBu)_2$ [7]. DFT calculations of ^{77}Se NMR chemical shifts tentatively indicate that this species is the eight-membered ring 1,3,5,7-$Se_4(N^tBu)_4$ (see Fig. 5.7). The amount of *cyclo*-Se_8 formed is consistent with the following transformation (Eq. 5.7).

$$ClSe[N(^tBu)Se]_2Cl + Me_3SiSiMe_3 \rightarrow {}^1\!/_2\ Se_4(N^tBu)_4 + 1/8\ Se_8 + 2\ Me_3SiCl \quad (5.7)$$

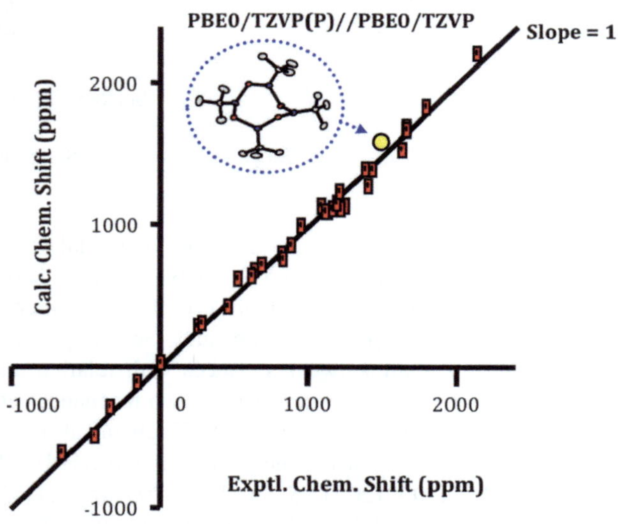

Fig. 5.7 The observed and calculated ^{77}Se chemical shifts of a number of selenium-nitrogen species. The data for the *red rectangular points* have been taken from Ref. [8] and that indicated by the *yellow circle* is taken from Ref. [47]

5.5.2 Metal Complexes

Although sulfur imides form *S*-coordinated complexes with metal centres, e.g., the sandwich complex $(S_4N_4H_4)_2{\cdot}AgClO_4$ [94] and the *S*-monodentate complexes $(S_4N_4H_4)M(CO)_5$ (M = Cr, W) [95], adducts of cyclic selenium imides have not yet been generated by direct treatment with metallic reagents. Interestingly, however, such complexes are formed when the selenium(IV) diimide $Se(N^tBu)_2$ is treated with one equivalent of $[PdCl_2(NCPh)_2]$ [46]. The *Se,Se′*-chelated complexes $[PdCl_2\{Se_4(N^tBu)_4\}]$ and $[PdCl_2\{Se_4(N^tBu)_3\}]$ were isolated as red and brown crystals, respectively, and identified by X-ray crystallography (see Fig. 5.8) [46]. The formation of the former complex provides further evidence for the presence of 1,3,5,7-$Se_4(N^tBu)_4$ among the decomposition products of $Se(N^tBu)_2$ (see Sect. 5.5.1). The latter complex embodies the novel seven-membered ring 1,3,5-$Se_4(N^tBu)_3$. The hydrolysis product $OSe(\mu\text{-}N^tBu)_2SeO$ (see Sect. 5.2.3) is also formed in this reaction.

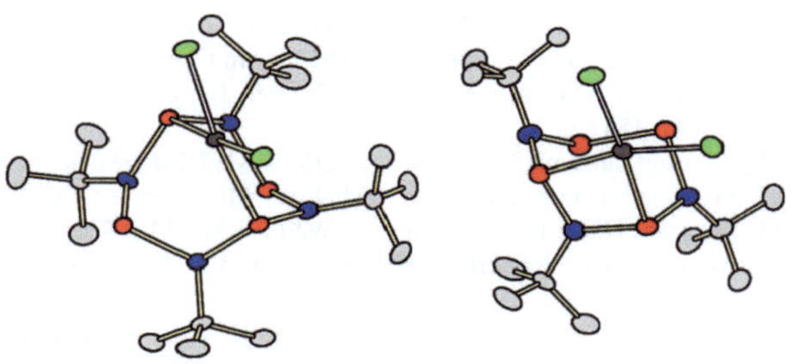

Fig. 5.8 Crystal structures of $[PdCl_2\{Se,Se'\text{-}Se_4(N^tBu)_n\}]$ ($n = 3, 4$) showing the eight- and seven-membered cyclic selenium imide ligands. (Redrawn from crystallographic data in Ref. [46])

5.6 Outlook

Selenium- and tellurium-nitrogen chemistry is progressing at an increasing rate paralleling the development in sulfur-nitrogen chemistry in the 1970s and 1980s. An understanding of the unusual structures, bonding and reactivities of imido-selenium and -tellurium compounds has been developed through a combination of experimental and theoretical investigations. This knowledge will be pivotal in future applications of these labile chalcogen-nitrogen species in areas ranging from materials science to organic synthesis. For example, the use of selenium or tellurium imides as a source of selenium or tellurium for the generation of metal chalcogenides as thin films or nanomaterials, via the thermodynamocally favourable elimination of a diazene RN = NR, merits consideration. The dearth of cyclic tellurium imides compared to the abundance of selenium imides in which selenium is in the +2 or lower oxidation states represents a challenge for synthetic chemists.

Acknowledgments Financial support from Academy of Finland and the Natural Sciences and Engineering Research Council (Canada) is gratefully acknowledged.

References

1. Chivers T (2005) A guide to chalcogen-nitrogen chemistry. World Scientific Publishing Co. Ltd, Singapore
2. Chivers T, Laitinen RS (2007) In: Devillanova F (ed) Handbook of chalcogen chemistry. RSC Publishing, Cambridge, pp 223–285
3. Chivers T (2005) In: King RB (ed) Encyclopedia of inorganic chemistry Vol. VIII, 2nd edn., Wiley, Chichester, pp. 5378–5403; Chivers T (2008) In: Crabtree RH (ed) Encyclopedia of inorganic chemistry, 2nd edn. Wiley, Chichester. doi:10.1002/0470862106.ia239.pub2
4. Laitinen RS, Oilunkaniemi R (2005) In: King RB (ed) Encyclopedia of inorganic chemistry Vol. IX, 2nd edn., Wiley, Chichester, pp. 5516–5539; Laitinen RS, Oilunkaniemi R (2008) In: Crabtree RH (ed) Encyclopedia of inorganic chemistry, 2nd edn. Wiley, Chichester. doi:10.1002/0470862106.ia239.pub2
5. Sharpless KB, Hori T, Truesdale LK, Dietrich CO (1976) J Am Chem Soc 98:269–271
6. Wrackmeyer B, Distler B, Gerstmann S, Herberhold M (1993) Z Naturforsch 48B:1307–1314
7. Maaninen T, Chivers T, Laitinen RS, Schatte G, Nissinen M (2000) Inorg Chem 39:5341–5347
8. Maaninen T, Tuononen HM, Kosunen K, Oilunkaniemi R, Hiitola J, Laitinen R, Chivers T (2004) Z Anorg Allg Chem 630:1947–1954
9. Chivers T, Sandblom N, Schatte G (2004) Inorg Synth 34:42–48
10. Chivers T, Gao X, Parvez M (1995) J Am Chem Soc 117:2359–2360
11. Tuononen HM, Suontamo RJ, Valkonen JU, Laitinen RS, Chivers T (2003) Inorg Chem 42: 2447–2454
12. Yu. Bagryanskaya I, Gatilov Y, Shakirov MM, Zibarev AV (1994) Mendeleev Commun 136–137
13. Yu. Bagryanskaya I, Gatilov Y, Shakirov MM, Zibarev AV (1994) Mendeleev Commun 167–168
14. Suenram RD, Lovas FJ, Stevens WJ (1985) J Mol Spectrosc 112:482–493
15. Anderson DG, Robertson HE, Rankin DWH, Woollins JD (1989) J Chem Soc Dalton Trans 859–862
16. Maaninen T, Laitinen R, Chivers T (2002) Chem Commun 1812–1813
17. Buschman J, Steudel W, personal communication cited in Borrmann T, Lork E, Mews R, Shakirov MM, Zibarev AV (2004) Eur J Inorg Chem 2452–2458
18. Stahl K, Legros J-P, Galy J (1992) Z Kristallogr 202:99–107
19. Orosel D, Leynaud O, Balog P, Jansen M (2004) Solid State Chem 177:1631–1638
20. Lindqvist O (1968) Acta Chem Scand 22:977–982
21. Ozin GA, Vander Voet A (1971) J Mol Struct 10:173–182
22. Beyer H (1967) Z Kristallogr 124:228–237
23. Champarnaud-Mesjard JC, Blanchadin S, Thomas P, Mitgorodsky A, Merle-Mejean T, Frit B (2000) J Phys Chem Solids 61:1499–1507
24. Chivers T, Gao X, Parvez M (1996) Inorg Chem 35:9–15
25. Sandblom N, Ziegler T, Chivers T (1998) Inorg Chem 37:354–359
26. Maaninen T, Tuononen HM, Schatte G, Suontamo R, Valkonen J, Laitinen R, Chivers T (2004) Inorg Chem 43:2097–2104
27. Laitinen RS (2005) Phosphorus Sulfur Silicon Relat Elem 180:777–782
28. Herberhold M, Distler H, Maisel H, Milius W, Wrackmeyer B, Zanello P (1996) Z Anorg Allg Chem 622:1515–1523
29. Vrieze K, van Koten G (1980) J R Neth Chem Soc 99:145–153
30. Herberhold M, Jellen W (1986) Z Naturforsch 41B:144–148
31. Kirchoff WH (1969) J Am Chem Soc 91:2437–2442
32. Beagly B, Chantrell SJ, Kirby RG, Schmidling DG (1975) J Mol Struct 25:319–327
33. Schatte G, Chivers T, Tuononen HM, Suontamo R, Laitinen R, Valkonen J (2005) Inorg Chem 44:443–451
34. Schatte G, Chivers T, Jaska C, Sandblom N (2000) Chem Commun 1657–1658
35. Li G, Chang HT, Sharpless KB (1996) Angew Chem Int Ed Engl 35:454–456
36. Elsegood MRJ, Kelly PF, Reid G, Staniland PM (2008) Dalton Trans 3798–3800

37. Chivers T, Parvez M, Schatte G (1996) Inorg Chem 35:4094–4095
38. Chivers T, Gao X, Parvez XM (1995) Angew Chem Int Ed Engl 34:2549–2550
39. Chivers T, Gao X, Parvez M (1996) Inorg Chem 35:553–554
40. Fleischer R, Freitag S, Stalke D (1998) J Chem Soc Dalton Trans 193–198
41. Chivers T, Parvez M, Schatte G (2001) Inorg Chem 40:540–545
42. Konu J, Chivers T, Schatte G, Parvez M, Laitinen RS (2005) Inorg Chem 44:2973–2982
43. Chivers T, Parvez M, Schatte G (1999) Inorg Chem 38:5171–5177
44. Hill AF (1994) Adv Organomet Chem 36:159–227
45. Gindl J, Björgvinsson M, Roesky HW, Freire-Erdbrügger C, Sheldrick GM (1993) J Chem Soc Dalton Trans 811–812
46. Risto M, Eironen A, Männistö E, Oilunkaniemi R, Laitinen RS, Chivers T (2009) Dalton Trans 8473–8745
47. Eironen A, Risto M, Takaluoma T, Oilunkaniemi R, Laitinen RS, Chivers T, unpublished results
48. Risto M, Konu J, Oilunkaniemi R, Laitinen RS, Chivers T (2010) Polyhedron 29:871–875
49. Wrackmeyer B, Köhler C, Milius W, Herberhold M (1995) Z Anorg Allg Chem 621: 1625–1631
50. Chivers T, Schatte G (2003) Can J Chem 81:1307–1314
51. Chivers T, Parvez M, Schatte G (1999) Angew Chem Int Ed 38:2217–2219
52. Risto M, Takaluoma TT, Bajorek T, Oilunkaniemi R, Laitinen RS, Chivers T (2009) Inorg Chem 48:6271–6279
53. Pyykkö P (1997) Chem Rev 97:597–636
54. Pyykkö P (2004) Angew Chem Int Ed Engl 43:4412–4456
55. Pyykkö P (2008) Chem Soc Rev 37:1967–1997
56. Van den Ancker TR, Bhargava SK, Mohr F, Papadopoulos S, Raston CL, Skelton BW, White AH (2001) J Chem Soc Dalton Trans 3069–3079
57. Merz KM, Hoffmann R (1988) Inorg Chem 27:2120–2127
58. Cotton FA, Feng X, Matusz M, Poli R (1988) J Am Chem Soc 110:7077–7083
59. Kölmel C, Aldrich R (1990) J Phys Chem 94:5536–5542
60. Fernández EJ, Lopez-de-Luzuriaga JM, Monge M, Crespo O, Gimeno MC, Laguna A, Jones PG (1998) Inorg Chem 37:6002–6006
61. Olson LP, Whitcomb DR, Rajeswaran M, Blanton TN, Stwertka BJ (2006) Chem Mater 18: 1667–1674
62. Glemser O, Mews R (1972) Adv Inorg Chem Radiochem 14:333–390
63. Roesky HW (1982) In: Senning A (ed) Sulfur in organic and inorganic chemistry, vol 4. Marcel Dekker, Inc, New York, p 15
64. Thrasher JS, Bauknight CW, Desmarteau DS (1985) Inorg Chem 24:1598–1599
65. Chivers T, Enright GD, Sandblom N, Schatte G, Parvez M (1999) Inorg Chem 38:5431–5436
66. Maaninen T, Chivers T, Laitinen RS, Wegelius E (2000) Chem Commun 759–760
67. Roesky HW, Münzenberg J, Noltemeyer M (1990) Angew Chem Int Ed Engl 29:61–63
68. Münzenberg J, Roesky HW, Besser S, Herbst-Irmer R, Sheldrick GM (1992) Inorg Chem 31: 2986–2987
69. Haas A, Pryka M (2002) J Organomet Chem 646:80–93
70. Haas A (1998) Adv Heterocycl Chem 71:115–144
71. Boese R, Dworak A, Haas A, Pryka M (1995) Chem Ber 128:477–480
72. Björgvinsson M, Roesky HW (1991) Polyhedron 10:2353–2370
73. Allan RE, Gornitzka H, Kärcher J, Parver MA, Rennie MA, Russell CA, Raithby PR, Stalke D, Steiner A, Wright DS (1996) J Chem Soc Dalton Trans 1727–1730
74. Björgvinsson M, Roesky HW, Pauer F, Stalke D, Sheldrick GM (1990) Inorg Chem 29:5140–5143
75. Siivari J, Maaninen A, Haapaniemi E, Laitinen RS, Chivers T (1995) Z Naturforsch 50B: 1575–1582
76. Konu J, Maaninen A, Paananen K, Ingman P, Laitinen RS, Chivers T, Valkonen J (2002) Inorg Chem 41:1430–1435

77. Maaninen A, Siivari J, Suontamo RJ, Konu J, Laitinen RS, Chivers T (1997) Inorg Chem 36: 2170–2177
78. Wolmershäuser G, Brulet CR, Street GB (1978) Inorg Chem 17:3586–3589
79. Maaninen A, Konu J, Laitinen RS, Chivers T, Schatte G, Pietikäinen J, Ahlgren M (2001) Inorg Chem 40:3539–3543
80. Maaninen A, Siivari J, Laitinen RS, Chivers T, Lawrence JD, Rauchfuss TB (2002) Inorg Synth 33:196–199
81. Siivari J, Chivers T, Laitinen RS (1993) Inorg Chem 32:1519–1520
82. Maaninen A, Laitinen RS, Chivers T, Pakkanen TA (1999) Inorg Chem 38:3450–3454
83. Siivari J, Chivers T, Laitinen RS (1993) Inorg Chem 32:4391–4395
84. Haas A, Pryka M, Schäfers M (1994) Chem Ber 127:1865–1870
85. Haas A, Kasparowski J, Pryka M (1992) J Chem Soc Chem Commun 1144–1145
86. Lidy W, Sundermeyer W, Verbeek W (1974) Z Anorg Allg Chem 406:228–234
87. Heal HG (1980) The inorganic heterocyclic chemistry of sulfur, nitrogen and phosphorus. Academic, London
88. Steudel R, Schumann O, Buschmann J, Luger P (1998) Angew Chem Int Ed Engl 37:492–493
89. Steudel R, Bergemann K, Buschmann J, Luger P (1996) Angew Chem Int Ed Engl 35: 2537–2538
90. Roesky HW, Weber KL, Bats JW (1984) Chem Ber 117:2686–2692
91. MacKinnon C (1997) Ph.D. Thesis, University of Guelph
92. Cordes AW, Glarum SH, Haddon RC, Hallford R, Hicks RG, Kennepohl DK, Oakley RT, Palstra TTM, Scott SR (1992) J Chem Soc Chem Commun 1265–1266
93. Maaninen A, Chivers T, Parvez M, Pietikäinen J, Laitinen RS (1999) Inorg Chem 38: 4093–4097
94. Hursthouse MB, Malik KMA, Nabi SN (1980) J Chem Soc Dalton Trans 355–359
95. Schmid G, Greese R, Boese R (1982) Z Naturforsch 37B:620–626

Chapter 6
A New Class of Paramagnetics: 1,2,5-Chalcogenadiazolidyl Salts as Potential Building Blocks for Molecular Magnets and Conductors

Andrey V. Zibarev and Rüdiger Mews

6.1 Introduction

The design, synthesis, and investigation of new molecular-based functional materials, especially of magnetic, conducting and superconducting materials, is a topical field of current scientific and technological progress. Of special interest as spin and/or charge carriers are open-shell molecules, both neutral and charged, *i.e.* radicals and radical ions.

In this field, chalcogen-nitrogen chemistry plays an important role [1, 2]. Particularly, π-heterocyclic thiazyl radicals (in individual state or as ligands in transition metal complexes) and radical cations were recognized as promising building blocks for molecular magnets and/or conductors (Refs. [3–30] and references therein; since relevant literature is too abundant to be cited completely, only selected recent references are given with special attention to review articles). Their common structural feature is the SN fragment, *i.e.* the same moiety that the molecular metal and low-temperature superconductor poly(sulfur nitride) $(SN)_x$ is composed of [31]. In particular, the neutral 4-R-1,2,3,5-dithiadiazolidyl radical (R = 4-NCC_6F_5) reveals spin-canted antiferromagnetism (weak ferromagnetism) with $T_C = 36$ K at normal pressure and $T_C = 65$ K at 16 kbar [32]. Some resonance-stabilized radicals of the dichalcogenazolyl type (chalcogen = S, Se) display properties of molecular conductors together with those of weak (spin-canted) ferromagnets (T_C = 17 and 28 K, depending on the structure), or even bulk

A.V. Zibarev (✉)
Institute of Organic Chemistry, Russian Academy of Sciences, 630090 Novosibirsk, Russia
and
Department of Physics, National Research University – Novosibirsk State University, 630090 Novosibirsk, Russia
e-mail: zibarev@nioch.nsc.ru

R. Mews
Institute for Inorganic and Physical Chemistry, University of Bremen, 28334 Bremen, Germany

J.D. Woollins and R.S. Laitinen (eds.), *Selenium and Tellurium Chemistry*,
DOI 10.1007/978-3-642-20699-3_6, © Springer-Verlag Berlin Heidelberg 2011

ferromagnets ($T_C = 12$ K) [8, 9]. In the solid state, some π-heterocyclic thiazyl radicals demonstrate spin-state crossover accompanied by magnetic hysteresis. The bistability arises from the coexistence over a temperature range of two solid-state phases, one based on paramagnetic radicals, the other on their weakly bonded diamagnetic π-dimers [33–36]. It is believed that magnetically bistable materials can find applications in molecular spintronic devices, *i.e.* in magneto-thermal switching and information storage devices.

The diversity of properties revealed by the open-shell thiazyl species covering magnetic ordering, magnetic bistability and electric conductivity is not observed for any other class of compounds.

The Se congeners of π-heterocyclic thiazyl radicals and radical cations are much less studied, whereas Te congeners are unknown [37]. This is due to experimental difficulties since variation of chalcogens in the chalcogen-nitrogen π-heterocycles does not involve conceptual challenges: S, Se and Te substitute each other isostructurally and isoelectronically, and 2,1,3-benzochalcogenadiazoles give representative example [38]. On the other hand, heavier-chalcogen congeners of open-shell thiazyl compounds are of enhanced interest especially in the context of the heavy atom effect on electrical and magnetic properties [3–11].

π-Heterocyclic thiazyl radical anions (RAs, known since 1960s from EPR experiments) [39–41] remained a missing link due to the lack of approaches to their isolation. This is also true for Se analogs, whereas Te derivatives were never observed even by EPR. The only related species are RAs of 1,2,5-thia(selena) diazoles fused with TCNQ which were isolated in the form of salts [42–45]. However, these RAs are normally considered as TCNQ derivatives [42]. The RAs of 1,2,5-thiadiazole 1,1-dioxide series studied recently by EPR [46] have no heterocyclic π-delocalization. Synthesis of authentic chalcogen-nitrogen π-heterocyclic RA salts is therefore of fundamental interest.

It should be noted that structural diversity of the known π-delocalized RAs suitable as spin and/or charge carrying units for molecular materials is rather limited. Besides the aforementioned RAs, they are represented mainly by those derived from N-heterocycles, such as mono- [47], di- [48, 49] and tetrazines [50, 51]; cyanocarbons, such as TCNE and TCNQ [52–62]; (semi)quinones [63–68]; and, perhaps, aromatics bearing strong electron withdrawing groups [69]. Ferromagnetic properties were observed for $[MCp^*_2]^+[TCNE]^-$ and $[MCp^*_2]^+[TCNQ]^-$ salts [M = Fe, Mn, Cr; Cp* = η^5-$C_5(CH_3)_5)_2$] with $T_C = 8.8$ K for $[MnCp^*_2]^+[TCNE]^-$ [52, 53, 55–62]; for a salt of 1,2,5-thiadiazole-TCNQ RA with $T_C = 107$ K [42]; and for the non-stoichiometric $[V]^{2+}[TCNE]_z^-[TCNE]_{1-z/2}^{2-}$ ($1 < z < 2$) salt with T_C exceeding room temperature [54]. The $[MCp_2]^+$ $[TCNE]^-$(M = Co, Fe) and some other relevant RA salts revealed also moderate electrical conductivity [43–45, 70]. These findings stimulated further investigation of RA salts.

This Chapter presents the results of the long-term Russian-German joint project focused on its current stage on chalcogen-nitrogen π-heterocyclic RA salts – derivatives of 1,2,5-chalcogenadiazole ring system (Chart 6.1), with special emphasis on their design, synthesis, and characterization including magnetic and electrical properties.

Chart 6.1 1,2,5-Chalcogenadiazoles and their RAs

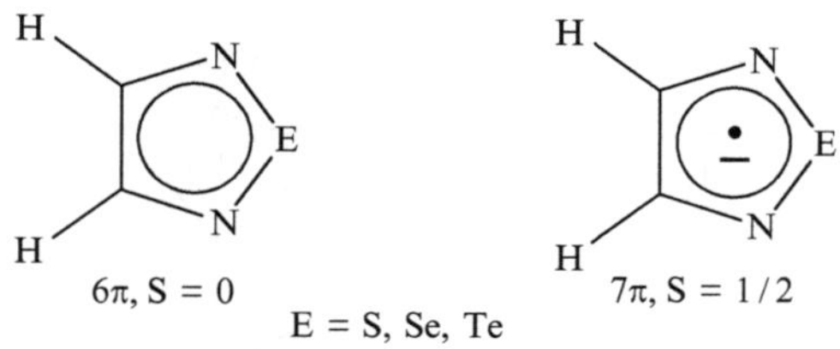

6.2 Neutral Precursors of the RAs

The most important property of neutral chalcogen-nitrogen π-heterocycles as precursors of RAs is their electron affinity (EA). According to the DFT calculations, the EA of 1,2,5-chalcogenadiazole derivatives is positive (Chart 6.2). Thermodynamically, it means that the target RAs are more stable than their neutral precursors. This provides a basis for the RAs isolation in the form of salts. At the same time, kinetically these open-shell and charged species should be (and they are) very active.

Within the heterocycles of interest, the EA varies in relatively broad range from ~1 to ~3 eV (Chart 6.2). At the same level of theory the EA of TCNE and TCNQ is 3.48 and 3.73 eV, respectively [71]. Of special interest is the dependence of the EA on the nature of the chalcogen atom and ring substituents. It follows from the calculations that in isostructural series the EA increases with atomic number of the chalcogen, *i.e.* from S to Te (Chart 6.2). Amongst simple ring substituents, F atoms and CN groups are, of course, the most effective, and, for example, for compounds **7–12** the EA is higher in the fluorocarbon series than in the hydrocarbon one (Chart 6.2) [71].

According to the calculations, the π-MOs of the discussed heterocycles are isolobal within the structural classes, *i.e.* their shapes are invariant to the nature of the chalcogen atom. In all cases the LUMO is a delocalized π-orbital, as it is shown in Fig. 6.1 for compound **9** as representative of the theoretically most interesting tellurium-nitrogen systems. Consequently, the target RAs are expected to be π-RAs [71].

Based on the EA values and synthetic availability, heterocycles **1–5, 7–12, 17, 19, 20** and **25** (Chart 6.2) [72–75] have been selected as starting materials. In the cases of compounds **2–5**, **7**, **8**, **10–12** and **25** original preparative approaches were used (Scheme 6.1) based in particular on ECl_4 (E = Se, Te) as source of the chalcogen [73–78], as well as on nucleophilic hetero ring-closure mediated by $[ArNSN]^-$ anions [72, 73, 77, 79–81]. The latter were isolated and characterized by X-ray diffraction (XRD). Because of the very short terminal S-N distance (1.44–1.48 Å) and the relatively long internal S-N distance (1.58–1.60 Å) these anions should be regarded as rather thiazylamides ($Ar\text{-}N^-\text{-}N{\equiv}S$) than the expected sulfur diimides ($Ar\text{-}N = S = N^-$), *i.e.* rare species containing a S≡N triple bond [82–85].

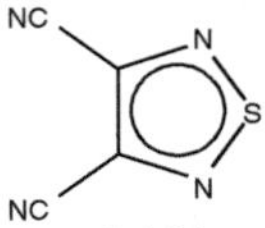

1, 1.84

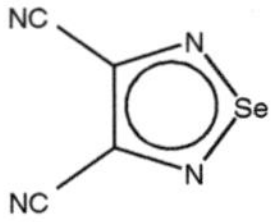

2, 1.94

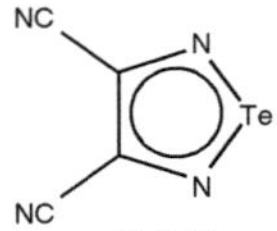

3, 2.04

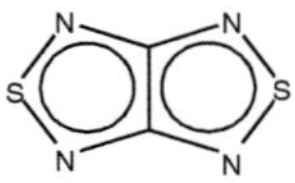

4, 2.14

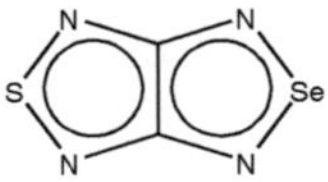

5, 2.19

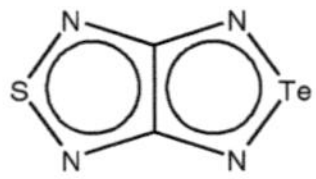

6, 2.21

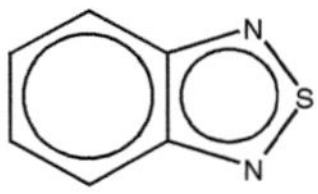

7, 0.95

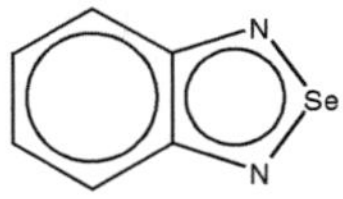

8, 1.06

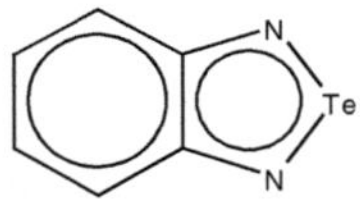

9, 1.15

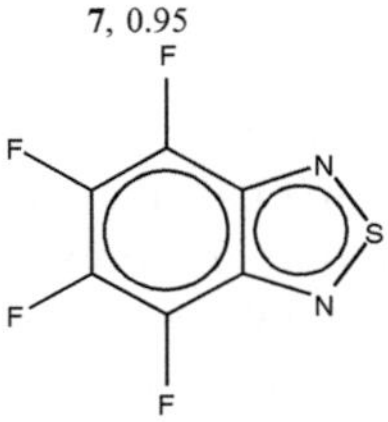

10, 1.57

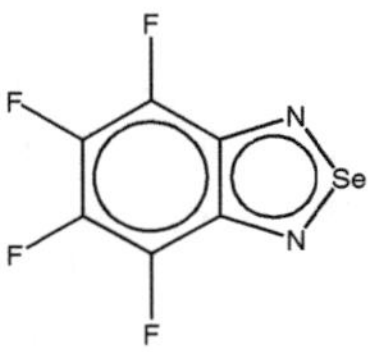

11, 1.67

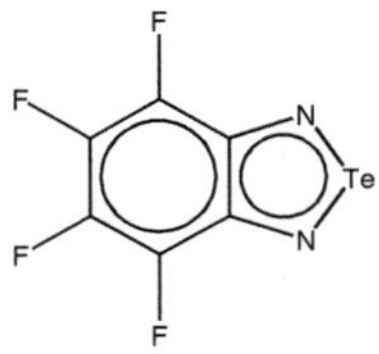

12, 1.83

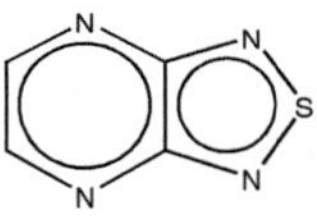

13, 1.75

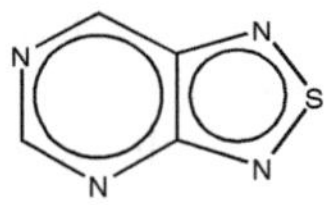

14, 1.75

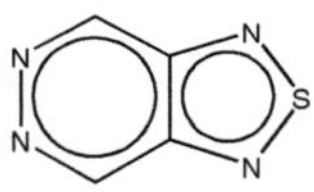

15, 1.79

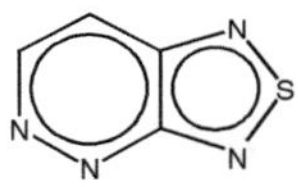

16, 1.88

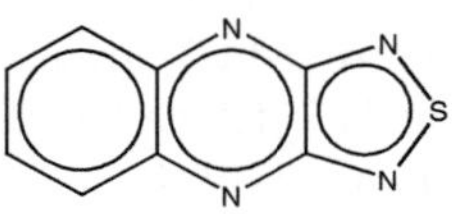

17, 2.21

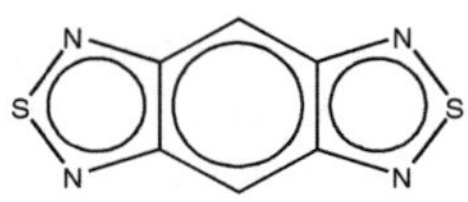

18, 2.26

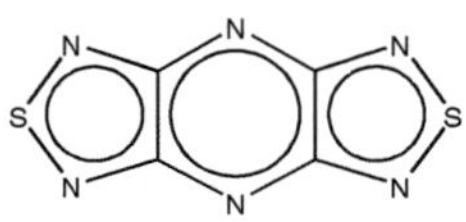

19, 3.04

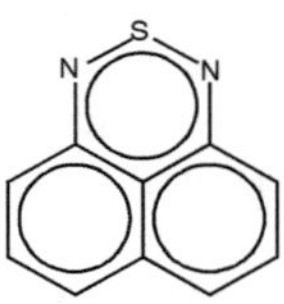

20, 1.77

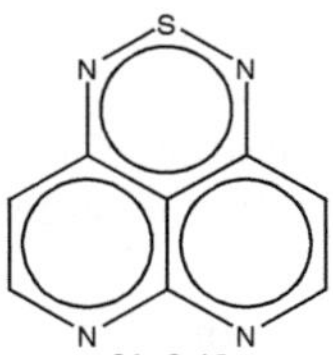

21, 2.45

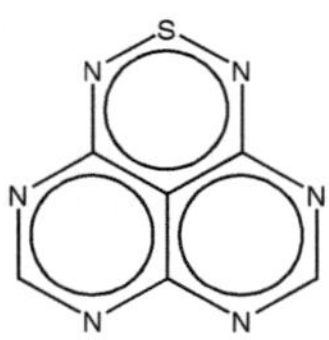

22, 2.99

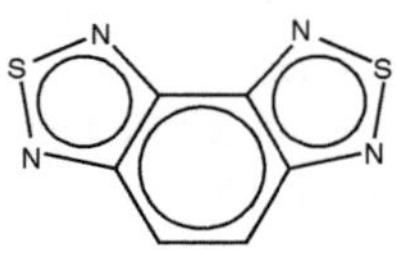

23, 1.24

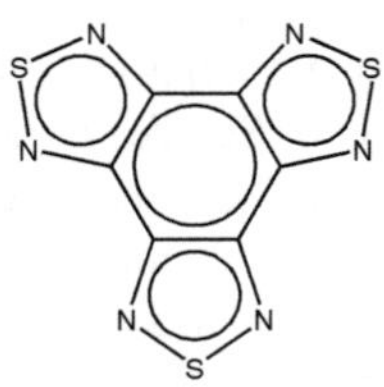

24, 1.70

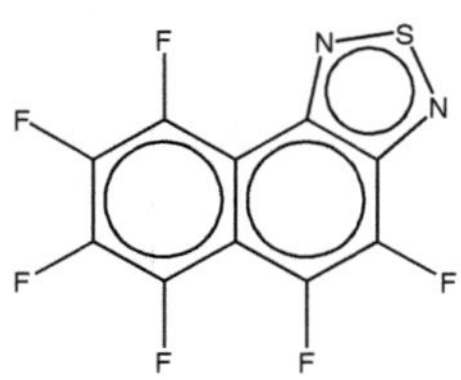

25, 1.67

Chart 6.2 Selected chalcogen-nitrogen π-heterocycles and their first adiabatic EA (eV) from (U)B3LYP/6–31 + G(d) (for Te: Def2-SVP with ECP) calculations [71]

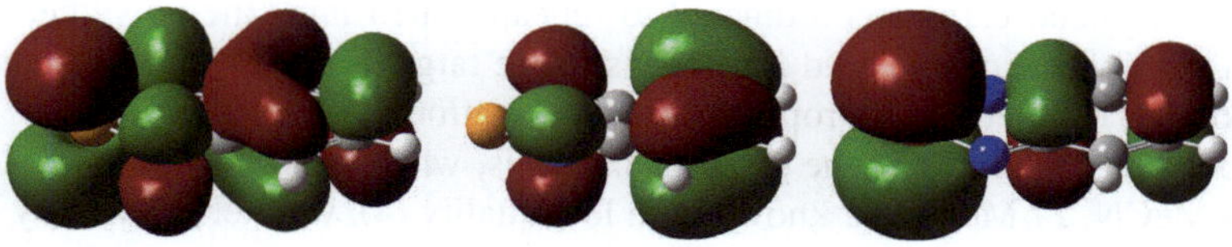

Fig. 6.1 Frontier MOs of compound **9** and their energies (eV) from B3LYP/6-31+G(d) (for Te: Def2-SVP with ECP) calculations [71]

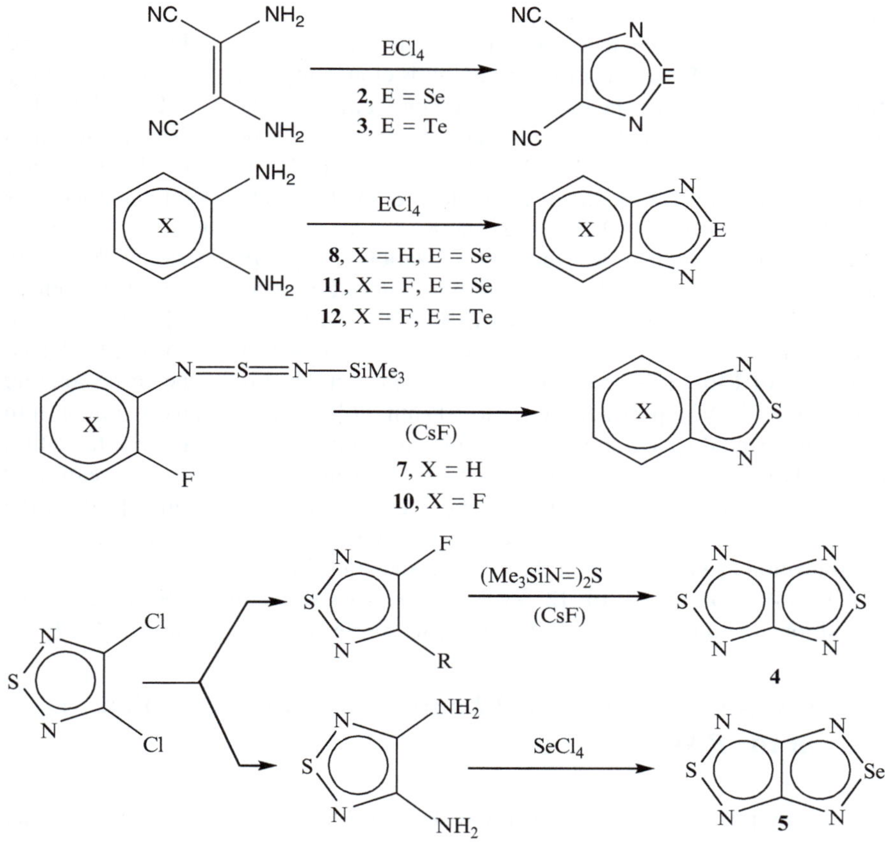

Scheme 6.1 Preparation of starting materials

These starting materials allow the preparation of p-homospin, $S_1 = 1/2$, RA salts in the case of diamagnetic, $S_2 = 0$, cations, as well as p-d-heterospin salts in the case of paramagnetic, $S_2 \neq 0$, cations of d-transition metals. For homospin salts antiferromagnetic (AF) exchange interactions (↑↓) are mostly expected whereas for heterospin salts ferromagnetic (FM) interactions (↑↑) are possible (see below). Additionally, derivatives of **7** and **8** fused at the positions 6 and 7 with 18-crown-6 (**26** and **27**, respectively) were synthesized from benzo-18-crown-6. They are potential precursors of more complex p-d-f-heterospin systems with paramagnetic f-transition metal cations coordinated to (encapsulated into) the polyether cycles [86]. Participation of unpaired f-electrons in the target RA salts' spin systems can lead to high magnetic anisotropy and slow relaxation of magnetization.

Molecular structures of the starting materials, where unknown (**1–3**, **5**, **17**, **24**, **26**, **26**·MeCN, **27**·MeCN) or known with low quality (**4**), were elucidated by XRD. This allowed direct experimental comparison of the structural effects of chalcogen variation, and well as of transformation of neutral precursors into the RAs. The latter effects are diagnostic toward structure of the SOMOs in the RAs (see below). The experimental XRD geometries of the RAs were well-reproduced by DFT calculations [76, 78, 80, 86–88].

In the neutral precursors, variation of the chalcogen leads to structural and spectral changes. Expected structural effect is elongating NX bonds and narrowing NXN bond angle of the heterocycle in going from X = S to X = Se and Te. For example, in going from **1** to **2** and **3** the experimental NX bond distance elongates from 1.623 (X = S) to 1.790 (X = Se) and 2.016 Å (X = Te) (in each case averaged over two NX bonds of the molecule); and the XNX bond angle narrows from 99.6 (X = S) to 93.0 (X = Se) and 82.7° (X = Te) [78, 86].

In the isostructural series, the first ionization energy (IE) in HeI UPS spectra reveals progressive lowering with growing atomic number of the chalcogen as observed for compounds **7** (8.98 eV), **8** (8.80) and **9** (8.57) [38]. The situation with electronic transition to the first excited state is however not so simple. Transformation of **1** to **2** is accompanied by long-wavelength shift of the corresponding band in the UV-Vis spectra from 273 to 304 nm [86]. On the contrary, in going from **4** to **5** this band displays short-wavelength shift from 353 to 319 nm. For the latter, the TD-DFT calculations identify two $\pi \rightarrow \pi^*$ transitions in each case appearing as one unresolved band in the experimental spectra of both compounds [76]. In the case of **10** and **11** variation of the chalcogen does practically not affect energy of the transition as emerged from the UV-Vis spectra [77].

^{19}F NMR spectra of **10–12** reveal progressive shielding of F nuclei in the o-positions in going from the S derivative to the Se and Te congeners [74, 77].

6.3 CV Generation and EPR and DFT Characterization of the RAs

The neutral precursors were reversibly reduced in MeCN and DMA solutions under CV conditions into long-lived RAs (Table 6.1; for a comprehensive discussion of the redox properties of chalcogen-nitrogen compounds, see Ref. [26]). In all cases

Table 6.1 First reduction potential (V, vs SCE) of chalcogen-nitrogen π-heterocycles

Compound	$-E_P^{1Red}$	Compound	$-E_P^{1Red}$
1	1.07	**11**	1.08
2	1.03	**17**	0.46 [26]
4	0.59	**19**	−0.10 [26]
5	0.53	**20**	0.96 [26]
7	1.56	**25**	1.29
8	1.40	**26**	1.80
10	1.21	**27**	1.66

the Se derivatives were reduced at lower potentials than their S congeners. The RAs were characterized by EPR spectroscopy (Fig. 6.2). Experimental hfc constants were in good agreement with those calculated by DFT methods (Fig. 6.2) [76, 80, 86, 89].

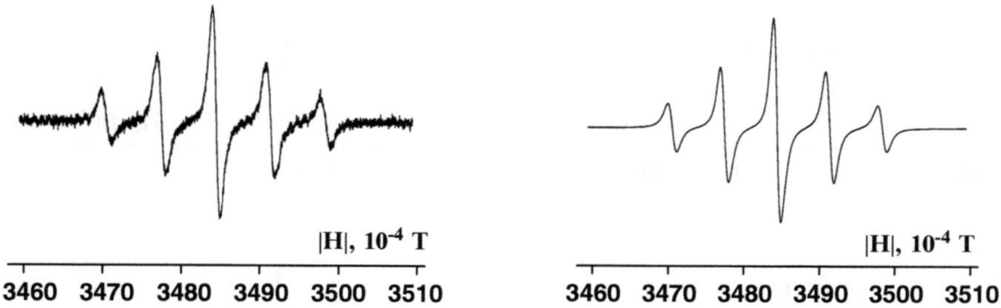

1, 0.698, 0.023 (0.741, 0.034)

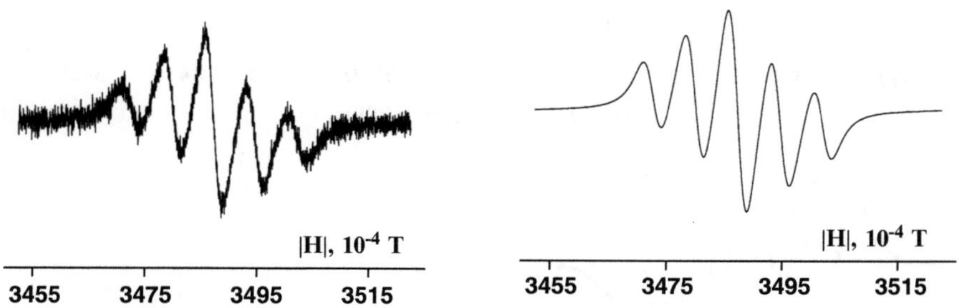

2, 0.730, 0.050 (0.762, 0.032)

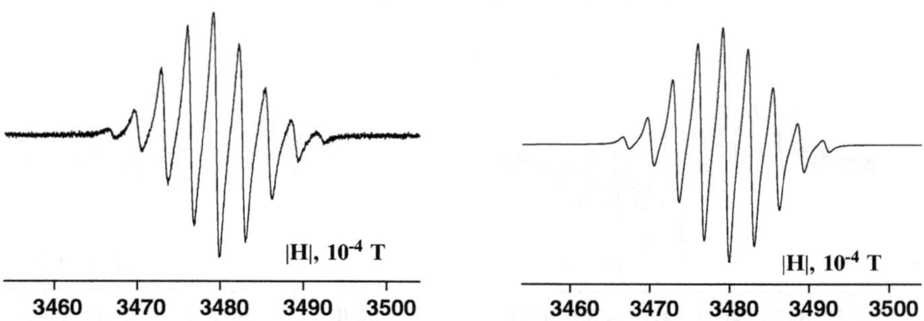

4, 0.314 (0.238)

Fig. 6.2 (continued)

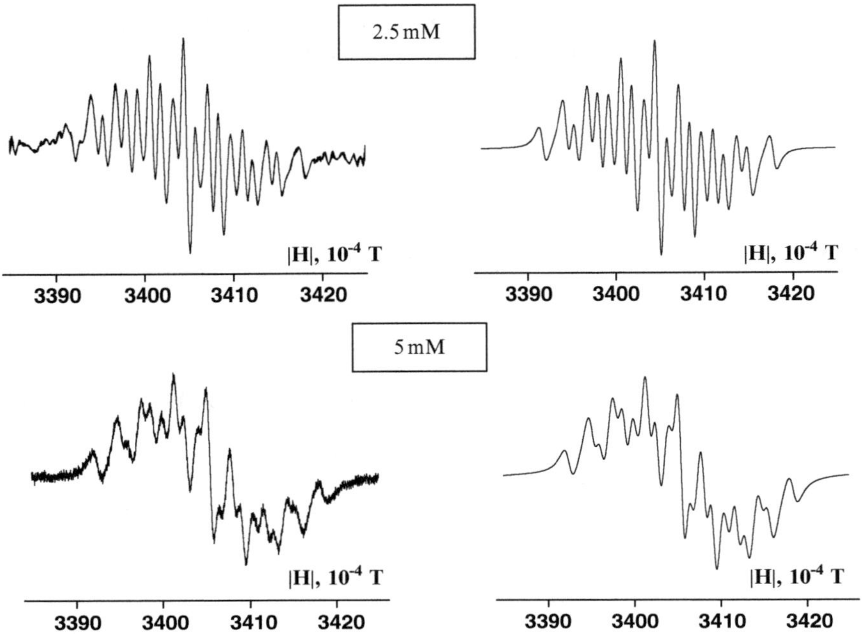

5, 0.379, 0.269 (0.360, 0.309)

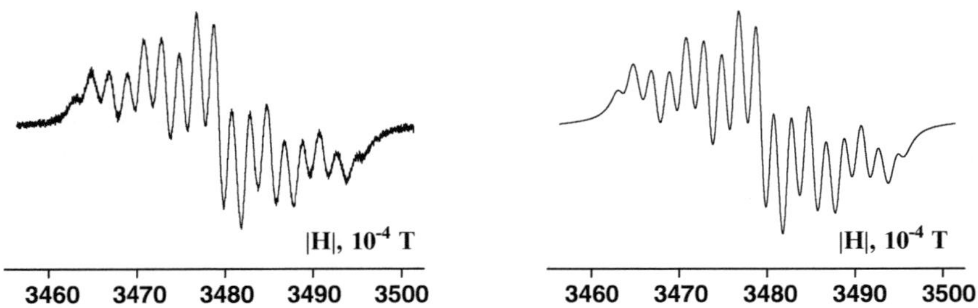

8, 0.594, 0.234, 0.167 (0.558, –0.339, –0.150)

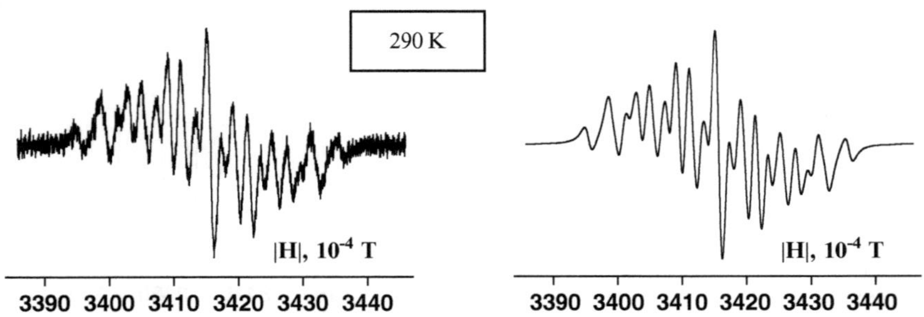

Fig. 6.2 (continued)

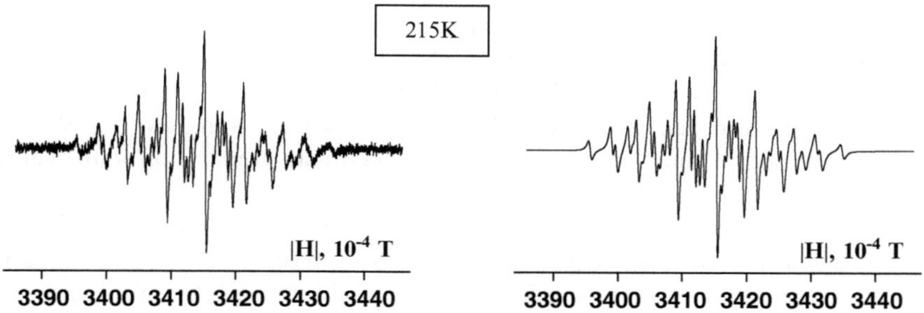

11, 0.613, 0.436, 0.361 (0.581, 0.423, 0.257)

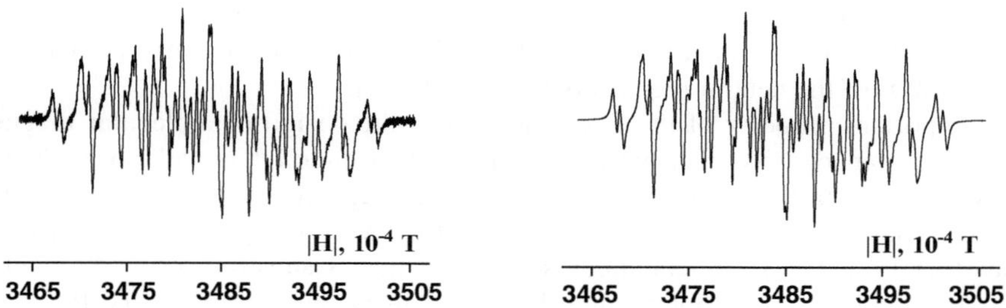

25, 0.533,0.493, 0.375, 0.313, 0.312, 0.280, 0.068, 0.023 (0.514, 0.506, 0.395, 0.366, 0.240, 0.175, 0.134, -0.057)

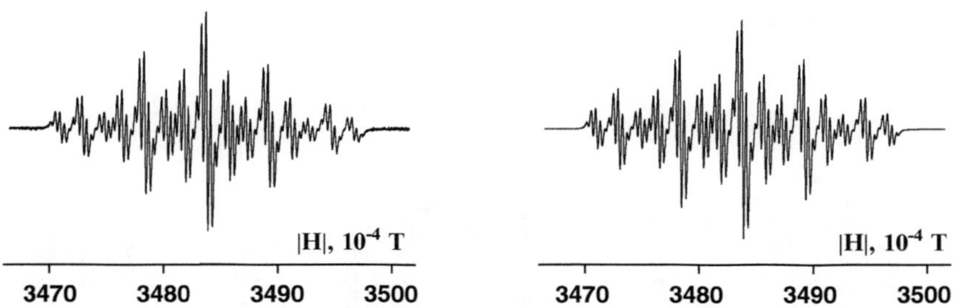

26, 0.542, 0.194, 0.039

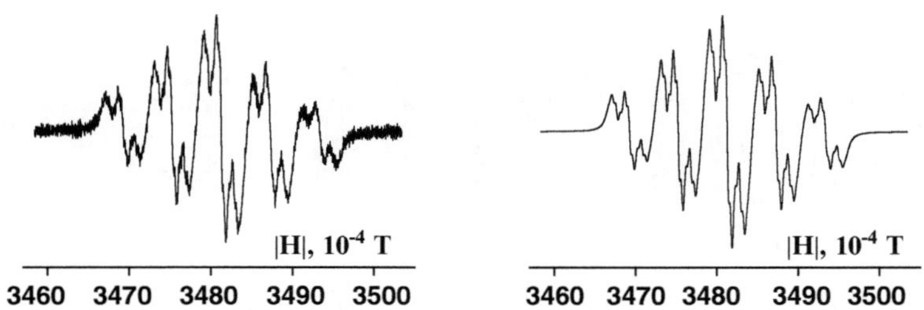

27, 0.602, 0.155, 0.046

Fig. 6.2 Experimental (*left*) and simulated (*right*) EPR spectra of chalcogen-nitrogen π-heterocycles, and experimental and calculated at the UB3LYP/6-31+G(d) level hfc constants (in *parenthesis*; mT)

Of special interest are polyfluorinated precursors. It was found that in contrast to the RAs of polyfluorinated aromatics (C_6F_6, $C_{10}F_8$) or simple heteroaromatics (C_5F_5N) which are highly unstable and cannot be observed in solution by conventional EPR (Ref. [90] and references therein), the RAs of chalcogen-nitrogen π-heterocycles are enormously long-lived [89]. This stability provides good prospects for their chemical preparation and eventual isolation.

At 295 K in MeCN solutions, the spectra of the Se containing RAs of **8** and **11** revealed line broadening (with the line widths of 0.084 and 0.105 mT, respectively), on average by a factor 2.5 as compared with the spectra of the S analogs **7** and **10**. In part, this may be due to the larger g-tensor anisotropy in the Se containing species as evident from their g-values which are higher than those for the S congeners [89].

Quantum chemical calculations at various levels of theory confirm the π-character of the RAs with their π-SOMOs (Fig. 6.3) being expectedly very similar to the π-LUMOs of the neutral precursors [71, 76, 80, 91].

Important property of the RAs regarded to the design of molecular magnets is their spin density distribution. With the McConnell I model, to achieve a FM ground state it is necessary to have paramagnetic moieties of two types, with a positive spin density at one interacting with a negative spin density at another [92, 93]. DFT calculations revealed that S substitution by Se, and H by F, practically does not change the spin density distribution in the RAs, and that the spin density on their van der Waals (VdW) surfaces is mostly positive (Fig. 6.4) [71, 79, 89]. Therefore, for homospin salts of the discussed RAs the AF exchange interactions are mainly expected. The FM interactions are possible in heterospin salts with paramagnetic cations possessing negative peripheral spin density.

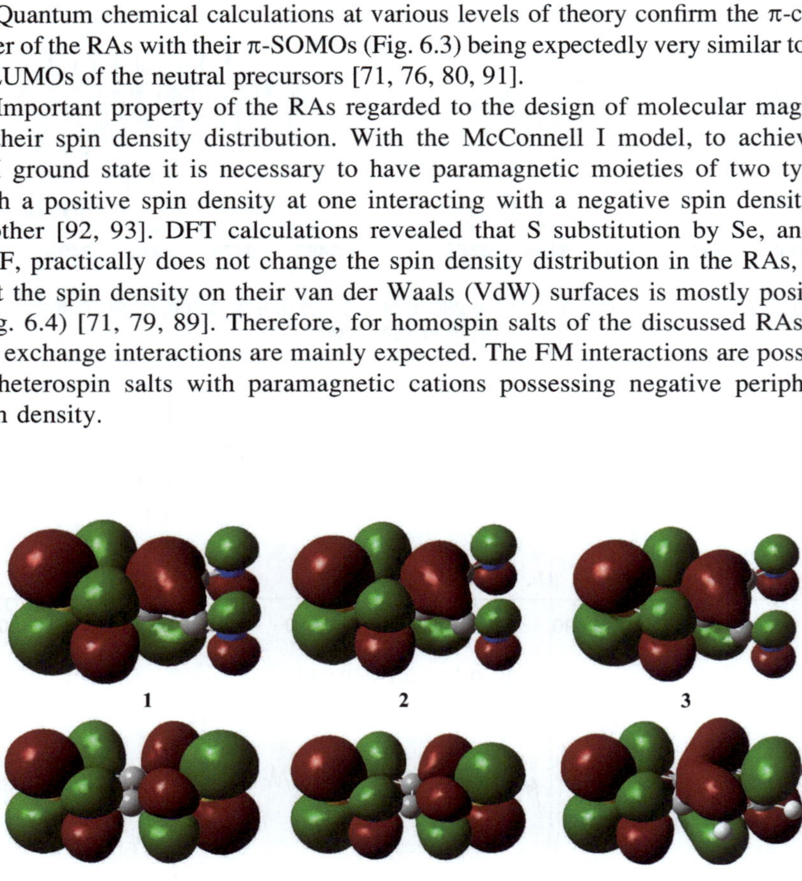

Fig. 6.3 π-SOMOs of the RAs from quantum chemical calculations

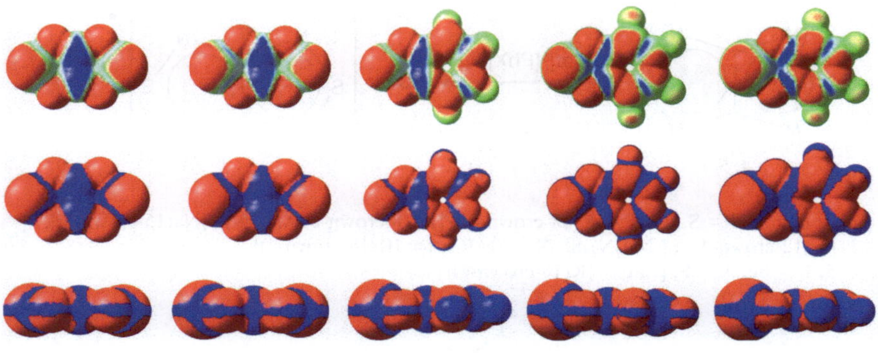

Fig. 6.4 Spin density (ρ) distribution on the VdW surfaces of the RAs from UB3LYP/6-31+G(d) calculations. First row: *red*, $\rho > 4 \times 10^{-4}$; *green*, $|\rho| < 4 \times 10^{-4}$; *blue*, $\rho < -4 \times 10^{-4}$. Second and third rows: *red*, $\rho > 0$; *blue*, $\rho < 0$

6.4 Synthesis and Characterization of the RA Salts

The discussed neutral precursors were preparatively reduced into RAs by various chemical reducing agents including thiophenolate and its Se congener, tetrakis (dimethylamino)ethene (TDAE), metallocenes, bis(arene)metals and elemental potassium.

Compounds **4** and **5** were reduced in MeCN solutions into their RAs by action of thiophenolate. This method allowed the high-yield preparation of early alkali metal (Li, Na, K) and tris(dimethylamino)sulfonium (TAS) cations' salts of the RAs. The metal cations were encapsulated into the corresponding crown-ethers for better solubility of the salts. Selenophenolate can also be used in these syntheses (Scheme 6.2) [76, 80, 94]. In all cases the cations were diamagnetic therefore the salts were homospin, S = 1/2.

The structures of all salts were confirmed by XRD. Like their neutral precursors, the RAs are planar. Coordination patterns in the crystals were different. In [K(18-crown-6)][**4**], [K(18-crown-6)][**5**] and [Na(15-crown-5)][**4**] the RAs act as bridging ligands forming 1D coordination chains, uniform in the K salts and non-uniform in the Na salt. In [K(18-crown-6)(MeCN)][**4**] the RA acts as chelating ligand. In [Li(12-crown-4)$_2$][**4**] the RA is not coordinated to the Li cation encapsulated into two crown-ether molecules and can be considered as "naked" anion. In [TAS][**4**] one of the two crystallographically independent RAs acts as bridging ligand via S…S contacts with cation, whilst the other is not coordinated to the cation (Fig. 6.5) [76, 80, 94].

With compounds **1** and **2** it was found that thiophenolate does not reduce the S derivative, whereas in the case of the Se congener hypercoordination of the nucleophile to the chalcogen center was observed (Scheme 6.2). This reaction is the first example of the nucleophilic hypercoordination to the chalcogen centers of 1,2,5-chalcogenadiazole derivatives for which electrophilic addition at the nitrogen centers is typical. The hypercoordination product is EPR-silent in both the solid

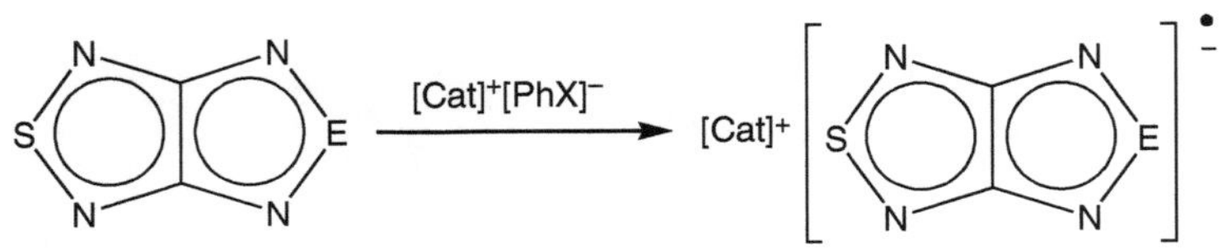

4, 5

4, E = S; X = S, [Cat] = [K(18-crown-6)], [K(18-crown-6)(MeCN)], [Na(15-crown-5)], [Li(12-crown-4)$_2$], [(Me$_2$N)$_3$S]; X = Se, [Cat] = [K(18-crown-6)]
5, E = Se, X = S, [Cat] = [K(18-crown-6)]

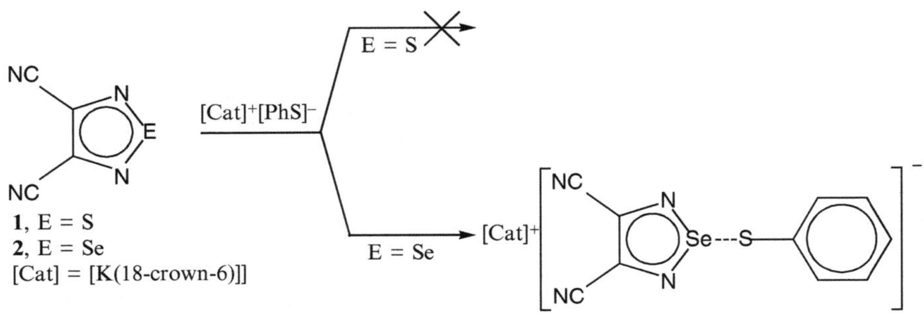

1, E = S
2, E = Se
[Cat] = [K(18-crown-6)]]

Scheme 6.2 Interaction of 1,2,5-chalcogenadiazoles with PhX$^-$ (X = S, Se)

state and a solution, and stable toward atmosphere. The structure of the product in the solid state was confirmed by XRD and in solution by UV-Vis spectroscopy. Its most interesting structural feature is the Se-S distance of 2.73 Å, which is ~0.5 Å longer than the sum of the covalent radii of these atoms, but ~1 Å shorter than the sum of their VdW radii. DFT calculations reproduced the XRD structure, IR and UV-Vis spectra of the hypercoordination product very well. According to the QTAIM and NBO calculations, the Se-S bond of the product can be considered as weak donor-acceptor bond whose formation leads to transfer of *ca.* 40% of negative charge from PhS$^-$ onto the heterocycle [95, 96]. The hypercoordination is reversible, and in very dilute solutions its dissociation to compound **2** and thiophenolate was observed by UV-Vis techniques.

DFT calculations with the polarizable continuum model to account for the THF solvent revealed that the experimental results (Scheme 6.2) are in agreement with thermodynamics of the reaction systems. For **4** and PhS$^-$, the reduction to RA with PhSSPh as by-product is thermodynamically favorable ($\Delta G = -7.3$ kcal.mol^{-1}) whereas the nucleophilic addition is not (3.7 kcal.mol^{-1}). For **1** and PhS$^-$, both reduction and addition are non-favorable (5.2 and 6.1 kcal.mol^{-1}, respectively). For **2** and PhS$^-$, the addition is thermodynamically favorable ($\Delta G = -0.8$ kcal.mol^{-1}) whereas the RA formation is not (8.1 kcal.mol^{-1}) [95, 96].

With compound **4** and TDAE, the dicationic RA salt [TDAE][**4**]$_2$ was obtained in quantitative yield (Scheme 6.3). Its structure was confirmed by XRD (Fig. 6.5).

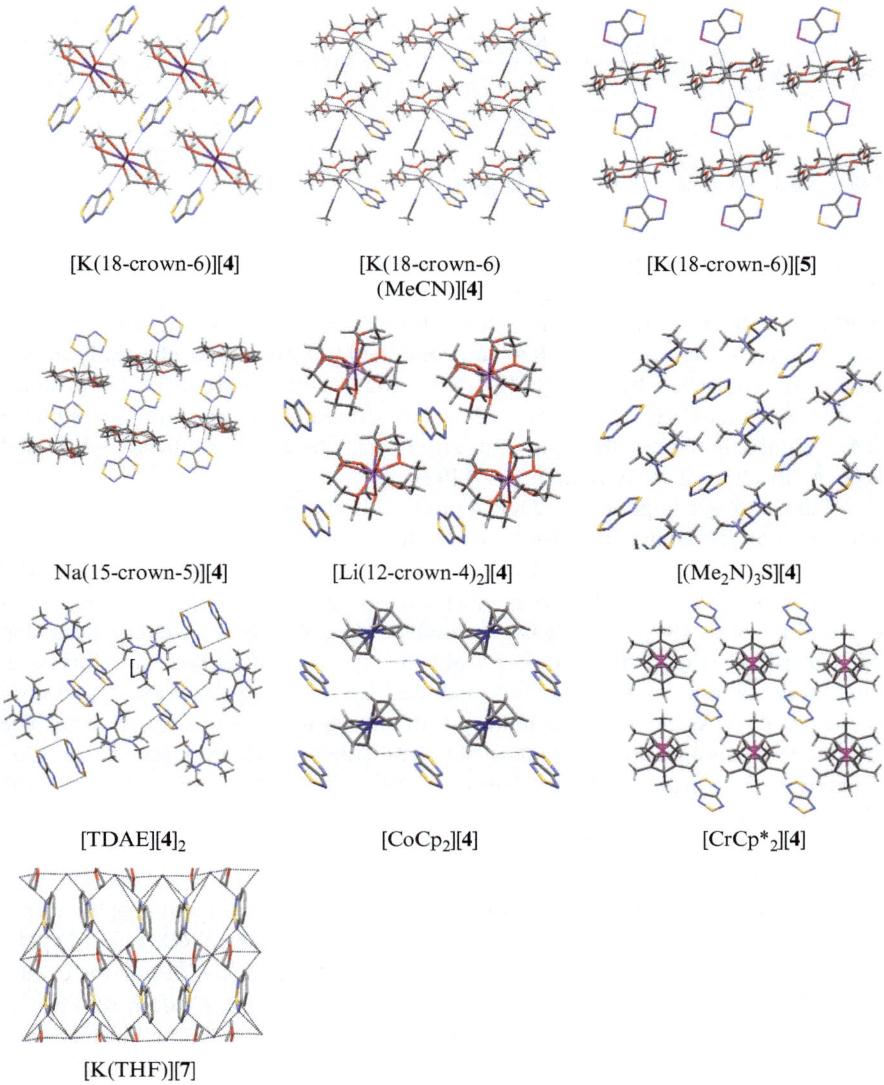

Fig. 6.5 XRD structures of the RA salts

The most interesting feature is π-dimers of the RAs possessing an ideally centrosymmetric structure. The interplanar separation of 3.25 Å is significantly shorter than the sum of the VdW radii of two S atoms of ~ 3.60 Å. The dimers are stable only in the solid state, in solution they immediately dissociate into the RAs. Due to π-dimers, this homospin RA salt is EPR-silent in the solid state [97].

Compound **4** and $CoCp_2$ gave homospin RA salt [$CoCp_2$][**4**] (Scheme 6.4) whose structure was elucidated by XRD (Fig. 6.5). In the crystal, the RAs form

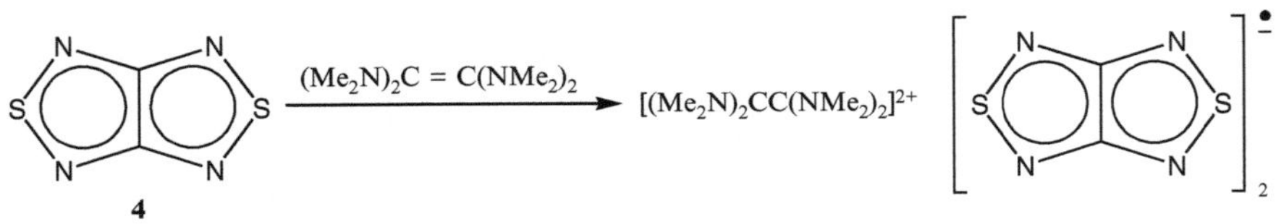

Scheme 6.3 Synthesis of RA salt [TDAE][**4**]$_2$

π...π contacts with Cp moieties of two neighboring cations. The angle between the corresponding best planes is 17.2°, the shortest interplanar contacts, 3.31 and 3.50 Å, are between the C atoms (sum of VdW radii 3.54 Å). Because of these π...π contacts, the cations and RAs are combined to form zigzag chains. Different chains are connected by weak C-H...N hydrogen bonds with an H...N distance of 2.60 Å and a C-H...N angle of 131°. The structure also displays layers of the RAs. Within the layers, the RAs form zigzag belts by S...N contacts of 3.58 and 3.66 Å (the sum of VdW radii 3.55 Å) [98].

With $CrCp^*_2$, compounds 1, **2** and **4** produced heterospin salts where both cation and anion are paramagnetic with $S = 3/2$ and $S = 1/2$, respectively (Scheme 6.4) [79, 86]. The structure of the [$CrCp^*_2$][**4**] salt was confirmed by single-crystal XRD (Fig. 6.5) [79] whereas salts [$CrCp^*_2$][**1**] and [$CrCp^*_2$][**2**] were obtained as microcrystalline solids [86]. The XRD structure of [$CrCp^*_2$][**4**] is different from that of [$CoCp_2$][**4**] (Fig. 6.5). In the crystal, the ions form a layered structure with alternating cationic and anionic layers shifted by x/2 along the *a* axis. As a consequence, the RAs are displaced in the pits of the cationic layer, and vice versa. The structure reveals a stacked orientation of two neighboring cations whose Cp*...Cp* separation, 3.77 Å, exceeds the sum of VdW radii of the C atoms, 3.54 Å; the Cr...Cr separation is 7.49 Å [79].

With $FeCp^*_2$, compounds **1**, **2** and **4** do not react, also **4** does not react with the Vaska's compound [$IrBr(CO)(PPh_3)_2$] [99].

Compounds **7**, **26** and **27** were reduced by elemental potassium to give homospin RA salts (Scheme 6.5). The salt [K(THF)][**7**] was identified by XRD (Fig. 6.5) [91], two other salts were microcrystalline solids not suitable to XRD [99]. In the crystal of [K(THF)][7], each N and S atom of the RA, as well the O atom of THF, is coordinated to two cations. Atoms S and K lie in the crystallographic plane *bc*, atoms K form chains along the *b* axis with K...K distances of 3.73 Å (effective ionic radius of K^+ is 1.38 Å for coordination number 6). Alternating RA and THF molecules form layers [91].

Compound **4** was also reduced by di(benzene)chromium but the structure of the amorphous solid obtained is still unknown [99].

Comparison of XRD structures of **4** and its RA in various salts revealed that in the RA the SN and CC bonds are elongated by ~0.03 Å, whereas the CN bonds and the NSN angles are insensitive to the charge/spin state [80, 94]. For **5** and its RA comparison is difficult because in the XRD structures chalcogens are statistically disordered [76]. In the case of **7** and its RA, the SN bonds are elongated in the latter by ~0.05 Å whilst the CN and CC bonds and the NSN angle are insensitive to the charge/spin state [91]. Elongation of the bonds directly indicates that the π-SOMOs of the RAs are antibonding toward them (*cf.* Fig. 6.3).

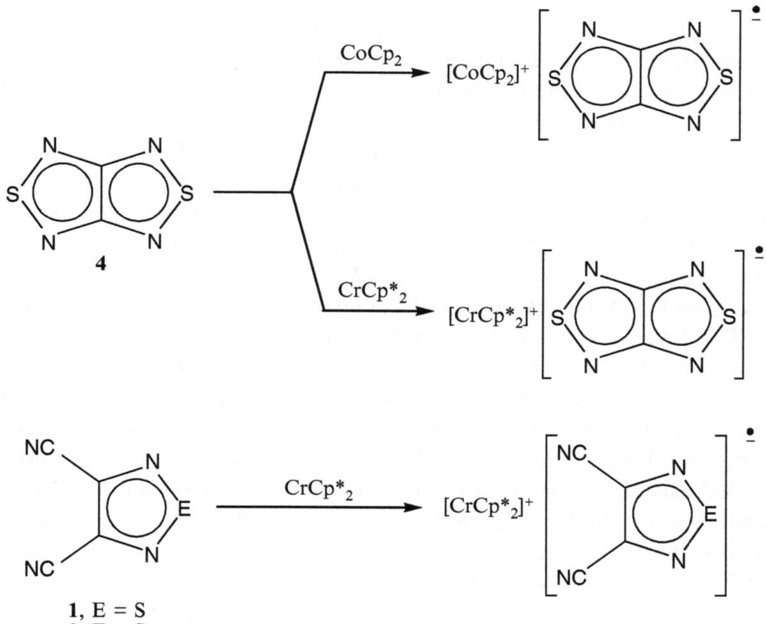

Scheme 6.4 Synthesis of RA salts [$CoCp_2$][**4**], [CrCp*2][**4**], [$CrCp^*_2$][**1**] and [$CrCp^*_2$][**2**]

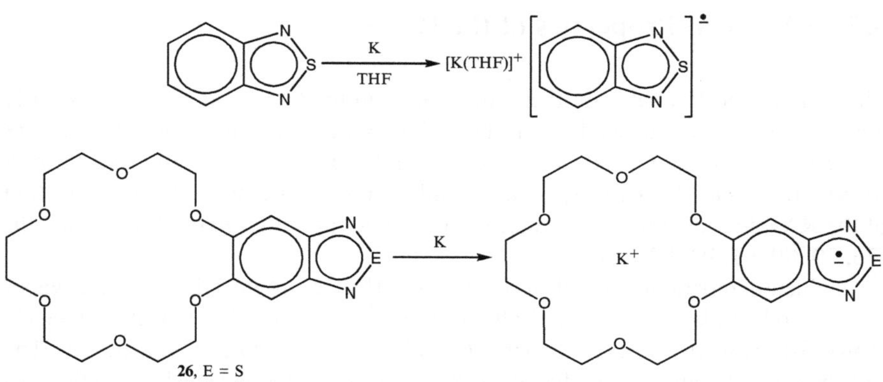

Scheme 6.5 Synthesis of RA salts [K(THF)][**7**], [K][**26**] and [K][**27**]

The paramagnetic nature of the synthesized salts was confirmed by solid-state and solution EPR. In most cases the solution spectra were very similar to those of electrochemically generated RAs [76, 80, 91, 94]. The crystalline salt [TDAE][**4**]$_2$

was found to be EPR-silent due to π-dimerization of the RAs (Fig. 6.5), its solutions however revealed strong signal of the RA [97]. At room temperature and with a conventional X-band EPR technique, all salts of the $[CrCp*_2]^+$ cation were EPR-silent in both the solid state and MeCN solutions [79, 86]. This can be associated with the huge zero-field splitting and fast relaxation of the cation [100, 101] provoking fast relaxation of the anion; in solution, the salts exist very likely in the ion-pair form typical of these types of RAs [39–41]. Differences between solution EPR spectrum of salt [K(18-crown-6)][**5**] and that of electrochemically generated RA of **5** also implied formation of ion contact pairs in the salt's solution [76].

Due to enhanced current interest to π-dimers of ions and radicals (Refs. [102, 103] and references therein), bonding situation in diamagnetic π-dimers observed for salt [TDAE][**4**]$_2$ (Fig. 6.5) was studied by means of CASSCF calculations. They predicted a singlet ground state for the $[\mathbf{4}_2]^{2-}$ dimer, with an S-T splitting of -10.9 kcal.mol^{-1} at the CASSSCF(14,12) level of theory and -8.5 kcal.mol^{-1} at the CASPT2 level. The ground state configuration makes the largest contribution of 71% to the wavefunction of the singlet ground state of the $[\mathbf{4}_2]^{2-}$. The next largest contribution of 12% comes from the doubly excited HOMO → LUMO configuration. Since the CASSCF natural orbital occupation numbers reveal significant deviation from 0 and 2 (at most 0.37 for the LUMO and 1.67 for the HOMO) one can conclude that the ground state of the $[\mathbf{4}_2]^{2-}$ dimer has noticeable (~22%) singlet biradical character. In the calculations, the energy difference between $[\mathbf{4}_2]^{2-}$ and two isolated RAs was always positive: the system of isolated RAs is more stable than $[\mathbf{4}_2]^{2-}$ because of Coulombic repulsion in the latter. Therefore the $[\mathbf{4}_2]^{2-}$ dimers do not exist in a gas phase and solution but can be stabilized by electrostatic (packing) forces of the crystal lattice [97].

6.5 Chemical Properties of the RA Salts

The synthesized RA salts are thermally stable up to 100°C and higher, especially metallocenium salts, depending on the cation. When protected from air, they are stable in the crystalline state for several months and in MeCN solutions at least for several weeks. [TDAE][**4**]$_2$ is kinetically stabilized because of π-dimerization of the RAs. In sealed tubes its solutions are stable for a minimum of a few months [76, 79, 80, 91, 94, 97, 98].

Ambient-temperature hydrolysis of solid salt [Na(15-crown-5)][**4**] by water vapor gave a product identified by XRD after crystallization from MeCN-THF as $[Na(15\text{-crown-}5)]_2[S(SO_3)_2]\cdot 2MeCN$ (Scheme 6.6, Fig. 6.6), *i.e.* a rare trithionate salt. The MeCN molecules are not coordinated to Na^+ cations, and the salt can be considered in this aspect as inclusion compound. In the crystal, each Na^+ center is surrounded by seven O atoms, five from the crown-ethers, and two from two different SO_3 groups of the same anion. This coordination pattern is very different from a previously studied trithionate salts. In particular, in K^+ and Cs^+ salts the interactions of the $[S(SO_3)_2]^{2-}$ with eight or nine metal centers, respectively, lead to 3D polymeric structures, whereas in $[Na(15\text{-crown-}5)]_2[S(SO_3)_2]$ the interaction of the anion with only two Na^+ centers results in the formation of isolated ion pairs [104].

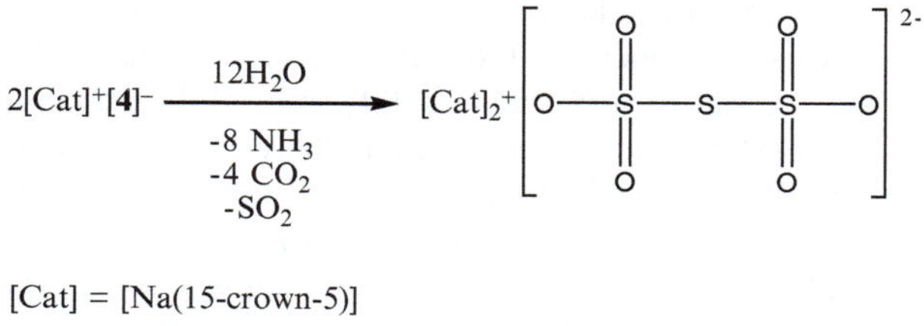

Scheme 6.6 Water vapor hydrolysis of crystalline RA salt [Na(15-crown-5)][**4**]

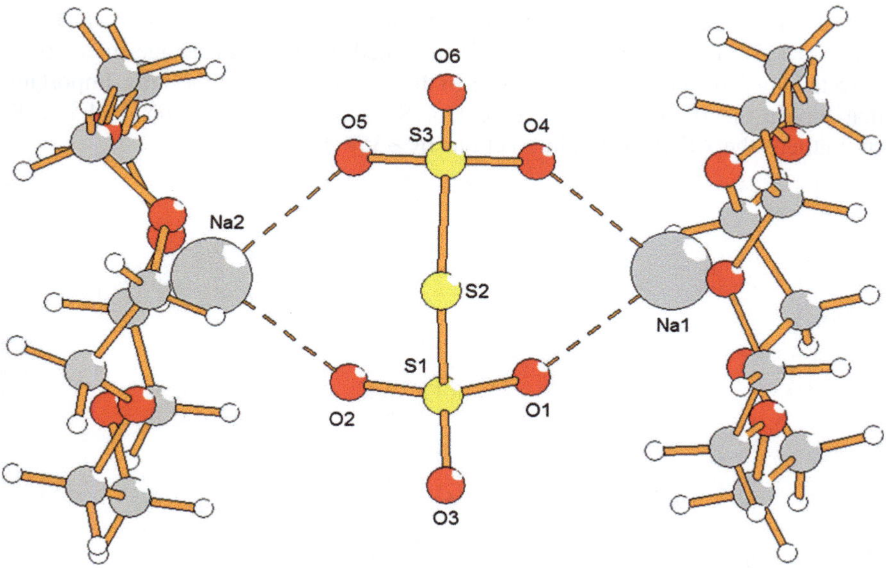

Fig. 6.6 XRD structure of the trithionate salt

6.6 Magnetic and Electrical Properties of the RA Salts

Magnetic properties of the synthesized RA salts were studied by experimental and theoretical methods. Particularly, temperature dependences of molar magnetic susceptibility (χ) and effective magnetic moment (μ_{eff}) were measured in the temperature range 2–300 K (Fig. 6.7).

For the homospin salt [TDAE][**4**]$_2$, paramagnetic contribution to χ(T) is constantly equal to zero in the 2–300 K range due to the solid-state π-dimerization of the RAs [97]. For such systems at higher temperatures, one may expect coexistence of diamagnetic and paramagnetic phases interesting in the context of switchable materials. However, according to the estimated energy of the S-T splitting, for [TDAE][**4**]$_2$ the corresponding temperature is *ca.* 2000 K whereas the salt is thermally stable only up to ~ 380 K.

In the 2–300 K temperature range, χ(T) of the homospin salts [K(18-crown-6)][**4**] and [K(18-crown-6)][**5**] increased steadily with lowering of the temperature (Fig. 6.7a), χ(T) of the salt [Na(15-crown-5)][**4**] showed the maximum at 8 K. To estimate the energy of the exchange interaction J between RAs, three phenomenological models were used at first: (1) the model of dimers, (2) the Bonner-Fisher (BF) uniform chain model with exchange parameter J equal along the 1D chain, and (3) the alternative chain model with exchange parameters J and αJ alternated along the 1D chain. The experimental χ(T) dependences were best fitted by the alternative chain model for the Na^+ salt, and by the BF model for the K^+ salts (Fig. 6.7a), in all cases in accordance with the character of the crystal packing (Fig. 6.5). Corresponding J and αJ values ([Na(15-crown-5)][**4**], −3,4, −1.1; [K(18-crown-6)][**4**], −1.2; [K(18-crown-6)][**5**], −1.7 cm^{-1}) indicated weak AF exchange interactions despite long distances between the RAs in the crystal lattices of the salts. Important, that within the BF model replacement of S by Se (transformation of **4** into **5**) strengthens the AF exchange interactions by ~35% [76, 94].

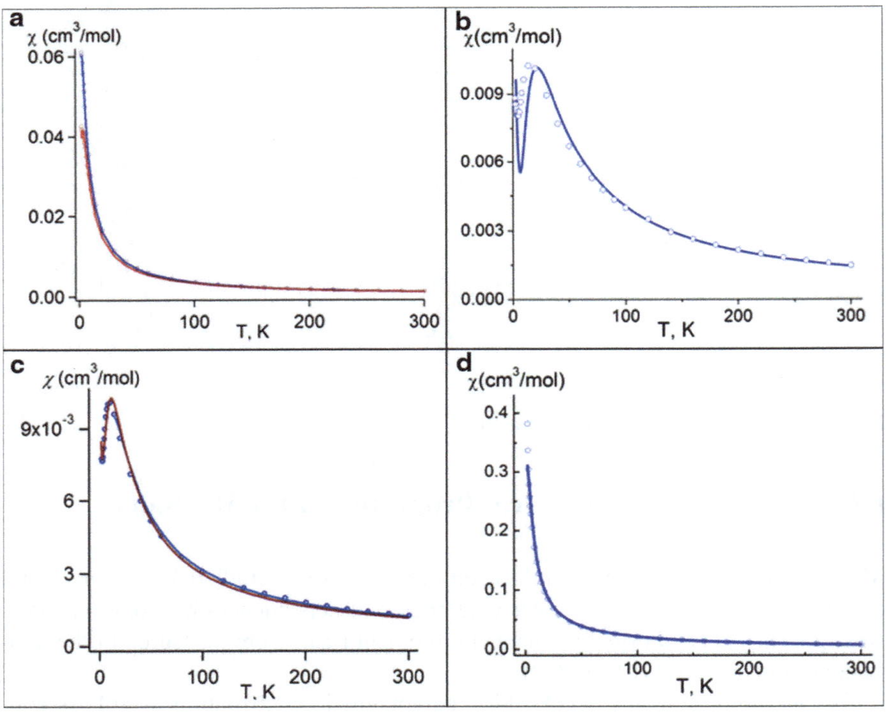

Fig. 6.7 Experimental χ(T) dependences (*points*) for the RA salts and their theoretical approximations (*solid curves*): (**a**) [K(18-crown-6)][**4**] (*blue*) and [K(18-crown-6)][**5**] (*red*) salts; theoretical curves were calculated for Bonner-Fisher uniform chain model with $J = -1.65$ cm^{-1} (*blue*) and $J = -1.22$ cm^{-1} (*red*); (**b**) [K(THF)][**7**] salt, theoretical curve is the twisted-to-tube [3 × 4] grid fit; (**c**) [$CoCp_2$][**4**] salt, theoretical curves are high-temperature series expansion fit (*blue*) and twisted-to-tube [3 × 4] grid fit (*red*); (**d**) [$CrCp^*_2$][**4**] salt, theoretical curve was calculated for the sample consistent of the AF coupled $[4]^{\bullet-}\ldots[4]^{\bullet-}$ ($J_1 = -40$ cm^{-1}) and $[CrCp^*_2]^+\ldots[CrCp^*_2]^+$ ($J_2 = -0.58$ cm^{-1}) pairs

For the K^+ salts, the spin-unrestricted (SU) broken-symmetry (BS) DFT calculations of pair exchange interactions performed at the UB3LYP/6–31 + G(d) level of theory for unique RA pairs indentified in the XRD structures of the salts gave AF exchange energies of the same order of magnitude as obtained with the discussed models but revealed 3D magnetic motifs. In these motifs, however, the exchange interaction in one unique RA pair was considerably stronger than in all other pairs. The main magnetic motif of the salts can therefore be approximately described as 1D chain in agreement with the BF model (Fig. 6.7a). For similar RA pairs, the calculated J values for Se-containing RAs of **5** were considerably larger than for S-containing RAs of **4**. This agrees with the result obtained from the fitting of the $\chi(T)$ dependences [76].

For the homospin salt [K(THF)][**7**], the SU BS DFT calculations based on its XRD structure suggested a complex 3D magnetic motif with 12 unique pair exchange interactions of both AF and FM types, varying from -8.1 to $7.2\ cm^{-1}$. The motif can be described as non-uniform chains of AF-coupled paramagnetic centers with FM interactions between the chains. The experimental $\chi(T)$ dependence (Fig. 6.7b) revealed that the AF interactions dominate the magnetic properties of the salt [91, 105].

For the homospin salt $[CoCp_2]$[**4**], μ_{eff} increased with temperature to the value of 1.74 BM at 300 K corresponding to one unpaired electron per molecule; the $\chi(T)$ dependence revealed the maximum at 10 K (Fig. 6.7c). The SU BS DFT calculations of pair exchange interactions between the RAs gave a 2D-rhomb magnetic motif which can be reduced to a 2D-square motif since one of two exchange integrals ($-2.8\ cm^{-1}$) was significantly larger than another ($-0.13\ cm^{-1}$). The resulting motif corresponds to the 2D version of the AF Heisenberg model for 1/2 spins. The low- and high-temperature approximations proposed for this model were used in analysis of the experimental $\chi(T)$ dependence (Fig. 6.7c). The obtained J value of $-4.2 \pm 0.7\ cm^{-1}$ was in agreement with the results of the SU BS DFT calculations. The $\chi(T)$ dependence was also modeled with the general form of the Van Fleck equation applied to planar and twisted to a tube [3×4] grid of spins 1/2. The best fit was obtained with $J = -5.8$ and $5.0\ cm^{-1}$, respectively [98].

The heterospin salts $[CrCp^*_2]$[**1**], $[CrCp^*_2]$[**2**] and $[CrCp^*_2]$[**4**] formally satisfy the conditions of the McConnell I model [92, 93]: whereas spin density at the VdW surface of the RAs is mostly positive (Fig. 6.4), the $[CrCp^*_2]^+$ cation (as well as its Fe and Mn congeners) possesses peripheral negative spin density [106–108]. Therefore anion-cation exchange interactions might lead to the FM ground state. For the salt $[CrCp^*_2]$[**4**], μ_{eff} increased with temperature; at 300 K its value of 4.27 BM was close to the spin-only value of 4.24 BM for an uncorrelated randomly-oriented two-spin system with $S = 3/2$ and $S = 1/2$ and $g = 2$ for both spins. The $\chi(T)$ increased steadily with a lowering of the temperature (Fig. 6.7d). The reciprocal $\chi(T)$ obeyed the Curie-Weiss law between 300 and 50 K and then deviated from linearity; extrapolation of the linear part onto axis T gave a negative Weiss constant of *ca.* -10 K. Altogether, $\mu_{eff}(T)$, $\chi(T)$ and $\chi^{-1}(T)$ dependences evidenced AF interactions in the salt's spin system [79]. The properties of $[CrCp^*_2]$[**1**] and $[CrCp^*_2]$[**2**] are similar [86, 105]. Thus, changing the spin state of the cation from S = 0 ([$CoCp_2$][**4**]) to S = 3/2 ([$CrCp^*_2$][**4**]) does not change the AF character of the exchange interactions in the spin system.

Theoretical analysis of the magnetic motif of [$CrCp^*_2$][**4**] within the CASSCF and SU BS DFT approaches also revealed dominance of AF interactions: significant between RAs, weak between cations, and very weak between RAs and cations. Only very weak FM interactions (0.08 cm^{-1}) between the RAs and cations were found. Experimental χ(T) and μ_{eff}(T) dependences were very well reproduced in the assumption of the AF coupled [**4**]$^-$...[**4**]$^-$ ($J_1 = -40 \pm 9$ cm^{-1}) and [$CrCp^*_2$]$^+$...[$CrCp^*_2$]$^+$ [$J_2 = -0.58 \pm 0.03$ cm^{-1}) pairs [for χ(T), see Fig. 6.7d]. The experimental J_1 value was in reasonable agreement with the values calculated using SU BS DFT (−61 cm^{-1}) and CASSCF(10,10) (−15.3 cm^{-1}) approaches. The experimental J_2 value was also in agreement with that calculated using the SU BS DFT technique (−0.33 cm^{-1}) [74]. Thus, the FM interactions in [$CrCp^*_2$]$^+$...[**4**]$^-$ pairs, satisfying the McConnell I model, belong to the weakest among theoretically calculated exchange interactions. The main reason is that the crystal packing of the salt (Fig. 6.5) is not well-suited to the required contacts between paramagnetic ions. In contrast to the crystal lattices of metallocenium [TCNE]$^-$ and [TCNQ]$^-$ salts with FM properties, featuring chains of alternating cations (donors, D^+) and anions (acceptors, A^-), ... $D^+A^-D^+$ A^-..., with interchain M...M separation of 10.7 ± 0.3 Å [52, 53, 55–62], that of [$CrCp^*_2$][**4**] (Fig. 6.5) reveals layered structure with alternating cationic and anionic layers (see above) [79].

Preliminary investigated electrical properties of the synthesized RA salts expectedly revealed that they can be classified as Mott insulators (paramagnetic dielectrics). For example, for polycrystalline sample of [$CoCp_2$][**4**] salt the room-temperature contactless estimations of bulk electrical conductivity and its thermal activation energy gave values of ~1.90 $Ohm^{-1}.cm^{-1}$ and ~0.5 eV, respectively [109]. These values have the same orders of magnitude as the previously studied RA salts of [MCp_2][TCNE] (M = Co, Fe) [70].

It should be noted that currently there is a strong interest to (molecular) Mott insulators since some of them undergo low-temperature transition from dielectric state directly to superconducting state (Refs. [110, 111] and references therein).

6.7 Attempted Preparation of Lanthanide Complexes of Chalcogen-Nitrogen π-Heterocycles

As preliminary stage of synthesis of p-d-f heterospin systems based on the chalcogen-nitrogen π-heterocyclic RA salts, complex formation from compounds **26** and **27** and lanthanide nitrates was studied [112].

It was known that aqueous nitrates of the early, larger, lanthanides form 12-coordinate directly complexed species [$Ln(NO)_3$-(18-crown-6)] (Ln = La-Nd) whereas those of mid-to-late, smaller, lanthanide crystallize as H-bonded second-sphere complexes [$Ln(NO_3)_3(OH_2)_3$]·18-crown-6 (Ln = Eu-Lu) [113].

Under the same conditions of crystallization from MeCN/MeOH mixture, however, compounds **26** and **27** formed with Ln(III) nitrides (**26**, Ln = Gd, Ho; **27**, Ln = Ce, Gd, Nd, Eu, Ho) only cocrystals whose XRD structures (Fig. 6.8) revealed the absence of coordination between Ln^{3+} cations and polyether fragments of **26** and **27**; instead, the Ln^{3+} centers were coordinated only to NO_3^- anions and H_2O molecules [112]. In the cases of Ce, Nd and Eu, the cocrystals contain ions $[Ln(NO_3)_2(H_2O)_5]^+$ and $[Ln(NO_3)_4(H_2O)_{2-m}(CH_3OH)_m]^-$ (m = 1, 2). In the case of Gd, the anions are NO_3^-. In the case of Ho, the cocrystal is composed of crown ether and $[Ho(NO_3)_3(H_2O)_3]$ building blocks. With all Ln, except Ln = Ho, two-link chains were observed in the cocrystals, being formed by O-H...H hydrogen bonds between water molecules coordinated to lanthanide-containing ion and oxygen atoms of the crown ether. With Ln = Ho, similar one-link chains were observed.

Further work with ultra dry Ln(III) chlorides is planned (*cf*. Ref. [114]).

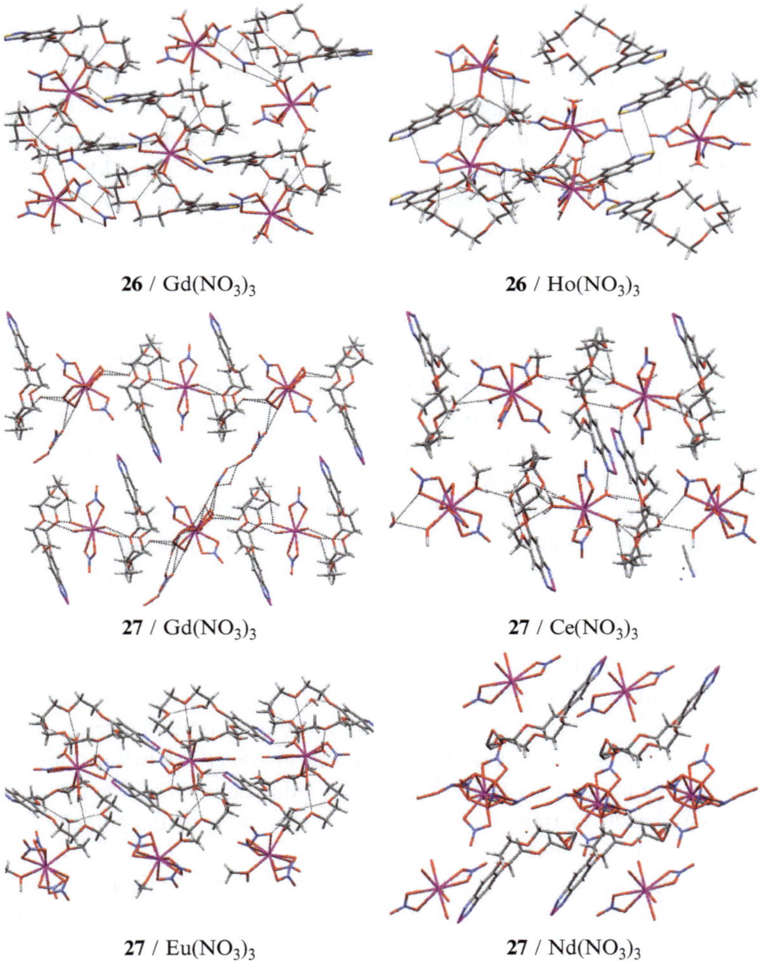

Fig. 6.8 XRD structures of co-crystals formed by aqueous Ln(III) nitrates and compounds **26** and **27** (crystallization from MeCN/MeOH, see the text for structural details)

6.8 Conclusions and Further Work

A new class of paramagnetics was found and characterized, namely, chalcogen-nitrogen π-heterocyclic RA salts.

Potentially, this class is very wide because it is possible to vary composition and structure of chalcogen-nitrogen π-heterocyclic RAs, as well as the chemical nature, charge (+1, +2, +3, etc.) and spin ($S = 0$, $S \neq 0$) state of cations.

Particularly, in neutral precursors one can vary chalcogens (S, Se, Te) together with sizes and numbers of hetero- and carbocycles and the ways of their fusion in molecules. Obviously, not only 1,2,5-chalcogenadiazole derivatives but also those of 1,2,5-chalcogenadiazine and some other heterocyclic systems can be involved including systems with larger numbers of heteroatoms. The EA of the precursors can be further enlarged in atom-economic way. For example, substitution of one S atom in compound **19** by O atom gives derivative with EA = 3.49 eV at the level of DFT theory specified in Chart 6.2, *i.e.* the same as the EA of TCNE [71].

Variation of the chemical nature of paramagnetic cations for the heterospin RA salts might be possible with MCp^*_2 and MBz^*_2 organometallics (M = Ti, Cr, Fe, Ni, Mo, W, Os) with low gas-phase IEs [115, 116] comparable with those of alkali metals (K, 4.34; Na, 5.14 eV) and TDAE (6.13 eV) [117]. For MCp_2^* derivatives, HeI UPS IE ranges from 4.93 (M = Cr) to 5.88 (M = Fe) and 6.68 (M = Os) eV, and for MBz_2 derivatives from 5.40 (M = W) to 5.52 (M = Mo) eV [115, 116] (MBz^*_2 should have lower IEs), allowing, in principle, exploitation of the whole block of 3d-5d transition metals. Transition metal carbonyls, $M(CO)_6$ (M = Cr, Mo, W), featuring IE of *ca.* 8.5 eV [118] are less promising.

Of special interest are heteroleptic MCp*Bz* compounds with even lower IEs, for example, 4.21 eV for M = Fe [119]. Despite the $[FeCp^*Bz^*]^+$ cation is diamagnetic, one can expect that the use of FeCp*Bz* will allow to generate RAs from a large number of various chalcogen-nitrogen π-heterocycles.

Promising reducing agents – sources of cations – are metal arylchalcogenolates $M(XAr)_n$ (X = S, Se; *cf.* Scheme 6.2), especially polyfluorinated, revealing good solubility in aprotic organic solvents (MeCN, THF) and known for s-, p-, d- and f-metals (Refs. [120–122] and references therein). In this context, aromatic tellurolates [123] should also be considered.

Metalloporphyrins and phthalocyanines featuring transition metals in lower oxidation states [124] form another interesting group of potential reducing agents. In this context, intramolecular redox processes in transition metal complexes of porphyrazines annulated with 1,2,5-thia(selena)diazoles [125], including photo-initiated, with electron transfer from metal center onto chalcogenadiazole ring, might be of special interest.

Already synthesized RA salts [76, 79, 80, 86, 91, 94, 97, 98] can also be used as reducing agents for neutral chalcogen-nitrogen π-heterocycles (Chart 6.2) whose EAs are higher those of the precursors of RAs in the starting salts. As relevant example, [K(18-crown-6)][**4**] and compound **19** can be considered.

Chalcogen-nitrogen π-heterocyclic RA salts can be considered as new building blocks in the design and synthesis of molecular magnets and/or conductors.

The design of magnets can be performed with the McConnell I, the Kahn, and the Dzyaloshinsky-Moriya models. The McConnell I model [92, 93] was discussed above in the context of the heterospin RA salts.

According to the Kahn model [93], orthogonality of magnetically-coupled SOMOs of paramagnetic species will lead to FM (↑↑) interactions, whereas non-orthogonality (spatial overlap) to AF (↑↓) interactions; the latter case resembles formation of electron pair of ordinary covalent bond. This model is applicable to both homo- and heterospin RA salts. With metalloporphyrins/phthalocyanines as reducing agents, two mutual orientations of planar cation and planar anion are the most expected in the solid state as face-to-edge and face-to-face orientations. It allows controlling the magnetic situation to some extent – if face-to-edge orientation leads to, say, AF interactions, the face-to-face one should lead to FM interactions, and vice versa.

According to the Dzyaloshinsky-Moriya model [93, 126], AF exchange interactions between paramagnetic centers of low-symmetry crystal lattice in combination with local (on-site) spin-orbit coupling can lead to weak ferromagnetism by virtue of spin canting. Similar to the Kahn model, this model is applicable to both homo- and heterospin RA salts being especially promising in the case of RAs containing Se and Te atoms because of stronger spin-orbit coupling.

The design of conductors requires transformation of the RAs' π-SOMOs into delocalized semi-occupied electron energy band in the solid state, which, in principle, is possible (*cf.* for example Ref. [127]). The obvious problem is on-site electrostatic repulsion. The [TDAE][**4**]$_2$ salt reveals that in some crystal lattices this repulsion can be overcome.

In all cases, the properties of new RA salts will critically depend on their crystal packing. Thus, for polymorphs of [MCp*$_2$][TCNE] and [MCp*$_2$][TCNQ] salts [52, 53, 55–62] and some other relevant systems [30], the typical situation is that only one has the FM properties whereas the other have not. For the non-stoichiometric $[V]^{2+}[TCNE]_z^{-}[TCNE]_{1-z/2}^{2-}$ salt only the non-crystalline kinetic phase reveals FM properties whereas the crystalline phase is not a magnetically-ordered material [54]. The crystal packing is a property which can hardly be predicted and/or controlled nowadays, especially in the case of complex system. Therefore, the further work will necessary include the trial-and-error procedure especially in the crystal engineering aspect.

The most interesting new results are expected for Se and Te derivatives. Progress in the field is very fast. In particular, after submitting this Chapter to the Editors, synthesis and XRD structure of compound **3** [128] and XRD structure of compound **12** (prepared earlier [74]) were published [129], and preparation of some other relevant derivatives was accomplished [130].[1] In this sense the work inherently belongs to the frontiers of selenium and tellurium chemistry.

[1]Compound **3** (synthetic protocol differs from that in Ref. [128]), its 1 : 1 complex with pyridine, as well as pyridinium and triethylammonium salts of 1-X-3,4-dicyano-1,2,5-telluradiazolide (X = Cl, Br) and 1,1-X_2-3,4-dicyano-1,2,5-telluradiazolide (X = Cl) anions are prepared and characterized by XRD.

Acknowledgments The authors are grateful to all their coauthors whose names are presented in corresponding references, especially to Prof. Nina P. Gritsan and Prof. Yuri V. Gatilov, as well as to Prof. Yuri N. Molin and Prof. Jens Beckmann for valuable discussions, and to Mr. Peter Brackmann for his contribution to XRD measurements. The authors are also grateful to the Deutsche Forschungsgemeinschaft (project 436 RUS 113/967/0-1 R), the Russian Foundation for Basic Research (projects 10-03-00735 and 10-03-08249), the Presidium of the Russian Academy of Sciences (project 18.17), and the Siberian Branch of the Russian Academy of Sciences (project 105) for financial support.

References

1. Chivers T, Laitinen RS, Devillanova FA (eds) (2007) Handbook of chalcogen chemistry. RSC Publishing, Cambridge, pp 223–285
2. Chivers T (2005) A guide to chalcogen-nitrogen chemistry. World Scientific Publishing Co. Ltd, Singapore
3. Tse JS, Leitch AA, Yu X, Bao X, Zhang S, Liu Q, Jin C, Secco RA, Desgreniers S, Ohish Y, Oakley RT (2010) J Am Chem Soc 132:4876–4886
4. Mito M, Komorida Y, Tsuruda H, Tse JS, Desgreniers S, Ohishi Y, Leitch AA, Cvrkalj K, Robertson CM, Oakley RT (2009) J Am Chem Soc 131:16012–16013
5. Leitch AA, Yu X, Winter SM, Secco RA, Dube PA, Oakley RT (2009) J Am Chem Soc 131: 7112–7125
6. Winter SM, Cvrkalj K, Dube PA, Robertson CMR, Probert MR, Howard JAK, Oakley RT (2009) Chem Commun 7306–7308
7. Robertson RC, Leitch AA, Cvrkalj K, Reed RW, Myles DJT, Dube PA, Oakley RT (2008) J Am Chem Soc 130:8414–8425
8. Robertson CM, Myles DJT, Leitch AA, Reed RW, Dooley BM, Frank NL, Dube PA, Thompson LK, Oakley RT (2007) J Am Chem Soc 129:12688–12689
9. Leitch AA, Brusso JL, Cvrkalj K, Reed RW, Robertson CM, Dube PA, Oakley RT (2007) Chem Commun 3368–3370
10. Leitch AA, Reed RW, Robertson CM, Britten JF, Yu X, Secco RA, Oakley RT (2007) J Am Chem Soc 129:7903–7914
11. Brusso JL, Cvrkalj K, Leitch AA, Oakley RT, Reed RW, Robertson CM (2006) J Am Chem Soc 128:15080–15081
12. Alberola A, Eisler D, Less RJ, Navarro-Moratello E, Rawson JM (2010) Chem Commun 46:6114–6116
13. Deumal M, Rawson JM, Goetha AE, Howard JAK, Copley RCB, Robb MA, Novoa JJ (2010) Chem Eur J 16:2741–2750
14. Rawson JM, Clarke CS, Bruce DW (2008) Magn Reson Chem 47:3–8
15. Deumal M, LeRoux S, Rawson JM, Robb MA, Novoa JJ (2007) Polyhedron 26:1949–1958
16. Rawson JM, Alberola A, Whalley A (2006) J Mater Chem 16:2560–2575
17. Rawson JM, Luzon P, Palacio F (2005) Coord Chem Rev 249:2631–2641
18. Rawson JM, Palacio F (2001) Struct Bonding (Berlin) 100:93–128
19. Rawson JM, MacManus GD (1999) Coord Chem Rev 189:135–168
20. Cameron TS, Decken A, Grein F, Knapp C, Passmore J, Rautiainen JM, Shuvaev KV, Thompson RC, Wood DJ (2010) Inorg Chem 49:7861–7879
21. Shuvaev KV, Decken A, Grein F, Abeldin TMS, Thompson LK, Passmore J (2008) Dalton Trans 4029–4037
22. Cameron TS, Decken A, Kowalczyk RM, McInnes EJL, Passmore J, Rawson JM, Shuvaev KV, Thompson LK (2006) Chem Commun 2277–2279
23. Decken A, Matar SM, Passmore J, Shuvaev KV, Thompson LK (2006) Inorg Chem 45: 3878–3886

24. Cameron TS, Lemaire MT, Passmore J, Rawson JM, Shuvaev KV, Thompson LK (2005) Inorg Chem 44:2576–2578
25. Boere RT, Tuonomen HM, Chivers T, Roemmele TL (2007) J Organomet Chem 692: 2683–2696
26. Boere RT, Roemmele TL (2000) Coord Chem Rev 210:369–445
27. Preuss KE (2007) Dalton Trans 2357–2369
28. Knapp C, Lork E, Gupta K, Mews R (2005) Z Anorg Allg Chem 631:1640–1644
29. Zienkiewicz J, Kaszynski P, Young VG (2004) J Org Chem 69:7525–7536
30. Fujita W, Kikuchi K (2009) Chem Asian J 4:400–405
31. Banister AJ, Gorrell IB (1998) Adv Mater 10:1415–1429
32. Mito M, Kawae T, Takeda K, Takagi S, Matsushida Y, Daguchi H, Rawson JM, Palacio F (2001) Polyhedron 20:1509–1512
33. Shuku Y, Suizu R, Awaga K, Sato U (2009) Cryst Eng Comm 11:2065–2068
34. Awaga K, Tanaka T, Shirai T, Umezono Y, Fujita W (2007) C R Chim 10:52–57
35. Fujita W, Awaga K (2003) Synth Met 137:1263–1265
36. Fujita W, Awaga K (1999) Science 286:261–262
37. Risto M, Assoud A, Winter SM, Oilunkaniemi R, Laitinen RS, Oakley RT (2008) Inorg Chem 47:10100–10109
38. Cozzolino AF, Gruhn NE, Lichtenberger DL, Vargas-Baca I (2008) Inorg Chem 47:6220–6226
39. Hanson P (1980) Adv Heterocycl Chem 27:31–149
40. Bock H, Haenel P, Neidlein R (1988) Phosphorus Sulfur Silicon Relat Elem 39:235–252
41. Kwan CL, Carmack M, Kochi JK (1976) J Phys Chem 80:1786–1792
42. Motokawa N, Miyasaka H, Yamashita M, Dunbar KR (2008) Angew Chem Int Ed 47: 7760–7763
43. Yamashida Y, Tomura T (1998) J Mater Chem 8:1933–1944
44. Tsubata Y, Suzuki T, Yamashita Y, Mikai T, Miyashi T (1992) Heterocycles 33:337–348
45. Yamashita Y, Mikai T, Miyashi T, Saito G (1988) Bull Chem Soc Japan 61:483–493
46. Mirifico MV, Caram JA, Gennaro AM, Cobos CJ, Vasini EJ (2009) J Phys Org Chem 22: 964–970
47. Denning MS, Irwin M, Golcoechea JM (2008) Inorg Chem 47:6118–6120
48. Choua S, Djukic JP, Dallery J, Bieber A, Welter R, Gisselbrecht JP, Turek P, Ricard L (2009) Inorg Chem 48:149–163
49. Bock H, John A, Naether C, Ruppert K (1994) Helv Chim Acta 77:1505–1509
50. Gloekle M, Hueber K, Kuemmerer HJ, Denninger G, Kaim W (2001) Inorg Chem 40: 2263–2269
51. Schwach M, Hausen HD, Kaim W (1999) Inorg Chem 38:2242–2243
52. Miller JS (2010) J Mater Chem 20:1846–1857
53. Her JH, Stephens PW, Ribas-Arino J, Novoa JJ, Shum WW, Miller JS (2009) Inorg Chem 48: 3296–3307
54. Miller JS (2009) Polyhedron 28:1596–1605
55. Miller JS (2006) Dalton Trans 2742–2749
56. Miller JS (2000) Inorg Chem 39:4392–4408
57. Miller JS, Epstein AJ (1994) Angew Chem Int Ed 33:385–415
58. Miller JS, Epstein AJ, Reiff WM (1988) Chem Rev 88:201–220
59. Koizumi K, Shoji M, Ritawaga Y, Takeda R, Yamanaka S, Kawakami T, Okumura M, Yamaguchi K (2007) Polyhedron 26:2135–2141
60. Broderick WE, Eichhorn DM, Liu X, Toscano PJ, Owens S, Hoffman BM (1995) J Am Chem Soc 117:3641–3642
61. Eichhorn DM, Skee DC, Broderick WE, Hoffman BM (1993) Inorg Chem 32:491–492
62. Broderick WE, Hoffman BM (1991) J Am Chem Soc 113:6334–6335
63. Lu JM, Rosokha SV, Neretin IS, Kochi JK (2006) J Am Chem Soc 128:16708–16719
64. Bencini A, Daul CA, Dei A, Mariotti F, Lee H, Shulz DA, Sorace L (2001) Inorg Chem 40: 1582–1590

65. Shulz DA, Bodnar SH, Vostrikova KE, Kampf JW (2000) Inorg Chem 39:6091–6093
66. Pierpoint CG, Lange CW (1994) Prog Inorg Chem 41:331–442
67. Adams M, Dei A, Rheingold AL, Hendrickson DN (1993) Angew Chem Int Ed 32:880–882
68. Dei A, Gatteschi D (1992) Inorg Chim Acta 198–200:813–822
69. Gallardo I, Guirardo G, Marquet J, Vila N (2007) Angew Chem Int Ed 46:1321–1325
70. Brandon RL, Osiecki JH, Ottenberg A (1966) J Org Chem 31:1214–1217
71. Gritsan NP, Lonchakov AV, Rakitin OA, Semenov NA, Suturina EA, Zibarev AV (2010) work in progress
72. Blockhuys F, Gritsan NP, Makarov AYu, Tersago K, Zibarev AV (2008) Eur J Inorg Chem 2008:655–672
73. Zibarev AV, Beregovaya IV (1992) Rev Heteroatom Chem 7:171–190
74. Kovtonyuk VN, Makarov AYu, Shakirov MM, Zibarev AV (1996) Chem Commun 1991–1992
75. Cozzolino AF, Britten JF, Vargas-Baca I (2006) Cryst Growth Des 6:181–186
76. Bagryanskaya IYu, Gatilov VYu, Gritsan NP, Ikorskii VN, Irtegova IG, Lonchakov AV, Lork E, Mews R, Ovcharenko VI, Semenov NA, Vasilieva NV, Zibarev AV (2007) Eur J Inorg Chem 2007:4751–4761
77. Zibarev AV, Miller AO (1990) J Fluorine Chem 50:359–363
78. Semenov NA, Lork E, Mews R, Zibarev AV (2010) work in progress
79. Semenov NA, Pushkarevsky NA, Lonchakov AV, Bogomyakov AS, Pritchina EA, Suturina EA, Gritsan NP, Konchenko SN, Mews R, Ovcharenko VI, Zibarev AV (2010) Inorg Chem 49:7558–7564
80. Makarov AYu, Irtegova IG, Vasilieva NV, Bagryanskaya IYu, Borrmann T, Gatilov YuV, Lork E, Mews R, Stohrer WD, Zibarev AV (2005) Inorg Chem 44:7194–7199
81. Lork E, Mews R, Shakirov MM, Watson PG, Zibarev AV (2001) Eur J Inorg Chem 2123–2134
82. Borrmann T, Lork E, Mews R, Shakirov MM, Zibarev AV (2004) Eur J Inorg Chem 2452–2458
83. Borrmann T, Lork E, Mews R, Stohrer WD, Watson PG, Zibarev AV (2001) Chem Eur J 7: 3504–3510
84. Borrmann T, Zibarev AV, Lork E, Knitter G, Chen SJ, Watson PG, Cutin E, Shakirov MM, Stohrer WD, Mews R (2000) Inorg Chem 39:3999–4005
85. Zibarev AV, Lork E, Mews R (1998) Chem Commun 991–992
86. Semenov NA (2009) Diploma work, Novosibirsk State University
87. Bagryanskaya IYu, Gatilov YuV, Lork E, Mews R, Rakitin OA, Semenov NA, Zibarev AV (2010) work in progress
88. Makarov AYu, Zhivonitko VV, Makarov AG, Zikirin SB, Bagryanskaya IYu, Bagryansky VA, Irtegova IG, Gatilov VYu, Shakirov MM, Zibarev AV (2011) Inorg Chem 50:3017–3027
89. Vasilieva NV, Irtegova IG, Gritsan NP, Lonchakov AV, Makarov AYu, Shundrin LA, Zibarev AV (2010) J Phys Org Chem 23:536–543
90. Barlukova MM, Beregovaya IV, Vysotsky VP, Shchegoleva LN, Bagryansky VA, Molin YuN (2005) J Chem Phys A 109:4404–4409
91. Konchenko SN, Gritsan NP, Lonchakov AV, Radius U, Zibarev AV (2009) Mendeleev Commun 19:7–9
92. Novoa JJ, Deumal M (2001) Struct Bonding (Berlin) 100:33–60
93. Kahn O (1993) Molecular magnetism. VCH Publishers, New York
94. Ikorskii VN, Irtegova IG, Lork E, Makarov AYu, Mews R, Ovcharenko VI, Zibarev AV (2006) Eur J Inorg Chem 2006:3061–3067
95. Semenov NA, Suturina EA, Lonchakov AV, Gritsan NP, Bagryanskaya IYu, Gatilov YuV, Irtegova IG, Vasilieva NV, Lork E, Mews R, Zibarev AV (2011) J Phys Chem A 115:4851–4860
96. Suturina EA (2010) Theoretical study of reactions between chalcogen-nitrogen π-heterocycles and thiophenolate and selenophenolate anions, B Phys, Dissertation, Novosibirsk State University

97. Gritsan NP, Lonchakov AV, Lork E, Mews R, Pritchina EA, Zibarev AV (2008) Eur J Inorg Chem 1994–1998
98. Konchenko SN, Gritsan NP, Lonchakov AV, Irtegova IG, Mews R, Ovcharenko VI, Radius U, Zibarev AV (2008) Eur J Inorg Chem 3833–3838
99. Mews R, Zibarev AV (2008) work in progress
100. Solodovnikov SP (1982) Russ Chem Rev 51:961–975
101. Warren KD (1976) Struct Bonding (Berlin) 27:45–159
102. Mota F, Miller JS, Novoa JJ (2009) J Am Chem Soc 131:7699–7707
103. Rosokha SV, Zhang J, Kochi JK (2010) J Phys Org Chem 23:395–399
104. Lork E, Mews R, Zibarev AV (2009) Mendeleev Commun 19:147–148
105. Bogomyakov AS, Pushkarevsky NA, Semenov NA (2009) work in progress
106. Kaupp M, Koehler FH (2009) Coord Chem Rev 253:2376–2386
107. Heise H, Koehler FH, Herker M, Hiller W (2002) J Am Chem Soc 124:10823–10832
108. Kollmar C, Kahn O (1992) J Chem Phys 96:2988–2997
109. Vorobiev AB, Ya Prinz V, Zibarev AV (2008) work in progress
110. Lee PA, Nagaosa N, Wen XG (2006) Rev Mod Phys 78:17–86
111. Seo H, Hotta C, Fukuyama H (2004) Chem Rev 104:5005–5036
112. Gatilov YuV, Semenov NA, Lork E, Mews R, Zibarev AV (2009) Abstracts of papers, National crystallochemical conference, Kazan, Russia, C-30
113. Rogers RD, Rollins AN (1994) J Chem Crystallogr 24:321–329
114. Mishra S (2008) Coord Chem Rev 252:1996–2025
115. Green JC (1981) Struct Bonding (Berlin) 43:37–112
116. O'Hare D, Green JC, Chadwick TP, Miller JS (1988) Organometallics 7:1335–1342
117. Bock H, Ruppert K, Naetzer C, Havlas Z, Herrmann HF, Arad C, Goebel I, John A, Meurel J, Nick S, Rauschenbach A, Seitz W, Vaupel T, Solouki B (1992) Angew Chem Int Ed 31: 550–581
118. Fukuda R, Hayaki S, Hiroshi N (209) J Chem Phys 131:174303-1–174303-10
119. Green JC, Kelly MR, Payne MP, Seddon EA, Astruc D, Hamon JR, Michaud P (1983) Organometallics 2:211–218
120. Eichhoefer A, Wood PT, Viswanath R, Mole RA (2007) Eur J Inorg Chem 2007:4794–4799
121. Emge TJ, Romanelly MD, Moore BF, Brennan JG (2010) Inorg Chem 49:7304–7312
122. Fuller AL, Knight FR, Slawin AMZ, Woollins JD (2010) Eur J Inorg Chem 2010:4034–4043
123. Sadekov ID, Minkin VI (1997) J Sulfur Chem 19:285–348
124. Ishikawa N (2010) Struct Bonding (Berlin) 135:211–228
125. Donzello MP, Ercolani C, Stuzhin PA (2006) Coord Chem Rev 250:1530–1561
126. Moriya T (1960) Phys Rev 120:91–98
127. Cordes AW, Haddon RC, Oakley RT (2004) Phosphorus Sulfur Silicon 179:673–684
128. Cozzolino AF, Yang Q, Vargas-Baca I (2010) Crystal Growth Des 10:4959–4964
129. Cozzolino AF, Whitfield PS, Vargas-Baca I (2010) J Am Chem Soc 132:17265–17270
130. Semenov NA, Pushkarevsky NA, Beckmann J, Lork E, Mews R, Zibarev AV (2011) work in progress

Chapter 7
Organotelluroxanes

Jens Beckmann and Pamela Finke

7.1 Introduction

Sulfoxides, R_2SO, and Sulfones, R_2SO_2, as well as sulfinic acids, RS(O)OH, and sulfonic acids, $RS(O)_2OH$, are well-known organosulfur compound classes that have found numerous applications in organic synthesis and other branches of chemistry. These applications reach from day-to-day laboratory use of commodities, such as dimethylsulfoxide (DMSO) and trifluoromethane sulfonic acid (triflic acid, TfOH) to the utility of more sophisticated reagents, such as, chiral sulfoxides [1, 2] as well as acetylenyl, allenyl [3], vinyl [4], and α-amido sulfones [5] in contemporary organic synthesis. Moreover, sulfones are increasingly employed in the emerging field of organo catalysis [6, 7]. Cysteine sulfinic acid (CSA) is an important intermediate in cysteine metabolism formed by the Cysteine dioxygenase catalyzed conversion of L-cysteine [8, 9]. Sulfinates and sulfonates, the conjugated bases of sulfinic and sulfonic acids are of paramount importance for many industrial and pharmacological products. Arylsulfinate salts serve as photoinitiator systems for the rapid polymerization of dental adhesive resins [10]. Linear Alkylbenzene Sulfonate (LAS) is the most widely used surfactant in the world, primarily for laundry detergents and cleaning products [11, 12], whereas sodium polystyrene sulfonate (SPS) is frequently employed as ion exchange resin [13, 14]. Sulfonate moieties are often introduced to increase the water-solubility of organic substrates. For instance in organometallic chemistry, sulfonated triphenylphosphane (TPPTS) was recently used as ligand for the preparation of water-soluble transition metal complexes [15]. Previously, much attention has been focused on metal alkylsulfinates, that were obtained by the insertion of SO_2 into metal alkyls [16], whereas more recently, metal arylsulfonates are under

J. Beckmann (✉) • P. Finke
Fachbereich 2: Biologie/Chemie, Institut für Anorganische und Physikalische Chemie, Universität Bremen, Leobener Straße, D-28359 Bremen, Germany
e-mail: j.beckmann@uni-bremen.de

J.D. Woollins and R.S. Laitinen (eds.), *Selenium and Tellurium Chemistry*,
DOI 10.1007/978-3-642-20699-3_7, © Springer-Verlag Berlin Heidelberg 2011

investigation for the preparation of mesoporous metal organic frameworks (MOFs) [17, 18]. Although the chemistry of the related selenium congeners is developed to a much smaller extent and for applications less significant, there is a large resemblance in the structure and reactivity to sulfur, perhaps with the exception, that selenium(IV) and (VI) species are potentially more oxidizing. However, compared to sulfur and selenium, the chemistry of telluroxanes including diorganotellurium oxides, diorganotellurones, organotellurinic acids and organotelluronic acids is still in its infancy. This holds particularly true for structural information of these compound classes. Until the year 2000, the only X-ray structure report was that of diphenyltellurium oxide Ph_2TeO (**1**) [19]. However, using the kinetic stabilization of bulky organic substituents or the electronic stabilization of intramolecularly coordinating N-donor ligands, in the last 10 years a number of well-defined telluroxanes were obtained, the structure and reactivity of which are described in this review.

7.2 Discussion

7.2.1 Diorganotellurium Oxides

The synthesis of the first diaryltellurium oxide, namely Ph_2TeO (**1**) dates back to the work of Krafft and Lyons in 1894 [20], but it was nearly 20 years later when Lederer undertook the first systematic studies on the synthesis and reactivity of this compound class [21–23]. There are two principle routes for the preparation of diaryltellurium oxides R_2TeO (**1**, R = Ph; **2**, R = 4-MeC_6H_4; **3**, R = 4-$MeOC_6H_4$; **4**, R = 4-$Me_2NC_6H_4$; **5**, R = Mes; **6**, R = 2,4,6-i-$Pr_3C_6H_2$; **7**, R = C_6F_5): the complete base hydrolysis of appropriate diaryltellurium dihalides, R_2TeX_2 (X = Cl, Br) or the oxidation of diaryltellurides, R_2Te (Scheme 7.1).

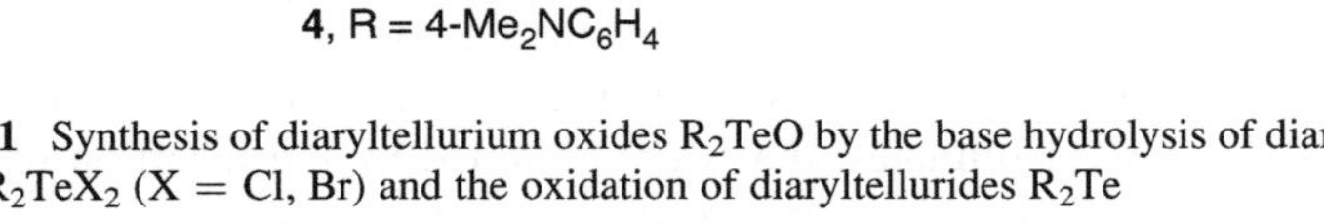

1, R = Ph
2, R = 4-MeC_6H_4
3, R = 4-$MeOC_6H_4$
4, R = 4-$Me_2NC_6H_4$
5, R = Mes
6, R = 2,4,6-i-$Pr_3C_6H_2$
7, R = C_6F_5

Scheme 7.1 Synthesis of diaryltellurium oxides R_2TeO by the base hydrolysis of diaryltellurium dihalides R_2TeX_2 (X = Cl, Br) and the oxidation of diaryltellurides R_2Te

The preparation of diaryltellurium oxides, R_2TeO, by the hydrolysis route usually requires an excess of base, e.g. NaOH or NH_3 in aqueous ethanol, from which the products precipitate as crystalline solids. The incomplete hydrolysis of diaryltellurium dihalides R_2TeX_2 using water gives rise to either diaryltellurium hydroxy halides $R_2Te(OH)X$ or their formal condensation products, tetraarylditelluroxane dihalides $XR_2TeOTeR_2X$, which could not be distinguished at that time (R = Ph, 4-MeC_6H_4; X = Cl, Br, I) [21]. A better way of preparing the tetraarylditelluroxane dihalides $XR_2TeOTeR_2X$ (**8**, R = Ph; **9**, R = 4-$MeOC_6H_4$; **10**, R = 4-$Me_2NC_6H_4$; X = Cl; **11**, R = 4-$MeOC_6H_4$, X = I), involves the redistribution reaction of diaryltellurium oxides, R_2TeO, and diaryltellurium dihalides, R_2TeX_2 [24, 25]. The oxidation of diaryltellurides, R_2Te, to give diaryltellurium oxides, R_2TeO, was achieved using *N*-chlorosuccinimide (NCS), *tert*-butyl hypochlorite, *t*-BuOCl [26], or stoichiometric amounts of sodiumperiodate, $NaIO_4$ [27]. Larger amounts of $NaIO_4$ sometimes cause the oxidation to the corresponding diaryltellurones (see below, Scheme 7.13). Very recently, it was also demonstrated that diaryltellurides can be smoothly converted into diaryltellurium oxides, R_2TeO, by the aerobic oxygenation in the presence of the photosensitizer Rose Bengal [28, 29]. It is interesting to note that the corresponding dialkyltellurium oxides are yet unknown. The reaction of Me_2TeI_2 and *cy*-$(CH_2)_4TeI_2$ with a suspension of Ag_2O in water gives rise to basic solutions of the dialkyltellurium dihydroxides $Me_2Te(OH)_2$ (**12**)[30] and *cy*-$(CH_2)_4Te(OH)_2$ (**13**) [31], respectively. The dehydration of $Me_2Te(OH)_2$ proceeds with methyl group migration and apparent formation of the hygroscopic salt $[Me_3Te]^+[MeTeO_2]^-$ (**14**), which was converted into crystalline $[Me_3Te]^+[MeTeI_4]^-$ upon reaction with HI [30, 32, 33]. The incomplete hydrolysis of Me_2TeI_2 and *cy*-$(CH_2)_4TeI_2$ produced the tetramethylditelluroxane diiodide $IMe_2TeOTeMe_2I$ (**15**) and the hexaalkyltritelluroxane diiodide I*cy*-$(CH_2)_4$TeO*cy*-$(CH_2)_4$TeOTe*cy*-$(CH_2)_4$I (**16**), respectively [25, 34]. Compound **15** is also accessible by the equimolar redistribution reaction of $Me_2Te(OH)_2$ (**12**) and Me_2TeI_2, while the same reaction in a ratio of 3:1 produced the alkyltelluroxane $(Me_2Te)_2O(I)OH{\cdot}H_2O$ (**17**) having a complex structure [25]. It is worth mentioning that early attempts to prepare dialkyltellurium oxides by oxidation of dialkyltellurides using air and hydrogen peroxide occurred with alkyl group cleavage and provided ill-defined alkyltellurinic acids or peroxides [35, 36].

While diaryltellurium oxides, R_2TeO, are known for a long time, little attention was directed to their structural characterization. The first X-ray structure analysis of a diaryltellurium oxide, namely Ph_2TeO (**1**), was reported by Alcock and Harrison 1982 [19]. Ph_2TeO (**1**) comprises an unsymmetric four-membered Te_2O_2 ring structure with two crystallographically independent monomers, having a (formal) elongated Te–O double bond (av. 1.89(1) Å).[1] These monomers are associated by two short secondary Te···O interactions (av. 2.55(2) Å), a term coined by Alcock for bond lengths that are longer than the sum of covalent radii (2.04 Å), but shorter or reasonably close to the sum of van-der-Waals radii (3.58 Å) [37–39]. The spatial arrangement of the tellurium atoms of **1** is completed by even longer secondary Te···O interactions (av. 3.77(2) Å) that slightly exceed the sum of van-der-Waals radii giving rise to the formation of an one-dimensional coordination polymer (Fig. 7.1a). This structural motif is in contrast to the related sulfur and selenium oxides Ph_2SO and Ph_2SeO, which lack any secondary E···O interactions (E = S, Se) [40, 41].

---

[1] To account for the polar bond character, the (formal) Te–O double bond might in fact be better described as bipolar single bond Te^+–O^-, in which Coulomb forces might play an important role.

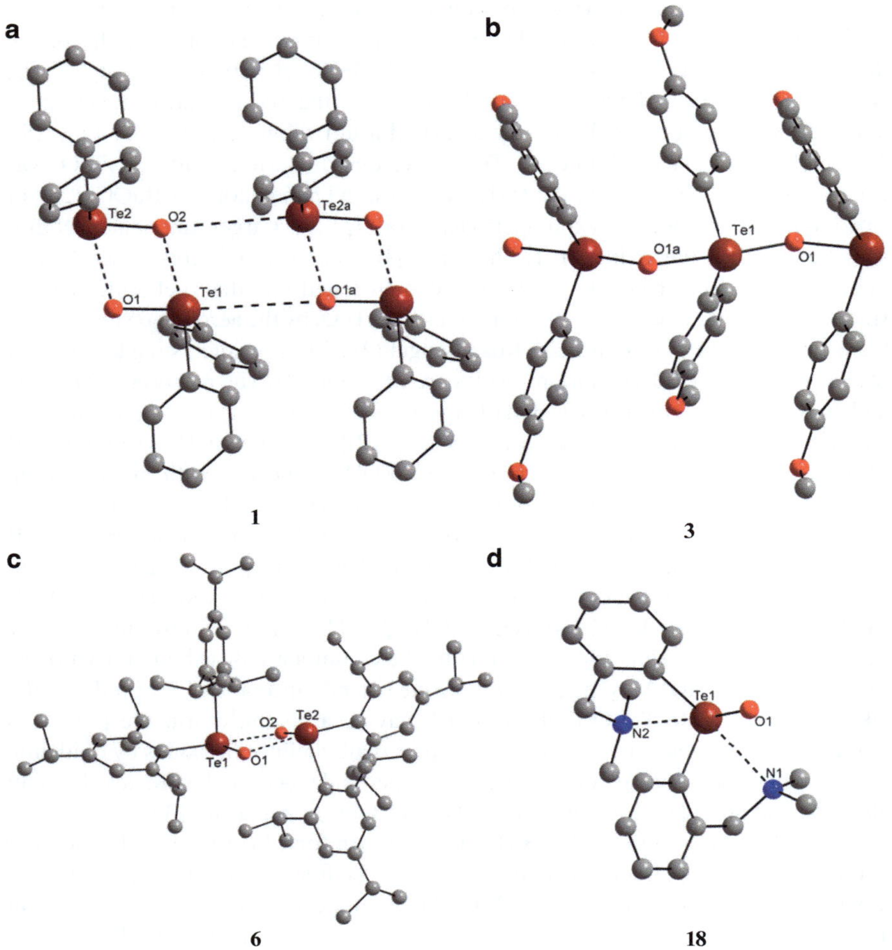

Fig. 7.1 (**a**) Molecular structure of Ph_2TeO (**1**). Selected bond lengths (Å): Te1–O1 1.87(2), Te2–O2 1.909(8), Te1···O2 2.56(2), Te2···O1 2.55(2), Te1···O1a 3.76(2), Te2a···O2 3.780(8); (**b**) Crystal structure of polymeric $(4\text{-}MeOC_6H_4)_2TeO$ (**3**). Selected bond lengths (Å): Te1–O1 2.100 (2), Te1–O1a 2.025(2); (**c**) Molecular structure of $(2,4,6\text{-}i\text{-}Pr_3C_6H_2)_2TeO$ (**6**). Selected bond lengths (Å): Te1–O1 1.853(5), Te2–O2 1.861(5), Te1···O2 2.536(5), Te2···O1 2.518(4); (**d**) Molecular structure of $(2\text{–}Me_2NCH_2C_6H_4)_2TeO$ (**18**). Selected bond lengths (Å): Te1–O1 1.829 (1), Te1···N1 2.755(6), Te1···N2 2.566(5)

The X-ray structures of $(2,4,6\text{-}i\text{-}Pr_3C_6H_2)_2TeO$ (**6**)[29] and $(C_6F_5)_2TeO$ (**7**) [42, 43] also contain unsymmetric four-membered Te_2O_2 ring structures with (formal) elongated Te–O double bonds and short secondary Te···O interactions. However, these compounds lack any longer secondary Te···O contacts (Fig. 7.1c). An intermediate structure is found in Mes_2TeO (**5**) [44], which consists of unsymmetric four-membered Te_2O_2 rings. Two of these rings are associated by longer secondary

Te···O interactions, but further aggregation is restricted by the shielding of the bulky mesityl groups [29].

A completely different structural motif is found in $(4\text{-MeOC}_6H_4)_2TeO$ (**3**) [45]. It comprises a one-dimensional polymeric Te–O structure with alternating Te–O single bonds (av. 2.063(2) Å), but no secondary Te···O interactions were observed (Fig. 7.1b). Based on a comparison of ^{125}Te MAS NMR parameters of **3** it was proposed that $(4\text{-Me}_2NC_6H_4)_2TeO$ (**4**) adopts a similar polymeric structure [24]. It is to note that the complex solid-state structure of $(Me_2Te)_2O(I)OH{\cdot}H_2O$ (**17**) also contains a polymeric $(Me_2TeO)_n$ chain, reminiscent to that of **3** [25].

In solution, diaryltellurium oxides appear to be mostly monomeric. Osmometric molecular weight determinations of Ph_2TeO (**1**) and $(4\text{-MeOC}_6H_4)_2TeO$ (**3**) revealed a degree of aggregation of 1.2 and 1.3, respectively [45].

The first diaryltellurium oxide that is also monomeric in the solid-state, namely $(2\text{-Me}_2NCH_2C_6H_4)_2TeO$ (**18**), was accidently obtained by the slow hydrolysis of the intramolecularly coordinated diaryltellurium diazide $(2\text{-Me}_2NCH_2C_6H_4)_2Te(N_3)_2$ in a loosely closed vessel (Scheme 7.2) [46].

18

Scheme 7.2 Accidental formation of the diaryltellurium oxide $(2\text{-Me}_2NCH_2C_6H_4)_2TeO$ (**18**)

It comprises a (formal) Te–O double bond (1.829(1) Å), that is substantially shorter than those of Ph_2TeO (**1**, 1.89(1) Å), $(C_6F_5)_2TeO$ (**7**, 1.87(1) Å), $(2,4,6\text{-}i\text{-Pr}_3C_6H_2)_2TeO$ (**6**, 1.857(5) Å) and Mes_2TeO (**5**, 1.851(4) Å). The lack of secondary Te···O interactions in $(2\text{-Me}_2NCH_2C_6H_4)_2TeO$ (**18**) is most likely due to the intramolecularly coordinating N-donor ligands that give rise to two intramolecular Te···N interactions (2.755(6) and 2.566(5) Å) (Fig. 7.1d).

Unlike the relatively inert diorganosulfoxides, R_2SO, and diorganoselenium oxides, R_2SeO, diaryltellurium oxides, R_2TeO, undergo a number of reactions involving the (formal) Te–O double bond. The reaction of diaryltellurium oxides, R_2TeO [24, 47–49], or dialkyltellurium dihydroxides, $R_2Te(OH)_2$ [25, 31], with proton acids HX, produces the four principle products, namely, R_2TeX_2, $R_2Te(OH)X$, $XR_2TeOTeR_2X$ and $XR_2TeOR_2TeOTeR_2X$. This is exemplified in the reaction of R_2TeO (**1**, R = Ph; **3**, R = 4-MeOC$_6$H$_4$; **4**, R = 4-Me$_2$NC$_6$H$_4$) with phenol and 2-nitrophenol, which provide $R_2Te(OPh)_2$ (**19**, R = Ph; **20**, R = 4-MeOC$_6$H$_4$; **21**, R = 4-Me$_2$NC$_6$H$_4$), $R_2Te(OH)(OPh)$ (**22**, R = Ph; **23**, R = 4-MeOC$_6$H$_4$; **24**, R = 4-Me$_2$NC$_6$H$_4$), $(R'O)R_2TeOTeR_2(OR')$ (**25**, R = 4-MeOC$_6$H$_4$; **26**, R = 4-Me$_2$NC$_6$H$_4$; R′ = 2-O$_2$NC$_6$H$_4$) and $(R'O)Ph_2TeOPh_2TeOTePh(OR')$ (**27**, R′ = 2-O$_2$NC$_6$H$_4$) depending on the reaction conditions applied (Scheme 7.3) [24].

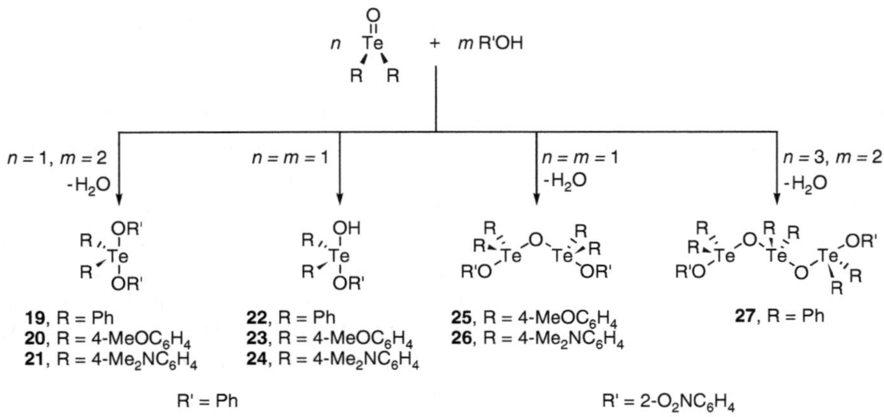

Scheme 7.3 Reactions of diorganotellurium oxides R_2TeO with phenols R′OH

Protonation of $(4\text{-}MeOC_6H_4)_2TeO$ (**3**) with triflic acid, TfOH, in acetonitrile gave rise to the formation of the diarylhydroxytelluronium cation $[(4\text{-}MeOC_6H_4)_2TeOH]^+$ (**28**, counterion TfO^-) in solution, which, however, underwent reversible condensation to afford the tetraarylditelluroxane $TfOR_2TeOTeR_2OTf$ (**29**, R = 4-$MeOC_6H_4$) in the solid state (Scheme 7.4) [48].

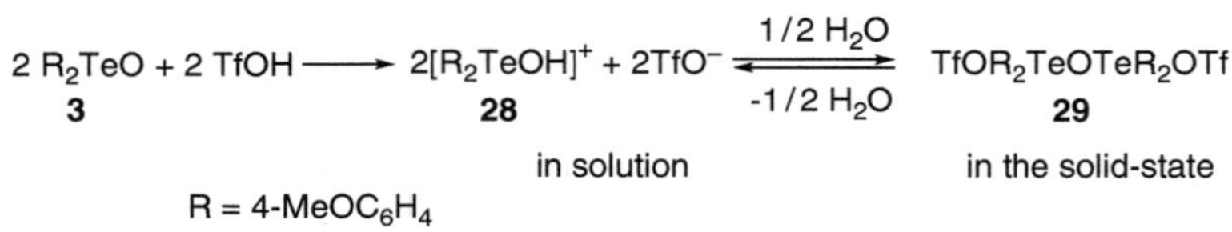

Scheme 7.4 Synthesis of the tetraorganoditelluroxane ditriflate $TfOR_2TeOTeR_2OTf$ (**29**, R = 4-$MeOC_6H_4$)

Notably, the closely related diaryltelluroxane ditriflate $TfOR_2TeOTeR_2OTf$ (**30**, R = 4-MeC_6H_4) was obtained by the air oxidation of $(4\text{-}MeC_6H_4)_2Te$ in the presence of triflic anhydride [50], whereas the reaction of $(4\text{-}MeC_6H_4)_2TeO$ (**2**) with triflic anhydride at low temperatures afforded the diaryltriflatotelluronium cation $[(4\text{-}MeC_6H_4)_2TeOTf]^+$ (**31**, counterion TfO^-) *in situ.* [51] The structure of $TfOR_2TeOTeR_2OTf$ (**29**, R = 4-$MeOC_6H_4$) features two $[R_2TeOTeR_2]^{2+}$ cations with short Te–O single bonds (av. 1.968(6) Å), that are associated by four triflate ions *via* secondary Te···O interactions (av. 2.619(6) Å) (Fig. 7.2b). The latter bond length is significantly longer than those of the related tetraarylditelluroxanes $(NO_3)Ph_2TeOTePh_2(NO_3)\cdot Ph_2Te(OH)NO_3$ (**32**, 2.452(3) Å) [47], $(Ph_2PO_2)R_2TeOTeR_2(O_2PPh_2)\cdot 2Ph_2PO_2H$ (**33**, 2.440(2) Å; R = 4-$MeOC_6H_4$) [48] $(F_3CCO_2)Ph_2TeOTePh_2(O_2CCF_3)\cdot 0.5H_2O$ (**34**, 2.35(1) Å) [52], and (R′O)$R_2TeOTeR_2$(OR′) (**25**, 2.240(2) Å; R = 4-$MeOC_6H_4$; R′ = 2-$O_2NC_6H_4$) [24], respectively. This structural motif of **29** is strong reminiscent to that of the diaryltelluroxane diiodide $IR_2TeOTeR_2I$ (**11**, R = 4-$MeOC_6H_4$) (Fig. 7.2a) [25].

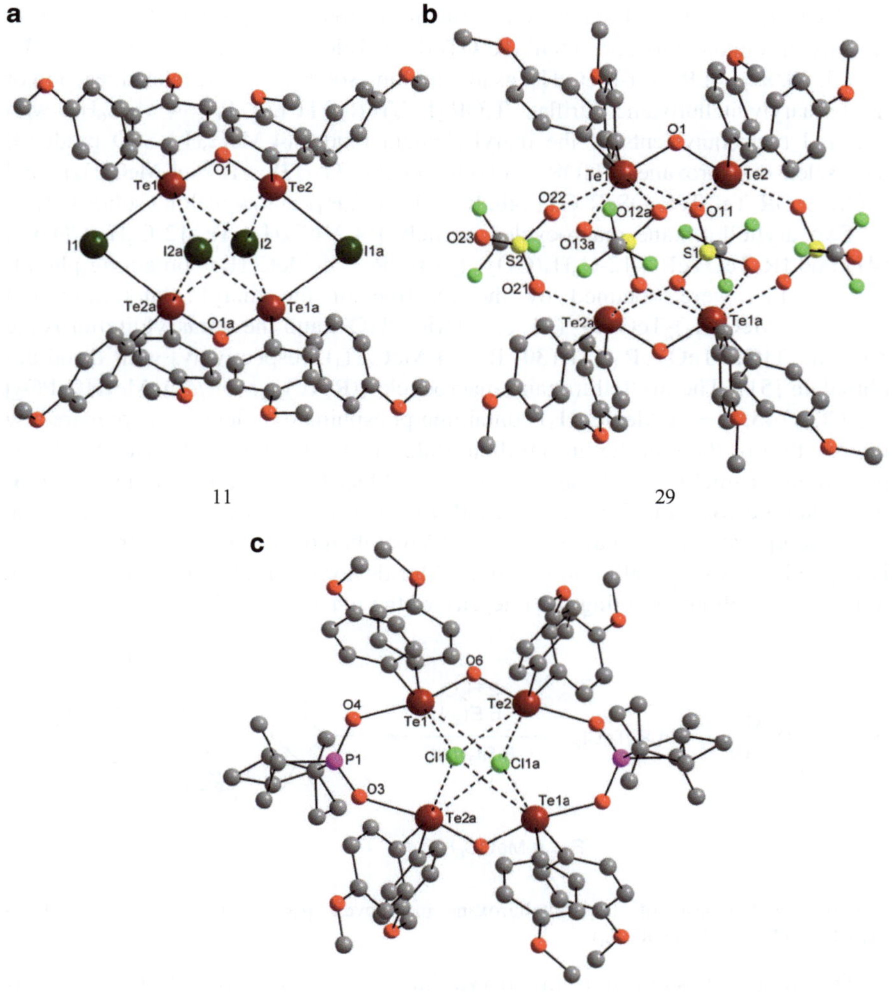

Fig. 7.2 (**a**) Molecular structure of $IR_2TeOTeR_2I$ (**11**, R = 4-MeOC$_6$H$_4$). Selected bond lengths (Å): Te1–O1 1.989(6), Te1–I1 3.294(2), Te1···I2 3.499(2), Te1···I2a 3.696(1), Te2–O1 1.982(6), Te2–I1a 3.262(2), Te2···I2 3.740(2), Te2···I2a 3.513(2); (**b**) Molecular structure of $TfOR_2TeOTeR_2OTf$ (**29**, R = 4-MeOC$_6$H$_4$). Selected bond lengths (Å): Te1–O1 1.974(6), Te2–O1 1.972(6), Te1···O22 2.621(6), Te2a···O21 2.572(7), Te1···O13a 3.28(1), Te1···O12a 3.323(8), Te1···O11 2.880(6), Te2···O12a 2.94(1), Te2···O11 2.971(7); (**c**) Molecular structure of $[(R_2Te)_2(\mu\text{-}O)(\mu\text{-}cy\text{-}Me_5HC_3PO_2)(\mu_4\text{-}Cl)]_2$ (**43**, R = 4-MeOC$_6$H$_4$). Selected bond lengths (Å): Te1–O6 1.988(3), Te2-O6 1.980(3), Te1-O4 2.313(3), Te2a–O3 2.371(3), Te1···Cl1 3.330(1), Te1···Cl1a 3.586(2), Te2···Cl1 3.349 (1), Te2···Cl1a 3.199(1)

The tetraarylditelluroxane ditriflate $TfOR_2TeOTeR_2OTf$ (**30**, R = 4-MeC$_6$H$_4$) is the precursor for the preparation of even longer oligotelluroxanes, which can be regarded as short segments of the polymeric solid-state structure of the diaryltellurium oxide

$(4\text{-MeOC}_6H_4)_2TeO$ (**3**). The reaction of **30** with various amounts of $(4\text{-MeC}_6H_4)_2TeO$ (**2**) gave the oligotelluroxanes $TfOR_2Te(OTeR_2)_nOTeR_2OTf$ (**35**, $n = 1$; **36**, $n = 2$, **37**, $n = 3$; **38**, $n = 4$; R = 4-MeC_6H_4) as low-melting solids [53]. The related reaction of the tetraaryLyditelluroxane ditriflate $TfOR_2TeOTeR_2OTf$ (**30**, R = 4-MeC_6H_4) with one and two equivalents of the diarylselenium oxide, $(4\text{-MeC}_6H_4)_2SeO$ produced the selenaditelluroxanes $TfOR_2TeOTeR_2OSeR_2OTf$ (**39**, R = 4-MeC_6H_4) and $TfOR_2SeOR_2TeOTeR_2OSeR_2OTf$ (**40**, R = 4-MeC_6H_4) as low melting solids [54].

Two aryltelluroxane macrocycles, namely $[(4\text{-MeC}_6H_4)_2Te]_3[2\text{-C}_6H_4(CO_2)_2]_3$ (**41**) and $[R_2TeOTeR_2]_2[2\text{-C}_6H_4(CO_2)_2]_2$ (**42**, R = 4-MeC_6H_4) containing phthalate moieties were obtained by the reaction of the diaryltriflatotelluronium cation $[(4\text{-MeC}_6H_4)_2TeOTf]^+$ (**31**, counterion TfO^-) and the tetraarylditelluroxane ditriflate $TfOR_2TeOTeR_2OTf$ (**30**, R = 4-MeC_6H_4), respectively, with disodium phthalate [51]. The aryltelluroxane macrocycle $[(R_2Te)_2(\mu\text{-O})(\mu\text{-}cy\text{-Me}_5HC_3PO_2)(\mu_4\text{-Cl})]_2$ (**43**, R = 4-MeOC_6H_4) containing phosphinate moieties was prepared by the reaction of the diaryltellurium dichloride $(4\text{-MeOC}_6H_4)_2TeCl_2$ with 1,1,2,3,3-pentamethyltrimethylenephosphinic acid, $cy\text{-Me}_5HC_3PO_2H$, in the presence of triethylamine (Scheme 7.5). Compound **43** comprises two tetraarylditelluroxane units incorporated into an almost planar 12-membered $Te_4P_2O_6$ ring structure that is capped by two loosely coordinating chloride ions, which can be replaced by iodide ions without effecting the integrity of the macrocycle (Fig. 7.2c) [55].

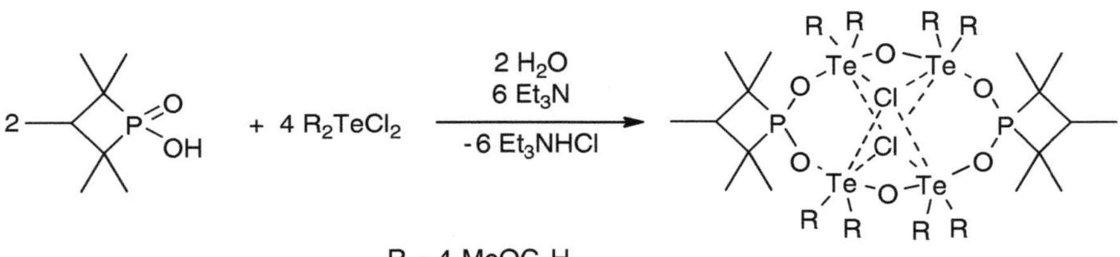

43

Scheme 7.5 Synthesis of the aryltelluroxane macrocycle $[(R_2Te)_2(\mu\text{-O})(\mu\text{-}cy\text{-Me}_5HC_3PO_2)(\mu_4\text{-Cl})]_2$ (**43**, R = 4-MeOC_6H_4)

The first persistent diarylhydroxytelluronium cation, $[(2\text{-Me}_2NCH_2C_6H_4)_2TeOH]^+$ (**44**, counterion SiF_6^{2-}), was recently obtained by the accidental hydrolysis of $(2\text{-Me}_2NCH_2C_6H_4)_2TeF_2$ in a loosely closed glass vessel during a period of 3 months (Scheme 7.6, Fig. 7.3a) [56].

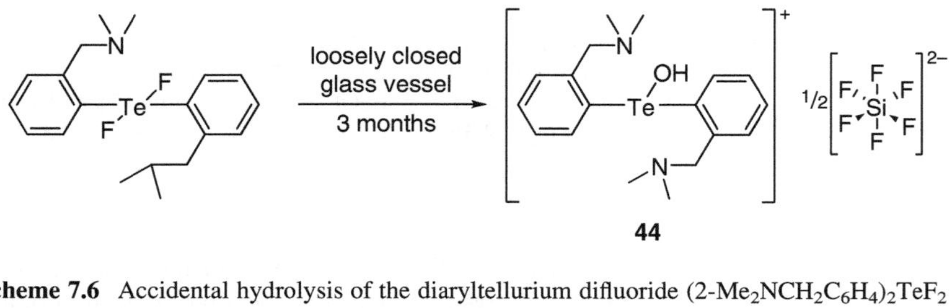

44

Scheme 7.6 Accidental hydrolysis of the diaryltellurium difluoride $(2\text{-Me}_2NCH_2C_6H_4)_2TeF_2$

The synthesis of the related persistent diarylhydroxytelluronium cation $[(8\text{-}Me_2NC_{10}H_6)_2TeOH]^+$ (**45**, counterion TfO^-) was achieved by the protonation of $(8\text{-}Me_2NC_{10}H_6)_2TeO$ (**46**) using triflic acid, TfOH (Scheme 7.7, Fig. 7.3b) [57].

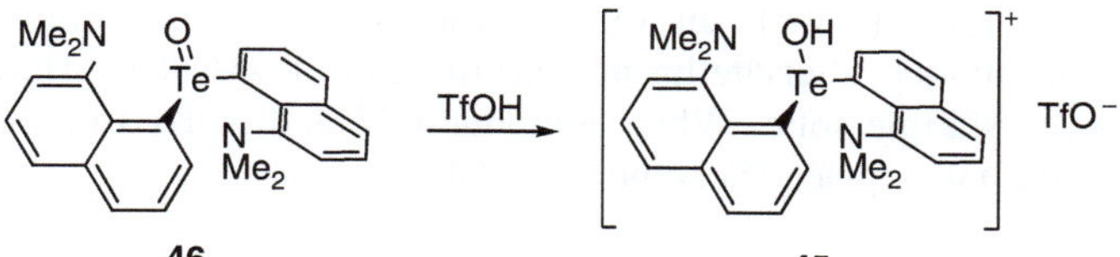

Scheme 7.7 Synthesis of the diarylhydroxytelluronium cation $[(8\text{-}Me_2NC_{10}H_6)_2TeOH]^+$ (**45**, counterion TfO^-)

The stability of the diarylhydroxytelluronium cations **44** and **45** is arguably due to the effect of the intramolecularly coordinating N-donor substituents that give rise to significant Te···N interactions (2.557(2) and 2.719(3) Å for **44**; 2.591(5) and 2.706(6) Å for **45**). The shorter of the Te···N contacts is associated to the N-donor situated in *trans*-position to the hydroxy group. The Te–O bond lengths (1.961(2) Å for **44**; 1.957(4) Å for **45**) are in the typical range for tellurium oxygen single bonds (Fig. 7.3).

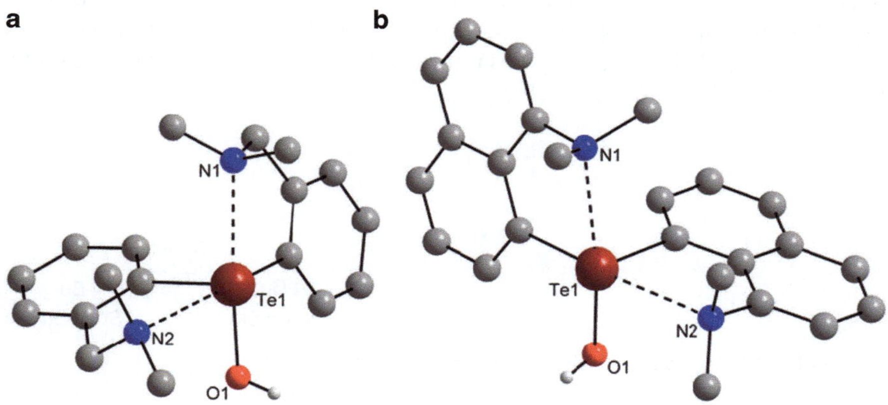

44 **45**

Fig. 7.3 (**a**) Molecular structure of $[(2\text{-}Me_2NCH_2C_6H_4)_2TeOH]^+$ (**44**, counterion SiF_6^{2-}). Selected bond lengths (Å): Te1–O1 1.961 (2), Te1···N1 2.557(2), Te1···N2 2.719(3); (**b**) Molecular structure of $[(8\text{-}Me_2NC_{10}H_6)_2TeOH]^+$ (**45**, counterion OTf^-). Selected bond lengths (Å): Te1–O1 1.957(4), Te1···N1 2.591(5), Te1···N2 2.706(6)

Diaryltellurium oxides, R_2TeO (**1**, R = Ph; **3**, R = 4-$MeOC_6H_4$), have attracted some interest as mild and selective transfer reagents for 'oxygen atoms' for a variety of substrates including thiols, phosphines [58], silanes [59], and metal carbonyls [60–63]. While in these reactions the oxygen transfer is irreversible, the intramolecularly coordinated diaryltellurium oxide (8-$Me_2NC_{10}H_6)_2TeO$ (**46**) surprisingly reacts reversibly with acetonitrile, MeCN. This reaction proceeds with an oxygen transfer from tellurium to carbon providing the unique diaryltellurium acetimidate (8-$Me_2NC_{10}H_6)_2TeNC(O)CH_3$ (**47**) as crystalline solid. When compound **47** is dissolved in solution, the reverse reaction takes place (Scheme 7.8) [64].

Scheme 7.8 Reversible formation of the diaryltellurium acetimidate (8-$Me_2NC_{10}H_6)_2Te$ NC(O) CH_3 (**47**)

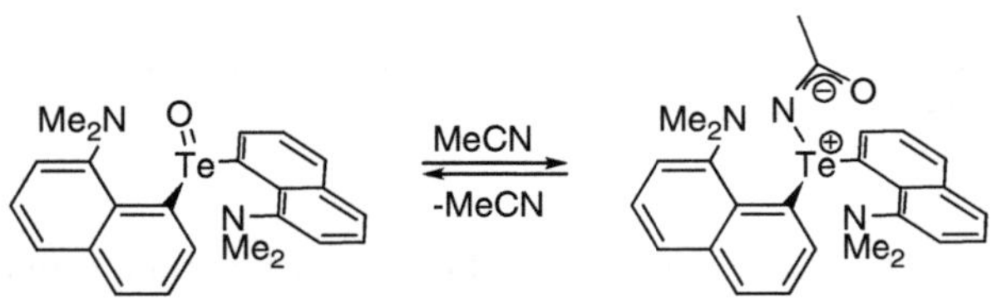

46 **47**

Metallatelluroxanes, compounds containing Te–O–M linkages (whereby M is a main group element), are assessable by the condensation reaction of diaryltellurium oxides, R_2TeO, with metal or element oxides and hydroxides. Reaction of the diaryltellurium oxides R_2TeO (**3**, R = 4-$MeOC_6H_4$; **4**, R = 4-$Me_2NC_6H_4$) with the organosilanols *t*-$Bu_2Si(OH)_2$, *t*-$BuSi(OH)_3$, [*t*-$BuSi(OH)_2]_2O$ and $[Ph_2Si(OH)]_2O$ afforded the macrocycles [(4-$MeOC_6H_4)_2TeOSi$*t*-$Bu_2O]_3$ (**48**) [65], $(R_2TeOSiPh_2OSiPh_2O)_2$ (**49**, R = 4-$MeOC_6H_4$; **50**, R = 4-$Me_2NC_6H_4$) and cages [(4-$Me_2NC_6H_4)_2Te]_6$(*t*-$BuSi)_4O_{12}$ (**51**), [(4-$MeOC_6H_4)_2Te]_6$(*t*-$BuSi)_6O_{15}$ (**52**) [66], respectively (Schemes 7.9 and 7.10).

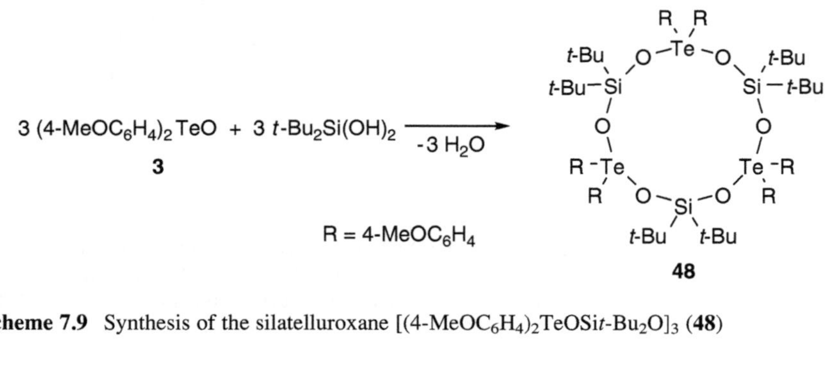

48

Scheme 7.9 Synthesis of the silatelluroxane [(4-$MeOC_6H_4)_2TeOSi$*t*-$Bu_2O]_3$ (**48**)

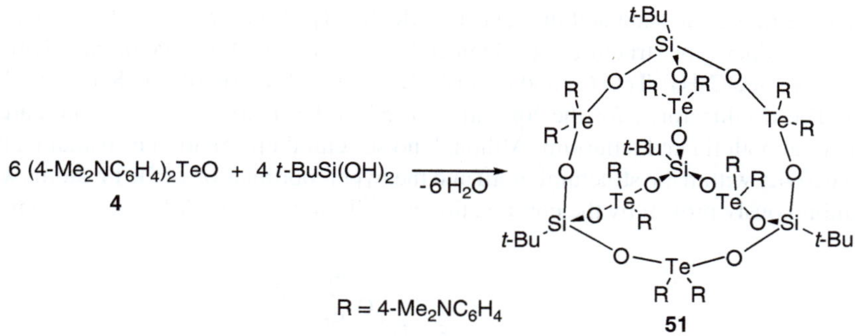

Scheme 7.10 Synthesis of the silatelluroxane cage $[(4\text{-}Me_2NC_6H_4)_2Te]_6(t\text{-}BuSi)_4O_{12}$ (**51**)

While the silatelluroxane **48** contains a slightly puckered 12-membered $Te_3O_6Si_3$ ring structure, compound **51** possesses a cage structure in which the silicon atoms are located at the angles of a tetrahedron and the six OR_2TeO units occupy the corresponding edges (Fig. 7.4). No secondary Te···O interactions were observed for **48** and **51**.

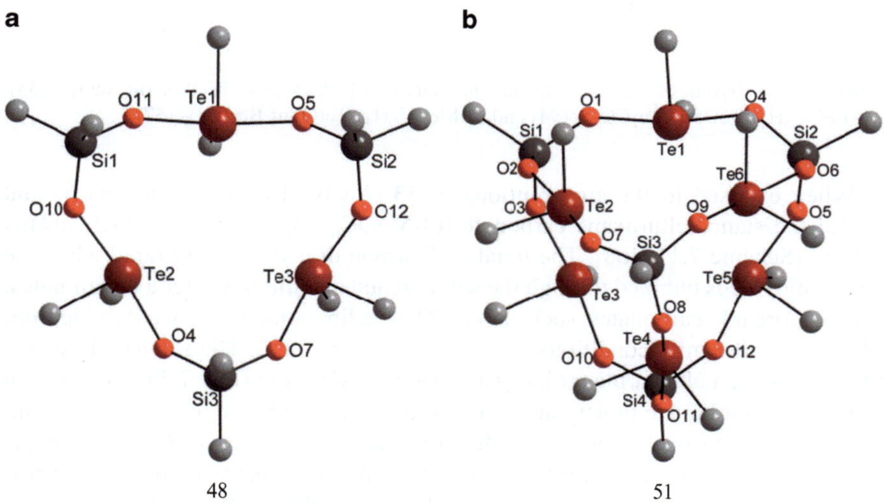

Fig. 7.4 (**a**) Molecular structure of $[(4\text{-}MeOC_6H_4)_2TeOSi\textit{t}\text{-}Bu_2O]_3$ (**48**) (for clarity only the ipso-carbons are shown). Selected bond lengths (Å): Te–O 2.051(3)–2.064(3), Si–O 1.617(4)–1.630(2); (**b**) Molecular structure of $[(4\text{-}Me_2NC_6H_4)_2Te]_6(t\text{-}BuSi)_4O_{12}$ (**51**) (for clarity only the ipso-carbons are shown). Selected bond lengths (Å): Te–O 2.027(5)–2.061(4), Si–O 1.598(6)–1.638(6)

The tetranuclear stannatelluroxane $[(4\text{-}MeOC_6H_4)_2TeOSn\textit{t}\text{-}Bu_2SnO]_2$ (**53**) containing a ladder-type structure was obtained by the reaction of the two organoelement oxides $(4\text{-}MeOC_6H_4)_2TeO$ (**3**) and $(\textit{t}\text{-}Bu_2SnO)_3$ at a Te:Sn ratio of 1:1 (Scheme 7.11) [67]. The driving force for the formation of **53** is the increase of the coordination number at both tin and tellurium. Although no structural information is available, the connectivity within the structural motif and the hypercoordination of the tin atoms was unambiguously proven by the combination of ^{125}Te and ^{119}Sn NMR spectroscopy.

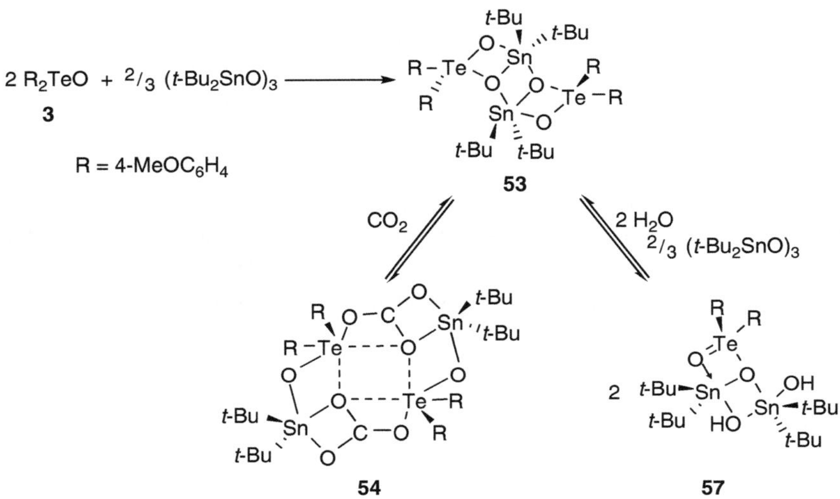

Scheme 7.11 Synthesis of the stannatelluroxanes $[(4\text{-}MeOC_6H_4)_2TeOSn\textit{t}\text{-}Bu_2SnO]_2$ (**53**), $[(4\text{-}MeOC_6H_4)_2TeOSn\textit{t}\text{-}Bu_2CO_3]_2$ (**54**) and $(4\text{-}MeOC_6H_4)_2Te(OSn\textit{t}\text{-}Bu_2OH)_2$ (**57**)

When exposed to the air, solutions of **53** slowly absorb carbon dioxide and the related stannatelluroxane carbonate $[(4\text{-}MeOC_6H_4)_2TeOSn\textit{t}\text{-}Bu_2CO_3]_2$ (**54**) is formed (Scheme 7.11) [68]. The uptake of carbon dioxide is more rapid when the carbon dioxide is bubbled through the solution and quantitative after a few minutes. Unlike inorganic carbonates, such as Na_2CO_3, the liberation of carbon dioxide from **54** occurs under mild conditions between 90°C and 145°C. The stannatelluroxane **54** comprises an almost planar inorganic $Te_2Sn_2C_2O_8$ core (Fig. 7.5a). All oxygen atoms of the carbonate moiety are involved in the coordination to tin and tellurium atoms. Owing to the different coordination modes, the three C–O bond lengths (1.259(3), 1.278(3) and 1.329(2) Å) vary substantially; a potential useful observation when perusing bond activation in carbonates. While $(\textit{t}\text{-}Bu_2SnO)_3$ reacts only very slowly with carbon dioxide in the presence of moisture to give the trinuclear stannoxane carbonate $OC(OSn\textit{t}\text{-}Bu_2)_2O\textit{t}\text{-}Bu_2Sn(OH)_2$ [69, 70], the diaryltellurium oxide $(4\text{-}MeOC_6H_4)_2TeO$ (**3**) shows no reactivity towards CO_2. However, the basic solutions of the dialkyltellurium dihydroxides $Me_2Te(OH)_2$ (**12**) and $cy\text{-}(CH_2)_4Te(OH)_2$ (**13**), readily absorb carbon dioxide affording the crystalline alkyltelluroxane carbonates $(Me_2TeOTeMe_2CO_3)_n$ (**55**) and HO*cy*-$(CH_2)_4$TeO Te*cy*-$(CH_2)_4CO_3$*cy*-$(CH_2)_4$TeOH (**56**) [71]. When the reaction of the two organo-element oxides $(4\text{-MeOC}_6H_4)_2TeO$ (**3**) and $(t\text{-Bu}_2SnO)_3$ was repeated at different Te:Sn ratios, no other stannatelluroxanes than **53** were observed by NMR spectroscopy [67]. However, when a solution of $(4\text{-MeOC}_6H_4)_2TeO$ (**3**) and $(t\text{-Bu}_2SnO)_3$ with a Te:Sn ratio of 1:2 was exposed to moist air, another trinuclear stannatelluroxane $(4\text{-MeOC}_6H_4)_2Te(OSn\textit{t}\text{-Bu}_2OH)_2$ (**57**) containing hydroxy groups crystallized from solution, which gives rise to a loosely associated dimer *via* secondary Te···O interactions (2.756(2) Å) in the solid-state (Scheme 7.11, Fig. 7.5b) [67].

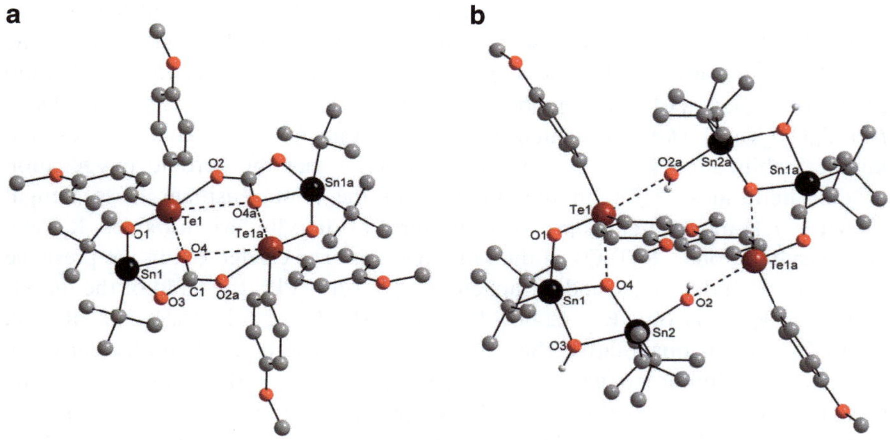

54 57

Fig. 7.5 (**a**) Molecular structure of $[(4\text{-MeOC}_6H_4)_2TeOSn\textit{t}\text{-Bu}_2CO_3]_2$ (**54**). Selected bond lengths (Å): Te1–O1 1.921(2), Te1–O2 2.481(2), Te1···O4 3.279(2), Te1···O4a 3.059(2), Sn1–O1 2.043(2), Sn1–O4 2.094(2), Sn1–O3 2.307(2), C1–O3 1.278(3), C1–O4 1.329(3), C1–O2a 1.259(3); (**b**) Molecular structure of $[(4\text{-MeOC}_6H_4)_2Te(OSn\textit{t}\text{-Bu}_2OH)_2]_2$ (**57**). Selected bond lengths (Å): Te1–O1 1.898(2), Te1···O2a 2.756(2), Te1···O4 2.595(2), Sn1–O1 2.145(2), Sn1–O3 2.167(2), Sn1–O4 2.007(2), Sn2–O2 2.106(2), Sn2–O3 2.300(2), Sn2–O4 1.996(2)

7.2.2 Diorganotellurones

In principle, diaryltellurones, R_2TeO_2, can be prepared by the simple oxidation of diaryltellurides, R_2Te, or diaryltellurium oxides, R_2TeO, using the right oxidizing agent (Scheme 7.12). Early attempts at preparing diaryltellurones using hydrogen peroxide most likely provided only hydroxyl-perhydrates [72]. However, the use of sodium periodate, $NaIO_4$ [72, 73], and sodium hypochlorite, NaOCl [74], afforded the diaryltellurones, R_2TeO_2 (**58**, R = 4-MeC_6H_4; **59**, R = 4-$MeOC_6H_4$; **60**, R = 2,4,6-*i*-$Pr_3C_6H_2$).

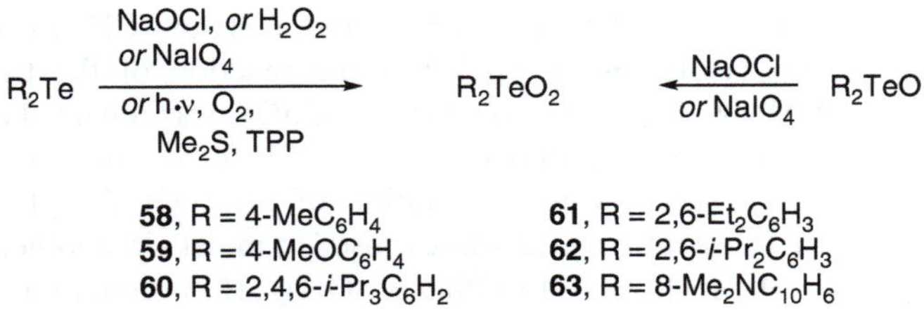

Scheme 7.12 Synthesis of diaryltellurones, R_2TeO_2, by the oxidation of diaryltellurides, R_2Te, or diaryltelluriumoxides, R_2TeO

The first diaryltellurone, namely, $(4\text{-}MeOC_6H_4)_2TeO_2$ (**59**), reported in 1982 by Engman and Cava [72], possesses a very high melting point and the ^1H, ^{13}C and ^{125}Te NMR spectra give rise to extremely broad signals, which lead to the speculation that it might be polymeric [73]. Only recently, Nishiyama and Ando *et al.* described the first arguably monomeric diaryltellurone, namely $(2,4,6\text{-}i\text{-}Pr_3C_6H_2)_2TeO_2$ (**60**), in which the bulky organic substituents provide the kinetic stabilization [73]. It is worth mentioning that the aerobic oxygenation of diaryltellurides, R_2Te, in the presence of the photosensitizer Rose Bengal (Scheme 7.1) in some cases provided mixtures of diaryltellurium oxides, R_2TeO, and diaryltellurones, R_2TeO_2. If the same reaction was carried out in the presence of dimethylsulfide, Me_2S, and tetraphenylporphyrin (TPP), the yield of the diaryltellurones R_2TeO_2 (**60**, R = $2,4,6\text{-}i\text{-}Pr_3C_6H_2$; **61**, R = $2,6\text{-}Et_2C_6H_3$; **62**, R = $2,6\text{-}i\text{-}Pr_2C_6H_3$) was quantitative (Scheme 7.12) [29]. The first intramolecularly coordinated diaryltellurone, namely $(8\text{-}Me_2NC_{10}H_6)_2TeO_2$ (**63**) was prepared by the oxidation of the diaryltelluride $(8\text{-}Me_2NC_{10}H_6)_2Te$ using an excess of aqueous hydrogen peroxide. Compound **63** crystallizes as hydrate perhydrate and dihydrate from the reaction mixture [57]. The diaryltellurones $(2,4,6\text{-}i\text{-}Pr_3C_6H_2)_2TeO_2$ (**60**) [73] and $(8\text{-}Me_2NC_{10}H_6)_2TeO_2$ (**63**) [57], shown in Fig. 7.6, possess two (formal) Te–O double bonds (av. 1.802(3) Å for **60** and av. 1.822(6) Å for **63**) that are significantly shorter than the sum of the covalent radii (2.04 Å) [38].

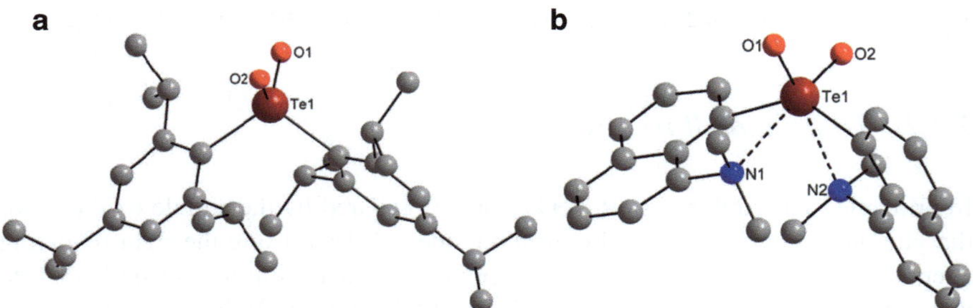

60 63

Fig. 7.6 (**a**) Molecular structure of $(2,4,6\text{-}i\text{-}Pr_3C_6H_2)_2TeO_2$ (**60**). Selected bond lengths (Å): Te1–O1 1.801(2), Te1–O2 1.802(3). (**b**) Molecular structure of $(8\text{-}Me_2NC_{10}H_6)_2TeO_2$ (**63**). Selected bond lengths (Å): Te1–O1 1.811(5), Te1–O2 1.832(6), Te1···N1 2.709(7), Te1···N2 2.762(7)

Little is known about the reactivity of diaryltellurones. Unlike $(8\text{-}Me_2NC_{10}H_6)_2TeO_2$ (**63**), which appears to have no oxidizing power, $(2,4,6\text{-}i\text{-}Pr_3C_6H_2)_2TeO_2$ (**60**) is a strong oxidizing agent towards primary and secondary alcohols that are converted to the corresponding aldehydes and ketones [73]. The symproportionation reaction of the diaryltelluride, $(8\text{-}Me_2NC_{10}H_6)_2Te$, and the diaryltellurone $(8\text{-}Me_2NC_{10}H_6)_2TeO_2$ (**63**) provided the intramolecularly coordinated diaryltellurium oxide $(8\text{-}Me_2NC_{10}H_6)_2TeO$ (**46**) (Scheme 7.13) [57]. Surprisingly, $(8\text{-}Me_2NC_{10}H_6)_2TeO$ (**46**) is sensitive towards air oxidation, which slowly converts it to $(8\text{-}Me_2NC_{10}H_6)_2TeO_2$ (**63**).

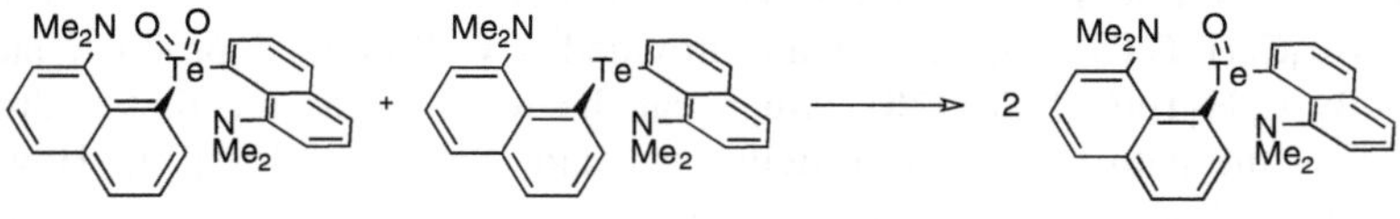

63 **46**

Scheme 7.13 Synthesis of the diaryltellurium oxide $(8\text{-}Me_2NC_{10}H_6)_2TeO$ (**46**)

7.2.3 Organotellurinic Acids

The first aryltellurinic acid, namely $[PhTe(O)OH]_n$ (**64**), was obtained by Lederer in 1915, by the neutralization of $[PhTe(O)NO_3]_n$ (**65**) which was prepared upon oxidation of the diarylditelluride, PhTeTePh, with nitric acid [75]. A more general route for the preparation of aryltellurinic acids, RTe(O)OH, involves the base hydrolysis of aryltellurium trichlorides, $RTeCl_3$ (Scheme 7.14). Like $[PhTe(O)OH]_n$ (**64**), most previously known aryltellurinic acids $[RTe(O)OH]_n$ (**66**, R = $4\text{-}MeOC_6H_4$; **67**, R = $4\text{-}EtOC_6H_4$) or their corresponding anhydrides $[RTeO_{1.5}]_n$ (**68**, R = $4\text{-}MeOC_6H_4$; **69**, R = $4\text{-}EtOC_6H_4$) [76–78], prepared by the hydrolysis route, are presumably amorphous and characterized by a high melting point and a very poor solubility in common organic solvents, and therefore most likely comprise polymers both in solution and in the solid-state. This is supported by

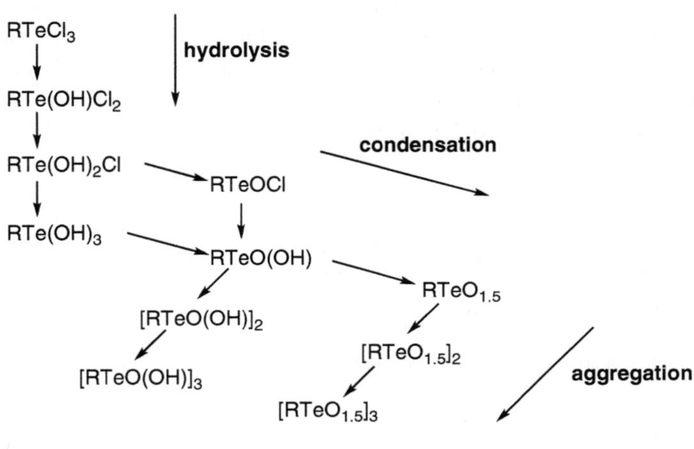

Scheme 7.14 Hydrolysis pathway of aryltellurium trichlorides

the X-ray structure determination of $[PhTe(O)NO_3]_n$ (**65**), which also contains an oxygen-bridged polymeric structure [47]. The formation of polymeric aryltellurinic acids $[RTe(O)OH]_n$ or their corresponding anhydrids, $[RTeO_{1.5}]_n$ can be attributed to concurrent condensation and aggregation processes that occur along the hydrolysis pathway (Scheme 7.14).

However, the use of bulky substituents and intramolecularly coordinating N-donor substituents allows the kinetic and electronic stabilization of molecular aryltellurinic acids as well as related oligomeric and polymeric aryltelluroxanes that possess well-defined structures. Thus, the solid-state hydrolysis of the bulky 2,4,6-*t*-$Bu_3C_6H_2TeCl_3$ with moist air provided a well-defined product, namely 2,4,6-*t*-$Bu_3C_6H_2Te(OH)Cl_2$ (**70**), which can be regarded as the first hydrolysis product on the pathway to an aryltellurinic acid (Scheme 7.15) [79]. However, the base hydrolysis of 2,4,6-*t*-$Bu_3C_6H_2TeCl_3$ in solution once again only produced an amorphous ill-defined aryltellurinic acid, [2,4,6-*t*-$Bu_3C_6H_2Te(O)OH]_n$ (**71**) [80].

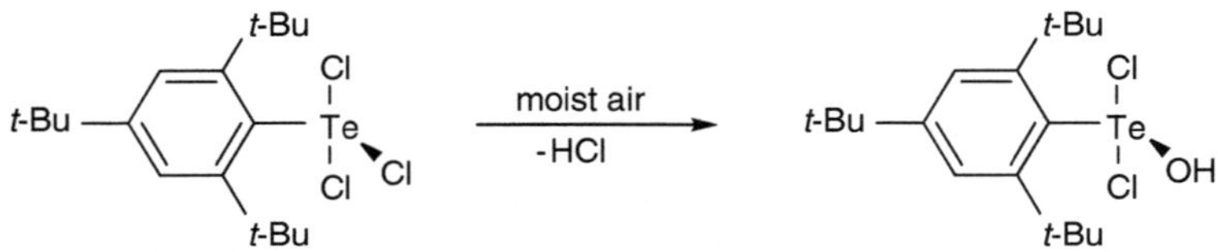

70

Scheme 7.15 Solid-state hydrolysis of 2,4,6-*t*-$Bu_3C_6H_2TeCl_3$

The aryltellurinic acid [2,4,6-*i*-$Pr_3C_6H_2Te(O)OH]_n$ (**72**) possesses a slightly less bulky organic substituent then **71**, but has a small solubility in hexane in which it appears to be monomeric. Given the stereochemically active lone pair of the tellurium (IV) atoms, tetravalent *meta*-aryltellurinic acids are chiral molecules. Partial resolution of racemic **72** by chiral chromatography to give enantiomerically enriched **72** supports the idea of the tetravalent *meta*-aryltellurinic acid 2,4,6-*i*-$Pr_3C_6H_2Te(O)OH$ in high dilution. However, the configurational stability of this enantiomerically enriched **72** is lower than those of the corresponding arylsulfinic acids and arylseleninic acids. Within minutes epimerization takes place, which is accelerated in the presence of water. This observation leads to propose that the racemization mechanism involves an achiral *ortho*-aryltellurinic acid 2,4,6-*i*-$Pr_3C_6H_2Te(OH)_3$ (**73**) (Scheme 7.16) [81, 82].

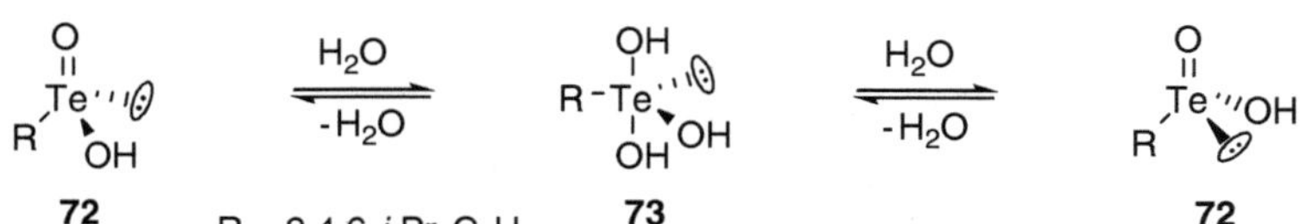

72 R = 2,4,6-*i*-$Pr_3C_6H_2$ **73** **72**

meta-aryltellurinic acid *ortho*-aryltellurinic acid *meta*-aryltellurinic acid

Scheme 7.16 Proposed mechanism for the epimerization of monomeric tellurinic acids in the presence of water

However, the controlled base hydrolysis of 2,6-$Mes_2C_6H_3TeCl_3$ proceeds in two steps and afforded the dinuclear aryltellurinic chloride [2,6-$Mes_2C_6H_3Te(O)Cl]_2$ (**74**) and the dinuclear aryltellurinic acid [2,6-$Mes_2C_6H_3Te(O)OH]_2$ (**75**) as highly crystalline materials (Scheme 7.17, Fig. 7.7) [83].

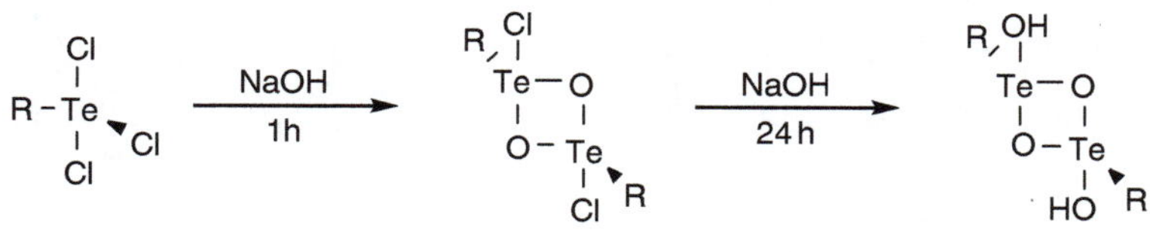

74 **75**

Scheme 7.17 Synthesis of the aryltellurinic acid [2,6-$Mes_2C_6H_3Te(O)OH]_2$ (**75**)

The molecular structure of **75** differs from those of the lighter congeners, namely the mononuclear sulfinic and seleninic acids possessing (formal) S–O and Se–O double bonds (E = S, Se), and features an asymmetric four-membered Te_2O_2 ring structure with Te–O single bonds (Fig. 7.6b). The four-membered Te_2O_2 ring is reminiscent to that in Ph_2TeO (**1**), however, the Te–O bond lengths of **75** (1.897(5) and 2.232(4) Å) are shorter than those of **1** (1.89(1) and 2.55(2) Å).

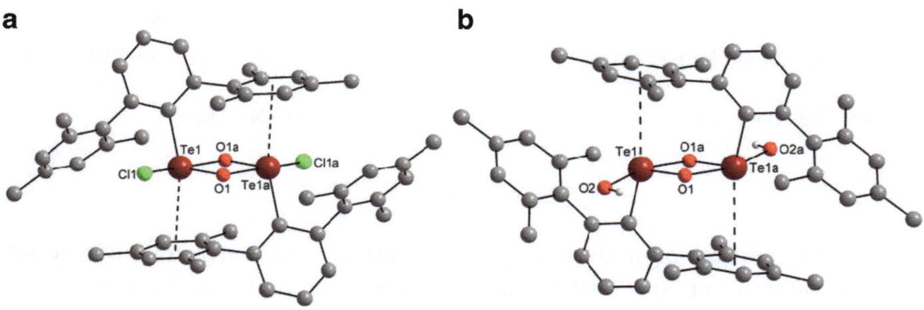

Fig. 7.7 (**a**) Molecular structure of [2,6-$Mes_2C_6H_3Te(O)Cl]_2$ (**74**). Selected bond lengths (Å): Te1–O1 1.912(4), Te1–O1a 2.098(4), Te1–Cl1 2.478(2); (**b**) Molecular structure of [2,6-$Mes_2C_6H_3Te(O)OH]_2$ (**75**). Selected bond lengths (Å): Te1–O1 1.897(5), Te1–O2 2.232(4), Te1–O1a 2.143(5)

The hydrolysis of the intramolecularly coordinated aryltellurium trichloride 8-$Me_2NC_{10}H_6TeCl_3$ using water and various amounts of base provided the aryltelluroxanes (8-$Me_2NC_{10}H_6Te)_2OCl_4$ (**76**), (8-$Me_2NC_{10}H_6Te)_6O_5Cl_8$ (**77**), (8-$Me_2NC_{10}H_6Te)_6O_8Cl_2$ (**78**), [(8-$Me_2NC_{10}H_6Te)_2O_3]_n$ (**79**), and

$(8\text{-}Me_2NC_{10}H_6Te)_6O_8(OH)_2$ (**80**) as highly crystalline materials (Scheme 7.18, Figs. 7.8 and 7.9a) [84].

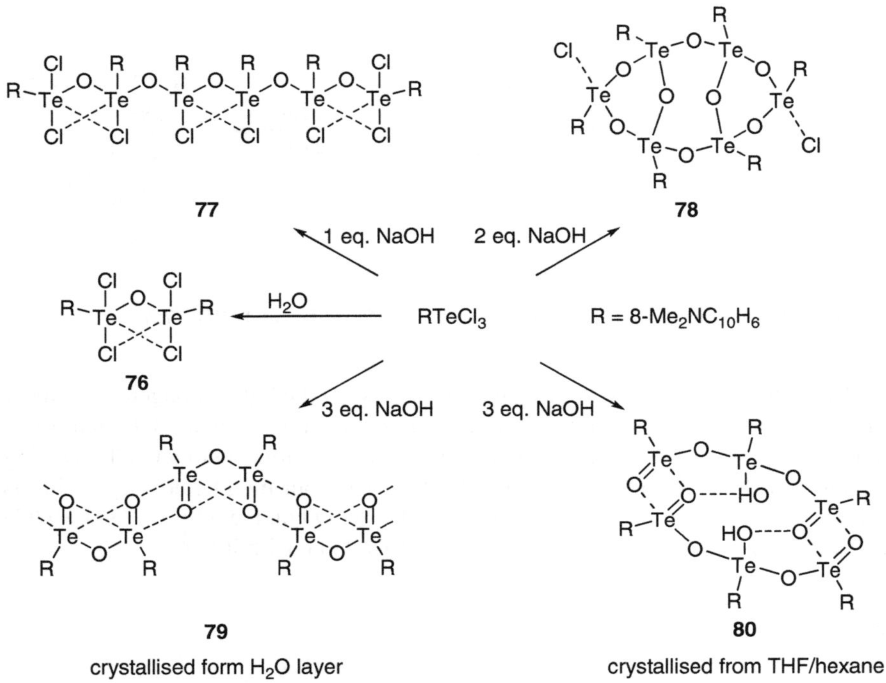

Scheme 7.18 Base hydrolysis of the aryltellurium trichloride $8\text{-}Me_2NC_{10}H_6TeCl_3$

The chlorine-free aryltelluroxanes **79** and **80** are interconvertible by crystallization from water and THF/hexane, respectively, and can be regarded as (partial) anhydrides of the elusive tellurinic acid (Fig. 7.8). In aqueous and organic solutions, **79** and **80** gave identical spectra and the degree of aggregation determined by osmometric molecular weight measurements was found to be 2.5 in water. Perhaps the dinuclear aryltelluroxane $[(8\text{-}Me_2NC_{10}H_6Te)_2O_3]_2$ (**81**) or a related hydrate exists in equilibrium with higher aryltelluroxane oligomers in aqueous solutions. The Te–O bond lengths of **79** span a wide range from (formal) elongated Te–O double bonds (1.890(5) Å), single bonds (1.976(3) Å), shorter (2.311(5) Å) and longer (3.319(5) Å) secondary Te···O interactions (Fig. 7.8a).

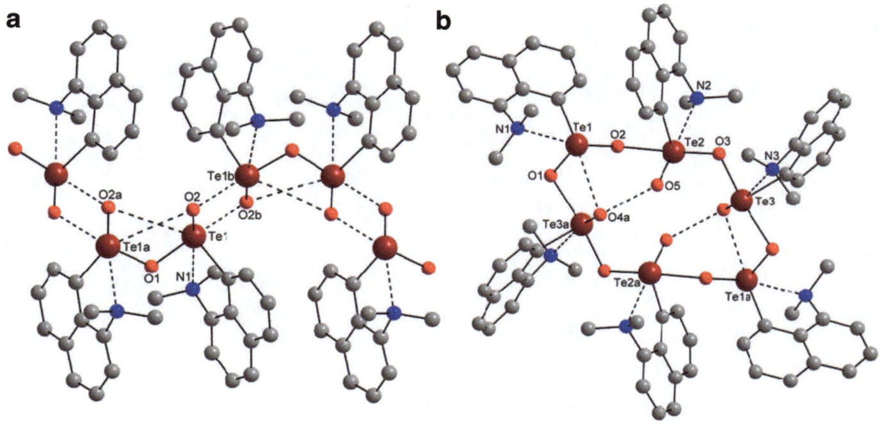

79 80

Fig. 7.8 (**a**) Crystal structure of $[(8\text{-}Me_2NC_{10}H_6Te)_2O_3]_n$ (**79**). Selected bond lengths (Å): Te1–O1 1.976(3), Te1–O2 1.890(5), Te1···O2a 3.319(5), Te1···O2b 2.311(5), Te1···N1 2.718(6); (**b**) Molecular structure of $(8\text{-}Me_2NC_{10}H_6Te)_6O_8(OH)_2$ (**80**). Selected bond lengths (Å): Te1–O1 1.875(5), Te1–O2 1.929(6), Te2–O2 2.257(5), Te2–O3 1.978(5), Te3–O3 2.010(6), Te3a–O1 2.273(6), Te1–O4a 2.896(6), Te2–O5 1.922(6), Te3a–O4a 1.877(5), Te1···N1 2.598(8), Te2···N2 2.639(7), Te3···N3 2.703(9), O4a···O5 2.532(7)

The accidental base co-hydrolysis of $8\text{-}Me_2NC_{10}H_6TeCl_3$ with $TeCl_4$ gave rise to the formation of the octanuclear aryltelluroxane cluster $(8\text{-}Me_2NC_{10}H_6Te)_6Te_2O_{12}Cl_2$ (**82**) containing six aryltellurium(IV) sites and two inorganic tellurium (IV) atoms (Fig. 7.9c). A more rational synthesis of **82** involved the condensation of $(8\text{-}Me_2NC_{10}H_6Te)_6O_8(OH)_2$ (**80**) with $(i\text{-}PrO)_3TeCl$ in the presence of water (Scheme 7.19) [84]. The heptanuclear aryltelluroxane cluster $[2\text{-}(PhNN)C_6H_4Te]_6TeO_{11}$ (**83**) containing six aryltellurium(IV) sites and one inorganic tellurium(IV) atom was probably obtained in a similar way by the accidental base co-hydrolysis of the intramolecularly coordinated aryltellurium trichloride $2\text{-}(PhNN)C_6H_4TeCl_3$ with $TeCl_4$ (Fig. 7.9b) [85].

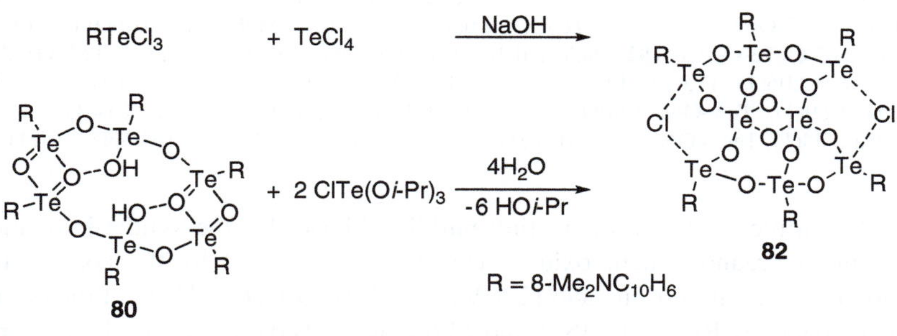

Scheme 7.19 Synthesis of the aryltelluroxane $(8\text{-}Me_2NC_{10}H_6Te)_6Te_2O_{12}Cl_2$ (**81**)

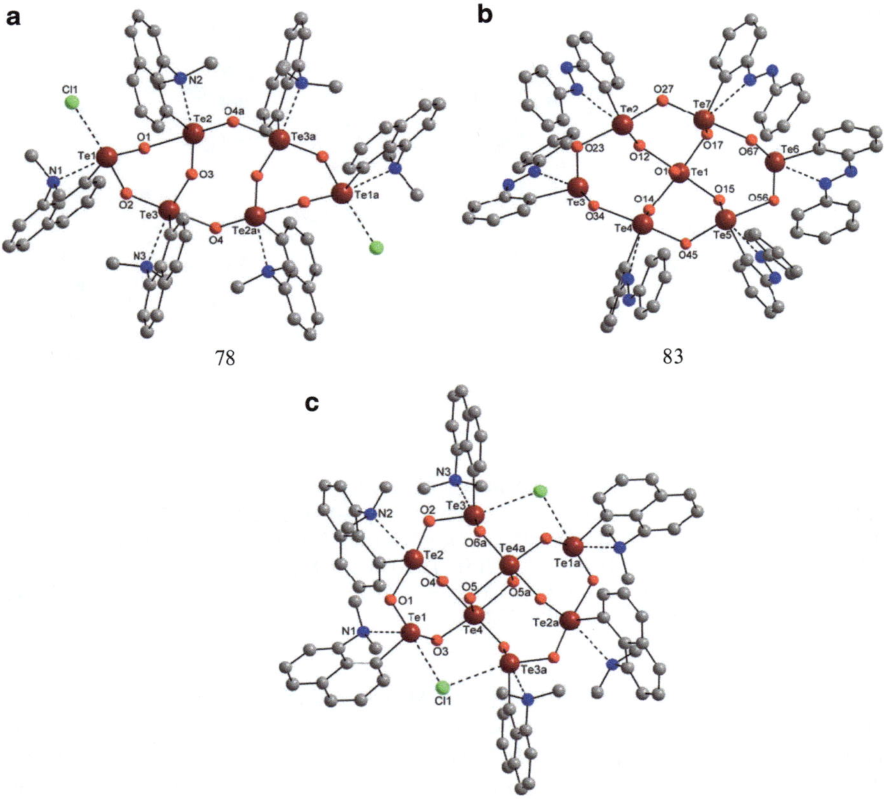

78 83

82

Fig. 7.9 (**a**) Molecular structure of $(8\text{-}Me_2NC_{10}H_6Te)_6O_8Cl_2$ (**78**). Selected bond lengths (Å): Te1–O1 1.902(3), Te1–O2 1.936(3), Te1···Cl1 3.074(2), Te1···N1 2.601(6), Te2–O1 2.243(3), Te2–O3 1.994(3), Te2–O4a 1.971(3), Te2···N2 2.570(5), Te3–O2 2.107(3), Te3–O3 1.964(3), Te3–O4 2.096(3), Te3···N3 2.619(3); (**b**) Molecular structure of $[2\text{-}(PhNN)C_6H_4Te]_6TeO_{11}$ (**83**). Selected bond lengths (Å): Te1–O1 1.876(2), Te1–O12 2.063(2), Te1–O14 2.084(2), Te1–O15 2.106(2), Te1–O17 2.083(2), Te2–O12 1.911(2), Te4–O14 1.903(2), Te5–O15 1.913(2), Te7–O17 1.910(2), Te2–O23 2.249(2), Te3–O34 1.874(3), Te4–O45 1.987(2), Te5–O56 2.193(2), Te6–O67 1.891(3), Te7–O67 2.174(2), Te···N 2.666–2.980; (**c**) Molecular structure of $(8\text{-}Me_2NC_{10}H_6Te)_6Te_2O_{12}Cl_2$ (**82**). Selected bond lengths (Å): Te1–O1 1.892(9), Te2–O1 2.166 (9), Te2–O2 2.06(1), Te3–O2 1.90(1), Te1···Cl1 3.051(8), Te3a···Cl1 3.08(1), Te1–O3 1.94(1), Te2–O4 1.89(1), Te3–O6a 1.908(9), Te4–O3 1.98(1), Te4–O4 2.104(9), Te4–O5 1.92(1), Te4–O5a 2.24(1), Te4a–O6a 2.073(9), Te1···N1 2.60(1), Te2···N2 2.71(2), Te3···N3 2.61(1)

Aryltellurinic acids are interesting building blocks for the synthesis of molecular binary organoelement oxides. This was exemplified by the condensation reaction of the aryltellurinic acid $[2,6\text{-}Mes_2C_6H_3Te(O)OH]_2$ (**75**) with the organoelement oxides $(t\text{-}Bu_2SnO)_3$ [86] and $[2,6\text{-}(Me_2NCH_2)_2C_6H_3BiO]_2$ [87], respectively, giving rise to the formation of the tetranuclear stannatelluroxane cluster

$(t\text{-}Bu_2Sn)_2(2,6\text{-}Mes_2C_6H_3Te)_2O_5$ (**84**) and the pentanuclear bismatelluroxane cluster $[2,6\text{-}(Me_2NCH_2)_2C_6H_3Bi](2,6\text{-}Mes_2C_6H_3Te)_4O_7$ (**85**) (Scheme 7.20, Fig. 7.10).

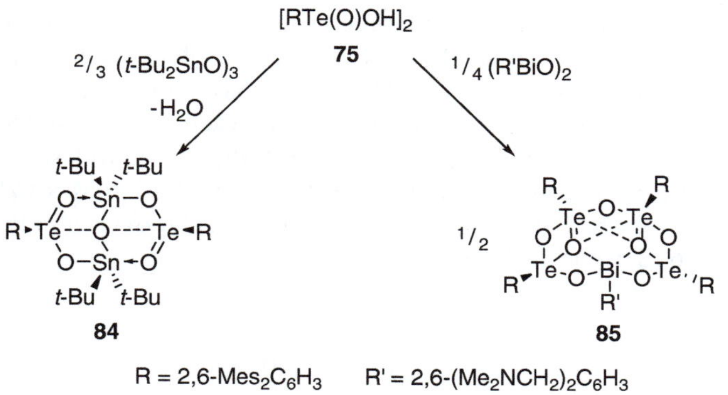

Scheme 7.20 Synthesis of the metallatelluroxanes $(t\text{-}Bu_2Sn)_2(2,6\text{-}Mes_2C_6H_3Te)_2O_5$ (**84**) and $[2,6\text{-}(Me_2NCH_2)_2C_6H_3Bi](2,6\text{-}Mes_2C_6H_3Te)_4O_7$ (**85**)

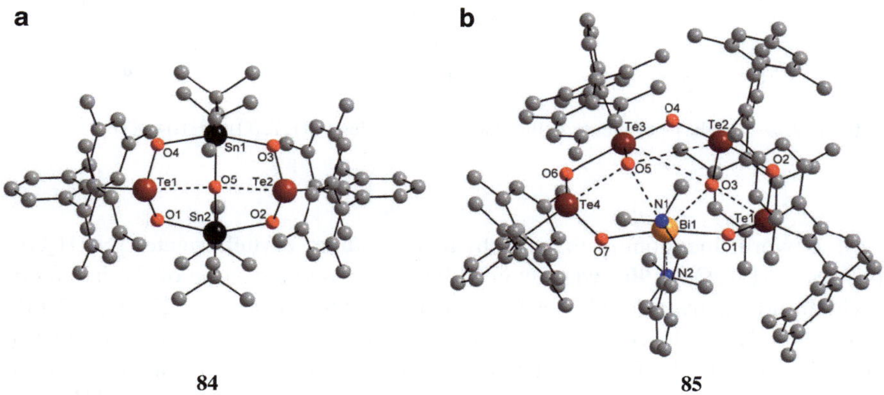

Fig. 7.10 (**a**) Molecular structure of $(t\text{-}Bu_2Sn)_2(2,6\text{-}Mes_2C_6H_3Te)_2O_5$ (**84**). Selected bond lengths (Å): Te1–O1 1.869(7), Te1–O4 1.857(6), Sn1–O4 2.174(6), Sn1–O3 2.165(6), Te2–O3 1.858(7), Te2–O2 1.874(7), Sn2–O2 2.142(7), Sn2–O1 2.162(7), Te1···O5 2.588(7), Sn1–O5 1.997(5), Te2···O5 2.581(7), Sn2–O5 2.016(5); (**b**) Molecular structure of $[2,6\text{-}(Me_2NCH_2)_2C_6H_3Bi](2,6\text{-}Mes_2C_6H_3Te)_4O_7$ (**85**). Selected bond lengths (Å): Te1–O1 1.857(5), Te1–O2 1.874(5), Te2–O2 2.310(5), Te2–O4 2.012(5), Te3–O4 1.977(5), Te3–O6 2.390(5), Te4–O6 1.856(4), Te4–O7 1.877 (4), Bi1–O7 2.342(5), Bi1–O1 2.385(4), Te1···O3 2.608(6), Te2–O3 1.845(5), Te2···O5 3.605(5), Te3···O3 3.558(5), Te3–O5 1.847(4), Te4···O5 2.695(5), Bi1···O5 2.787(4), Bi1···O3 2.839(4), Bi1···N1 2.501(6), Bi1···N2 2.557(6)

Notably, the related dinuclear stannatelluroxanes $RTeO_2Sn\,t\text{-}Bu_2Cl$ (**86**, R = 2,4,6-t-$Bu_3C_6H_2$; **87**, R = 8-$Me_2NC_{10}H_6$) containing four-membered ring structures were obtained by the reaction of $(t\text{-}Bu_2SnO)_3$ with the aryltellurium trichlorides $RTeCl_3$ (R = 2,4,6-t-$Bu_3C_6H_2$ or 8-$Me_2NC_{10}H_6$) [86].

7.2.4 Organotelluronic Acids

Up to very recently, little was known about organotelluronic acids. The attempted oxidation of the dinuclear aryltellurinic acid $[2,6\text{-}Mes_2C_6H_3Te(O)OH]_2$ (**75**) using strong oxidants, such as H_2O_2, $KMnO_4$ and $NaIO_4$, failed to provide the aryltelluronic acid $[2,6\text{-}Mes_2C_6H_3Te(O)(OH)_3]_2$ (**88**). However, the synthesis of **88** was achieved by the following two consecutive reactions. At first, the dinuclear aryltellurinic acid $[2,6\text{-}Mes_2C_6H_3Te(O)OH]_2$ (**75**) was reacted with sodium hydride to give the tetranuclear sodium aryltellurinate cluster $[Na(2,6\text{-}Mes_2C_6H_3TeO_2)]_4$ (**89**) (Fig. 7.11a), which was then oxidized with molecular oxygen in the presence of 18-crown-6 (Scheme 7.21) [88].

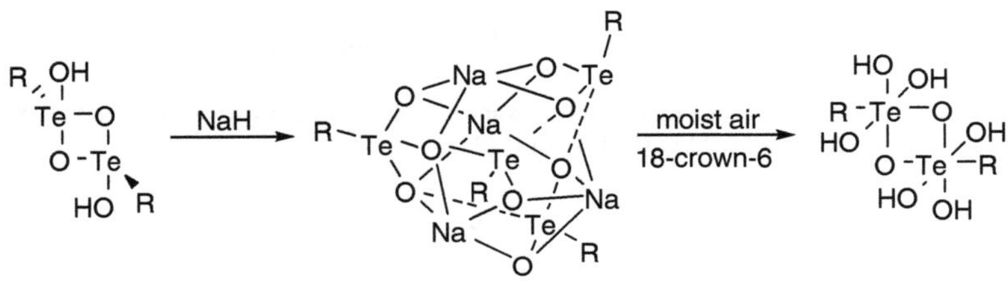

75 **89** **88**

Scheme 7.21 Synthesis of the aryltelluronic acid $[2,6\text{-}Mes_2C_6H_3Te(O)(OH)_3]_2$ (**88**)

It is worth mentioning that the hydrated sodium aryltellurinate, $[Na(H_2O)_4]$ $(4\text{-}MeOC_6H_4TeO_2)$, (**90**), was obtained by the depolymerisation of the polymeric aryltellurinic anhydride $[4\text{-}MeOC_6H_4TeO_{1.5}]_n$ (**68**) (Scheme 7.22, Fig. 7.11b) [89]. A related dodecanuclear lithium alkyltelluroxanate cluster $[Li(THF)_4]$ $\{(i\text{-}PrTe)_{12}O_{16}Br_4[Li(THF)Br]_4Br\}\cdot 2THF$ (**91**) was accidentally obtained by the air oxidation of isopropyl hex-1-ynyl telluride in the presence of LiBr in THF [90]. The organotelluroxanes **81, 82, 89** and **91** hold potential as molecular models for alkali tellurite glasses $(M_2O)_x(TeO_2)_{1\text{-}x}$ (M = alkali metal; $0.10 < x < 0.35$) for which only little structural information is available [91–95].

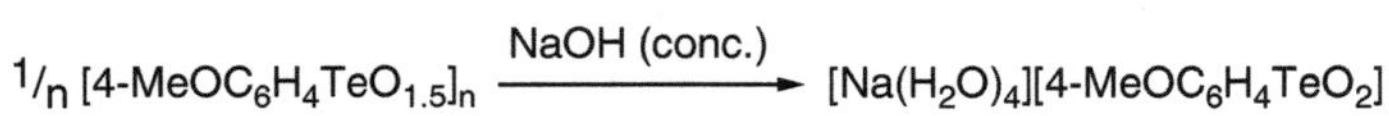

68 **90**

Scheme 7.22 Depolymerisation of the aryltellurinic anhydride $[4\text{-}MeOC_6H_4TeO_{1.5}]_n$ (**68**) with strong base

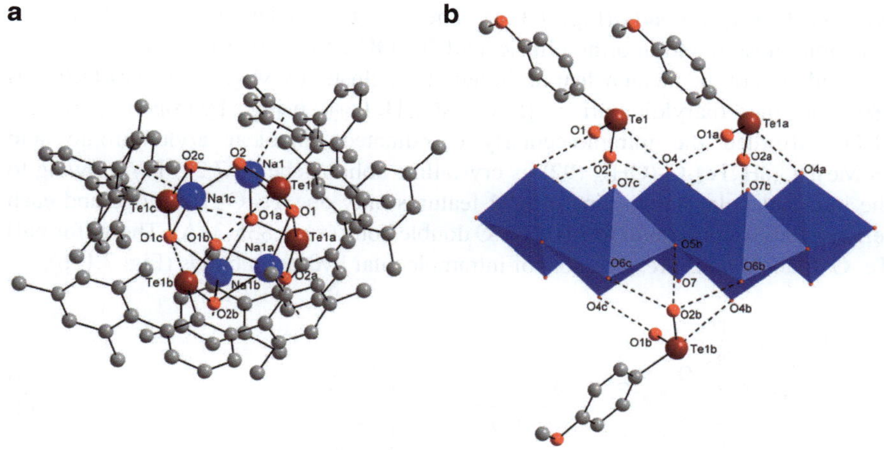

89 **90**

Fig. 7.11 (**a**) Molecular structure of $[Na(2,6\text{-}Mes_2C_6H_3TeO_2)]_4$ (**89**). Selected bond lengths (Å): Te1–O1 1.850(4), Te1···O1b 3.195(4), Te1–O2 1.832(4), Na1–O1 2.568(5), Na1–O1a 2.357(4), Na1–O2 2.469(5), Na1–O2c 2.261(4); (**b**) Crystal structure of $[Na(H_2O)_4](4\text{-}MeOC_6H_4TeO_2)$ (**90**). Selected bond lengths (Å): Te1–O1 1.823(4), Te1–O2 1.868(3), Te1···O4 3.031(4), O4···O1a 2.647(6), O2b···O6b 2.927(7), O2b···O5b 2.738(4), O2b···O6c 2.854(7)

The molecular structure of the dinuclear aryltelluronic acid $[2,6\text{-}Mes_2C_6H_3Te(O)(OH)_3]_2$ (**88**) differs from those of the lighter congeners, namely the mononuclear arylsulfonic and arylselenonic acids possessing (formal) S–O and Se–O double bonds (E = S, Se), and features an asymmetric four-membered Te_2O_2 ring structure

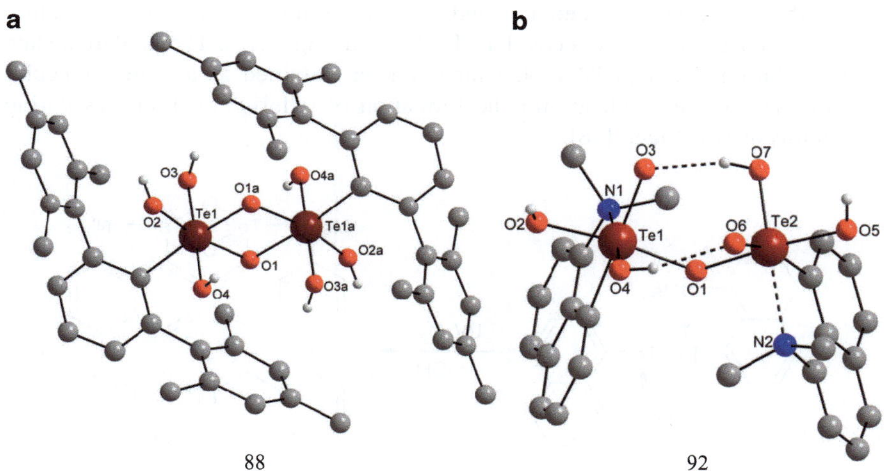

88 92

Fig. 7.12 (**a**) Molecular structure of $[2,6\text{-}Mes_2C_6H_3Te(O)(OH)_3]_2$ (**88**). Selected bond lengths (Å): Te1–O1 1.943(2), Te1–O1a 1.971(2), Te1–O2 1.918(2), Te1–O3 1.938(3), Te1–O4 1.918(3); (**b**) Molecular structure of $(8\text{-}Me_2NC_{10}H_6Te)_2O_3(OH)_4$ (**92**). Selected bond lengths (Å): Te1–O1 1.969(3), Te1–O3 1.836(3), Te1–O4 1.903(4), Te1–O2 1.921(4), Te2–O1 1.983(3), Te2–O6 1.834 (3), Te2–O7 1.912(4), Te2–O5 1.928(3), Te1···N1 2.376(4), Te2···N2 2.366(4) with Te–O single bonds (Fig. 7.12a). The Te–O bond lengths (av. 1.938(3) Å) resemble those found in orthotelluric acid, $Te(OH)_6$ (av. 1.913(6) Å) [96].

Oxidation of the hexanuclear aryltelluroxane cluster $(8\text{-}Me_2NC_{10}H_6Te)_6O_8(OH)_2$ (**80**) or the diarylditelluride $(8\text{-}Me_2NC_{10}H_6Te)_2$ using hydrogen peroxide, H_2O_2, afforded the intramolecularly coordinated dinuclear aryltelluronic acid $(8\text{-}Me_2NC_{10}H_6Te)_2O_3(OH)_4$ (**92**) as crystalline solid (Scheme 7.23) [97]. Owing to the intramolecular N-coordination, **92** features only one Te–O–Te bridge and each tellurium atom contains a (formal) Te–O double bond (av. 1.835(3) Å). These (formal) Te–O double bonds are acceptors of intramolecular hydrogen bonds (Fig. 7.12b).

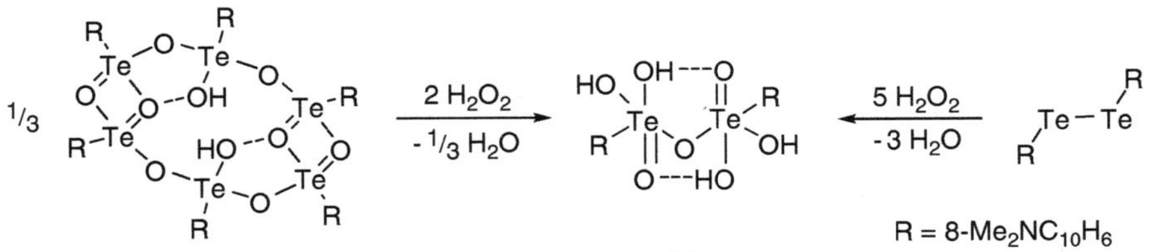

80 **92**

Scheme 7.23 Synthesis of the aryltelluronic acid $(8\text{-}Me_2NC_{10}H_6Te)_2O_3(OH)_4$ (**92**)

The photooxidation of the diarylditelluride $(8\text{-}Me_2NC_{10}H_6Te)_2$ afforded a complex reaction mixture from which the tetranuclear aryltelluroxane cluster $(8\text{-}MeNC_{10}H_5TeO)_4$ (**93**) was isolated as crystalline solid (Scheme 7.24). The key feature of **93** are the two 5,5′-binaphthyl moieties, which were formed upon oxidation of the activated C5–H bonds of the two naphthyl groups giving rise to a C5–C5′ cross-coupling. Concurrently, one of the N-methyl groups was cleaved and presumably converted into methanol and at the same time the remaining coordinative Te···N bond turned into a covalent Te–N bond (Fig. 7.13). The oxidation state of the tellurium atom of **93** is 4, which can be classified as an intramolecular tellurinic amide. No evidence for the formation of tellurium(VI) species during the photolysis was found [98].

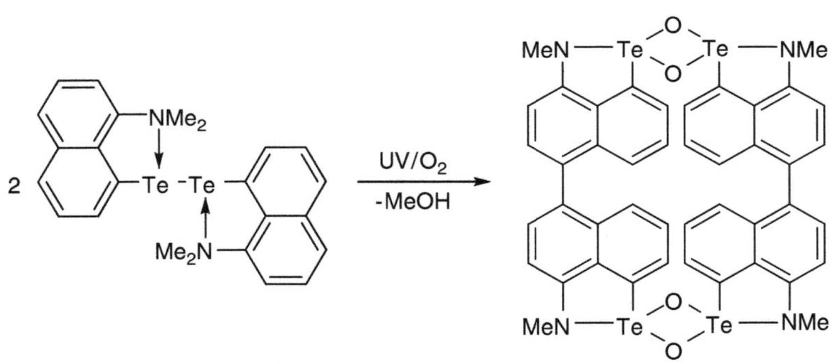

93

Scheme 7.24 Photooxidation of the diaryltelluride $(8\text{-}Me_2NC_{10}H_6Te)_2$ in the presence of air

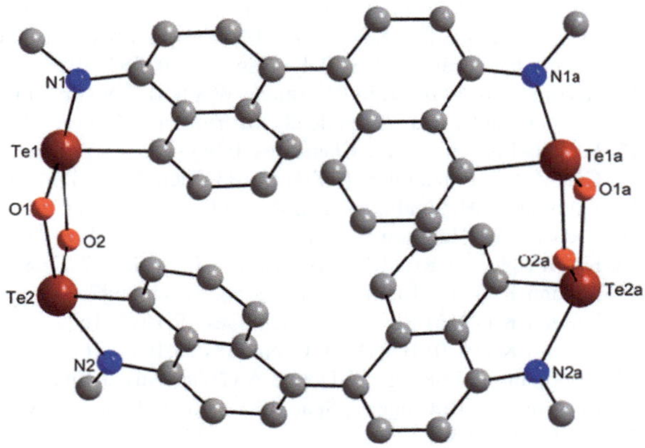

93

Fig. 7.13 Molecular structure of (8-MeNC$_{10}$H$_5$TeO)$_4$ (**93**). Selected distances (Å): Te1–O1 1.908(7),Te1–O2 2.195(6), Te1–N1 2.076(7), Te2–O1 2.155(6), Te2–O2 1.891(6), Te2–N2 2.103(8)

7.3 Conclusion

The last 10 years have witnessed significant progress in the field of organotelluroxanes. Using the kinetic stabilization of bulky organic substituents and the electronic stabilization of intramolecularly coordinating N-donor ligands, novel diorganotellurium oxides, diorganotellurones, organotellurinic acids as well as organotelluronic acids were prepared and fully characterized. The structural motifs observed for these compound classes show great diversity and differ from those of the lighter sulfur and selenium congeners. Frequently, structurally directing secondary Te···O bonds are observed, that are responsible for a high degree of aggregation compared to their lighter group 16 counterparts. Telluroxanes may be used for the preparation of binary metallatelluroxanes containing Te–O–M linkages (M = main group element), the reversible fixation of carbon dioxide and the reversible 'oxygen' transfer to small molecules such, as acetonitrile.

References

1. Pellissier H (2006) Tetrahedron 62:5559–5601
2. Fernandez I, Khiar N (2003) Chem Rev 103:3651–3706
3. Back TG (2001) Tetrahedron 57:5263–5301
4. Simpkins NS (1990) Tetrahedron 46:6951–6984
5. Petrini M (2005) Chem Rev 105:3949–3977
6. Alba AR, Companyo X (2010) Chem Soc Rev 39:2018–2033
7. Nielsen M, Jacobsen CB, Holub N, Paixão MW, Jørgensen KA (2010) Angew Chem Int Ed 49:2668–2679
8. Jacob C, Holme AL, Fry FH (2004) Org Biomol Chem 2:1953–1956
9. Joseph CA, Maroney MJ (2007) Chem Commun 32:3338–3349
10. Ikemura K, Endo T (1999) J Appl Polym Sci 72:1655–1668
11. Adami I (2009) In: Zoller U (ed) Handbook of detergents, vol F. CRC Press, Boca Raton

12. Vora B, Pujado P, Imai T, Fritsch T (1990) Chem Ind (London):187–191
13. Mahore JG, Wadher KJ, Umekar MJ, Bhoyar PK (2010) Int J Pharm Sci Rev Res 1:8–13
14. Kucera F, Jancár J (1998) Polym Eng Sci 38:783–792
15. Herrmann WA (2000) In: Herrmann WA (ed) Transition metals Part 3 of synthetic methods of organometallic and inorganic chemistry, vol 9. George Thieme, Stuttgart
16. Wojcicki A (1971) Acc Chem Res 4:344–352
17. Côté AP, Shimizu GKH (2003) Coord Chem Rev 245:49–64
18. Shimizu GKH, Vaidhyanathan R, Taylor JM (2009) Chem Soc Rev 38:1430–1449
19. Alcock NW, Harrison WD (1982) J Chem Soc Dalton Trans:709–712
20. Krafft F, Lyons RE (1894) Ber Dtsch Chem Ges 27:1768–1773
21. Lederer K (1912) Liebigs Ann Chem 391:326–347
22. Lederer K (1916) Ber Dtsch Chem Ges 49:1615–1622
23. Lederer K (1920) Ber Dtsch Chem Ges 53:1674–1680
24. Beckmann J, Bolsinger J, Duthie A (2008) Aust J Chem 61:172–182
25. Beckmann J, Bolsinger J, Spandl J (2008) J Organomet Chem 693:957–964
26. Detty MR (1980) J Org Chem 45:274–279
27. Akiba M, Lakshmikantham MV, Jen KY, Cava MP (1984) J Org Chem 49:4819–4821
28. Oba M, Endo M, Nishiyama K, Ouchi A, Ando W (2004) Chem Commun 14:1672–1673
29. Oba M, Okada Y, Endo M, Tanaka K, Nishiyama K, Shimada S, Ando W (2010) Inorg Chem 49:10680–10686
30. Vernon RH (1920) J Chem Soc 117:86–98
31. Srivastava PC, Bajpai S, Ram C, Kumar R, Butcher RJ (2007) J Organomet Chem 692: 2482–2490
32. Drew HDK (1929) J Chem Soc:560–569
33. Einstein F, Trotter J, Williston C (1967) J Chem Soc A:2018–2023
34. Vernon RH (1921) J chem Soc Trans 119:687–697
35. Balfe MP, Chaplin CA, Phillips H (1938) J Chem Soc:341–347
36. Balfe MP, Nandi KN (1941) J Chem Soc:70–72
37. Alcock NW (1972) Adv Inorg Chem Radiochem 15:1–58
38. Cordero B, Gomez V, Platero-Prats AE, Reves M, Echeverria J, Cremades E, Barragan F, Alvarez S (2008) Dalton Trans 21:2832–2838
39. Bondi A (1964) J Phys Chem 68:441–451
40. Abrahams SC (1957) Acta Crystallogr 10:417–422
41. Szymaniak A, Borowiak T (1976) Rocz Chem 50:803–804
42. Naumann D, Tyrra W, Hermannn R, Pantenburg I, Wickleder MS (2002) Z Anorg Allg Chem 628:833–842
43. Klapoetke TM, Krumm B, Mayer P, Piotrowski H, Ruscitti OP (2002) Z Naturforsch B 57: 145–150
44. Lederer K (1916) Ber Dtsch Chem Ges 49:345–349
45. Beckmann J, Dakternieks D, Duthie A, Ribot F, Schuermann M, Lewcenko NA (2003) Organometallics 22:3257–3261
46. Klapoetke TM, Krumm B, Scherr M (2009) Phosphorus Sulfur Silicon 184:1347–1354
47. Alcock NW, Harrison WD (1982) J Chem Soc Dalton Trans:1421–1428
48. Beckmann J, Dakternieks D, Duthie A, Lewcenko NA, Mitchell C, Schuermann M (2005) Z Anorg Allg Chem 631:1856–1862
49. Chandrasekhar V, Kumar A (2009) J Organomet Chem 694:2628–2635
50. Kobayashi K, Deguchi N, Horn E, Furukawa N (1998) Angew Chem Int Ed 37:984–986
51. Kobayashi K, Izawa H, Yamaguchi H, Horn E, Furukawa N (2001) Chem Commun:1428–1429
52. Alcock NW, Harrison WD, Howes C (1984) J Chem Soc Dalton Trans:1709–1716
53. Kobayashi K, Deguchi N, Takahashi O, Tanaka K, Horn E, Kikuchi O, Furukawa N (1999) Angew Chem Int Ed 38:1638–1640
54. Kobayashi K, Tanaka K, Izawa H, Arai Y, Furukawa N (2001) Chem Eur J 7:4272–4729
55. Chandrasekhar V, Thirumoorthi R (2009) Inorg Chem 48:10330–10337

56. Klapoetke TM, Krumm B, Scherr M (2007) Acta Crystallogr Sect E 63E:o4189
57. Beckmann J, Bolsinger J, Duthie A, Finke P. In preparation
58. Ley SV, Meerholz CA, Barton DHR (1981) Tetrahedron 37:213–223
59. Okada Y, Oba M, Arai A, Tanaka K, Nishiyama K, Ando W (2010) Inorg Chem 49:383–385
60. Shen JK, Gao Y, Shi Q, Rheingold AL, Basolo F (1991) Inorg Chem 30:1868–1873
61. Ming X, Yi-Ci G, Jian-Kun S, Qi-Zhen S, Basolo F (1993) Inorg Chim Acta 207:207–212
62. Liu X, Gao Y, Su Z, Wang Y, Shi Q (1999) Transit Met Chem 24:666–668
63. Song L, Li Q, Hu Q, Dong Y (2001) J Organomet Chem 619:194–203
64. Beckmann J, Bolsinger J, Finke P, Grabowsky S, Hesse M, Luger P, Mebs S. In preparation
65. Beckmann J, Bolsinger J (2007) Organometallics 26:3601–3603
66. Chandrasekhar V, Thirumoorthi R (2009) Inorg Chem 48:6236–6241
67. Beckmann J, Dakternieks D, Duthie A, Mitchell C (2005) Dalton Trans:1563–1564
68. Beckmann J, Dakternieks D, Duthie A, Lewcenko NA, Mitchell C (2004) Angew Chem Int Ed 43:6683–6685
69. Reuter H (1987) Spektroskopische und röntgenographische Untersuchungen zur Hydrolyse von Organozinnhalogeniden. Dissertation, Universität Bonn, Bonn
70. Ballivet-Tkatchenko D, Burgat R, Chambrey S, Plasseraud L, Richard P (2006) J Organomet Chem 691:1498–1504
71. Beckmann J, Bolsinger J, Duthie A (2010) Z Anorg Allg Chem 636:765–769
72. Engman L, Cava MP (1982) J Chem Soc Chem Commun:164–165
73. Oba M, Okada Y, Nishiyama K, Shimada S, Ando W (2008) Chem Commun 42:5378–5380
74. Khurana JM, Kandpal BM, Chauhan YK (2003) Phosphorus Sulfur Silicon Relat Elem 178: 1369–1375
75. Lederer K (1915) Ber Dtsch Chem Ges 48:1345–1350
76. Reichel L, Kirschbaum E (1936) Liebigs Ann Chem 523:211–223
77. Thavornyutikarn P, McWhinnie WR (1973) J Organomet Chem 50:135–143
78. Barton DHR, Finet J, Thomas M (1986) Tetrahedron 42:2319–2324
79. Beckmann J, Heitz S, Hesse M (2007) Inorg Chem 46:3275–3282
80. Beckmann J, Hesse M. In preparation
81. Nakashima Y, Shimizu T, Hirabayashi K, Kamigata N (2004) Org Lett 6:2575–2577
82. Nakashima Y, Shimizu T, Hirabayashi K, Yasui M, Nakazato M, Iwasaki F, Kamigata N (2004) Tetrahedron Asymm 15:3791–3797
83. Beckmann J, Finke P, Hesse M, Wettig B (2008) Angew Chem Int Ed 47:9982–9984
84. Beckmann J, Bolsinger J, Duthie A (2011) Chem Eur J 17:930–940
85. Srivastava K, Sharma S, Singh HB, Singh UP, Butcher RJ (2010) Chem Commun 46: 1130–1132
86. Beckmann J, Bolsinger J, Hesse M (2009) Organometallics 28:4225–4228
87. Beckmann J, Dostál L, Svoboda T, Hesse M. In preparation
88. Beckmann J, Bolsinger J, Finke P, Hesse M (2010) Angew Chem Int Ed 49:8030–8032
89. Beckmann J, Koehne T. In preparation
90. Citeau H, Kirschbaum K, Conrad O, Giolando DM (2001) Chem Commun:2006–2007
91. El-Mallawany R (1998) Mater Chem Phys 53:93–120
92. El-Mallawany R (1999) Mater Chem Phys 60:103–131
93. El-Mallawany R (2000) Mater Chem Phys 63:109–115
94. McLaughlin JC, Tagg SL, Zwanziger JW (2001) J Phys Chem B 105:67–75
95. McLaughlin JC, Tagg SL, Zwanziger JW, Haeffner DR, Shastri SD (2000) J Non-Cryst Solids 274:1–8
96. Falck L, Lindqvist O (1978) Acta Crystallogr B 34B:3145–3146
97. Beckmann J, Bolsinger J, Finke P. In preparation
98. Beckmann J, Bolsinger J (2011) Z Anorg Allg Chem 637:29–30

Chapter 8
Recent Developments in the Lewis Acidic Chemistry of Selenium and Tellurium Halides and *Pseudo*-Halides

Jason L. Dutton and Paul J. Ragogna

8.1 Redox Reactions Between Heavy Chalcogen Tetrahalides and Neutral 2-Electron Ligands

The coordination chemistry of the chalcogen elements is underdeveloped when compared to the rest of the periodic table (the halogens and noble gasses excepted), especially considering the existence of the selenium and tellurium tetrahalides as stable, easily handled, commercially available Lewis acids. The available reports focus on TeX_4, primarily involving the coordination of two monodentate or one bidentate sulfur or oxygen based ligands to tellurium (e.g. **1–3**) [1–7]. These studies are primarily structural in nature and concern themselves with the presence or absence of stereochemical activity for the lone pair of electrons at tellurium, as well as the *cis vs. trans* orientation of the ligands. Typically *pseudo*-octahedral complexes are observed (Fig. 8.1), with the lone pair being stereochemically inactive, due to the inert s-pair effect. It should be noted that virtually no onward reactivity has been examined with these compounds. There is only one report of $SeCl_4$ being used as a Lewis acid in a coordination complex, addition of pyridine was initially reported to give ionic complex **4a** based on conductivity studies [8]. However, subsequent researchers were unable to reproduce these results and proposed a neutral octahedral complex (**4b**), but no single crystal X-ray structure was obtained as confirmation (Chart 8.1) [9].

J.L. Dutton • P.J. Ragogna (✉)
Department of Chemistry, The University of Western Ontario, 1151 Richmond Street, London, ON, Canada, N6A 5B7
e-mail: pragogna@uwo.ca

J.D. Woollins and R.S. Laitinen (eds.), *Selenium and Tellurium Chemistry*,
DOI 10.1007/978-3-642-20699-3_8,

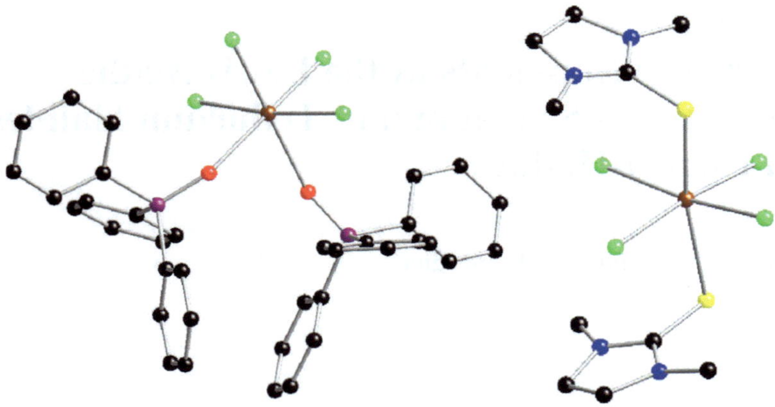

Fig. 8.1 Octahedral coordination complexes of $TeCl_4$

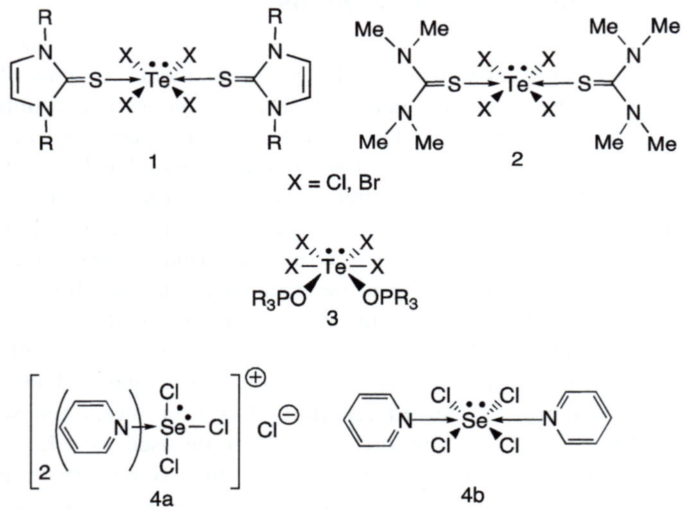

Chart 8.1 Examples of ChX_4 acting as Lewis acids in coordination complexes

A major reason for the lack of activity in this area is the propensity for the selenium and tellurium tetrahalides to participate in redox reactions in the presence of a variety of common two-electron ligands. For example, the reaction of SeX_4 (X = Cl, Br) with R_3P: (R = ^{n}Bu, Ph) results in reduction of the chalcogen centre, and formation of the P(V) containing $[R_3PX]^+$ cation paired with the Se(II) centred anion $[SeX_3]^-$ (**5**), rather than a P-Ch coordination complex [10]. Phosphorus-selenium bond formation only occurs after addition of three equivalents of phosphine, generating $R_3P = Se$ with selenium fully reduced to the −2 oxidation state. Similar reactivity was observed with tellurium; the addition of Ph_3P: to $TeCl_4$ immediately results in the generation of tellurium metal and the P(V) salt $[Ph_3PCl]_2[TeCl_6]$ (**6**). Addition of excess triphenylphosphine does not further reduce the Te^0 (Scheme 8.1).

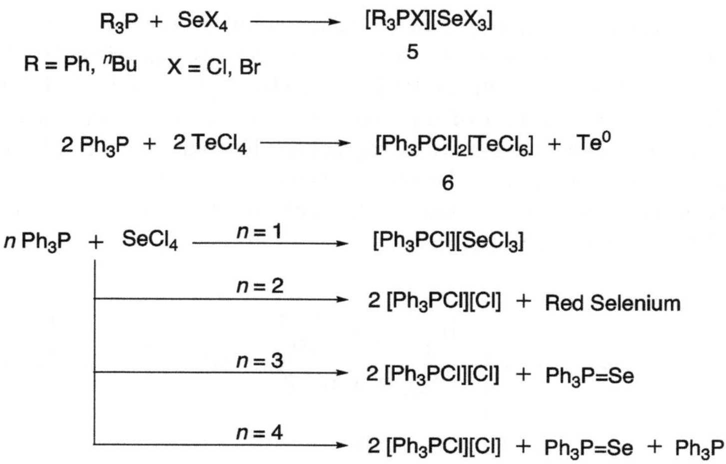

Scheme 8.1 Reduction of selenium and tellurium halides using triorganophosphine ligands

The reaction of N-heterocyclic carbene (NHC) ligands with ChX_4 does result in coordination, again with a reduction at the chalcogen centre to the +2 oxidation state. The oxidation by-product is molecular X_2, a highly reactive species, that can lead to undesirable side products and lowered yields. In cases where I_2 is eliminated, this is less of a problem, but for the more reactive lighter halides complex mixtures are often obtained due to halogenation reactions. For example, Kuhn et al. isolated the TeI_2-NHC complex **7** from the reaction of TeI_4 with NHC, but the same reaction with $TeCl_4$ or $TeBr_4$ gave many products [11]. For $SeCl_4$, the eliminated Cl_2 was sequestered by free NHC in solution, giving a chloroimidazolium salt paired with a $[SeCl_6]^{2-}$ dianion [10] This byproduct was easily separated from the $SeCl_2$ coordination compound (**8**), but consumed half of the NHC and one third of the $SeCl_4$ introduced into the reaction, thus resulting in a lowered yield of **8** (Scheme 8.2). In the reaction involving $SeBr_4$ a number of byproducts were generated in the redox reaction, again the $SeBr_2$•NHC coordination compound could be isolated by simple solvent extraction, but only in a moderate yield.

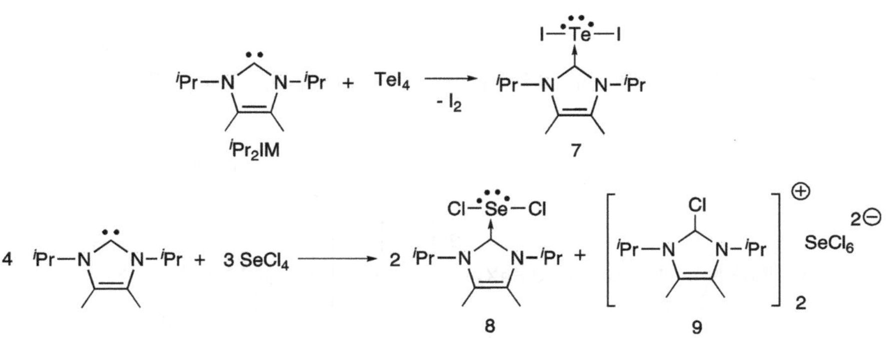

Scheme 8.2 Reactions of the NHC ^{i}Pr$_2$IM with ChX_4 giving Ch(II)-NHC coordination complexes with reductive elimination of X_2.

The spontaneous reduction of p-block halides in the presence of N-heterocyclic carbene ligands has also been observed for the phosphorus trihalides. In these reactions, two NHCs bind a cationic Pn(I) centre (**10**), again with reductive elimination of X_2 [12]. The reaction of organophosphines with phosphorus and arsenic trihalides gives similar complexes as tripnictogenium cations (**11**) [13–15]. For reactions involving the lighter phosphorus halides (X = Cl, Br), the addition of cyclohexene as a trap for the eliminated, highly reactive X_2 was found to give improved results (Scheme 8.3) [16].

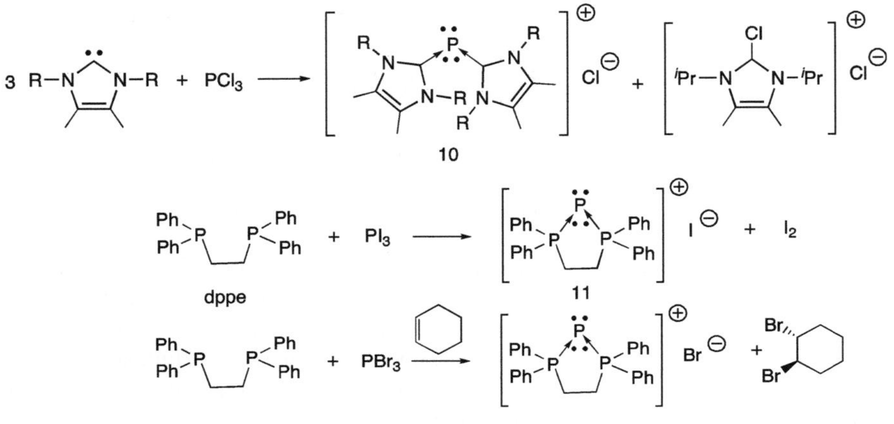

Scheme 8.3 Redox reactions between neutral donors NHC and dppe and phosphorus trihalides

Use of bidentante ligands with the selenium tetrahalides also resulted in reduction of selenium; the reactions of the bidentate α-diimine ^{t}Bu$_2$DAB (DAB = diazabutadiene) ligand with SeX_4 also resulted in a two reduction of selenium, the loss of a *tert*-butyl side group, and elimination of X_2 in the formation of unique 1,2,5-selenadiazolium cations (**11**) in low 30–35% yields [17]. As was observed for reactions involving phosphorus trihalides, it was demonstrated that the addition of a reagent to trap X_2 (cyclohexene) resulted in a dramatic increase in yield (80–90%) for the heterocyclic compound (Scheme 8.4) [18].

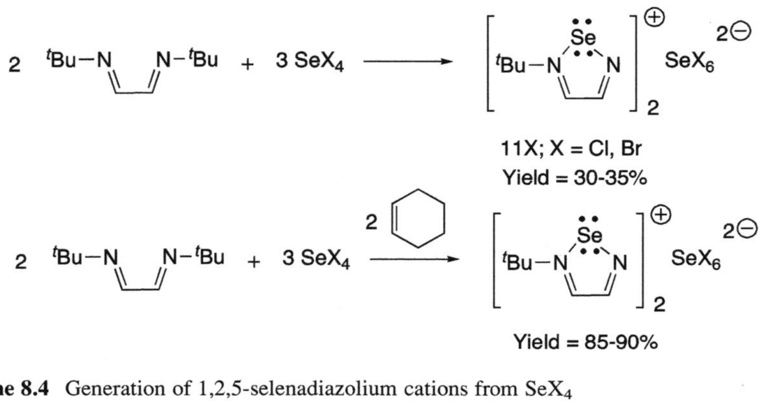

Scheme 8.4 Generation of 1,2,5-selenadiazolium cations from SeX_4

Richards et al. also noted the elimination of X_2 and reduction of selenium in reactions of SeX_4 with the related β-diketiminate ligand [19]. A methyl group on the ligand backbone was found bound to selenium, resulting from attack by the eneamine tautamer of the ligand, which has also been observed in other, related systems. The yields of the isolated compounds (**12**, **13**) were low and the ligand framework was halogenated in a variety of unpredictable positions, underscoring the complication that this redox activity can cause. Similar transformations were observed for DAB ligands bearing methyl substituents in the backbone, where reaction of $Dipp_2MeDAB$ with $SeCl_4$ and $SeBr_4$ gave N,C bound Se(II) complexes in low (< 20%) yield (**14**, **15**), with halogenation of ligand framework in the case of $SeBr_4$ (Scheme 8.5) [20].

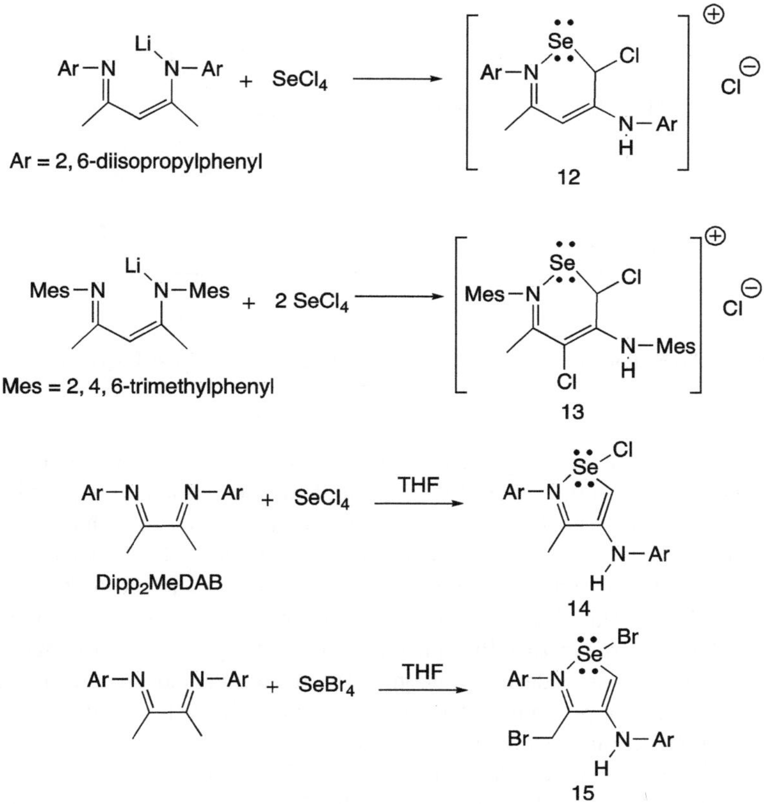

Scheme 8.5 Reactions of $SeCl_4$ with β-diketiminate and DAB ligands bearing reactive methyl groups on the ligand framework

The observation that reduction from the +4 to +2 oxidation state is a common occurrence during the reaction of ChX_4 with neutral ligands implies that use of the binary haldes (ChX_2) should give the same compounds, but circumvent the redox process and avoid the elimination of X_2. This was indeed found to be the case for reactions involving selenium and ^{t}Bu$_2$DAB, which gave improved yields and fewer byproducts as compared to using $SeCl_4$ [18]. In this reaction, the 1,2,5-selenadiazolium was isolated as the chloride, rather than paired with the $[SeCl_6]^{2-}$ dianion. The incorporation of hexachlorochalcogenate anions is common when using ChX_4, and this can present a problem as these types of anions are reactive, and can also act as a source of halides. The chlorinated compound **15** could easily undergo metathesis reactions to pair the cation with more inert anions (e.g. triflate, **16**), allowing for the Lewis acidic chemistry of the heterocycle to be explored (Scheme 8.6) [21]. Greatly improved results for the Dipp$_2$MeDAB system were also noted when $SeCl_2$ was used with **14** being generated in essentially quantitative yield.

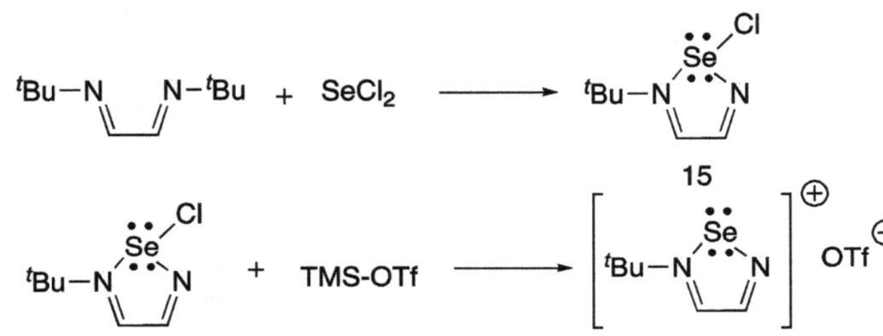

Scheme 8.6 Formation of 1,2,5-selenadiazolium species from $SeCl_2$ and ^{t}Bu$_2$DAB

The use of these same ligand sets in reactions with TeX_4 does not result in reduction of tellurium, rather Te(IV) complexes are isolated (e.g. **17**) [18]. Although, in the case of **17**, reduction occurs spontaneously upon halide abstraction, giving 1,2,5-telluradiazolium cations (**18**) analogous to 15 (Scheme 8.7) [21]. The tellurium tetrahalides appear be somewhat less susceptible to reduction in reactions with neutral ligands. However, when reduction does occur often it is often all the way down to tellurium metal, with the production of many other products, generally an undesired outcome. The reduction is also somewhat unpredictable, reactions of phosphine chalcogenides with TeX_4 were found to give either Te(IV) or Te(II) complexes with no apparent reason for the difference [6]. In cases where Te(IV) was obtained, the yields were >90%, for reactions involving reduction to Te(II) yields of ~30% were achieved, likely due to side products arising from the elimination of X_2. The binary tellurium halides are unknown for synthetic applications, having only been characterized spectroscopically in the gas phase, and therefore cannot be used as a source of Te(II) to circumvent the redox reactions [22, 23].

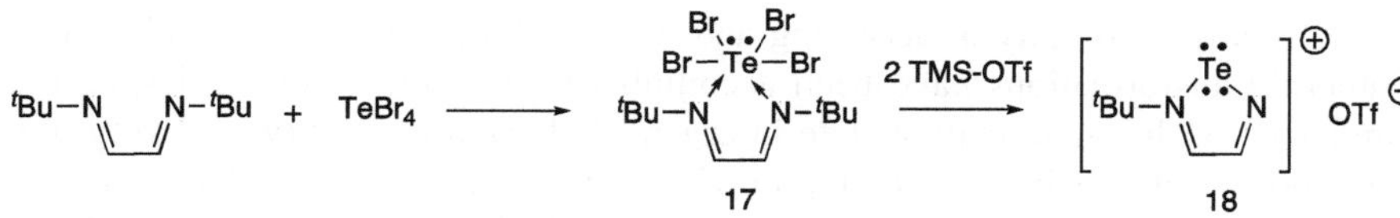

Scheme 8.7 Formation of 1,2,5-telluradiazolium cation *via* a halide abstraction reaction with spontaneous reduction of tellurium

8.2 Recent Developments in the Electrophilic Chemistry of SeX_2

The binary halides for selenium are also thermally unstable ($SeBr_2$ more so than $SeCl_2$), and cannot be isolated in a pure form, as they decompose *via* disproportionation [24]. Rather, they must be generated *in situ* and used immediately. There are several different methods for generating SeX_2, however the obvious route, halogenation of Se^0 using X_2, may not be used as this generates a mixture of the mono-, di-, and tetrahalides until excess X_2 eventually gives exclusively the tetrahalides. Therefore, stoichiometric methods must be employed. For $SeCl_2$, the synthesis of the "pure" material was not available until Chivers et al. reported the oxidation of Se^0 using SO_2Cl_2 in THF, which gives $SeCl_2$ and gaseous SO_2 as the only by-product [25]. Generally, the $SeCl_2$ is used *in situ*, removal of the THF solvent gives pure $SeCl_2$ as a highly unstable red oil, disproportionation is significantly slower in the THF solution. The primary drawback to this method is relatively large amounts of $SeCl_2$ must be generated, due to difficulty in handling extremely small amounts of liquid SO_2Cl_2. For smaller-scale (e.g. 20 mg) reactions $SeCl_2$ can be synthesized by reduction of $SeCl_4$ with Ph_3Sb, which gives Ph_3SbCl_2 as the by-product, this method may be used providing the chlorostiborane does not react with the substrates of interest and can be easily separated [26]. Selenium dibromide can be synthesized *via* a methathesis reaction using $SeCl_2$ and TMS-Br, or by the comproportionation reaction of Se^0 and $SeBr_4$ in THF or MeCN solution [27]. The disproportionation of $SeBr_2$ is substantially faster than $SeCl_2$, and these solutions must be used immediately to avoid significant contamination of Se_2Br_2 and Br_2 (Scheme 8.8).

$$3\,SeCl_2 \longrightarrow Se_2Cl_2 + SeCl_4$$

$$2\,SeBr_2 \longrightarrow Se_2Br_2 + Br_2$$

$$Se^0 + SO_2Cl_2 \xrightarrow{THF} SeCl_2 + SO_2\uparrow$$

$$SeCl_4 + Ph_3Sb \xrightarrow{THF} SeCl_2 + Ph_3SbCl_2$$

$$SeBr_4 + Se^0 \xrightarrow{THF} 2\,SeBr_2$$

Scheme 8.8 Syntheses and disproportionation reactions of SeX_2

The relative difficulty in accessing and using SeX_2 has meant that only a few synthetic transformations have been accomplished with this reagent, despite the dominance of the +2 oxidation state in organochalcogen chemistry, and very few were known before Chivers' key report of generated pure $SeCl_2$ in THF solution.

The most common type of reactions involving the binary halides are metathesis (s-block metal, $-Si(CH_3)_3$, $[H]^+$) reactions, to introduce selenium into some formally anionic moiety [28–30]. A simple example is the reaction of $SeCl_2$ with aryl Grignard reagents to generate diaryl selenides (Ar_2Se) reported by Zade et al. [31]. These transformations give Ar_2Se in good 70–80% yield. Interestingly, they were unable to synthesize ArSeCl type compounds using one stoichiometric equivalent of the Grignard reagent, which also gave Ar_2Se (in lowered yield).

Selenium dichloride has also been introduced to anionic transition metal complexes by metathesis reactions, which has been investigated by Song et al. For example, the reaction of an anionic butterfly compound with $SeCl_2$ gives the μ_4-Se double butterfly complex **19** (Scheme 8.9) [32, 33]. A variety of different –R groups bound to the tellurium atom were used, all reactions gave similarly low yields of ~20%.

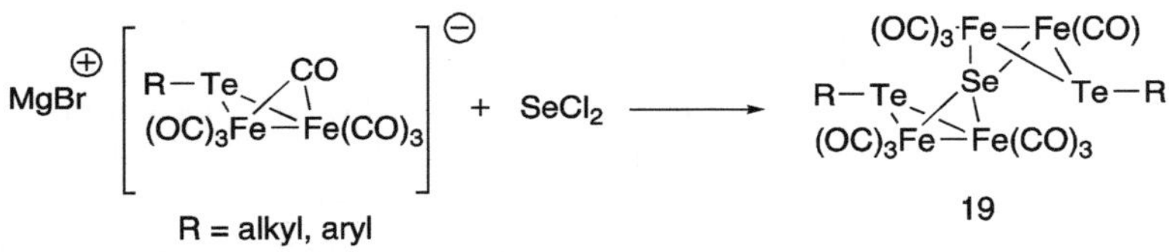

19

Scheme 8.9 Formation of double butterfly complexes from anionic metal precursor and $SeCl_2$

The other main class of reactions that can be preformed with SeX_2 is addition reactions with olefinic C-C bonds. The most common outcome of these transformations is the selenation of the α-carbon and halogenation of the β-carbon atom [34, 35]. Reaction of SeX_2 with acetylenes proceeds in the same fashion, but leaves a C=C bond functionality intact [36, 37]. Recently this strategy has been used for the synthesis of the monomer (**20**) for the conducting polymer poly(3,4-ethylenedioxyselenophene) (**21**), where the incorporation of selenium involves the addition of $SeCl_2$ to a 1,3-diene in the presence of NaOAc to remove the chloride atoms and protons (Scheme 8.10) [38].

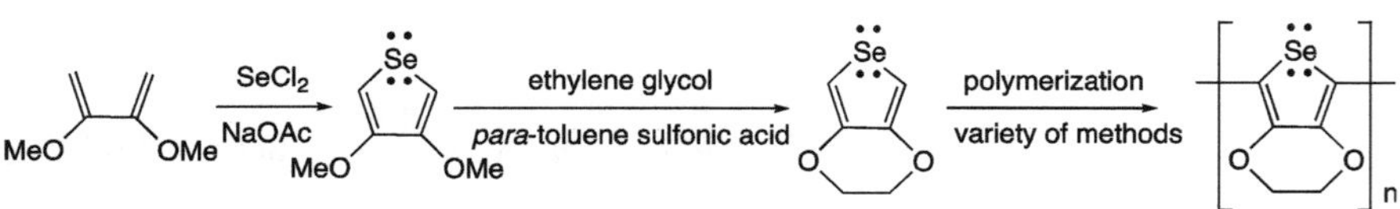

20 21

Scheme 8.10 Synthesis of poly(3,4-ethylenedioxyselenophene)

8.3 Base Stabilized Forms of ChX_2

A stable synthetic equivalent of SeX_2, which could be easily handled, stored and then delivered stoichiometrically would be a substantial advancement in the field, and open up new opportunities for the use of these reagents as sources of electrophilic Se(II).

Coordination by a Lewis base has been used by several groups to stabilize SeX_2 fragments in the past 40 years. Synthetic access to these compounds is either from the direct reaction of sulfur centered Lewis bases with SeX_2 [25, 39, 40], or by the oxidative addition of X_2 to lower oxidation state prescursors (e.g. phosphine selenides, selenoureas) [41–44].

A similar strategy is used to isolated base-stabilized TeX_2 complexes, except that "free" TeX_2 is synthetically inaccessible; coordination to the tellurium tetrahalides with spontaneous reduction of tellurium to the +2 oxidation state is the most common method, oxidative addition of X_2 to phosphine tellurides is also possible [5, 6, 11, 45–49]. The action of HCl on TeO_2 in the presence of thiourea was the historical route to the thiourea stabilized $TeCl_2$ complexes (Chart 8.2) [50, 51].

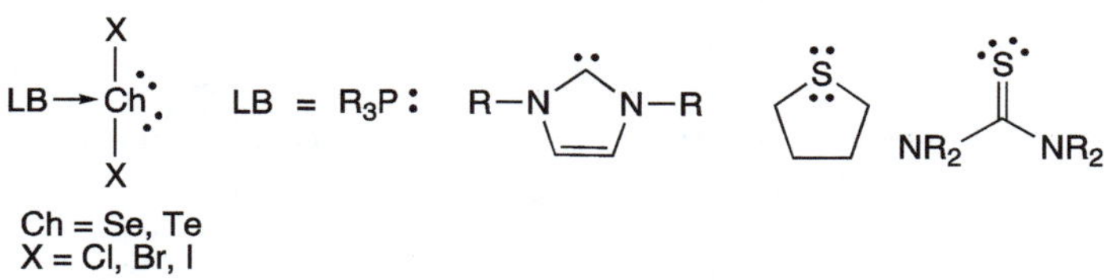

Chart 8.2 Examples of Lewis base stabilized ChX_2 species

Interestingly, for the phosphine-SeX_2 complexes, reaction of R_3P: with free SeX_2 does not give the coordination complex, instead reduction to Se^0 and the formation of the salt $[R_3PCl][Cl]$ is the outcome [49]. These compounds are, for the most part, easily handled stable solids, that may be structurally verified with single crystal X-ray diffraction studies (Fig. 8.2).

Fig. 8.2 T-shaped ChX_2 adducts with Et_3P, NHC and tmtu

Our contributions to this area have been the addition of nitrogen based ligands as stabilization agents for ChX_2. The 1:1 stoichiometric reactions of SeX_2 with $Dipp_2DAB$ or $Dipp_2BIAN$ (Dipp = 2,6-diisopropylphenyl; DAB = diazabutadiene; BIAN = bis(aryl)iminoacenapthene) resulted in the immediate generation of dark red solutions, and simple removal of the solvent allows for deeply coloured SeX_2•DAB complexes (**22**, **23**) to be isolated in nearly quantitative yields [52]. The solid-state structures, as determined by single crystal X-ray diffraction studies revealed compounds based on a square planar selenium centre, with long Se-N bonds (2.26–2.35 Å), reflective of the "dative" interaction between the DAB ligand and SeX_2. For comparison, the standard for a typical Se-N single bond is approximately 1.85 Å [53]. Cowley and Reeske reported a related compound with $Dipp_2BIAN$ stabilizing a TeI_2 fragment (**24**), generated from the reaction of $Dipp_2BIAN$ with TeI_4 by reductive elimination of I_2 (Scheme 8.11) [46].

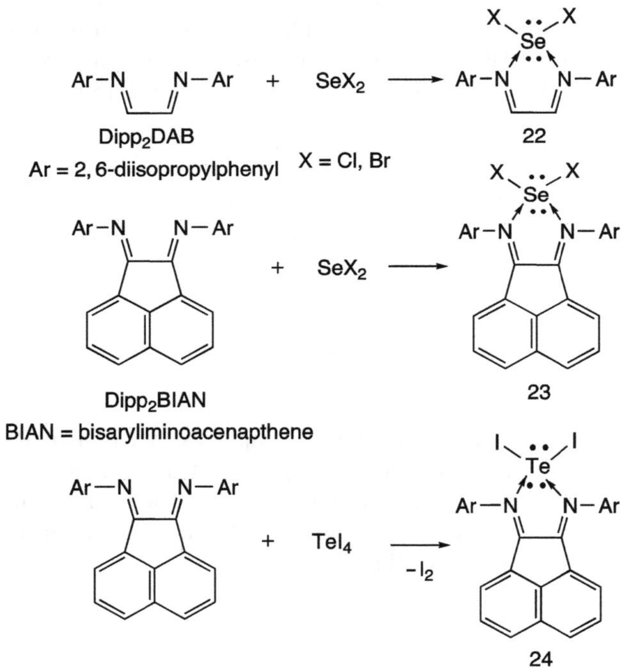

Scheme 8.11 Reaction of aryl substituted diazabutadiene ligands with chalcogen halides in the generation of DAB•ChX_2 complexes

The diazabutadiene complexes of $SeCl_2$ were found to be stable when exposed to the ambient atmosphere of the laboratory for periods of at least 2 months (as monitored by FT-Raman spectroscopy), and thus far have proven indefinitely stable (>1 year) when stored under N_2. This is in stark contrast with free $SeCl_2$, which decomposes almost immediately upon exposure to ambient conditions. Another difference from SeX_2 is that the DAB complexes are tolerant of a wide variety of organic solvents, whereas for free SeX_2 certain common solvents such as CH_2Cl_2 must be avoided (Fig. 8.3).

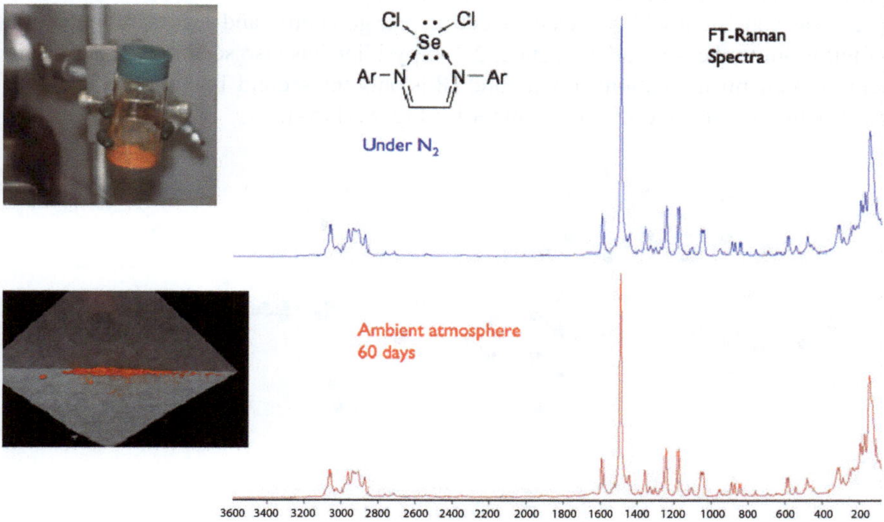

Fig. 8.3 Comparison of compound X stored under N_2 (*blue*) and left in the open (*red*)

The 2,2′-bipyridine (bipy) ligand was also used to stabilize SeX_2 *via* direct reaction [52]. Surprisingly, **25** represents the first selenium complexes using the ubiquitous bipy ligand. For tellurium (Scheme 8.12), it was found that reaction of TeX_4 with bipy

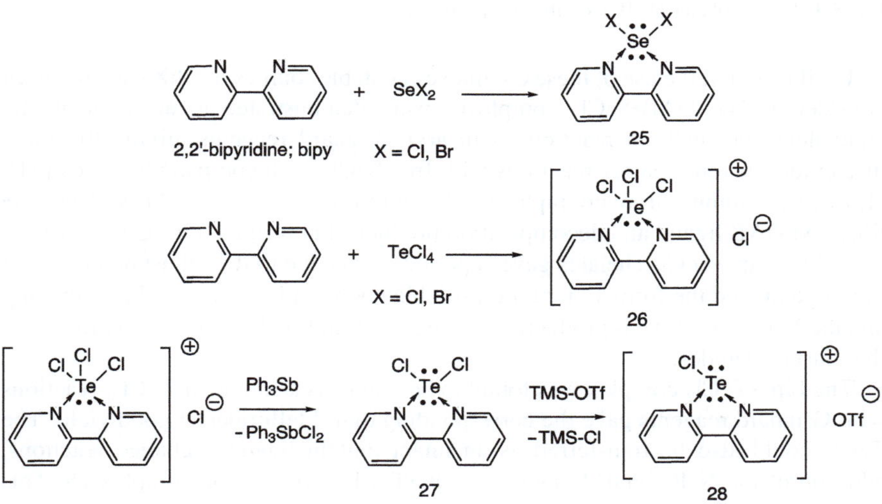

Scheme 8.12 Synthesis of bipy•ChX_2 complexes

gave Te(IV) coordination complexes (**26**), which were susceptible to facile reduction to TeX_2 compounds (**27**) using Ph_3Sb as the reductant. A halide abstraction reaction was employed to generate a cationic derivative (**28**), as single crystals could not be grown to confirm the identity of **27**. The T-shaped geometry about the tellurium centre in **28** was consistent with an AX_3E_2 electron pair geometry, and successful reduction of tellurium to the +2 oxidation state. 2,2′-bipyridine has also seen only scant use in tellurium chemistry, cationic compound **28** is only the second Te-bipy compound to be structurally characterized (Scheme 8.12; Fig. 8.4) [54].

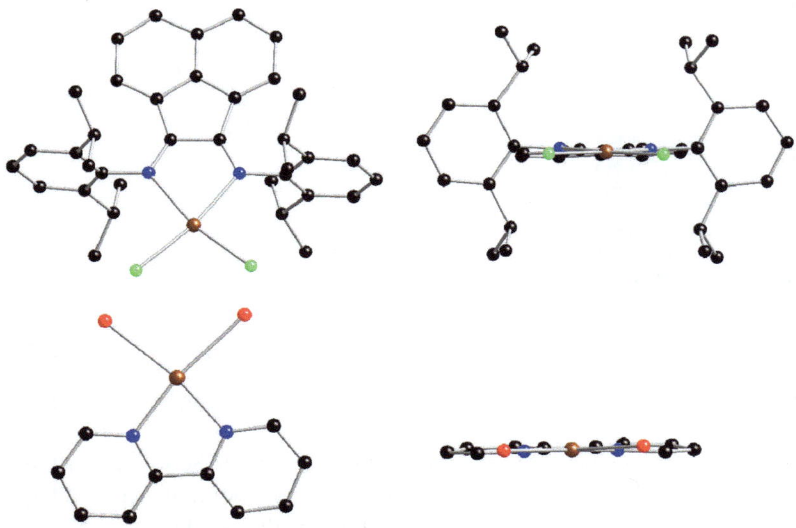

Fig. 8.4 Square planer DAB and bipy complexes of SeX_2

Until recently, the use of these complexes as stable sources of ChX_2 had not been considered. The DAB•$SeCl_2$ complexes were demonstrated to act as synthetic equivalents of $SeCl_2$ in reactions with aryl Grignard reagents, giving the same diselenides as reported for reactions with free $SeCl_2$ with comparable yields [31]. However, column chromatography was required to separate the Ar_2Se from the liberated DAB ligand and decomposition products that formed during the reaction. Use of the bipy•$SeCl_2$ chelates gave superior results due to the robust nature of the bipy ligand, and the formation of an insoluble coordination complex between bipy and the $MgCl_2$ metathesis product, obviating the need for chromatography to isolate the diaryl selenides.

The bipy•$TeCl_2$ complex was found to be viable as a source of $TeCl_2$, reactions with Grignard reagents gave the corresponding diaryl tellurides in good yield. The $TeCl_2$ could also be transferred as an intact unit in ligand exchange reactions, addition of the NHC $^{i}Pr_2IM$ gave the NHC•$TeCl_2$ coordination complex **29**. This transformation is significant, in that Kuhn et al. reported the synthesis of the TeI_2 analogue of **29** by the reaction of the NHC with TeI_4, with reductive elimination of I_2 [11]. However, they were unable to perform the synthesis using $TeCl_4$ or $TeBr_4$ due to the increased reactivity of the lighter halogens, giving a complex mixture. Use of a stabilized $TeCl_2$ fragment circumvents reduction of tellurium, and allows for **29** to be easily isolated. Chivers et al. recently used the $TeCl_2$•tmtu complex initially reported by Foss as a source of $TeCl_2$ [55]. Here, an unusual transformation then occurred, resulting in a disproportionation of the tellurium to Te^0 and Te(IV), where the Te^0 precipitated from the reaction mixture and Te(IV) was incorporated into a partially decomposed ligand framework (**30**) (Scheme 8.13).

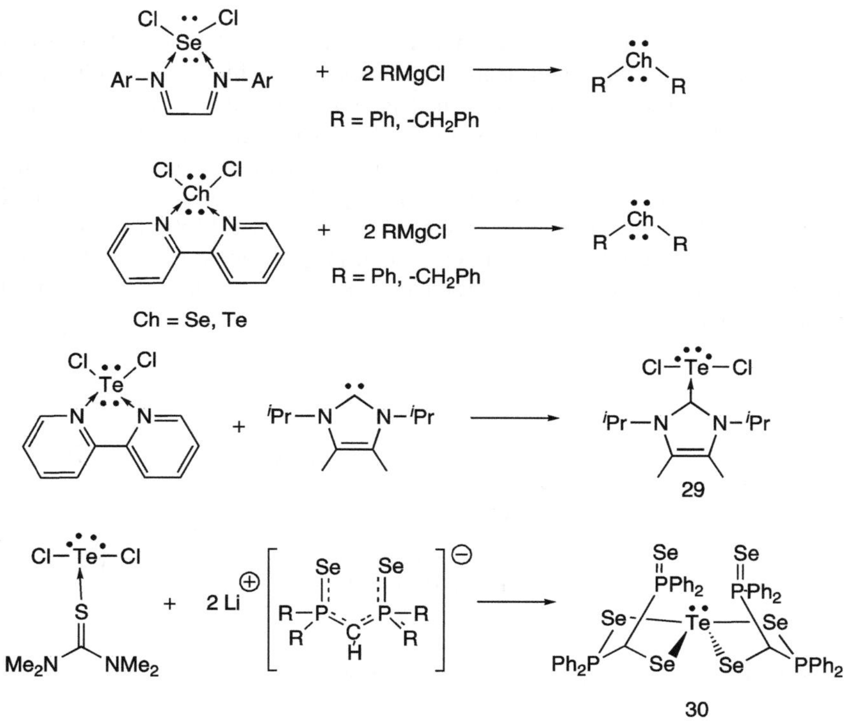

Scheme 8.13 Use of base stabilized ChX_2 complexes as synthetic equivalents of ChX_2

8.4 $[Ch]^{2+}$ Synthons

The selenium dihalides can retain some of the redox activity, which hampers the investigation of their chemistry, for example reduction and chlorination reactions upon the addition of organophospines. Therefore, materials containing a $[Ch(II)]^{2+}$ synthetic equivalent, without the reactivity of halides are desirable starting materials for the study of electrophilic group 16 chemistry. This chemistry is nascent, with only a few reports having appeared, although the area is currently undergoing growth as the usefulness of these synthons becomes apparent.

The first compound of this type could be considered the *pseudo*-halide $Te(CN)_2$, first reported in 1908 [56]. In a more recent publication, where its crystal structure was determined, Klapotke called $Te(CN)_2$ a "forgotten compound" [57]. Synthetic applications have yet to be realized. Selenium dicyanide is also known, but again its use as an electrophilic $[Se]^{2+}$ source has not been explored [58]. Interestingly, the binary cyandies are much more stable than the corresponding Ch(IV) tetracyanides, which decompose explosively to $Ch(CN)_2$ and cyanogens, in contrast with the halides where the higher oxidation state species are more stable [59]. This fact has implications for the use of *pseudo*-halides as superior alternatives to the binary halides for the study of electrophilic Ch(II) chemistry.

The use of $[Ch]^{2+}$ materials using a sulfur based support has started to develop, where the sulfur ligand is either neutral or monoanionic which were synthesized in an exhaustive series of studies published in *Acta Chemica Scandanavia* by the group of Foss from the 1950s to the 1980s, all of which are currently available as "open access". The first example of these complexes being used as a $[Ch]^{2+}$ synthon was by Woollins et al. in 2000 who used $[Te(tu)_4][Cl]_2$ in the generation of Te(II) centred imidophosphinate complexes (e.g. **31**) (Scheme 8.14) [60]; Ibers et al. also used this method to produce similar compounds [61]. It should be noted that many substrates might not tolerate the introduction of water into the system from this hydrated reagent.

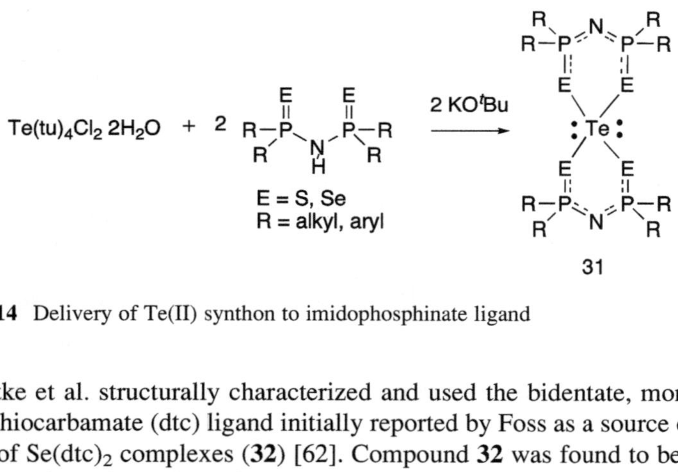

31

Scheme 8.14 Delivery of Te(II) synthon to imidophosphinate ligand

Klapötke et al. structurally characterized and used the bidentate, monoanionic diethyldithiocarbamate (dtc) ligand initially reported by Foss as a source of Se^{2+} in the form of $Se(dtc)_2$ complexes (**32**) [62]. Compound **32** was found to be versatile for the formation of a variety of diorganoselenides from Grignard reagents, which the researchers used as precursors for R_2SeF_2 compounds *via* oxidation by XeF_2, ultimately for the formation of $R_2Se(N_3)_2$ diorganoseleniumdiazides. $Ch(dtc)_2$ has also been used to introduce Ch(II) to organolithium reagents [63]. Despite being an anionic ligand, and therefore expected to make relatively strong bonds, dtc is inclined to release the $[Ch]^{2+}$ fragment due to the bite angle of the dtc, giving a highly distorted and therefore strained square planar arrangement about the chalcogen centre. The key consideration for its use is that a cation (e.g. Li^+, Mg^{2+}) must be available to replace the $[Ch]^{2+}$ centre, and drive the reaction (Fig. 8.5).

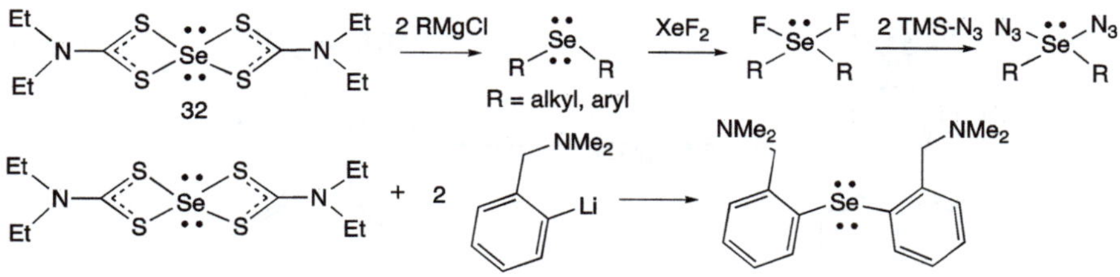

Fig. 8.5 Tetramethylthiourea-tellurium dication (*left*) and dtc$_2$(Se) complex used as sources of Ch(II)

The DAB•ChX$_2$ compounds can be used as precursors for halide abstraction and/or halide metathesis reactions to generate either base stabilized $[Ch]^{2+}$ or Ch (OTf)$_2$ complexes depending on the DAB ligand used. For example, the reaction of AgOTf with Cowley's Dipp$_2$BIAN stabilized TeI$_2$ results in the elimination of AgI and the formation of a Te(OTf)$_2$ complex (**33**) (Scheme 8.15) [64].

An examination of the metrical parameters for **33** obtained from single crystal X-ray diffraction studies reveal some interesting features. The Te-N bonds (2.15 Å) were found to be much shorter than those in the TeI$_2$ complex (2.4 Å) reflecting the greatly increased electrophilicity of the tellurium centre. There were also long Te-O bonds of 2.32–2.48 Å, much longer than a typical Te-O bond of ~2 Å. Despite the weak nucleophilicity of the triflate anion, the Te-O bonds are retained in solution based on $^{19}F\{^1H\}$ NMR studies. The chemical shift of the triflates was found to be $\delta = -78.2$ ppm, compared with $\delta = -79.0$ ppm (CH_2Cl_2) for a standard for purely ionic triflate ([NOct$_4$][OTf]). Thus, **33** can be considered to be a base stabilized form of Te(OTf)$_2$, which would be expected to be a powerful electrophile. Some evidence for this property can be inferred from the extremely downfield resonance observed in the $^{125}Te\{^1H\}$ NMR spectrum, with **33** having a chemical shift of $\delta = 2{,}853$ ppm.

Synthesis of a selenium analogue involved the action of TMS-OTf on Cy$_2$DAB•SeCl$_2$ gives a dicationic complex (**34**), with only weak Se-O interactions in the solid-state between the selenium centre and the triflate counterions, which are not retained in solution $^{19}F\{^1H\}$ NMR, $\delta = -78.5$ ppm (MeCN); [NOct$_4$][OTf]

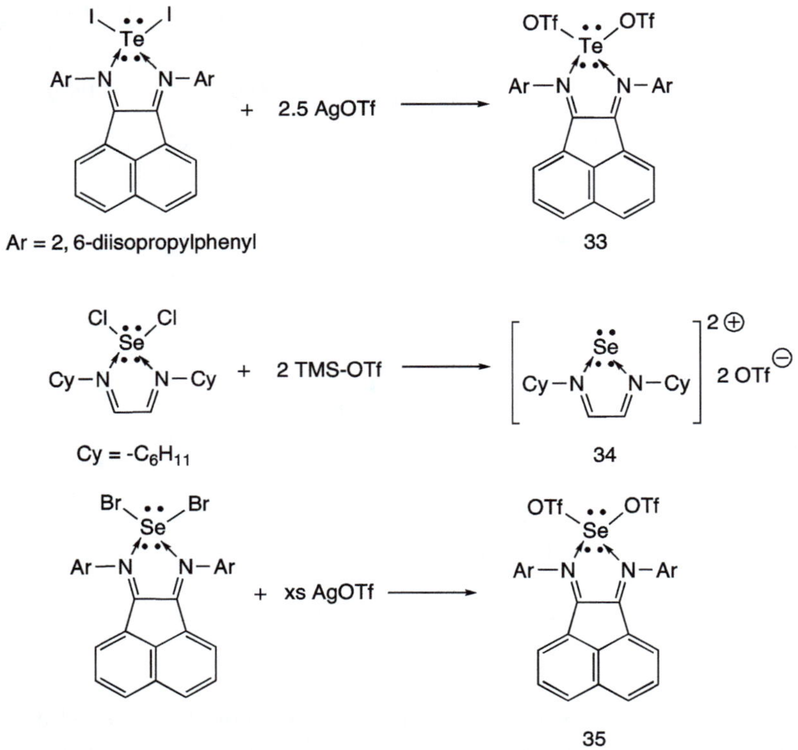

Scheme 8.15 Synthesis of bistriflate *pseudo*-halide selenium and tellurium complexes

δ = −78.5 ppm (MeCN) [65]. The $Dipp_2BIAN\bullet SeBr_2$ complex may also be used to generate a base stabilized $Se(OTf)_2$ species (**35**), however in this case single crystals of sufficient quality for X-ray diffraction studies could not be obtained (Fig. 8.6).

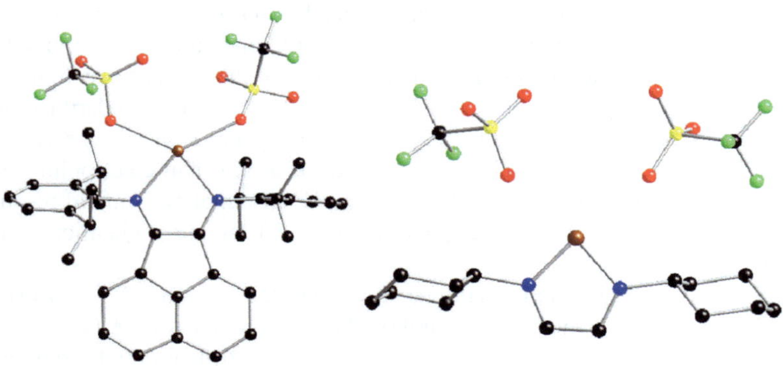

Fig. 8.6 Synthons of Te^{2+} (*left*) and Se^{2+} (*right*)

These complexes, with only dative, or weak covalent bonds to the chalcogen centres were found to susceptible to ligand exchange reactions with a variety of other Lewis bases, including 2,2′-bipyridine, 4-DMAP and the NHC $^{i}Pr_2IM$, which all gave selenium or tellurium centred dications. These transformations demonstrated the transferability of the dicationic chalcogen centres, indicating the ability of these compounds to act as sources of $[Ch]^{2+}$. Ligand exchange is an extremely rare mode of reactivity for the group 16 elements, likely due to the paucity of coordination complexes containing a Lewis acidic chalcogen centre (Scheme 8.16; Fig. 8.7).

In the case of 4-DMAP a 4-coordinate "pinwheel" coordination complexes (**36**) were obtained bearing nearly perfect D_{4h} symmetry for the dications. For the NHC, 2-coordinate dications were synthesized (**37**), which are structural analogues of the recently reported family of compounds known as the "bent allenes" or "carbodicarbenes" of Bertrand et al., centered about a C^0 atom [66, 67]. However,

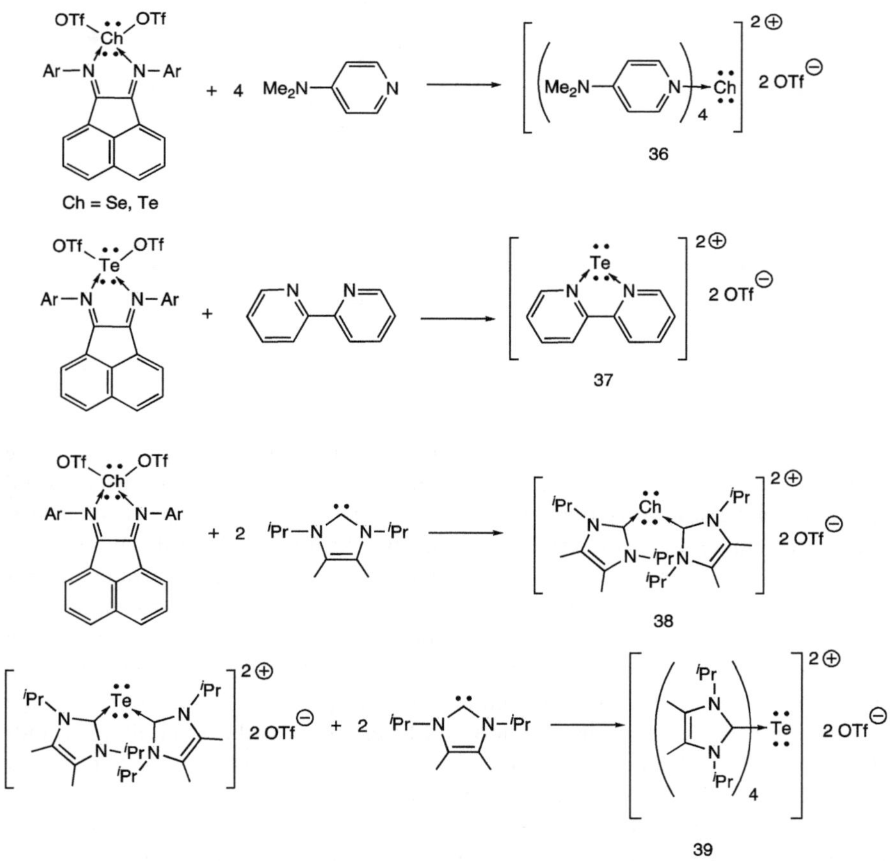

Scheme 8.16 Transfer of $[Ch]^{2+}$ from DAB supported $Ch(OTf)_2$ "Ch^{2+}" synthons

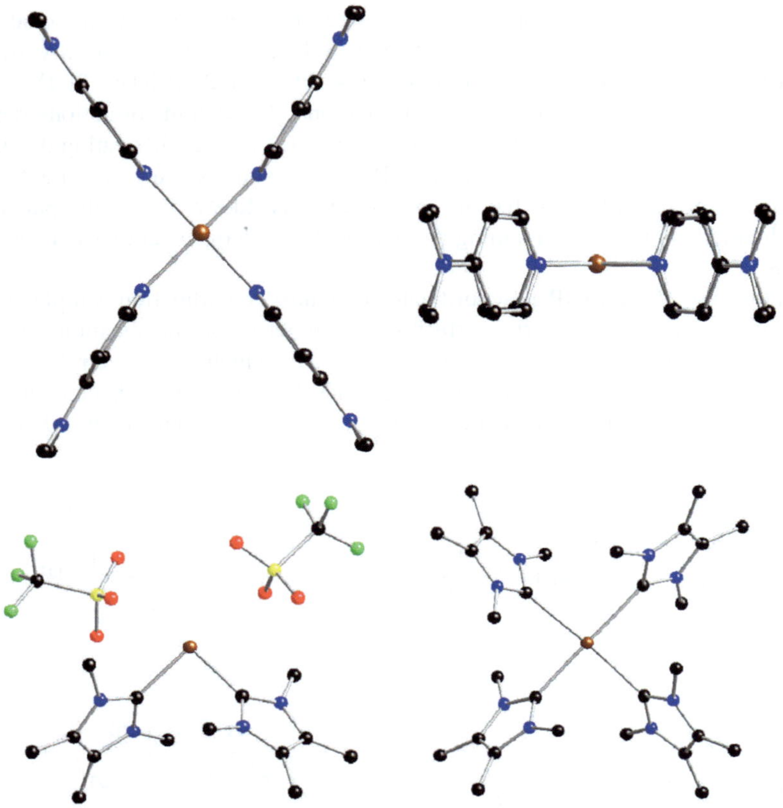

Fig. 8.7 Dicationic complexes

there are distinct differences in the electronic structures, which are borne out in the metrical parameters. The C-C-C bond angle in X is 134°, borne from the partial multiple bond character of the C-C bonds, while in the heavy group 16 congeners, the C-Ch-C angle is almost exactly 90° reflecting the use of unhybridized p-orbitals to form the Ch-C bonds (Scheme 8.16; Fig. 8.7). A similar bonding arrangement is found in the monocationic phosphorus analogues reported by Macdonald et al., although theoretical studies found some multiple character for the C-P bonds, in agreement with a larger C-P-C angle of 98° [12]. The carbodicarbene has been noted to be a strong nucleophile, and has been proposed for use as an even more powerful Lewis base than the ubiquitous N-heterocyclic carbene family of ligands [68]. In contrast, dicationic **37** acts as a Lewis acid, the addition of two further equivalents of NHC gives a square planar "paddlewheel" complex (**38**), bearing a D_4 symmetric "propeller" geometry. The strong *trans* influence of the NHCs rendered each Te-C bond extremely long and therefore labile, making the resulting complex much less stable than the 2-coordinate **39**.

8.5 Conclusions and Outlook

The spontaneous reduction of selenium and tellurium from the +4 to the +2 oxidation state is a common occurrence when examining the interaction of the stable tetrahalides of these elements with neutral donor ligands. Often, the pathway involves the reductive elimination of highly reactive X_2, which should be anticipated and dealt with in order to achieve reasonable yields of the desired compound. Alternatively, the binary halides may be used as +2 oxidation state starting materials, often giving better results due to the circumvention of the redox process. The binary halides are either unstable (Se) or unavailable (Te), however the formation of a coordination complex using a Lewis base has proven to be an effective method of trapping ChX_2 in a stable, solid form. Recently these complexes have found utility as transfer reagents for ChX_2. *Pseudo*-halides, or other forms of $[Ch]^{2+}$ also show great promise as new starting materials for Ch(II) chemistry, without suffering any of the detrimental reactivity resulting from the presence of $[X]^-$ in the system. Ideally these recent discoveries allow for an expansion in the area of electrophilic group 16 chemistry; largely unexplored are potential applications into organic syntheses and materials chemistry.

References

1. Husebye S, George JW (1969) Inorg Chem 8:313–319
2. Fleischer H, Schollmeyer D (2002) Acta Crystallogr E58:o901–o903
3. Katsaros N, George JW (1969) J Inorg Nucl Chem 31:3503–3508
4. Husebye S, Tornroos KW (2000) Acta Crystallogr C56:1242–1244
5. Williams DJ, Bevilacqua VLH, Morson PA, Pennington WT, Schimek GL, Kawai NT (2000) Inorg Chim Acta 308:129–134
6. Levason W, Reid G, Victor M, Zhang W (2009) Polyhedron 28:4010–4016
7. Carmalt CJ, Norman NC, Farrugia LJ (1995) Polyhedron 14:1405–1413
8. Cordes AW, Hughes TV (1964) Inorg Chem 3:1640–1641
9. Beattie IR, Milne M, Webster M, Blayden HE, Jones PJ, Killean RCG, Lawrence JL (1969) J Chem Soc 482–485
10. Dutton JL, Tabeshi R, Jennings MC, Lough A, Ragogna PJ (2007) Inorg Chem 46: 8594–8602
11. Kuhn N, Abu-Rayyan A, Piludu C, Steimann M (2005) Heteroatom Chem 16:316–319
12. Ellis BD, Dyker CA, Decken A, Macdonald CLB (2005) Chem Commun 1965–1967
13. Ellis BD, Carlesimo M, Macdonald CLB (2003) Chem Commun 1946–1947
14. Ellis BD, Macdonald CLB (2006) Inorg Chem 45:6864–6874
15. Schmidpeter A, Lochschmidt S, Sheldrick WS (1982) Angew Chem Int Ed 21:63–64
16. Norton EL, Szekely KLS, Dube JW, Bomben PG, Macdonald CLB (2008) Inorg Chem 47: 1196–1203
17. Dutton JL, Tindale JJ, Jennings MC, Ragogna PJ (2006) Chem Commun 2474–2476
18. Dutton JL, Sutrisno A, Schurko RW, Ragogna PJ (2008) Dalton Trans 3470–3477
19. Gushwa AF, Richards AF (2008) Eur J Inorg Chem 2008:728–736
20. Dutton JL, Martin CD, Sgro MJ, Jones ND, Ragogna PJ (2009) Inorg Chem 48:3239–3247
21. Dutton JL, Ragogna PJ (2009) Inorg Chem 48:1722–1790

22. Kovacs A, Konings RJM (1997) J Mol Struct 410–411:407–410
23. Fernholt L, Haaland A, Volden HV (1985) J Mol Struct 128:29–31
24. Milne J, Lamoureux M (1990) Polyhedron 9:589–595
25. Maaninen A, Chivers T, Parvez M, Pietikäinen J, Laitinen R (1999) Inorg Chem 38: 4093–4097
26. Del Bel Belluz P, Cordes AW, Kristof EM, Kristof PV, Liblong SW, Oakley RT (1989) J Am Chem Soc 111:9276–9278
27. Milne J (1985) Polyhedron 4:65–68
28. Konu J, Maaninen A, Paananen K, Ingman P, Laitinen RS, Chivers T, Valkonen J (2002) Inorg Chem 41:1430–1435
29. Maaninen T, Chivers T, Laitinen R, Schatte G, Nissinen M (2000) Inorg Chem 39:5341–5347
30. Konu J, Chivers T, Tuononen Inorg HM (2006) Chem 45:10678–10687
31. Zade SS, Panda S, Singh HB, Wolmershauser G (2005) Tet Lett 46:665–669
32. Song L, Hu Q, Fan H, Tang M, Yang Z, Lu G (2002) Organometallics 21:2468–2472
33. Song L, Fan H, Hu Q, Yang Z, Sun Y, Gong F (2003) Chem Eur J 9:170–180
34. Amosova SV, Penzik MV, Albanov AI, Potapov VA (2009) J Organomet Chem 694: 3369–3372
35. Potapov VA, Kurktov EO, Albanov AI, Amosova SV (2008) Russ J Org Chem 44:1547–1548
36. Braverman S, Jana R, Cherkinsky M, Gottlieb HE, Sprecher M (2007) Synlett 2007:2663–2666
37. Potapov VA, Amosova SV, Belozerova OV, Albanov AI, Yarosh OG, Voronkov MG (2003) Chem Heterocycl Comp 39:549–550
38. Patra A, Wijsboom YH, Zade SS, Li M, Sheynin Y, Leitus G, Bendikov M, (2008) J Am Chem Soc 130:6734–6736
39. Wynne KJ, Pearson PS (1971) J Chem Soc Chem Commun 293–294
40. Wynne KJ, Pearson PS, Newton MG, Golen J (1972) Inorg Chem 11:1192–1196
41. Godfrey SM, Jackson SL, McAuliffe CA, Pritchard RG (1998) Dalton Trans 4201–4204
42. Kuhn N, Kratz T, Henkel G (1994) Chem Ber 127:849–851
43. Aragoni MC, Arca M, Blake AJ, Devillanova FA, du Mont WW, Garau A, Isaia F, Lippolis V, Verani G, Wilson C (2001) Angew Chem Int Ed 40:4229–4232
44. Williams DJ, Wynne KJ (1976) Inorg Chem 15:1449–1451
45. Konu J, Chivers T (2006) Dalton Trans 3941–3946
46. Reeske G, Cowley AH (2006) Chem Commun 4856–4858
47. Hrib CG, Jones PG, du Mont WW, Lippolis V, Devillanova FA (2006) Eur J Inorg Chem 1294–1302
48. Foss O, Maartmann-Moe K (1987) Acta Chem Scand A41:121–129
49. Lee LM, Elder PJW, Cozzolino AF, Yang Q (2010) I Vargas Baca Main Group Chem 9: 117–133
50. Foss O, Hauge S (1959) Acta Chem Scand 13:1252–1253
51. Foss O, Hauge S (1961) Acta Chem Scand 15:1616–1617
52. Dutton JL, Farrar GJ, Sgro MJ, Battista TL, Ragogna PJ (2009) Chem Eur J 15:10263–10271
53. Roesky HW, Weber K, Seseke U, Pinkert W, Noltemeyer M, Clegg W, Sheldrick GM (1985) J Chem Soc Dalton Trans 565–571
54. Annan TA, Ozarowski A, Tian Z, Tuck DG (1992) J Chem Soc Dalton Trans 2931–2938
55. Konu J, Chivers T (2010) Chem Commun 46:1431–1433
56. Cocksedge HE (1908) J Chem Soc 93:2175–2177
57. Klapotke TM, Krumm B, Galvez-Ruiz JC, Noth H, Schwab I (2004) Eur J Inorg Chem 24: 4764–4769
58. Klapotke TM, Krumm B, Scheer M (2008) Inorg Chem 47:7025–7028
59. Lentz D, Szwak M (2005) Angew Chem Int Ed 44:5079–5082
60. Birdsall DJ, Novosad J, Slawin AMZ, Woollins JD (2000) J Chem Soc Dalton Trans 435–439
61. Sekar P, Ibers JA (2003) Inorg Chem 42:6294–6299
62. Klapotke TM, Krumm B, Scheer M (2008) Inorg Chem 47:4712–4722

63. Panda A, Mugesh G, Singh HB, Butcher RJ (1999) Organometallics 18:1986–1993
64. Dutton JL, Tuononen HM, Ragogna PJ (2009) Angew Chem Int Ed 48:4409–4413
65. Dutton JL, Battista TL, Sgro MJ, Ragogna PJ (2010) Chem Commun 46:1041–1043
66. Dyker CA, Lavallo V, Donnadieu B, Bertrand G (2008) Angew Chem Int Ed 47:3206–3209
67. Tonner R, Frenking G (2007) Angew Chem Int Ed 46:8695–8698
68. Tonner R, Frenking G (2008) Chem Eur J 14:3273–3289

Chapter 9
Selenium and Tellurium Containing Precursors for Semiconducting Materials

Mohammad Azad Malik, Karthik Ramasamy, and Paul O'Brien

9.1 Introduction

This chapter will provide a short overview of some of the chemistry that has been developed for the preparation of metal selenides or tellurides principally in the form of thin films or small particles; the later most often crystalline with critical dimensions of the order of nanometres. There are a number of related reviews of this general area of materials chemistry [1–4] and some older summaries [5–7]. However, no published work deals solely with the two heavier stable chalcogenides.

Metal chalcogenides have many potential applications: cadmium telluride thin films form one layer in CdS/CdTe polycrystalline solar cells [8], thin films and quantum dots in the CIGS family of materials, for our purposes copper indium or gallium selenide, are important collectors in thin film and hybrid solar cells [9], lead selenide quantum dots have been touted for use in polymer based solar cells [10, 11] and there are numerous other examples.

The chapter will emphasize the development of the approach in which both elements are delivered from a single molecule (the 'single-source' precursor route), the development of the suite of precursors available for such depositions and recent work in our own group. However the growth of group VI containing materials by such chemical methods can perhaps be traced to Manasevit's pioneering work. He used volatile metal alkyls such as dimethylcadmium (Me_2Cd) or dimethylzinc (Me_2Zn) in combination with H_2S, H_2Se, or Me_2Te delivered in dihydrogen [12], as a carrier gas, effectively inventing and defining the process of MOCVD. Many films were grown including ZnS, ZnSe, CdS, CdSe, and CdTe. The present review emphasizes more recent studies and approaches using single-molecular precursors'

M.A. Malik • K. Ramasamy • P. O'Brien (✉)
School of Chemistry, The University of Manchester, Oxford Road, Manchester M13 9PL, UK
e-mail: paul.obrien@manchester.ac.uk

J.D. Woollins and R.S. Laitinen (eds.), *Selenium and Tellurium Chemistry*,
DOI 10.1007/978-3-642-20699-3_9, © Springer-Verlag Berlin Heidelberg 2011

and to some extent our own work. The chapter is structured considering each material type in terms of precursors studied. The structures of the ligands used in such depositions are summarized in Fig. 9.1.

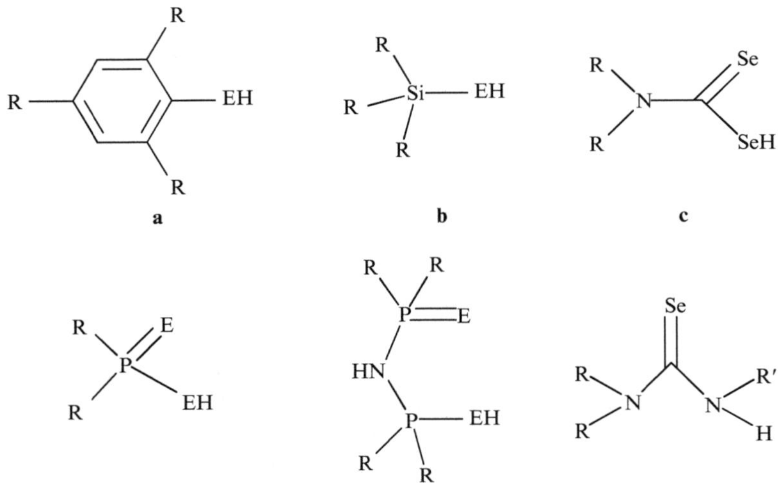

Fig. 9.1 Some common ligands used to prepare precursors: 2,4,6 substituted aryl chalcogenide R = ^{t}Bu (**a**); trimethylsilylchalcogenol; R = [Si(Me)$_3$] (**b**); diselenocarbamate, R and R′ are alkyls (**c**); diselenocarbamic, R is an alkyl E = E′ = Se, E = S and E′ = Se (**d**); dichalcogen-ophosphinic acid, R is an alkyl E = E′ = Se or Te, E = S E′ = Se or Te, E = Se E′ = Te (**e**); imidochalcogenophosphinic acid; (**f**) triorganylselenourea

9.2 II–VI Materials

9.2.1 Metal Selenides/Tellurides from Alkylchalcogenide Complexes

Steigerwald and coworkers [13, 14] reported a series of complexes with the general formula $M(ER)_2$ (M = Zn, Cd, Hg: E = S, Se, Te) in the form of adducts with 1,2-*bis* (diethylphosphino)ethane (depe). Complexes containing one or two mole equivalents of the phosphine have been isolated with 1:2 species being polymeric and the 1:1 as dimmers. Crystal structures for $[Cd_2(SeC_6H_5)_4(depe)]_n$ and $[Hg(SeC_6H_5)_2(depe)]_2$ have been reported [13, 14]. The decomposition of $[Cd_2(SeC_6H_5)_4(depe)]_n$ in 4-ethylpyridine, leads to the deposition of CdSe nanoclusters. Solid-state pyrolysis of the same complex in a vacuum sealed tube also provided CdSe. The decomposition of all polymeric or dimeric [15, 16] complexes result in the deposition of metal chalcogenides.

9.2.2 Metal Selenides/Tellurides from Arylchalcogenolates

A series of low-coordination metal complexes were prepared by Dilworth *et al.* [17] with 2,4,6-tri-*iso*-propylbenzenethiol(tipt). Bochmann *et al.* [18–21] produced a range of precursors for II–VI materials based on 2,4,6-tri-*tert*-butylphenylchalcogenolate. The general preparation method of one such complex is shown in Eqs. 9.1–9.3, and the structure of the product is given in Fig. 9.2.

(9.1)

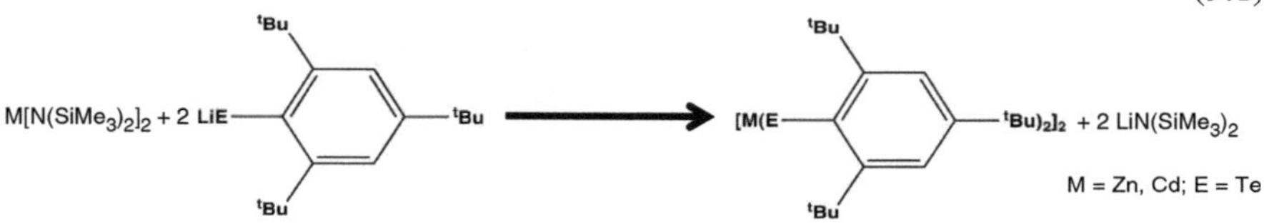

(9.2)

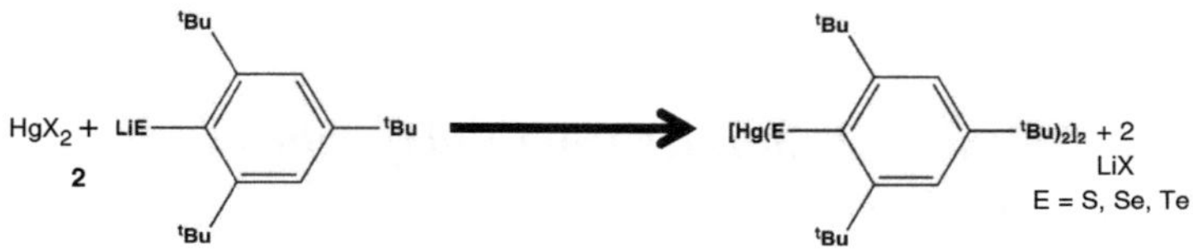

(9.3)

The compounds are dimeric even in the vapour phase and have been used for the deposition of sulfides or selenides by LP-MOCVD (Low Pressure-Metal Organic Chemical Vapour Deposition) [22, 23]. Mixed alkyl complexes [24, 25] with aryl chalcogenates RMSeR′ (SeR′ = $SeC_6H_2{}^iPr_3$-2,4,6; M = Zn, R = Me, Et, ^{n}Pr, ^{i}Pr; M = Cd, R = Me) have also been prepared which can be suitable precursors for AA-CVD (Aerosol Assisted-Chemical Vapour Deposition).

ZnTe nanocrystals were prepared by the thermolysis of $[Zn(TePh)_2][TMEDA]$ (TMEDA = tetramethylethylenediamine) in a mixture of trioctylamine and dodecylamine or trioctylamine and dimethylhexylamine [25]. X-Ray structure of $[Zn(TePh)_2][TMEDA]$ shows that it is monomeric and the zinc center adopts a distorted tetrahedral geometry. TGA (thermogravimetric analysis) of this complex showed the dissociation of the TMEDA in the first step and then clean residue of ZnTe at higher temperature [25].

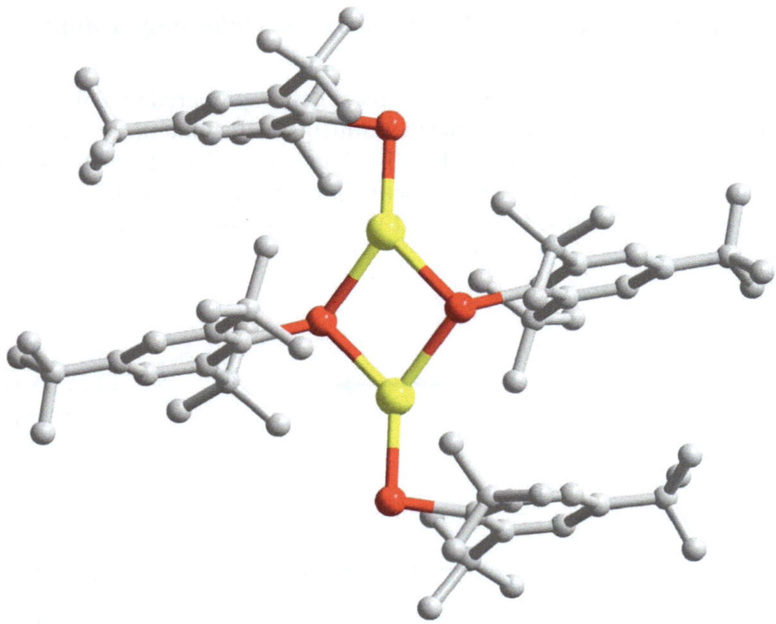

Fig. 9.2 Molecular structure of $[\{Cd(SeC_6H_2Bu^t_3\text{-}2,4,6)_2\}_2]$

9.2.3 Metal Selenides/Tellurides from Bulky Silicon Based Complexes

Arnold and coworkers used a series of precursors with the general formula $M[ESi(SiMe)_3]_2$ (M = Zn, Cd, Hg and E = S, Se or Te) for the deposition of chalcogenides (Fig. 9.3) [26, 27]. The most detailed work has been reported on the tellurides: thin films of the tellurides have been deposited by LP-MOCVD [28]. The tellurium-containing ligand, $HTeSi(SiMe_3)_3$, is termed HSitel and this reagent is potentially useful in the preparation of metal tellurolates [29–31]. Metal complexes of Sitel are generally prepared as illustrated below Eq. 9.4. The decomposition in the MOCVD proceeds *via* an elimination path as shown in Eq. 9.5.

$$[M(TeSi(SiMe_3)_3]_2(s) \rightarrow [M(TeSi(SiMe_3)_3]_2(g) \tag{9.4}$$

$$[M(TeSi(SiMe_3)_3]_2(g) \rightarrow \begin{array}{l} MTe + Te[Si(SiMe_3)_3]_2 \\ (M = Zn,\ Cd,\ Hg) \end{array} \tag{9.5}$$

Zinc telluride was deposited at temperatures between 250°C and 350°C onto quartz, silicon, InAs and GaSb substrates. The cadmium precursor showed the deposition of hexagonal CdSe films.

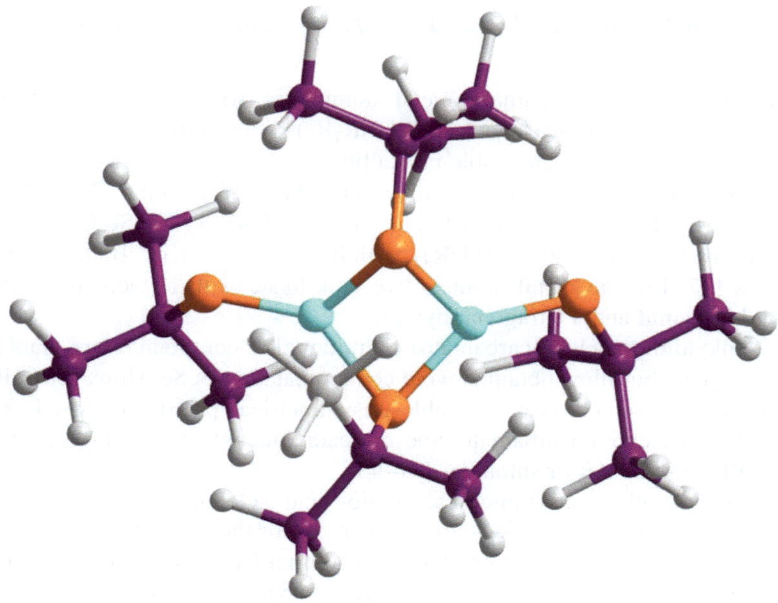

Fig. 9.3 Molecular structure of [Zn(TeSi(SiMe)$_3$)$_2$]

9.2.4 Metal Selenides from Diselenophosphinato Complexes

We have synthesized a series of metal complexes with the general formula [M(R$_2$PSe$_2$)$_n$] (M = Zn^{II}, Cd^{II}, Pb^{II}, In^{III}, Ga^{III}, Cu^{I}, Bi^{III}, Ni^{II}; R = ^{i}Pr, Ph) and Mo$_2$^{V}O$_2$Se$_2$(Se$_2$P^{i}Pr$_2$)$_2$ [32]. These complexes have been used for the deposition of metal selenide thin films by AA-CVD method [33]. The compound [R$_2$PSe$_2$]$_2$Se (R = ^{i}Pr, Ph) was prepared by the reaction of NEt$_3$ with ^{i}Pr$_2$PCl or ^{i}Pr$_2$PCl and HSiCl$_3$ in cold toluene. Further development of the synthesis method have led to a new type of diselenophosphinate ligand. With the notation that (R$_2$PSe$_2$)$^-$ anion may be stabilized, and crystallize as ionic compound if there is enough counter cation in reaction solution, an excess of HSiCl$_3$/NEt$_3$ have tendency of producing HNEt$_3$$^+$ cation. However this did not give expected result due to the fact that this combination led to the forming of (HNEt$_3$)(SiCl$_3$) precipitate. This problem was solved by using HSi(Et)$_3$ instead of HSiCl$_3$. As a result, the ionic compounds (HNEt$_3$)[R$_2$PSe$_2$] were obtained. The synthesis process is represented in Eq. 9.6.

$$\begin{aligned} &R_2PCl\ + 2\ HSiEt_3 + 2\ NEt_3 \xrightarrow[\text{toluene}]{\text{Se}} (HNEt_3)^+(R_2PSe_2)^- \\ &\quad + HNEt_3Cl + Si_2Et_6 \end{aligned} \tag{9.6}$$

Solid state structures of several of these complexes were determined by single crystal X-ray crystallography [33].

9.2.5 Metal Selenides from Dialkylselenocarbamato Complexes

Dialkyldithio-/diseleno-carbamato metal complexes with the general formula ($M(E_2CNR_2)_2$ (symmetrical) or $M(E_2CNR_1R_2)_2$ (unsymmetrical) R = alkyl, E = S, Se; M = Zn, Cd) are stable crystalline solids. Their solid state structures have been determined by X-ray crystallography. Most of them *e.g.* $Zn(S_2CNR_2)_2$, with R = Me, Et, or ^{i}Pr [34–36] and in $Cd(S_2CNR_2)_2$ $Zn(Se_2CNEt_2)_2$, Cd $(Se_2CNEt_2)_2$ (Fig. 9.4), $Zn(S_2CNMeR)_2$, with R = Et, ^{n}Pr, ^{i}Pr, or ^{n}Bu have dimeric structures [37–40], the metal atom is five coordinate with a geometry between trigonal bipyramid and a tetragonal pyramid.

Bis(dialkyldithio-/seleno-carbamato)-cadmium/zinc compounds are stable for periods of years but films obtained were contaminated with Se. However O'Brien *et al.* developed some novel air stable unsymmetrical precursors based on bis (methylalkyldiselenothiocarbamato)zinc or -cadmium which decompose cleanly in MOCVD to selenides or sulphides [41–43].

A series of other unsymmetrical dithio- and selenocarbamates were also synthesised to be used as single source precursors for the deposition of thin films. *Bis*(n-hexy(methyl)-dithio/selenocarbamato)cadmium/zinc were proved to be the best unsymmetrical derivatives for the growth of chalogenides [44].

A mechanistic study for the decomposition behavior reported by O'Brien *et al.* [43] showed that diethyldiselenocarbamato complexes of cadmium or zinc are poor sources for the deposition of ZnSe or CdSe films. Under similar reaction conditions the diethyldiselenocarbamates precursors give films of the metal selenide heavily contaminated with selenium [39]. However, the mixed alkyl diselenocarbamates complex Eq. 9.7 (Fig. 9.5) have been used successfully to deposit thin films of CdSe or ZnSe [45, 46].

$$Me_2M + M(Se_2CNEt_2)_2 \rightarrow 2\ MeM(Se_2CNEt_2) \quad M = Cd, Zn \tag{9.7}$$

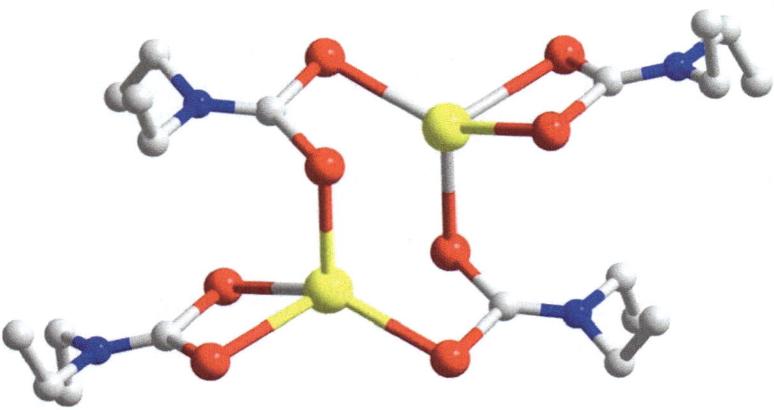

Fig. 9.4 Molecular structure of $[\{Cd(SeCNEt_2)_2\}_2]$

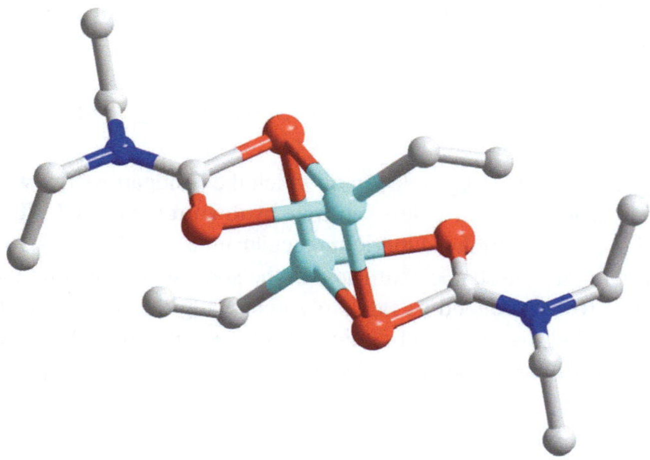

Fig. 9.5 Molecular structure of $[(Et)Zn(Se_2CNEt_2)_2]$

$Zn(Se_2CNMe^nHex)_2$ is monomeric in the solid phase [41] in contrast to the analogous diethyldiselenocarbamates and the mixed alkyl diselenocarbamates complexes which are both dimers.

9.2.6 Metal Selenides from Mixed Alkyldiselenocarbamato Complexes

Noltes [47] reported the first preparation of these compounds by a insertion reaction [48] (Eq. 9.8).

$$2\ MeZn(NEt_2) + 2\ CS_2 \rightarrow \left[MeZn(S_2CNEt_2)_2\right] \tag{9.8}$$

Another method for the preparation of these compounds is conproportionation (Eq. 9.9) [49–53].

$$Me_2Zn + Zn(S_2CNEt_2)_2 \rightarrow \left[MeZn(S_2CNEt_2)_2\right] \tag{9.9}$$

All these compounds are dimers [48] in the solid state and the parent dimeric structure has been confirmed for a wide range of compounds where R = Me, Et, ^{t}Bu or Me_3CCH_2, M = Zn or Cd, E = S or Se, and R′ = Me or Et [54, 55] single crystal X-ray structures of neopentylcadmium diethyldiselenocarbamate and ethylzincdiethyldithiocarbamate have been determined [56, 57].

An interesting application of the conproportionation reaction is the preparation of a mixed species such as methylcadmium/methylzinc diethyldiselenocarbamate [46] which is useful for the deposition of thin films of ternary solid solutions of

$Cd_{0.5}$,$Zn_{0.5}$Se. Thus the reaction of Me_2Zn with $Cd(Se_2CNEt_2)_2$ gave $Me_2CdZn(Se_2CNEt_2)_2$ as shown in Eq. 9.10.

$$Me_2Zn + Cd(Se_2CNEt_2)_2 \rightarrow [Me_2CdZn(Se_2CNEt_2)_2] \quad (9.10)$$

A polycrystalline $Cd_{0.5}Zn_{0.5}Se$ layer, of which the bandgap energy was estimated to be 2.1 eV, was deposited on a glass substrate by low-pressure MOCVD. The mixed metal complex showed similar dimeric molecular units, $[RM(Se_2CNEt_2)]_2$ to other alkylmetal dithio- and diselenocarbamates. In the solid-state structure, the cadmium and zinc atoms were modelled as randomly occupying the metal sites [46]. Many of these mixed alkyl diseleno- and dithio-carbamate compounds have been used to deposit thin films of metal chalcogenides by low-pressure MOCVD [45].

9.2.7 *Metal Selenides/Tellurides from Dichalcogenoimidodiphosphinato Complexes*

The dichalcogenoimidodiphosphinate anions were first synthesized by Schmidpeter *et al.* in the 1960 [51–53]. Woollins and co-workers initiated the chemistry of the selenium analogue [58]. Much of the early development of the coordination chemistry of these ligands with both main group [59] and transition metals [59, 60] was focused on the phenyl derivatives. Metal complexes incorporating the more volatile *iso*-propyl ligand were proved to be better precursors for the variety of binary metal selenides by CVD techniques [61–63].

Imido-diisopropylphosphineselenides are prepared by the oxidative insertion of elemental selenium [64–66]. The cadmium *bis*(imidodiisopropylphosphine selenide) compound, $Cd[N(SeP^iPr_2)_2]_2$ (Fig. 9.6) was first synthesised by Woollins *et al.* [67] from diisopropylchlorophosphine *via* a two step strategy. Improved yields for $Cd[N(SeP^iPr_2)_2]_2$ can be afforded by utilising $CdCl_2$/NaOMe conditions rather than metal carbonates [68, 69] (Eqs. 9.11–9.13).

$$NH(SiMe_3)_2 + {}^iPr_2PCl \xrightarrow[\text{Stirring (3 hours)}]{50-70^\circ C} NH(P^iPr_2)_2 + 2\ Me_3SiCl \quad (9.11)$$

$$NH(P^iPr_2)_2 \xrightarrow[\text{reflux}]{E\,=\,S,\,Se} NH(EP^iPr_2)_2 \quad (9.12)$$

$$2\ NH(ESiMe_3)_2 \xrightarrow[\text{MeOH}]{\text{NaOMe/MCl}_2} M[(EP^iPr_2)_2N]_2 + 2\ NaCl \quad (9.13)$$

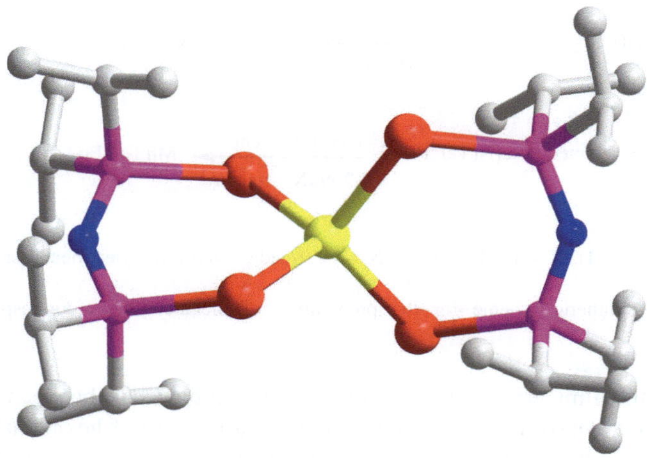

Fig. 9.6 Molecular structure of $[Cd(N(SeP^{i}Pr_2)_2)_2]$

The $NH(SeP^{i}Pr_2)_2$ ligand is more thermally stable than bulky selenolate ligands, such as $[SeSi(SiMe_3)_3]^-$, and thermolysis of its complexes produces cleaner products with reduced contamination caused by undesired ligand degradation reactions [14, 15].

$[M\{(EP^{i}Pr_2)_2N\}_2]$ (M = Cd, Zn and E = S, Se) and $[M\{(SePPh_2)_2N\}_2]$ (M = Cd, Zn) complexes have been used as precursors for zinc/cadmium selenide films by LP-MOCVD [68–70] at temperatures between 400°C and 500°C. All of these precursors have dimeric structures [70]. The complex $[MeCd\{(SeP^{i}Pr_2)_2N\}]_2$ was also prepared by the conproportionation of Me_2Cd and $Cd[(SeP^{i}Pr_2)_2N]_2$ in anhydrous toluene [71]. The structure of the compound [71] was determined by X-ray crystallography and consists of dimeric molecular units; each diselenoimidodiphosphinate chelates to one cadmium atom and bridges to the next. Each cadmium is four-coordinate and bound to three selenium atoms and one carbon. The compound is suitable for the deposition of cadmium selenide films by low pressure chemical vapour deposition.

The aerosol-assisted chemical vapour deposition (AACVD) of CdTe has been carried out using $Cd[(TeP^{i}Pr_2)_2N]_2$ at substrate temperatures between 375°C and 475°C [72]. The synthesis of the Te analogue of $Cd[(SeP^{i}Pr_2)_2N]_2$ could not be achieved by direct reaction of $NH(P^{i}Pr_2)_2$ with tellurium. An alternative approach involved metallation of $NH(PR)_2$ with NaH, prior to reaction with tellurium, which facilitates the preparation of $Na[N(TePR_2)_2]$ (R = Ph, ^{i}Pr) [73, 74]. This reagent was then used in metathetical reactions with metal halides to generate homoleptic complexes of the type $M[(TeP^{i}Pr_2)_2N]_2$ (M = Cd, Hg) (Scheme 9.1) [75].

Although crystalline samples of these metal complexes can be handled in air for short periods, extended exposure to moist air results in decomposition, especially for powdered samples. Consequently, these precursors have to be handled under the inert atmosphere of a glove box.

$$\mathrm{Na[(P^iPr_2)_2N] + 2Te \xrightarrow[\text{TMEDA}]{\text{Hot Toluene}} (tmeda)Na[N(TeP^iPr_2)_2]}$$

$$\mathrm{MX_2 + 2\ (tmeda)Na[N(TeP^iPr_2)_2] \xrightarrow[\text{-2 NaX, -tmeda}]{\text{Hot Toluene}} M[(TeP^iPr_2)_2N]_2}$$

M = Cd(1), X = I; M = Hg(2), X = Cl, tmeda = tetramethylethylenediamine

Scheme 9.1 Synthetic scheme for the preparation of dichalcogenoimidophosphinato-metal complexes

XRD of the films grown show the formation of cubic CdTe between 425°C and 475°C. At low deposition temperature (375°C), a mixture of hexagonal tellurium and cubic cadmium telluride is observed. SEM images reveal that the growth temperatures do not have a profound effect on the morphologies of films.

The AACVD of $Hg[(TeP^iPr_2)_2N]_2$ resulted in deposition of hexagonal tellurium [73] which may be due to reductive elimination of mercury at higher temperatures. Previously, mercury chalcogenide compounds have been shown to produce R_2E_2 (E = S, Se, Te) and Hg under CVD conditions rather than HgE [7, 76]. It is known that the anionic ligand $[(TeP^iPr_2)_2N]_2$ is readily oxidized to the ditelluride $(TeP^iPr_2N^iPr_2PTe_2)_2$ (which can be viewed as R_2E_2, where R = $TeP^iPr_2N^iPr_2P$ and E = Te) [77]. Thus reductive elimination of mercury with the concomitant formation of this ditelluride is a feasible pathway for the decomposition of $Hg[(TeP^iPr_2)_2N]_2$. The subsequent degradation of this ditelluride to give hexagonal Te films may account for the current observations. To confirm such a decomposition pathway, pyrolysis of the mercury precursor was carried out at 500°C. The black powder obtained was investigated by XRD which confirmed the presence of hexagonal Te along with cubic HgTe.

9.3 III–VI Materials

9.3.1 Metal Selenides from Selenolate Complexes

Indium selenide thin films have been prepared from the selenolate complexes [78]. Barron and co-workers prepared a number of dialkylindium selenolates and alkylindium selenides [79], and deposited indium selenide films by LP-MOCVD [80]. Whilst $[In^tBu_2(Se^tBu)]_2$ deposited indium rich films at temperatures between 230°C and 420°C, $[In(CEtMe_2)(\mu_3{-}Se)]_4$ gave crystalline films of InSe; however, the film quality depended on the growth temperature. Gysling *et al.* have also prepared thin films of indium selenide from $[InMe_2(SePh)]$ or $[In(SePh)_3]$, by a spray-assisted MOCVD technique, on GaAs(100) [78]. Different phases of InSe were grown at different substrate temperatures, with a cubic phase observed at deposition temperatures between 310°C and 365°C. Pyrolysis of [In(SePh)$_3$] gave hexagonal films of In_2Se_3 at temperatures of 470–530°C. In addition to these, a number of other potential precursors for the growth of indium selenide have been prepared with more bulky alkyl substituents [81].

9.3.2 Metal Selenides from Indium Diselenocarbamates

Indium diselenocarbamates have proved to be useful compounds for the deposition of indium selenide thin films. For example, [In(Se$_2$CNMe^{n}Hex)$_3$] has been used to deposit thin films of cubic In_2Se_3 [82]· At growth temperature of 450°C, films of cubic In_2Se_3 were deposited on glass substrates with a preferred (111) orientation. The results are similar to those of Arnold and co-workers who prepared cubic In_2Se_3 films from [In{SeC(SiMe$_3$)$_3$}$_3$] [83].

9.3.3 Metal Selenide/Tellurides from Cubane Selenide or Telluride Precursors

The hypothesis that molecular structure can influence the phase of the as-deposited films was investigated with the cubane selenide or telluride precursors [GaR (μ_3–E)]$_4$ (E = S, Se, Te; R = CMe$_3$, CEtMe$_2$, CEt$_2$Me or Et$_3$C) (Fig. 9.7) [83, 84]. However, in contrast to the sulfides, the hexagonal phase of GaSe was deposited by atmospheric pressure (AP)-MOCVD (≈350°C) and similarly metastable hexagonal GaTe was deposited by LP-MOCVD [85]. It was suggested that the formation of hexagonal GaSe and GaTe a consequence of the "controlled" cleavage of the appropriate Ga_4E_4 core during MOCVD. There are also examples of cubanes with indium chalcogenides [79]. However, [^{t}BuInSe]$_4$ was found to give only indium metal films by LP-MOCVD. The complex [(Me$_2$EtC)InSe]$_4$ gave hexagonal InSe in LP-MOCVD [80].

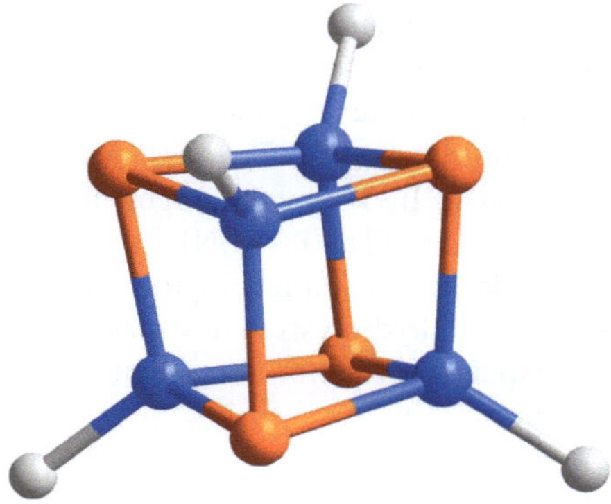

Fig. 9.7 Molecular structure of [(GaTe(μ_3-CMe))$_4$]

9.3.4 Metal Selenide/Tellurides from Dichalcogenoimidodiphosphinato Complexes

Dialkyl indium and gallium complexes (Fig. 9.8) with dichalcogenoimido-phosphinato ligands also proved useful compounds for MOCVD studies [86]. The acidic nature of the imino proton in $[NH(SeP^iPr_2)_2]$ makes it a good candidate for alkane elimination reactions with group 13 metal alkyls. Thermodynamic stable cubic Ga_2Se_3 could be grown on glass substrates by LP-MOCVD and AACVD. However, the quality of films interms of morphology is somewhat better in low pressure studies in comparison with AACVD.

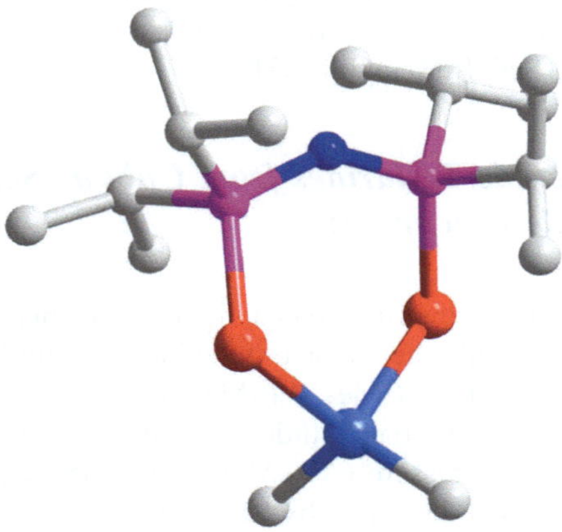

Fig. 9.8 Molecular structure of $[Me_2Ga(SeP^iPr_2)_2N]$

To prepare the indium complex of $[N(TeP^iPr_2)_2]^-$ anion, indium(I) chloride in the presence of elemental tellurium were used to give trimeric complex $\{In(\mu{-}Te)[N(TeP^iPr_2)_2]\}_3$ comprised of a central In_2Te_3 ring instead of expected octahedral complex (Fig. 9.9) [87]. The gallium analogue can be obtained in a similar manner by using GaI instead of InCl. AACVD studies of indium complex deposited cubic In_2Te_3 films whereas, gallium complexes yielded a mixture of Ga_2Te_3, GaTe and elemental tellurium on silicon substrates [87]. Mass spectrometric studies indicate that the indium complex deposits In_2Te_3 by the fragmentation of the central In_3Te_3 ring, with the remaining species $In[(TeP^iPr_2P)_2N]_2^+$ and $[N(TeP^iPr_2)_2N]_2^-$ staying intact. Gallium complex undergoes a similar fragmentation process; however it is significantly more susceptible to hydrolysis and oxidation than its indium counterpart. This is a possible explanation of the formation of Te during the AACVD deposition of the Ga complex.

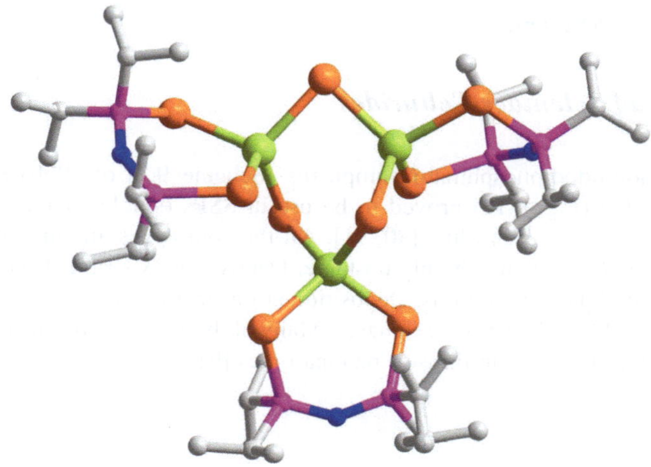

Fig. 9.9 Molecular structure of $[In(\mu\text{-}Te)[N(^{i}Pr_2PTe)_2]\}_3]$

9.3.5 Metal Selenides from Diselenophosphinato Complexes

Recently O'Brien *et al.* used tris(diisopropyldiselenophosphinato)-indium(III)/gallium(III) complexes [32, 33] for the deposition of GaSe and InSe thin films and nanoparticles [88, 89]. The indium complex show (Fig. 9.10a) the expected $[In(^{i}Pr_2PSe_2)_3]$ molecule. All three diselenophosphinate ligands are chelating to form three four membered rings (Se-P-Se-In). The geometry at indium is trigonally distorted octahedral due to the restricted bite angle of the chelating diselenophosphinate ligand. The structure of $[Ga(^{i}Pr_2PSe_2)_3]$ in Fig. 9.10b show a four coordinate gallium centre in contrast to six coordinate indium. The structure consist of one chelating and two pendant di-*iso*-propyldiseleno-phosphinate ligands in a distorted tetragonal geometry [33].

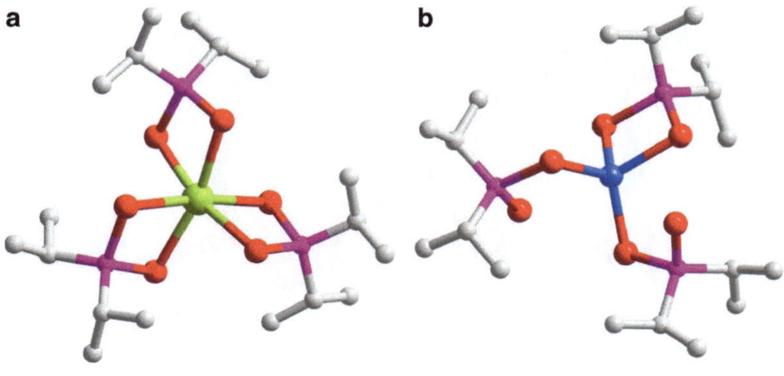

Fig. 9.10 Molecular structure of (**a**) $[In(Se_2P^{i}Pr_2)_3]$ and (**b**) $[Ga(Se_2P^{i}Pr_2)_3]$

9.4 IV–VI Materials

9.4.1 Lead Selenides/Tellurides

Dichalcogenoimidodiphosphinato complexes (Scheme 9.2) of $[Pb[(EP^{i}Pr_2)_2N]_2]$ (E = S, Se, Te) (Fig. 9.11) proved to be useful SSPs (single source precursors) for lead chalcogenide thin films [90, 91]. All the complexes are air-stable except $[Pb[(TeP^{i}Pr_2)_2N]_2]$ which is highly unstable. Deposition experiments conducted at 475°C produced pure PbTe films. Deposition at lower growth temperatures lead to the formation of both PbTe and Te phases. Many of these precursors have been used to deposit PbS or PbSe thin films or nanoparticles [91–99].

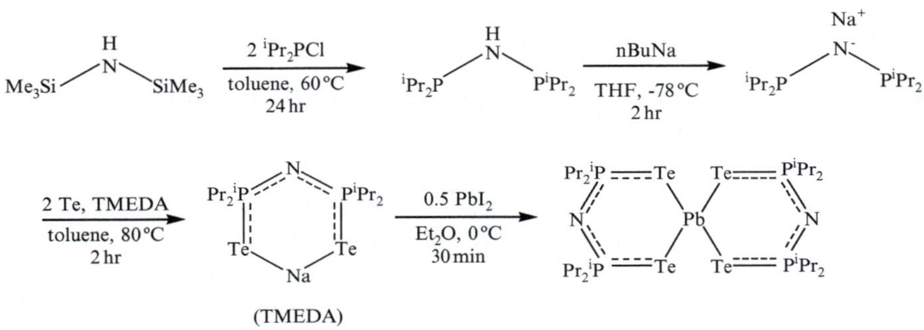

Scheme 9.2 Synthetic scheme for the preparation of *bis*(ditelluroimidophosphinato)lead(II)

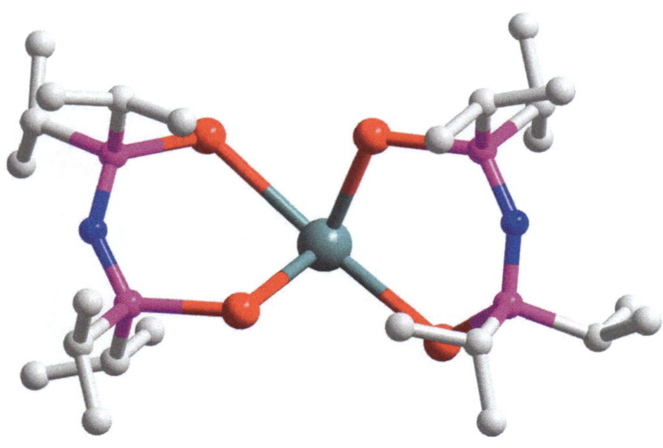

Fig. 9.11 Molecular structure of $[Pb[(SeP^{i}Pr_2)_2N]_2$

Bis(*di*-phenylthioselenophosphinato)lead(II) $[Pb\{(C_6H_5)_2PSSe)\}_2]$ has been synthesised and its single crystal X-ray structure determined (Fig. 9.12) [33]. Recently, this precursor has been used to deposit PbSe thin films and nanoparticles [100]. Figure 9.13 shows a HRTEM image of the PbSe nanocrystals prepared by thermolysis in a (1:1) mixture of oleylamine (OA): dodecanethiol (DDT) (1:1) at room temperature.

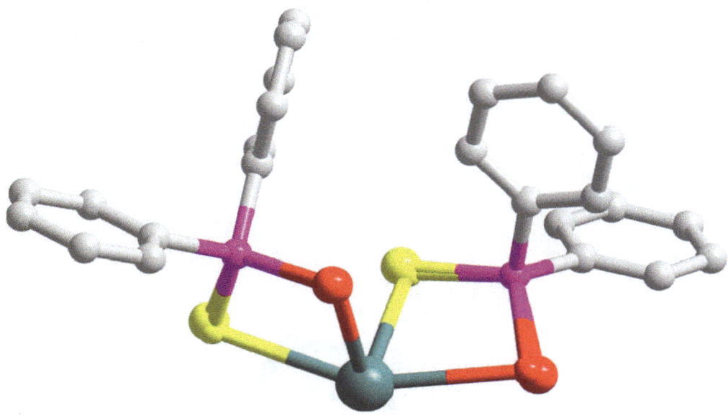

Fig. 9.12 Molecular structure of *bis*[di-phenylthioselenophosphinato)]lead(II)

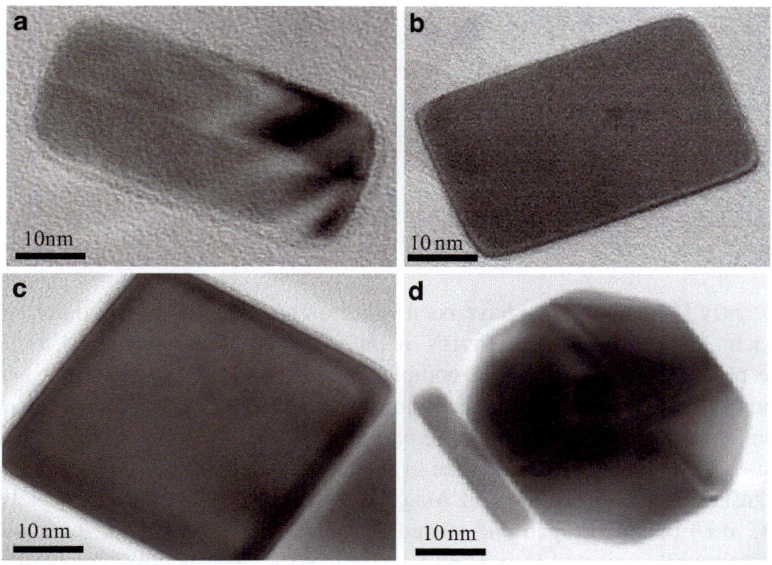

Fig. 9.13 TEM images (**a–d**) of some selected shape of PbSe nanoparticles prepared in Oleyal amine (OA): dodecanethiol(DDT) (1:1) at room temperature stirred for 15 min from $[Pb\{(C_6H_5)_2PSSe)\}_2]$

The deposition of PbSe thin films from diselenophosphinato- and imidodiselenodiphosphinato-lead complexes [90, 91] always results in contamination of phosphorous. In order to overcome this problem, O'Brien *et al.* have synthesized (4-nitro-*N,N-di-iso*-butyl-N-benzoylselenoureato)Pb(II) and *N,N*-diethyl-*N*-benzoylselenoureato) Pb(II) complexes to be used as SSP for the deposition of PbSe thin films [101]. The solid state structure of di-*iso*-butyl- *N*-(4-nitrobenzoylselenoureato)] lead(II), was determined by X-ray single crystallography (Fig. 9.14). The complex was used for the deposition of phosphorus free PbSe thin films by AACVD and nanoparticles by solution thermolysis. Figure 9.15 shows different shapes of nanoparticles obtained at different temperatures by thermolysis in a mixture of TOP/ oleic acid/octadecene.

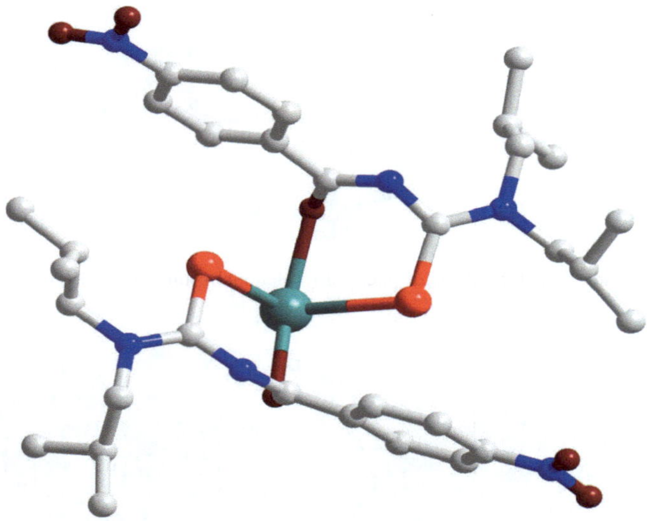

Fig. 9.14 Single crystal structure of *bis*[di-isobutyl-*N* -(4-nitrobenzoyl-selenoureato)]lead(II)

Recently PbSe nanowires have been synthesized by solution-liquid-solid (SLS) approach from [*bis*(*N,N*-di-ethyl-*N′*-naphthoylselenoureato)lead(II)] as single source precursor [102]. The lead complex was prepared by the reaction between aqueous $Pb(NO_3)_2$ and ethanolic solution of *N,N*-di-ethyl-*N′*-naphthoylselenourea at room temperature. Nanowires were prepared by heating of 5.5 g of TOPO under vacuum for 10 min at 130°C and then to 290–295°C under nitrogen. In separate two neck flask, 0.2 g of precursor and Au@Bi (40–120 μL) solution was dissolved in TOP (2 mL) under nitrogen and stirred at room temperature. This stock solution was then rapidly injected into TOPO at 290–295°C. The nanowires obtained are shown in Fig. 9.16.

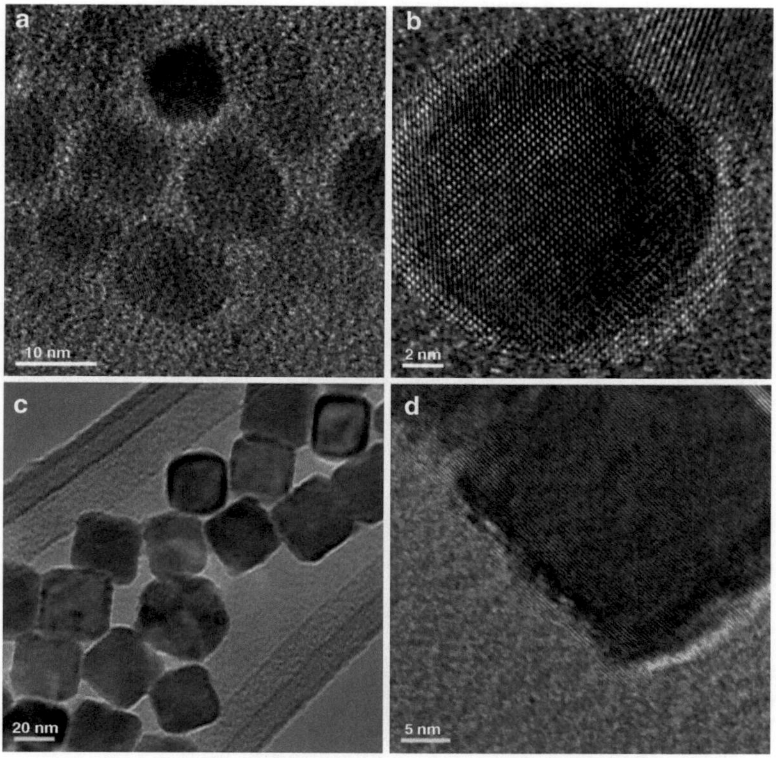

Fig. 9.15 (**a**, **b**) TEM image and HRTEM image of spherical shape PbSe nanoparticles prepared from *bis*[di-isobutyl-*N*-(4-nitrobenzoylseleno-ureato)]lead(II)at 200°C, (**c**, **d**) cubic shape PbSe nanoparticles prepared at 250°C (all in TOP/oleic acid/octadecene)

9.4.2 Tin Selenides/Tellurides

Dahmen and co-workers have used *bis*[*bis*(trimethylsilyl)methyl]tin chalcogenides, $[Sn\{CH(SiMe_3)_2\}_2(\mu{-}E)]_2$ (E = Se, Te) for the MOCVD of SnSe and SnTe thin films [103, 104]. These complexes are dimers in solution as well as in the solid state. The compounds are air-stable and volatile [105]. MOCVD experiments indicated that both compounds show strong selectivity in the decomposition reaction towards the metallic substrates, thus demonstrating the need for a seeding metallic layer to initiate the growth. In the case of non-metallic substrates, no film deposition was obtained under all experimental conditions.

The $[SnCl_4\{o\text{-}C_6H_4(CH_2SeMe)_2\}]$ deposits poor quality $SnSe_2$ films at 600°C. The Et_2Se derivative, $[SnCl_4(Et_2Se)_2]$ leads to uniform deposition of $SnSe_2$ with growth perpendicular to the substrate [106].

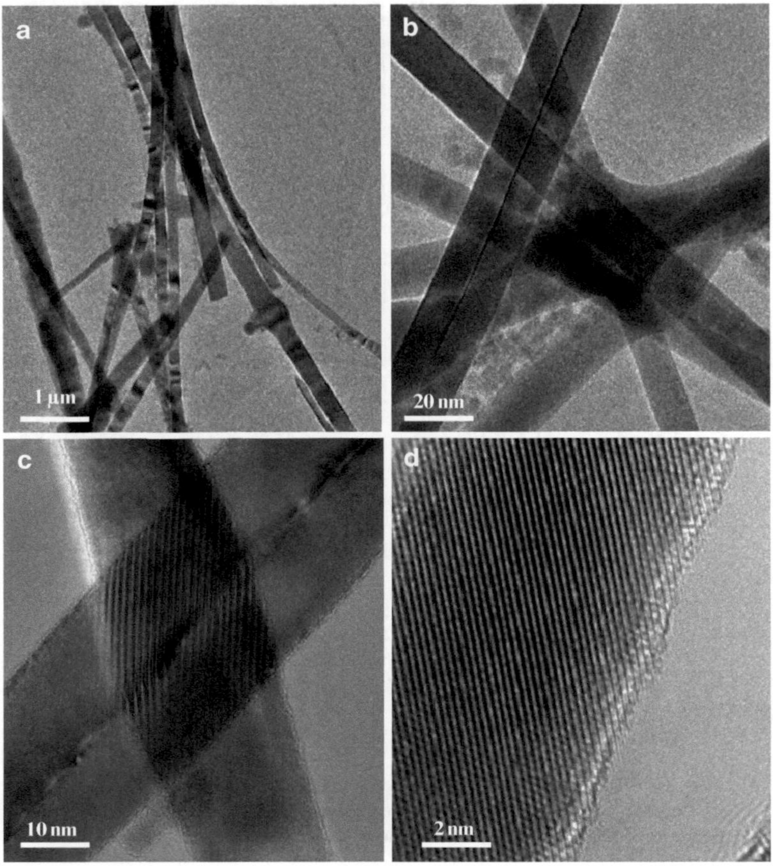

Fig. 9.16 TEM and HRTEM images of PbSe nanowires prepared using 0.2 g of [*bis*(*N*,*N*-di-ethyl-*N*′-naphthoylselenoureato)lead(II)] and 120 μL Au@Bi catalyst at 290°C

9.4.3 Germanium Tellurides

Reports on the deposition of germanium chalcogenide are extremely rare. Only recently, germanium telluride thin films are reported from a single-source route [107]. Insertion reaction between a germylene and dialkyl telluride leads to a stable alkyl tellurolato compound which has been used to deposit GeTe films in MOCVD studies.

9.5 V–VI Materials

9.5.1 *Antimony Selenides/Tellurides*

The V–VI materials attracted a considerable interest due to their thermo-electric and photovoltaic properties which find applications in thermo-electric and optoelectronic devices [108–111], television cameras [113, 114], and solar cells [115]. While there are several reports on V–VI metal sulfides [116–121], the work on metal selenides has recently started [122]. Previously the selenides have been prepared by heating elements (Sb/Bi and Se) at high temperatures [123–125]. Recently other techniques, such as solvothermal, hydrothermal, microwave assisted synthesis and chemical bath deposition have been used to prepare thin films and nanoparticles [126–130]. There are only few examples of using single source precursors for the preparation of V–VI materials [131–134]. The preparation of Sb_2Se_3 nanowires from $[Sb\{Se_2P(O^iPr)_2\}_3]$ under solvothermal conditions has been reported by Chang *et al.* [131]. In a later study the authors have studied the electrical and optical properties of a single Sb_2Se_3 nanorod with an average size of 70 nm in diameter and a length of ~1–2 μm [135]. Recently complexes with the general formula $[M\{Se–C_5H_3(R\text{-}3)N\}_3]$ (M = Sb or Bi) have been prepared by the reaction of $SbCl_3$ or $BiCl_3$ with $M'Se–C_5H_3(R\text{-}3)N$ (M′ = Li or Na; R = H or Me) and the molecular structures of both complexes have been determined by X-ray crystallography (Fig. 9.17a, b) [136].

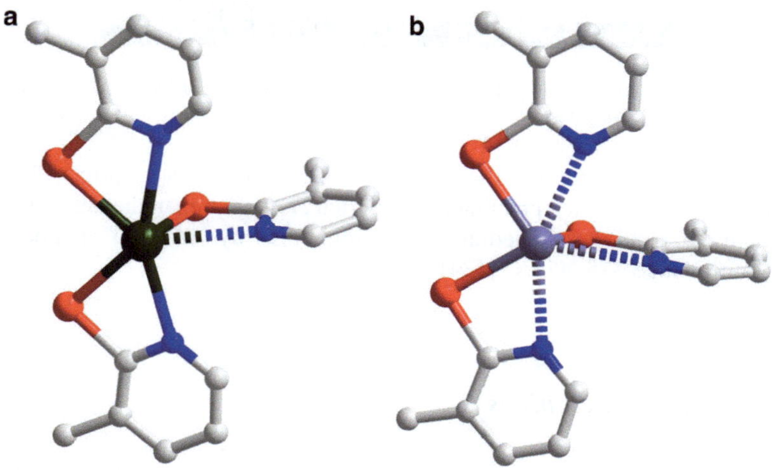

Fig. 9.17 Molecular structures of (**a**) $Sb\{Se–C_5H_3(Me\text{-}3)N\}_3]$; (**b**) $Bi\{Se–C_5H_3(Me\text{-}3)N\}_3]$

The structure of antimony complex (Fig. 9.17a) is monomer in which antimony atom is bonded to three selenolates of monodentate $Se–C_5H_3(Me\text{-}3)N$ ligands to acquire a trigonal pyramidal configuration. The void in the coordination sphere of antimony can be viewed conventionally as filled by a stereochemically active lone pair of electrons. The structure of the bismuth complex (Fig. 9.17b) is based on a distorted square pyramidal configuration by two chelating and a monodentate selenolate ligands. The monodentate ligand occupies the apical position and the position opposite is filled by stereochemically active lone pair of electrons, while the two chelating ligands occupy the corners of the square. Pyrolysis of antimony and bismuth complexes at 400°C and 450°C for 1 h under argon in a furnace produced black residues which were characterized by pXRD and SEM as orthorhombic Sb_2Se_3 rods (Fig. 9.18) and hexagonal BiSe flowers (Fig. 9.19) whereas thermolysis of these complexes in HDA provided orthorhombic Sb_2Se_3 and rhombohedral Bi_2Se_3 [136]. Thin films of these materials were also obtained by AACVD method.

Fig. 9.18 SEM images of Sb_2Se_3 nanorods obtained by pyrolysis of [Sb{Se–C_5H_3(Me-3)N}$_3$] (2) in a furnace at 400°C for 1 h

The homoleptic antimony(III) complex [Sb[(TeP^{i}Pr$_2$)$_2$N)$_2$]$_3$] (Fig. 9.20) have been used as single source precursor by O'Brien *et al.* to prepare hexagonal-shaped nanoplates of pure rhombohedral Sb_2Te_3 thin films by AACVD method in the temperature range 375–475°C [134].

9.5.2 *Bismuth Selenides*

Bismuth chalcogenide films have been reported from CVD studies of air-stable complexes of Bi[(EPR$_2$)$_2$N]$_3$ (E = S, Se; R = Ph, ^{i}Pr) [133]. The compounds were prepared using methodology developed by Woollins *et al.* (Scheme 9.3) [67]. These neutral ligands are readily deprotonated to the monoanions by treatment with bases such as sodium methoxide. Subsequent metathetical reactions with metal halides produce homoleptic complexes in good yields.

Fig. 9.19 (**a**–**d**) SEM images of different magnifications of BiSe flowers obtained by pyrolysis of [Bi{Se–C_5H_3(Me-3)N}$_3$] at 450°C for 1 h, respectively

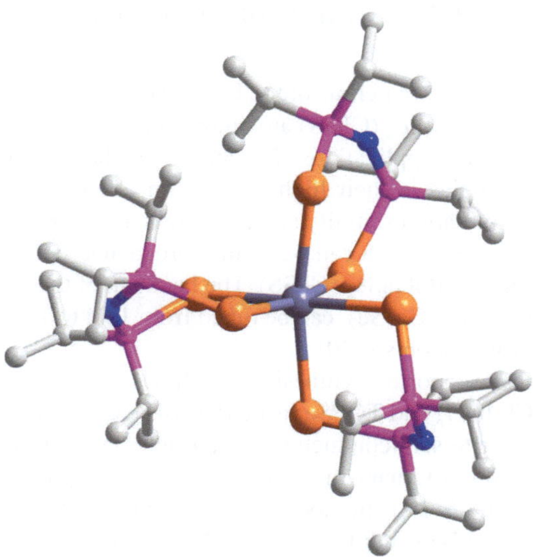

Fig. 9.20 Molecular structure of [Sb[(TePPri$_2$)$_2$N)$_2$]$_3$]

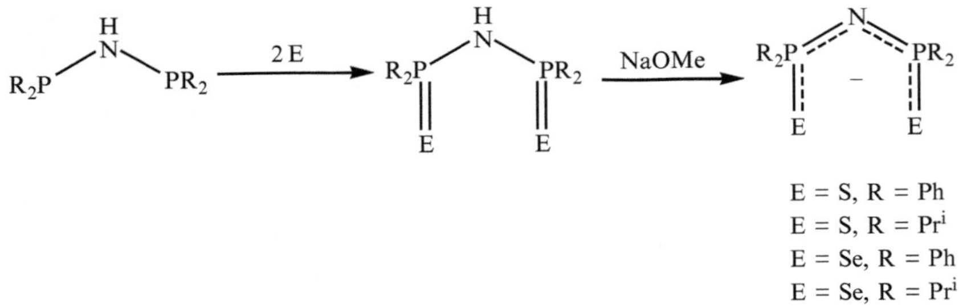

Scheme 9.3 Synthetic scheme for the preparation of dichalcogenoimidophosphinato ligand

Crystalline thin films of rhombohedral Bi_2Se_3 (from $Bi[(SeP^{i}Pr_2)_2N]_3$), hexagonal BiSe (from $Bi[(SePPh_2)_2N]_3$, and orthorhombic Bi_2S_3 (from $Bi[(SPR_2)_2N]_3$ have been deposited on glass substrates [133]. The suggested reason for the deposition of monophasic Bi_2Se_3 and BiSe from *iso*propyl and phenyl substituted precursors, respectively is that the difference in electron-donating character of the alkyl group on the P atoms affects the relative bond strengths within the structures and hence the decomposition profiles of the parent molecules.

Thin films composed of hexagonal Bi_2Se_3 nanoplates have also been deposited from MOCVD of bismuth diselenophosphato complex, $[Bi(Se_2P\{O^{i}Pr)_2\}_3]$ on modified and unmodified Si substrates [137]. The deposited Bi_2Se_3 nanoplates indicate a superior thermoelectric property over bulk Bi_2Se_3.

9.6 The CIS and CIGS Family of Materials

Compound semiconducting materials of the I-III-VI_2 family such as $CuInSe_2$ (CIS), $CuGaSe_2$ and $CuIn_{(1-x)}Ga_xSe_2$ (CIGS) are amongst the leading candidate absorber materials for photovoltaic applications. The attraction chalcopyrite based I-III-VI_2 photovoltaic materials lies in their high conversion efficiency, higher absorption coefficients, photo-irradiation stability and lower toxicity [138, 139]. $CuInSe_2$ has a direct bandgap of 1.04 eV which can be further broadened by incorporation of Ga leading to formation of $CuInGaSe_2$ (CIGS). The bandgap of the materials, generally as represented by $CuIn_{(1-x)}Ga_xSe_2$, can be tuned from 1.04 to 1.68 eV by controlled substitution of Ga for In atoms [140].

Previously we have reported synthesis of $CuInSe_2$ nanoparticles by a two step reaction using CuCl, $InCl_3$ and TOPSe in TOPO [141]. Castro and co-workers have synthesized chalcopyrite semiconductor nanoparticles of $CuInS_2$ and $CuInSe_2$ by thermal decomposition of ternary molecular single source precursors $(PPh_3)_2CuIn(SEt)_4$ and $(PPh_3)_2CuIn(SePh)_4$ in dioctyl phthalate at temperatures from 200°C to 300°C [142]. Allan and Bawendi have reported the synthesis of Cu-In-Se quantum dots of varying stoichiometric compositions, which exhibited high photoluminescence quantum yields (PLQY) from the red to near infrared (NIR) region [143].

Their approach involved the use of the metal halides of copper and indium precursors and bis(trimethylsilyl)selenide [$(Me_3Si)_2Se$] as the chalcogenide precursor, whereas a combination of tri-*n*-octylphosphine (TOP) and oleylamine (OA) was used as a coordinating solvent and capping ligand. However, the stoichiometric compositions obtained were not $CuInSe_2$ but materials of stoichiometry $CuIn_5Se_8$, $CuIn_{2.4}Se_4$, and $CuIn_{1.5}Se_3$.

$CuInSe_2$, $CuGaSe_2$ and $CuIn_{(1-x)}Ga_xSe_2$ (CIGS) nanoparticles were deposited from the diisopropyldiselenophosphinatometal complexes $M_x[^iPr_2PSe_2]_y$ (M = Cu, In, Ga) by thermal decomposition of precursors in HDA/TOP system at 120–210°C and 250°C, respectively [89]. The semiconductor nanoparticles obtained were characterized by X-ray diffraction (XRD), scanning electron microscopy (SEM), transmission electron microscopy (TEM) and energy dispersive X-ray (EDX) analysis. Figure 9.21 shows the TEM images and SAED pattern and a histogram representing size distribution of tetragonal $CuIn_{0.7}Ga_{0.3}Se_2$ nanoparticles.

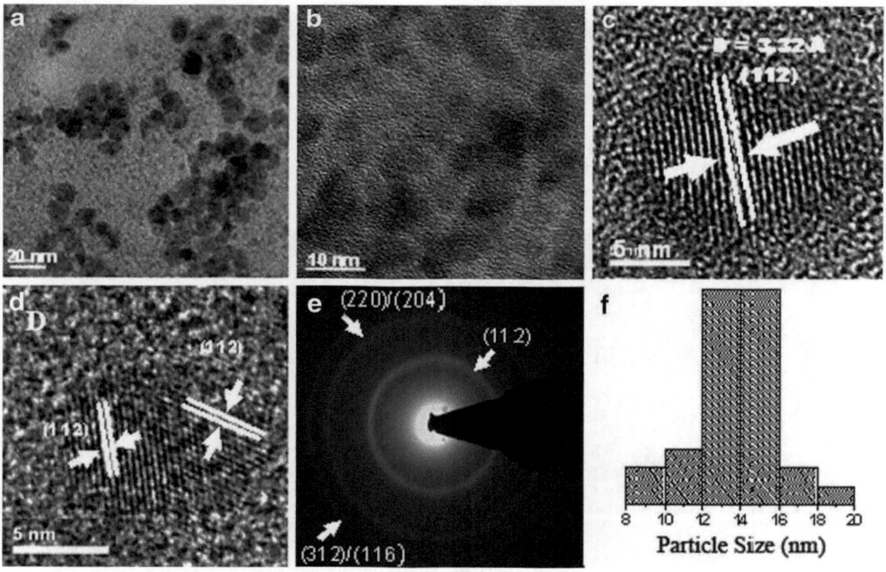

Fig. 9.21 (**a–d**) TEM images of tetragonal $CuIn_{0.7}Ga_{0.3}Se_2$ quantum dots (**e**) SAED pattern of the $CuIn_{0.7}Ga_{0.3}Se_2$ nanoparticles and (**f**) histogram representing size distribution for $CuIn_{0.7}Ga_{0.3}Se_2$ quantum dots

The corresponding diisopropyldiselenophosphinatometal complexes $M_x[^iPr_2PSe_2]_y$ (M = Cu, In, Ga) were also used to deposit CuSe, γ-In_2Se_3, $CuInSe_2$, $CuGaSe_2$, and $CuIn_{0.7}Ga_{0.3}Se_2$, thin films by AA-CVD method. Figure 9.22 shows SEM images of as deposited (a) γ-In_2Se_3 thin film at 450°C (b) $CuGaSe_2$ thin at 300°C (b) 500°C (d) 3D AFM image of $CuGaSe_2$ thin films at 450°C from 1:4 equivalent of $[Cu_4(^iPr_2P_2Se_2)_4]$ and $[Ga(^iPr_2PSe_2)_3]$. Figure 9.23 shows SEM images of as

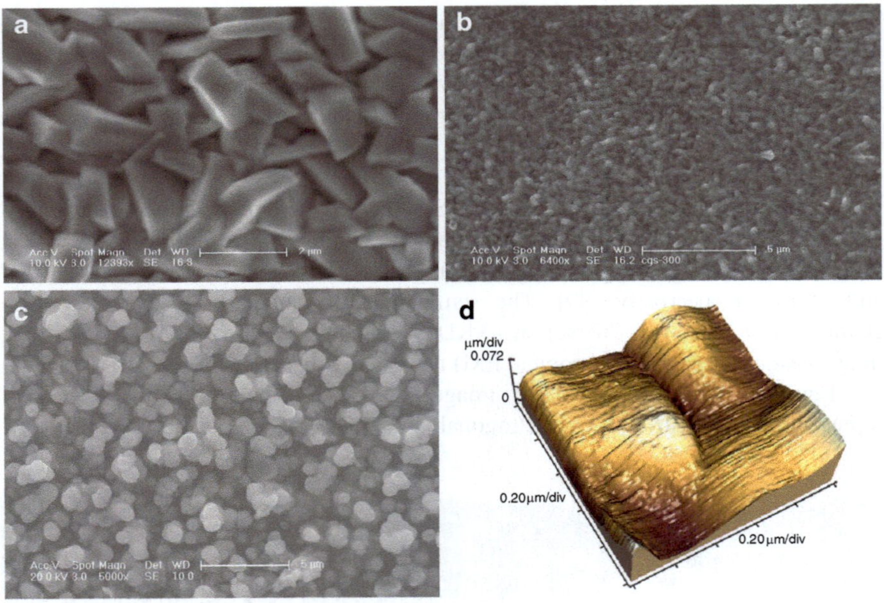

Fig. 9.22 SEM images of as deposited (**a**) γ-In_2Se_3 thin film at 450°C (**b**) $CuGaSe_2$ thin at 300°C (**b**) 500°C (**d**) 3D AFM image of as grown $CuGaSe_2$ thin films at 450°C from 1:4 equivalent of $[Cu_4(^iPr_2P_2Se_2)_4]$ and $[Ga(^iPr_2PSe_2)_3]$

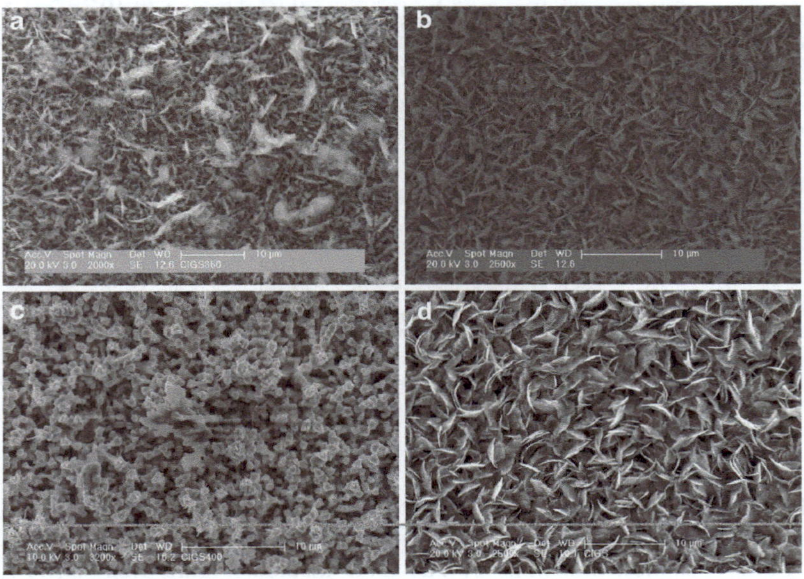

Fig. 9.23 SEM images of as deposited $CuIn_{0.7}Ga_{0.3}Se_2$ at (**a**) 300°C, (**b**) 350°C, (**c**) 400°C and (**d**) 450°C from 1:4:4 equivalent of $[Cu_4(^iPr_2P_2Se_2)_4]$, $[In(^iPr_2PSe_2)_3]$ and $[Ga(^iPr_2PSe_2)_3]$, respectively

deposited $CuIn_{0.7}Ga_{0.3}Se_2$ at different temperatures from 1:4:4 equivalent of $[Cu_4(^iPr_2P_2Se_2)_4]$, $[In(^iPr_2PSe_2)_3]$ and $[Ga(^iPr_2PSe_2)_3]$ respectively.

9.7 Transition Metal Selenides and Tellurides

9.7.1 Nickel and Cobalt Selenides

Air-stable, metal-organic compounds of the type $[M\{^iPr_2P(S)NP(Se)^iPr_2\}_2]$, (M = Ni, Co) were synthesised by the metathetical reaction between the sodium salt of the S/Se containing imidodiphosphinate ligand with the appropriate metal salt in methanol and the structure of cobalt complex was determined by X-ray crystallography [144]. Both compounds are monomeric and isomorphous with tetrahedral coordination geometry at the metal centre. These compounds have been used as single-source precursors to deposit thin films of metal phosphide or selenide by AA-CVD and LP-CVD methods. The mechanism of the unique formation of phosphide or selenide films from the same precursor was studied by pyrolysis GC-MS and by *ab initio* computational studies [144].

Diisopropylimidodiselenophosphinato complexes $[Ni[(SeP^iPr_2)_2N]_2]$, $[Co[(SeP^iPr_2)_2N)_2]$ and diisopropyldiselenophosphinato $[Ni(Se_2P^iPr_2)_2]$ and $[Co(Se_2P^iPr_2)_2]$ were synthesized and the single crystal X-ray structures of $Ni[(SeP^iPr_2)_2N]_2$ and $[Ni(Se_2P^iPr_2)_2]$ were determined [145]. All these compounds were used to deposit thin films of nickel and cobalt selenide by CVD.

The recrystallisation of $[Ni[(SeP^iPr_2)_2N)]_2]$ from THF solution at room temperature gave deep red and green colour crystals. Single crystal X-ray structure of the red crystals showed the nickel in tetrahedral geometry whereas in square planar geometry in green crystals (Fig. 9.24a, b). The square planar geometry also showed the presence of a molecule of water in partial occupancy in the crystal lattice.

The diselenophosphinato nickel complexes are square planar (Fig. 9.25). The Ni atoms are surrounded by a distorted square of four Se atoms belonging to two bidentate chelating ligands as in the $^iPr_2PSe_2$ complex.

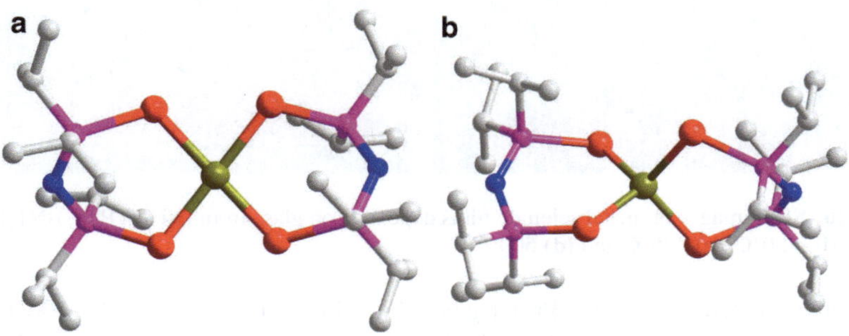

Fig. 9.24 X-ray crystal structure of $Ni[(SeP^iPr_2)_2N]_2$ (**a**) tetrahedral and (**b**) square planar

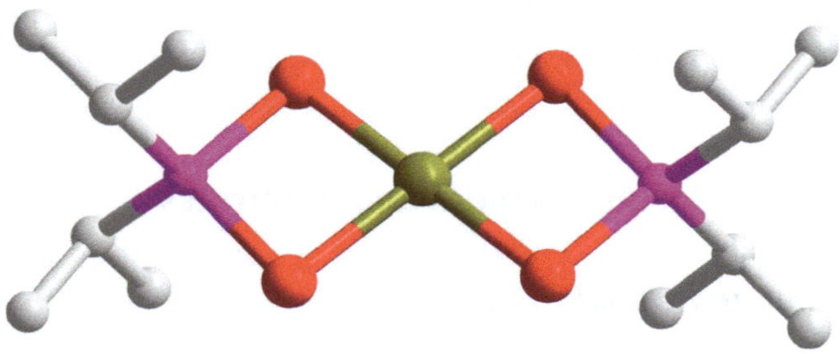

Fig. 9.25 X-ray crystal structure of [Ni(^{i}Pr$_2$PSe$_2$)$_2$]

The deposition by AA-MOCVD from Ni[(SeP^{i}Pr$_2$)$_2$N]$_2$ produced uniform black films at temperatures between 400°C and 500°C. XRD studies show that in each case, monophasic hexagonal $Ni_{0.85}Se$ was deposited, with preferred orientation in the (101) direction. SEM studies revealed that the films comprise of mixtures of fibrous wires and rods. The morphology of the films change from thin and short wires to thicker and longer wires as the deposition temperatures increases (Fig. 9.26a–d).

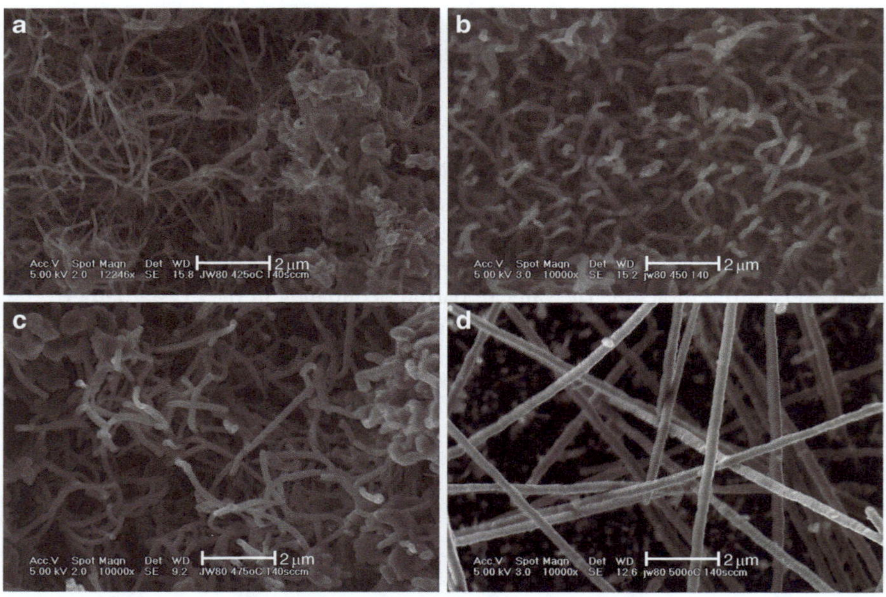

Fig. 9.26 SEM images of nickel selenide films deposited on glass from [Ni{(SeP^{i}Pr$_2$)$_2$N}$_2$] at (**a**) 425°C, (**b**) 450°C, (**c**) 475°C and (**d**) 500°C

Deposition from [Ni(Se$_2$P^{i}Pr$_2$)$_2$] gave thick black films at 350°C, 400°C and 450°C. XRD results showed that all films were composed of hexagonal $Ni_{0.85}Se$ with preferred growth along (101) plane. SEM images showed that morphologies varied from a mixture of cubic crystallites and flakes at 350°C to spherical particles connected by wires at 400°C. Deposition from $[Co(Se_2P^iPr_2)_2]$ by AA-MOCVD gave thick black films at 450°C. No deposition was obtained at lower temperatures. The films consits of a mixture of Co_9Se_8 and CoP phases as revealed by XRD analysis. The morphology appears to be irregular grains of varing sizes. Deposition of thin films was also carried out from selenophosphinato nickel/cobalt(II) and the results were compared to imidoselenophosphinato nickel/cobalt(II) and selenophophinato nickel/cobalt(II).

9.7.2 Copper/Silver Selenides/Tellurides

A series of dichalcogenophosphinato and imidodichalcogenodiphosphinato silver (I) complexes including $[Ag(Se_2P^iPr_2)]$, $[Ag_4(SSeP^iPr_2)_4]$, $[Ag\{^iPr_2P(S)NP(Se)^iPr_2\}]_3$ and $[Ag(SeP^iPr_2)_2N]_3$ have been synthesised and structures of $[Ag_4(SSeP^iPr_2)_4]$ and $[Ag\{^iPr_2P(S)NP(Se)^iPr_2\}]_3$ were determined by X-ray single crystallography [146]. The structure of $[Ag_4(SSeP^iPr_2)_4]$ consists of tetrahedral array of four Ag(I) atoms capped by four bidentate $[^iPr_2PSSe]$ ligands coordinated to the Ag atoms through S or Se atoms (Fig. 9.27).

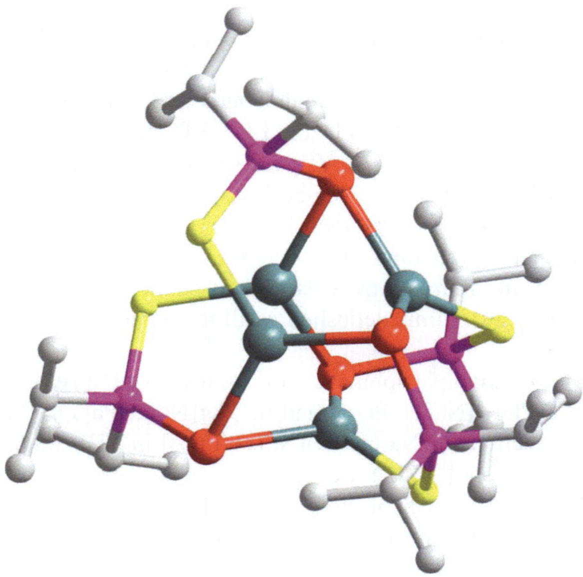

Fig. 9.27 Molecular structure of $[Ag_4(SSeP^iPr_2)_4]$

Each ligand caps one triangular face of the Ag(I) tetrahedrons by one bridging and one terminal chalcogen atom. Each Ag atom has coordination number of five with distorted trigonal distribution. The molecular structure of $[Ag\{^iPr_2P(S)NP$

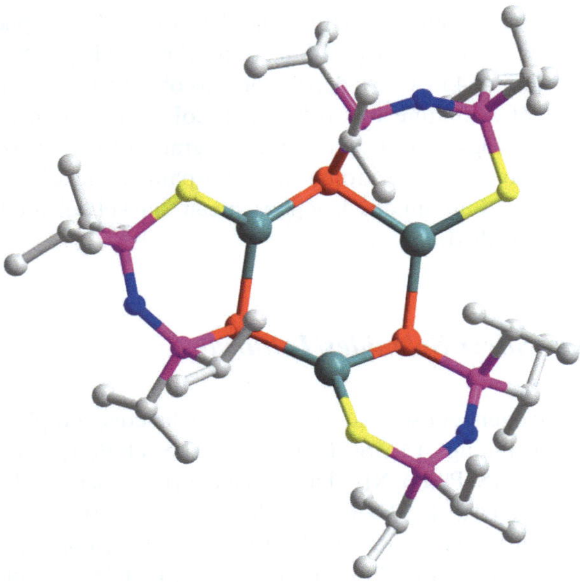

Fig. 9.28 Molecular structure of $[Ag\{^iPr_2P(S)NP(Se)^iPr_2\}]_3$

$(Se)^iPr_2\}]_3$ shows a trimeric unit along with toluene as solvate (Fig. 9.28) similar to that of the silver-selenium analogue $[Ag(SeP^iPr_2)_2N]_3$ [63].

All four compounds were used as single-source precursors for the deposition of silver selenide thin films by AA-CVD and LP-CVD methods. All precursors gave silver selenide (Ag_2Se) films by AACVD; whereas only $[Ag\{^iPr_2P(S)NP(Se)^iPr_2\}]_3$ and $[Ag(SeP^iPr_2)_2N]_3$ gave silver selenide films by LPCVD.

Films deposited from $[Ag\{^iPr_2P(S)NP(Se)^iPr_2\}]_3$ at 325°C were mainly composed of monodispersed spherical silver selenide nanoparticles (0.05–0.1 μm) but a significant change in morphology occurred at higher deposition temperatures (375–475°C) (Fig. 9.29). Films deposited at all temperatures were silver deficient and selenium rich.

Metal ditelluroimidodiphosphinate complexes $\{Cu[N(TeP^iPr_2)_2]\}_3$, $\{Ag[N(TeP^iPr_2)_2]\}_6$, $Au(PPh_3)[N(TeP^iPr_2)_2]$ and in $\{Ag[N(TePPh_2)_2\}_4{\cdot}2THF$ have been prepared by the reactions of $Na(tmeda)[N(TeP^iPr_2)_2]$ in THF solution with CuCl, AgI or AuCl respectively [147]. X-ray structures single crystal structure of $\{Cu[N(TeP^iPr_2)_2]\}$ (Fig.9.30) consists of three six membered $CuTe_2P_2N$ rings, which are linked together to form two central, highly distorted Cu_3Te_3 rings.

The distortion of these central rings results in surprisingly short copper–copper distances of 2.626(1) and 2.637(2) Å. Two of the ligands are equivalent with similar metrical parameters, each containing one tellurium which is complexed to one copper centre and one tellurium bridging two copper centres.

Structure of $\{Ag[N(TeP^iPr_2)_2]\}_6$ consists of six $Ag[N(TeP^iPr_2)_2]$ units to give the hexamer (Fig. 9.31). The molecule lies on an inversion centre so the two halves

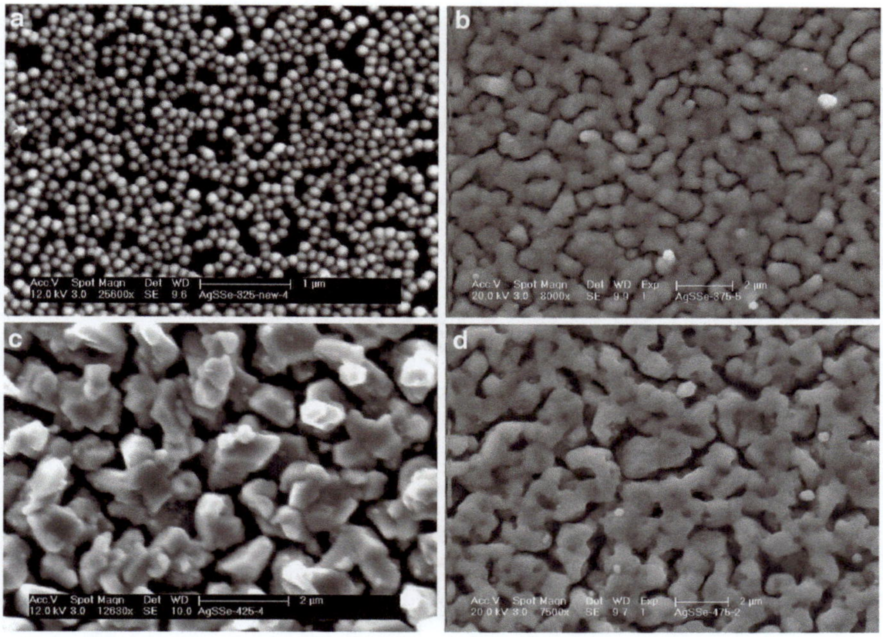

Fig. 9.29 SEM images of silver selenide films deposited on glass by AACVD from $[Ag\{^iPr_2P(S)NP(Se)^iPr_2\}]_3$ at flow rate of 160 sccm and growth temperatures of (**a**) 325°C, (**b**) 375°C, (**c**) 425°C and (**d**) 475°C

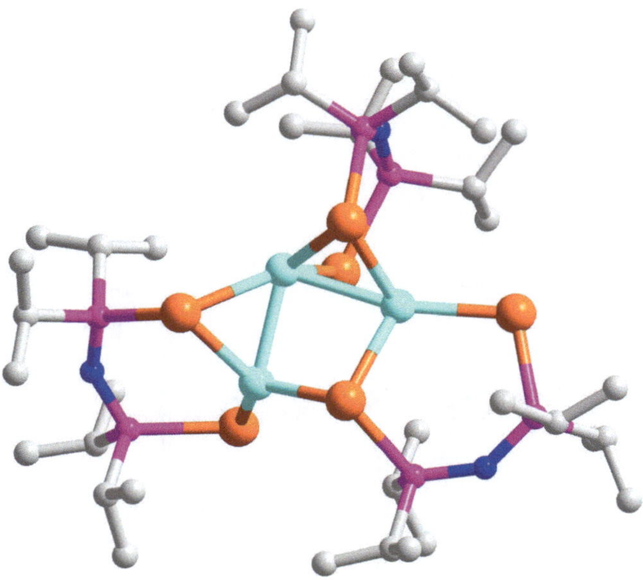

Fig. 9.30 Molecular structure of $[Cu\{N(^iPr_2PTe)_2\}]_3$

of the hexamer are crystallographically equivalent. The silver centres are each bound to two tellurium centres from one ligand and one tellurium from an adjacent $Ag[N(TeP^{i}Pr_2)_2]$ moiety. Each silver centre is coordinated equally between two bridging tellurium centres of adjacent $[N(TeP^{i}Pr_2)_2]$ ligands forming a central

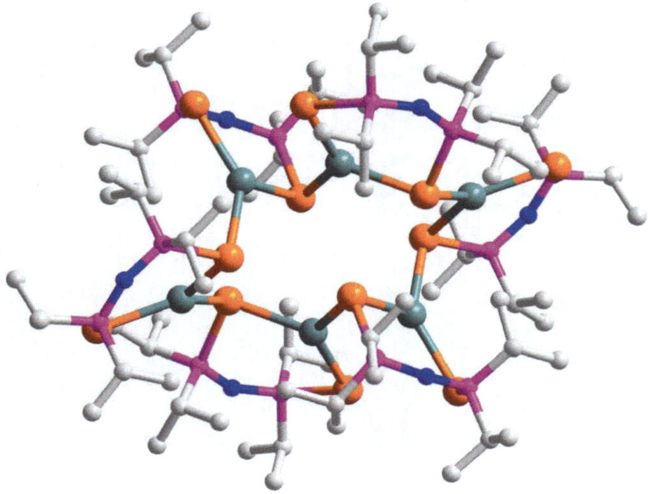

Fig. 9.31 Molecular structure of $[Ag\{N(^{i}Pr_2PTe)_2\}]_6$

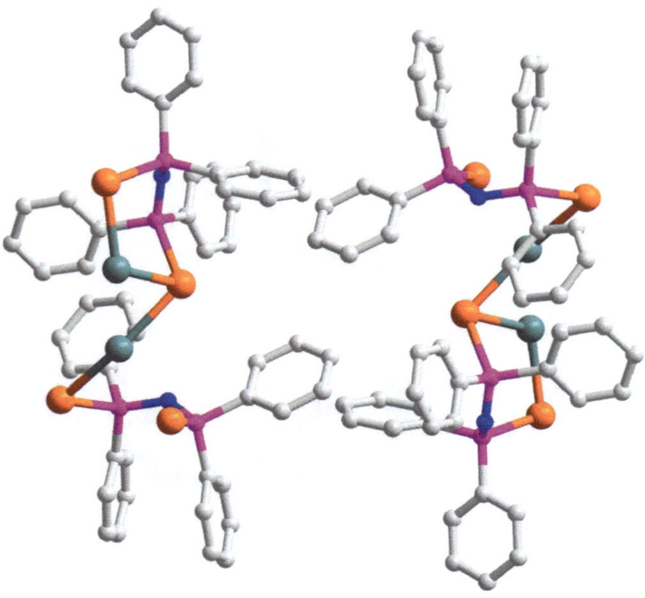

Fig. 9.32 Molecular structure of $\{Ag[N(Ph_2PTe)_2]\}_4$

twelve membered Ag_6Te_6 ring. The replacement of the ^{i}Pr groups on phosphorus by Ph results in a tetramer with a boat-shaped Ag_4Te_4 ring in $[Ag\{N(Ph_2PTe)_2\}_4]\cdot 2THF$ (Fig. 9.32).

The structure of gold complex $[Au(PPh_3)\{N(TeP^iPr_2)_2\}]$ based on a monomer with a trigonal planar gold(I) centre coordinated to one $[N(TeP^iPr_2)_2]$ ligand and

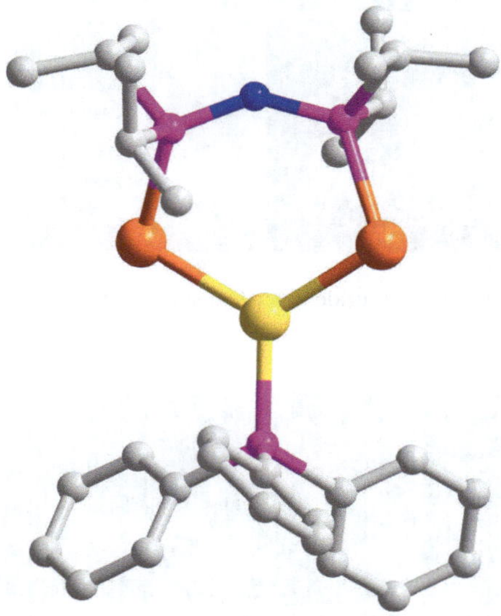

Fig. 9.33 Molecular structure of $[Au(PPh_3)\{N(TeP^iPr_2)_2\}]$

one molecule of triphenylphosphine. The gold centre is bound between the two tellurium centres of the ditelluroimido-diphosphinate ligand to form the six-membered $AuTe_2P_2N$ metallacycle (Fig. 9.33).

The complexes $\{Cu[N(TeP^iPr_2)_2]\}_3$, $\{Ag[N(TeP^iPr_2)_2]\}_6$, and $[Au(PPh_3)\{N(TeP^iPr_2)_2\}]$ were used to deposit thin films by AA-CVD method to give CuTe, Ag_7Te_4, $AuTe_2$ and Au films respectively. The films were grown at temperatures of 300–500°C and characterized XRD, SEM, and EDAX. Figures 9.34 and 9.35 show the different morphology of the copper telluride and Ag_7Te_4 films deposited at different temperatures respectively.

9.7.3 Palladium Selenides

A series of organochalcogenolato complexes including: monoselenocarboxylates, benzylselenolates, allylpalladium complexes, and palladium/Platinum sulphido/seleno bridged complexes have been prepared by Jain *et al.* [148–153]. They have investigated the decomposition of such complexes by pyrolysis.

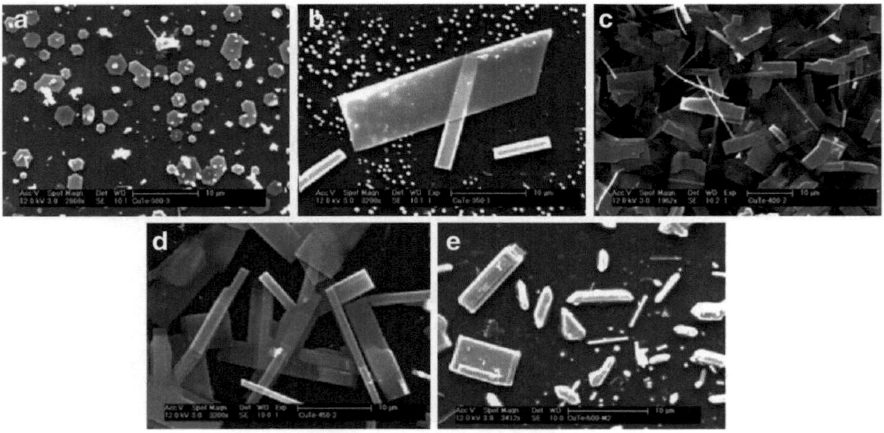

Fig. 9.34 SEM images of copper telluride films deposited at (**a**) 300°C, (**b**) 350°C, (**c**) 400°C, (**d**) 450°C and (**e**) 500°C

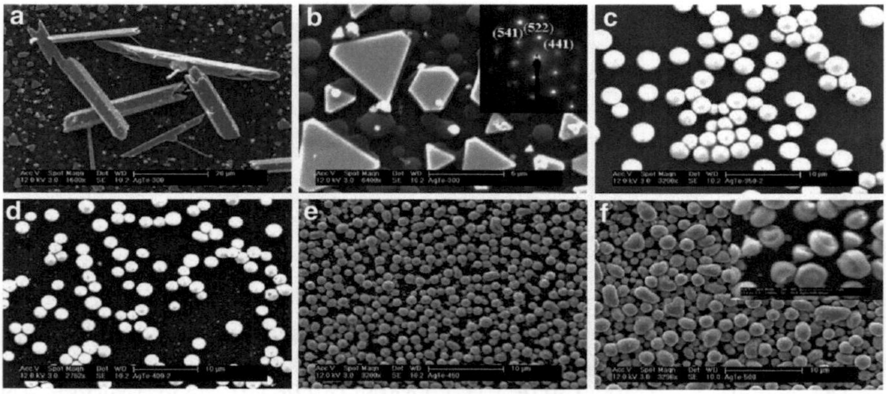

Fig. 9.35 SEM images of Ag_7Te_4 films deposited at (**a**) and (**b**) 300°C, (**c**) 350°C, (**d**) 400°C, (**e**) 450°C and (**f**) 500°C. The inset in (**b**) shows the SAED of truncated hexagonal plates. The inset in (**f**) shows the morphology of the film at 450°C after etching

The decomposition of $[M_2Cl_2(\mu{-}Se^nPr)_2(P^nPr_3)_2]$ in tributylphosphate at 195°C gave $Pd_{17}Se_{15}$ whereas the pyrolysis of related allylpalladium complexes in xylene gave Pd_4Se_{19} [154–156]. In another study $Pd(acac)_2$ (or $PdCl_2$) was reacted with nanowires of trigonal selenium (t-Se, as a template) in refluxing ethanol to generate $Pd_{17}Se_{15}$ on the surface of each t-Se nanowire. Recently for the first time a single source precursor [*bis*(*N*,*N*-diethyl-*N*′-naphthoylselenoureato)palladium(II)] was used to deposit thin films by AACVD at deposition temperatures of 400–500°C and nanoparticles by colloidal method [157].

XRD of the films showed the formation of the cubic form of palladium selenide (palladisite $Pd_{17}Se_{15}$) at 400°C, 450°C and 500°C. Thermolysis of the precursor in dodecanethiol (DDSH): oleylamine (OA) 1:1 gave cubic and hexagonal shaped nanoparticles (Fig. 9.36).

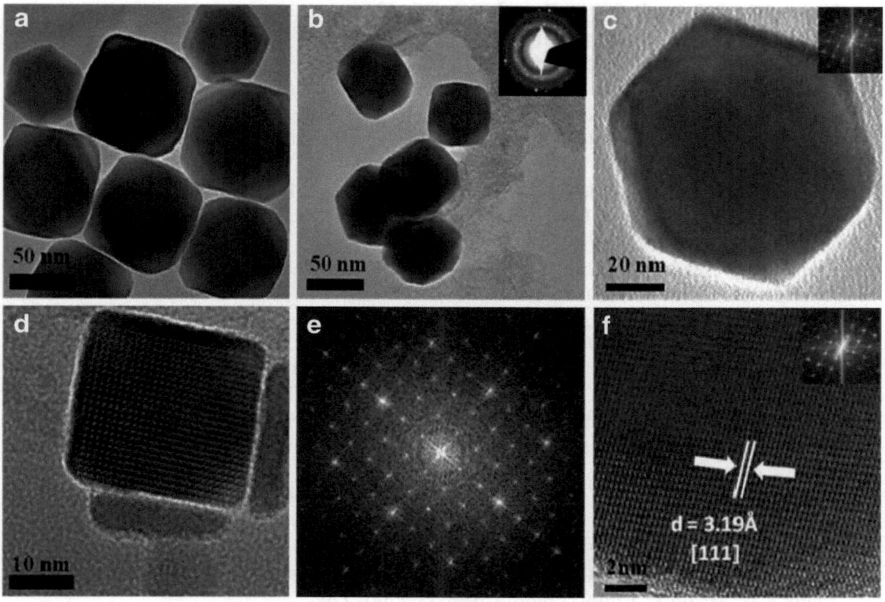

Fig. 9.36 TEM images of palladium selenide nanoparticles prepared in dodecanethiol (DDSH): oleyl amine (OA) 1:1 at 240°C for 1 h (**a**)and (**b**) TEM images with inset in (**b**) showing selected area electron diffraction (SAED) pattern (**c**) and (**d**) HRTEM images of nanocrystals with inset in (**c**) showing FFT image. (**e**) FFT image of cubic nanocrystals (**f**) HRTEM image with d-spacing and FFT inset

9.8 Conclusion

Metal chalcogenide materials have diverse applications such as in satellite TV receivers, optical fibre communications, compact disc players, bar-code readers, full color advertising displays and solar cells. This chapter gives an overview of the evolving chemistry of metal selenium/-tellurium complexes and their potential use as single source precursors for the preparation of metal chalcogenide thin films and nanoparticles. The relationship between the type of material produced and the design of the precursor is also discussed. The most common method used for the preparation of nanoparticles involves the thermolysis of the precursor in a high boiling solvent whereas the thin films are deposited by CVD method.

References

1. Malik MA, Afzaal M, O'Brien P (2010) Chem Rev 110:4417–4446
2. Malik MA, Afzaal M, O'Brien P (2010) J Mater Chem 20:4031–4040
3. Trindade T, O'Brien P, Pickett NL (2001) Chem Mater 13:3843–3858
4. Pickett NL, O'Brien P (2001) Chem Rec 1:467–479
5. O'Brien P (1991) Chemtronics 5:61–70
6. O'Brien P, Nomura R (1995) J Mater Chem 5:1761–1773
7. Bochmann M (1996) Chem Vap Deposition 2:85–96
8. Boyle DS, Hearne S, Johnson DR, O'Brien P (1999) J Mater Chem 9:2879–2884
9. Park JH, Afzaal M, Kemmler M, O'Brien P, Otway DJ, Raftery J, Waters J (2003) J Mater Chem 13:1942–1949
10. Ahmad K, Afzaal M, O'Brien P, Hua GH, Woollins DJ (2010) Chem Mater 22:4619–4624
11. Ma W, Luther JM, Zheng H, Wu Y, Alivisatos P (2009) Nano Lett 9:1699–1703
12. Manasevi HM (1981) J Cryst Growth 55:10–23
13. Brennan JG, Segrist T, Carroll PJ, Stuczynski SM, Reynders P, Brus LE, Steigerwald ML (1989) J Am Chem Soc 111:4141–4143
14. Brennan JG, Siegrist T, Carroll PJ, Stuczynski SM, Reynders P, Brus LE, Steigerwald ML (1990) Chem Mater 2:403–409
15. Osakada K, Yamamoto T (1987) J Chem Soc Chem Commun 1117–1118
16. Steigerwald ML, Sprinkle CR (1987) J Am Chem Soc 109:7200–7201
17. Zeng D, Hampden-Smith MJ, Densler EN (1994) Inorg Chem 33:5376–5377
18. Bochmann M, Webb K, Harman M, Hursthouse MB (1990) Angew Chem Int Ed Engl 29:638–639
19. Bochmann M, Webb KJ, Hursthouse MB, Mazid M (1991) J Chem Soc Dalton Trans 2317–2323
20. Bochmann M, Webb KJ (1991) J Chem Soc Dalton Trans 9:2325–2329
21. Bochmann M, Bwembya GC, Grinter R, Powell AK, Webb KJ, Hursthouse MB, Abdel Malik KM, Mazid MA (1994) Inorg Chem 33:2290–2296
22. Bochmann M, Webb KJ (1991) Mater Res Soc Symp Proc 204:149–154, Chem. Abstr 1992, 116, 33194h
23. Bochmann M, Webb KJ, Hails J-E, Wolverson D (1992) Eur J Solid State Inorg Chem 29:155–166
24. Bochmann M, Coleman AP, Powell AK (1992) Polyhedron 11:507–512
25. Jun YW, Choi CS, Cheon J (2001) Chem Commum 101–102
26. Dabbousi BO, Bonasia PJ, Arnold J (1991) J Am Chem Soc 113:3186–3188
27. Bonasi PJ, Arnold J (1992) Inorg Chem 31:2508–2514
28. Arnold J, Walker JM, Yu KM, Bonasia PJ, Seligson AL, Bourret ED (1992) J Cryst Growth 124:647–653
29. Bonasia PJ, Gindelberger DE, Dabbousi BO, Arnold J (1992) J Am Chem Soc 114:5209–5214
30. Bonasia PJ, Mitchell GP, Hollander FJ, Arnold J (1994) Inorg Chem 33:1797–1802
31. Arnold J, Bonasia PJ (1992) US Pat 5157136; Chem. Abstr. 1993, 118, 213295s
32. Adeogun A, Nguyen CQ, Afzaal M, Malik MA, O'Brien P (2006) Chem Commun 2179–2181
33. Nguyen CQ, Adeogun A, Afzaal M, Malik MA, O'Brien P (2006) Chem Commun 2182–2184
34. Klug HP (1966) Acta Crystallogr 21:536–546
35. Bonamico M, Mazzone G, Vaciago A, Zambonelli L (1965) Acta Crystallogr 19:898–909
36. Miyame H, Ito M, Iwasaki H (1979) Acta Crystallogr B B35:1480–1482
37. Domenicano A, Torelli L, Vaciago A, Zambonelli L (1968) J Chem Soc A 1351–1361
38. Bonamico M, Dessy G (1971) J Chem Soc A 264–269
39. Hursthouse MB, Malik MA, Motevalli M, O'Brien P (1992) Polyhedron 11:45–48

40. Motevalli M, O'Brien P, Walsh JR, Watson IM (1996) Polyhedron 15:2801–2808
41. Chunggaze M, McAleese J, O'Brien P, Otway DJ (1998) J Chem Soc Chem Commun 7:833–834
42. O'Brien P, Otway DJ, Walsh JR (1997) Adv Mater CVD 3:227–229
43. Chunggaz M, Malik MA, O'Brien P (1999) J Mater Chem 9:2433–2437
44. Ludolph B, Malik MA, O'Brien P, Revaprasadu N (1998) Chem Commun 17:1849–1850
45. Malik MA, O'Brien P (1994) Adv Mater Opt Electron 3:171–175
46. Hursthouse MB, Malik MA, Motevalli M, O'Brien P (1992) J Mater Chem 9:949–955
47. Noltes JG (1965) Recl Trav Chim Pay-B 84:126
48. Hursthouse MB, Malik MA, Motevalli M, O'Brien P (1991) Organometallics 10:730–732
49. Nyman M, Hampden-Smith MJ, Duesler E (1996) Chem Vap Deposition 2:171–174
50. Nyman M, Jenkins K, Hampden-Smith MJ, Kodas TT, Duesler EN, Rheingold AL, Liable-Sands ML (1998) Chem Mater 10:914–921
51. Schmidpeter A, Bohm R, Groeger H (1964) Angew Chem Int Ed Engl 3:704
52. Schmidpeter A, Stoll K (1967) Angew Chem Int Ed Engl 6:252–253
53. Schmidpeter A, Stoll K (1968) Angew Chem Int Ed Engl 7:549–550
54. Malik MA, O'Brien P (1991) Chem Mater 3:999–1000
55. Malik MA, Motevalli M, O'Brien P, Walsh JR (1992) Organometallics 11:3136–3139
56. Abrahams I, Malik MA, Motevalli M, O'Brien P (1994) J Organomet Chem 465:73–77
57. Malik MA, Motevalli M, O'Brien P (1996) Acta Crystallogr C C52:1931–1933
58. Bhattacharyya P, Slawin AMZ, Williams DJ, Woollins JD (1995) J Chem Soc, Dalton Trans 2489–2495
59. Silvestru C, Drake JE (2001) Coord Chem Rev 223:117–216
60. Ly TQ, Woollins JD (1998) Polyhedron 176:451–480
61. Afzaal M, Crouch D, Malik MA, Motevalli M, O'Brien P, Park J-H, Woollins JD (2004) Eur J Inorg Chem 171–177
62. Crouch DJ, O'Brien P, Malik MA, Skabara PJ, Wright SP (2003) Chem Commun 1454–1455
63. Afzaal M, Crouch DJ, O'Brien P, Raftery J, Skabara PJ, White AJP, Williams DJ (2004) J Mater Chem 14:233–237
64. Dance IG, Choy A, Scudder ML (1984) J Am Chem Soc 106:6285–6295
65. Lee GSH, Fisher KJ, Craig DC, Scudder M, Dance IG (1990) J Am Chem Soc 112:6435–6437
66. Cumberland SL, Hanif KM, Javier A, Khitrov GA, Strouse GF, Woessner SM, Yun CS (2002) Chem Mater 14:1576–1584
67. Cupertino D, Birdsall DJ, Slawin AMZ, Woollins JD (1999) Inorg Chim Acta 290:1–7
68. Afzaal M, Aucott SM, Crouch D, O'Brien P, Woollins JD, Park JH (2002) Chem Vap Deposition 8:187–189
69. Afzaal M, Crouch D, O'Brien P, Park JH (2002) Mater Res Soc Symp Proc 692–698
70. Afzaal M, Crouch D, Malik MA, Motevalli M, O'Brien P, Park JH, Woollins JD (2004) Eur J Inorg Chem 171–177
71. Afzaal M, Crouch D, Malik MA, Motevalli M, O'Brien P, Park JH (2003) J Mater Chem 13:639–640
72. Garje SS, Ritch JS, Eisler DJ, Afzaal M, O'Brien P, Chivers T (2006) J Mater Chem 16:966–969
73. Briand GG, Chivers T, Parvez M (2002) Angew Chem Int Ed 41:3468–3470
74. Chivers T, Eisler DJ, Ritch JS (2005) Dalton Trans 2675–2677
75. Green M, Wakefield G, Dobson PJ (2003) J Mater Chem 13:1076–1078
76. Okamoto Y, Yano T (1971) J Organomet Chem 29:99–103
77. Chivers T, Eisler DJ, Ritch JS, Tuononen HM (2005) Angew Chem Int Ed 44:4953–4956
78. Gysling HJ, Wernberg AA, Blanton TN (1992) Chem Mater 4:900–905
79. Stoll SL, Bott SG, Barron AR (1997) J Chem Soc, Dalton Trans 1315–1322
80. Stoll SL, Barron AR (1998) Chem Mater 10:650–657

81. Rahbarnoohi W, Wells RL, Liable-Sands LM, Yap GPA, Rheingold AL (1997) Organometallics 16:3959–3964
82. O'Brien P, Otway DJ, Walsh JR (1997) Chem Vap Deposition 3:227–229
83. Cheon J, Arnold J, Yu K-M, Bourret ED (1995) Chem Mater 7:2273–2276
84. Gillan EG, Barron AR (1997) Chem Mater 9:3037–3048
85. Gillan EG, Bott SG, Barron AR (1997) Chem Mater 9:796–806
86. Pernot P, Barron AR (1995) Chem Vap Deposition 1:75–78
87. Park J-H, Afzaal M, Helliwell M, Malik MA, O'Brien P, Raftery J (2003) Chem Mater 15:4205–4210
88. Malik S, Mehboob S, Malik MA, O'Brien P (2011) J Mater Chem to be submitted
89. Malik S, Mehboob S, Malik MA, O'Brien P, Raftery J (2011) J Mater Chem to be submitted
90. Afzaal M, Ellwood K, Pickett NL, O'Brien P, Raftery J, Waters J (2004) J Mater Chem 14:1310–1315
91. Ritch JS, Ahmad K, Afzaal M, Chivers T, O'Brien P (2010) Inorg Chem 49:1158–1205
92. Trindade T, O'Brien P, Zhang X-M, Motevalli M (1997) J Mater Chem 7:1011–1016
93. Trindade T, Monteiro OC, O'Brien P, Motevalli M (1999) Polyhedron 18:1171–1175
94. Lee SM, Jun YW, Cho SN, Cheon J (2002) J Am Chem Soc 124:11244–11245
95. Duan T, Lou W, Wang X, Xue Q (2007) Colloid Surface A 310:86–93
96. Lee SM, Cho SN, Cheon J (2003) Adv Mater 15:441–444
97. Zhihua Z, Lee SH, Vittal JJ, Chin WS (2006) J Phys Chem B 110:6649–6654
98. Pradhan N, Katz B, Efrima S (2003) J Phys Chem B 107:13843–13854
99. Acharya S, Gautam UJ, Sasaki T, Bando Y, Golan Y, Ariga K (2008) J Am Chem Soc 130:4594–4595
100. Akhtar J, Malik MA, Stubbs SK, O'Brien P, Helliwell M, Binksv DJ (2011) Inorg Chem submitted for publication
101. Akhtar J, Bruce JC, Malik MA, Klaus RK, Afzaal M, O'Brien P (2009) Mater Res Soc Symp Proc 1148-PP12-08
102. Akhtar J, Malik MA, O'Brien P (2011) Eur J Inorg Chem submitted for publication
103. Chuprakov IS, Dahmen K-H, Schneider JJ, Hagen J (1998) Chem Mater 10:3467–3470
104. Chuprakov IS, Dahmen K-H (1999) J Phys IV 9:313–319
105. Schneider JJ, Hagen J, Heinemann O, Bruckmann J, Krüger C (1997) Thin Solid Films 304:144–148
106. Reid SD, Hector AL, Levason W, Reid G, Waller BJ, Webster M (2007) Dalton Trans 42:4769–4777
107. Chen T, Hunks W, Chen PS, Stauf GT, Cameron TM, Xu C, DiPasuale AG, Rheingold AL (2009) Eur J Inorg Chem 2047–2049
108. Rowe DM (1995) CRC handbook of thermoelectric. CRS Press, Inc, Boca Raton
109. Bayaz AA, Giani A, Foucaran A, Delannoy FP, Boyer A (2003) Thin Solid Films 441:1–5
110. Kim IH (2000) Mater Lett 43:221–224
111. Zheng X, Xie Y, Zhu L, Jiang X, Jia Y, Song W, Sun Y (2002) Inorg Chem 41:455–461
112. Cope D (1959) U S Pat 2875359
113. Rajpure KY, Bhosale CH (2000) Mater Chem Phys 64:70–74
114. Kalyanasundaram K (1985) Sol Cells 15:93–156
115. Chaudhari KR, Wadawale AP, Ghoshal S, Chopade SM, Sagoria VS, Jain VK (2009) Inorg Chim Acta 362:1819–1824
116. Chaudhari KR, Yadav N, Wadawale A, Jain VK, Bohra R (2010) Inorg Chim Acta 363:375–380
117. Crouch DJ, Helliwell M, O'Brien P, Park JH, Waters J, Williams DJ (2003) Dalton Trans 1500–1504
118. Rodriguez-Castro J, Dale P, Mahon MF, Molloy KC, Peter LM (2007) Chem Mater 19:3219–3226
119. Koh YW, Lai CS, Du AY, Tiekink ERT, Loh KP (2003) Chem Mater 15:4544–4554
120. Zhang H, Ji Y, Ma X, Yang D (2003) Nanotechnology 14:974–977

121. Ye C, Meng G, Jiang Z, Wang Y, Wang G, Zhang L (2002) J Am Chem Soc 124:15180–15181
122. Ko TY, Sun KW (2009) J Lumin 129:1747–1749
123. Arivuoli D, Gnanam FD, Ramasamy P (1988) J Mater Sci Lett 7:711–713
124. Okamoto H (1994) J Phase Equilib 15:195–201
125. Sankapal BR, Lokhande CD (2002) Mater Chem Phys 73:151–155
126. Harpeness R, Gedanken A (2003) New J Chem 27:1191–1193
127. Xie Q, Liu Z, Shao M, Kong L, Yu W, Qian Y (2003) J Cryst Growth 252:570–574
128. Sigman MB, Korgel BA (2005) Chem Mater 17:1655–1660
129. Liu H, Cui H, Han F, Li X, Wang J, Boughton RI (2005) Cryst Growth Des 5:1711–1714
130. Ma J, Wang Y, Wang Y, Chen Q, Lian J, Zheng W (2009) J Phys Chem C 113:13588–13592
131. Chang HW, Sarkar B, Liu CW (2007) Cryst Growth Des 7:2691–2695
132. Monterio OC, Trindade T, Paz FAA, Klinowski J, Watersand J, O'Brien P (2003) J Mater Chem 13:3006–3010
133. Waters J, Crouch D, Raftery J, O'Brien P (2004) Chem Mater 16:3289–3298
134. Garje SS, Eisler DJ, Ritch JS, Afzaal M, O'Brien P, Chivers T (2006) J Am Chem Soc 128:3120–3121
135. Ko TY, Yang CH, Sun KW, Chang HW, Sarker B, Liu CW (2009) Cent Eur J Chem 7:197–204
136. Sharma RK, Kedarnath G, Jain VK, Wadawale A, Nalliath M, Pillai CGS, Vishwanadh B (2010) Dalton Trans 39:8779–8787
137. Lin Y-F, Chang H-W, Lu S-Y, Liu CW (2007) J Phys Chem C 111:18538–18544
138. Kessler F, Rudmann D (2004) Sol Energy 77:685–695
139. Ramanathan K, Contreras MA, Perkins CL, Asher S, Hasson FS, Keane J, Young D, Romero M, Metzger W, Noufi R, Ward J, Duda A (2003) Prog Photovolt Res Appl 11:225–230
140. Yamaguchi T, Matsufusa J, Yoshida A (1992) Jpn J Appl Phys 31:L703–L705
141. Malik MA, O'Brien P, Revaprasadu N (1999) Adv Mater 11:1441–1444
142. Castro SL, Bailey SG, Rafello RF, Banger KK, Hepp AF (2003) Chem Mater 15:3142–3147
143. Allen PM, Bawendi MG (2008) J Am Chem Soc 130:9240–9241
144. Panneerselvam A, Periyasamy G, Ramasamy K, Afzaal M, Malik MA, O'Brien P, Burton NA, Waters J, van Dongen BE (2010) Dalton Trans 39:6080–6091
145. Panneerselvam A, Nguyen CQ, Waters J, Malik MA, O'Brien P, Raftery J, Helliwell M (2008) Dalton Trans 33:4499–4506
146. Panneerselvam A, Nguyen CQ, Malik MA, O'Brien P, Raftery J (2009) J Mater Chem 19:419–427
147. Copsey MC, Panneerselvam A, Afzaal M, Chivers T, O'Brien P (2007) Dalton Trans 15:1528–1538
148. Singhal A, Jain VK, Mishra R, Varghese B (2000) J Mater Chem 10:1121–1124
149. Kumbhare LB, Jain VK, Phadnis PP, Nethaji M (2007) J Organomet Chem 692:1546–1556
150. Dey S, Jain VK, Varghese B (2001) J Organomet Chem 623:48–55
151. Dey S, Jain VK, Chaudhury S, Knoedler A, Lissner F, Kaim W (2001) J Chem Soc Dalton Trans 5:723–728
152. Dey S, Jain VK, Singh J, Trehan V, Bhasin KK, Varghese B (2003) Eur J Inorg Chem 4:744–750
153. Kumbhare BL, Jain VK, Varghese B (2006) Inorg Chim Acta 359:409–416
154. Dey S, Narayan S, Singhal A, Jain VK (2000) Indian Acad Sci Chem Sci 112:187–196
155. Jain VK, Kumbhare BL, Dey S, Ghavale ND (2008) Phosphorus Sulfur 183:1003–1008
156. Kumbhare BL, Wadawale AP, Jain VK, Kolay S, Nethaji M (2009) J Organomet Chem 694:3892–3901
157. Akhtar J, Mehmood RF, Malik MA, Iqbal N, O'Brien P, Raftery J (2010) Chem Commun. doi:10.1039/c0cc03079a

Chapter 10
Synthesis and Transformations of 2- and 3-hydroxy-Selenophenes and 2- and 3-Amino-Selenophenes

G. Kirsch, E. Perspicace, and S. Hesse

10.1 Introduction

Several monographs, series and articles have been published on selenophene [1–4]. These give a general overview of the nucleus synthesis. In this chapter we focus on the synthesis of hydroxy and aminoselenophenes and also describe examples of their reactivity or their transformation to different condensed heterocyclic systems.

10.2 Synthesis of Hydroxy Selenophenes and Their Transformation Studies

10.2.1 Synthesis of 2-hydroxyselenophenes

The first synthesis of a 2-hydroxyselenophene was made through an acid catalysed dealkylation of 2-tertiobutoxyselenophene obtained by lithiation of a substituted selenophene followed by reaction with magnesium bromide and tertiobutylperbenzoate (Scheme 10.1) [5, 6].

2-Hydroxyselenophenes exist in equilibrium between two lactonic forms (Fig. 10.1) [7].

Another route to the 2-hydroxy derivatives was the hydrogen peroxide oxidation of 2-selenophene diethylboronates (Scheme 10.2) [8]. The boronates were obtained from the reaction of triethyl borate on 2-lithiated selenophenes.

G. Kirsch (✉) • E. Perspicace • S. Hesse
Laboratoire d'Ingénierie Moléculaire et Biochimie Pharmacologique, Institut Jean Barriol, Université Paul Verlaine Metz, Metz, France
e-mail: kirsch@univ-metz.fr

J.D. Woollins and R.S. Laitinen (eds.), *Selenium and Tellurium Chemistry*,
DOI 10.1007/978-3-642-20699-3_10,

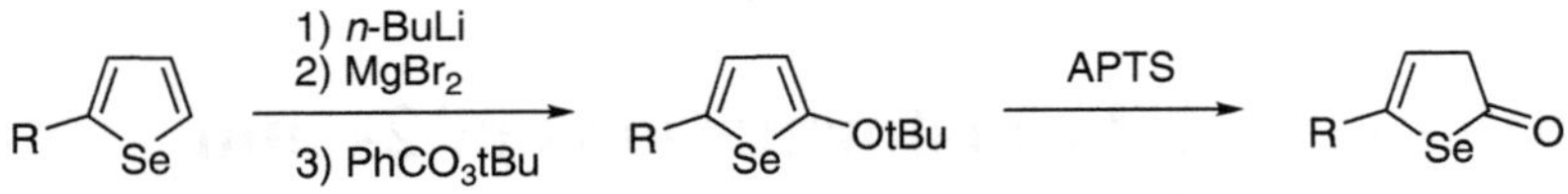

Scheme 10.1

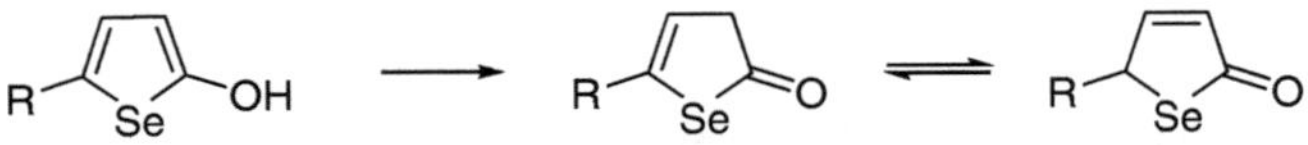

Fig. 10.1 Lactonic forms of 2-hydroxy selenophene

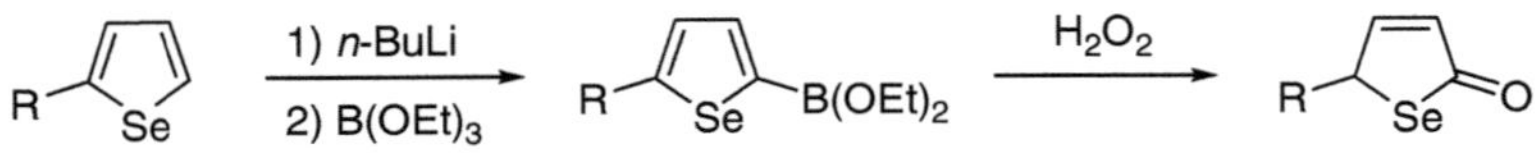

Scheme 10.2

10.2.2 Synthesis of 3-Hydroxy and 3, 4-Dihydroxyselenophenes

As for the 2-hydroxy derivatives, the literature is scarce on the 3-hydroxyisomers. Using the same approach as for the synthesis of the 2-hydroxyselenophenes, the compounds are obtained by acidic dealkylation of 3-tertiobutoxyselenophenes or oxidation of 3-selenophene boronic acids substituted in the 2-position by either a formyl or an acetyl group [9]. In the case of 2,5-disubstituted 3-hydroxyselenophenes [10] as well as in the case of the formyl and acetyl derivatives, the equilibrium between hydroxyl and keto form exists (Fig. 10.2).

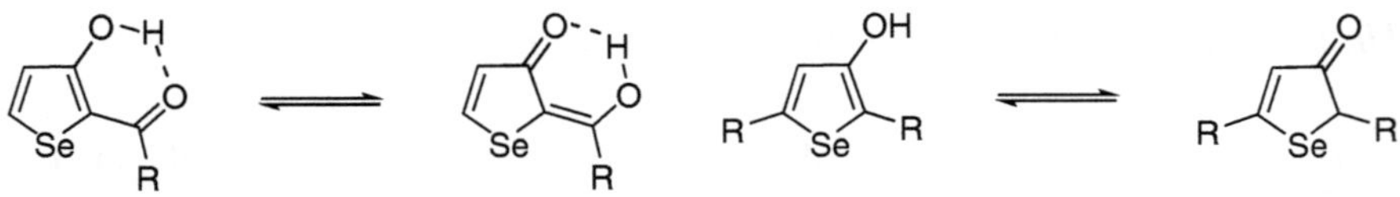

Fig. 10.2 Hydroxy-keto equilibrium in 3-hydroxyselenophenes

The dimethylketene dithioacetal, prepared from diethylmalonate, when treated with sodium selenide and phenacylbromide afforded after cyclisation the fully substituted 3-hydroxy- selenophene (Scheme 10.3) [11].

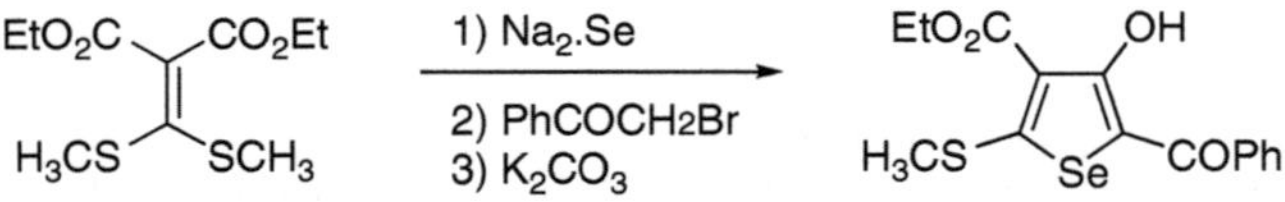

Scheme 10.3

The reaction of isoselenocyanates with ethyl γ-chloroacetate in presence of a base gave an intermediate keteneaminoselenolate which cyclised directly to the 3-hydroxy derivative. Spectroscopic and crystallographic studies have shown that the compounds exist in the keto form (Scheme 10.4) [12].

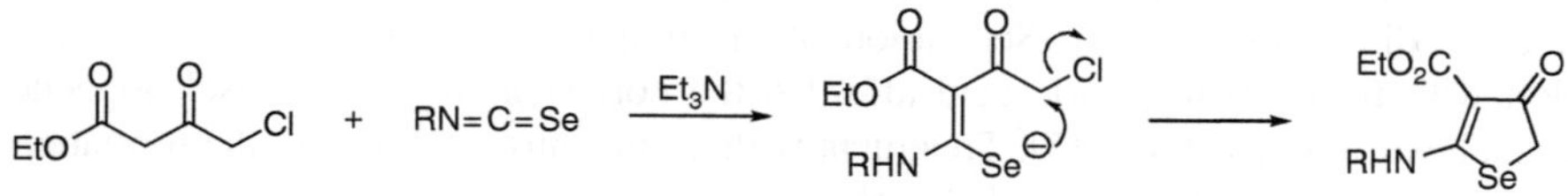

Scheme 10.4

Looking for a way to oxidize tetraketones, 1,6-diphenyl-1,3,4,6-hexanetetrone was treated with selenium dioxide. The compound isolated was in fact 2,5-dibenzoyl-3,4-dihyroxyselenophene [13]. Using Hinsberg's method [14], diethyl 3,4-dihydroxy 2,5-selenophene dicarboxylate could be prepared [15] (Scheme 10.5).

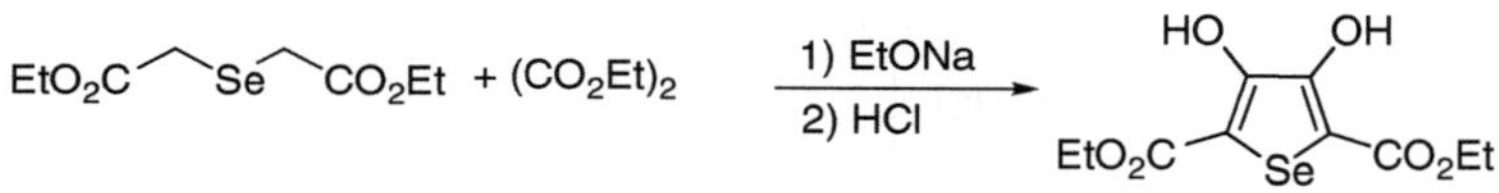

Scheme 10.5

10.2.3 Reactivity and Further Transformations of Hydroxyselenophenes

2- or 3-hydroxyselenophenes react under etherification conditions by C- or O-alkylation. Depending on the hardness of the alkylation reagent, methyl iodide or dimethylsulfate, the reaction gave respectively C or O alkylation [16]. Unsubstituted 2-hydroxyselenophene underwent Michael addition on one of its keto form [16] (Scheme 10.6).

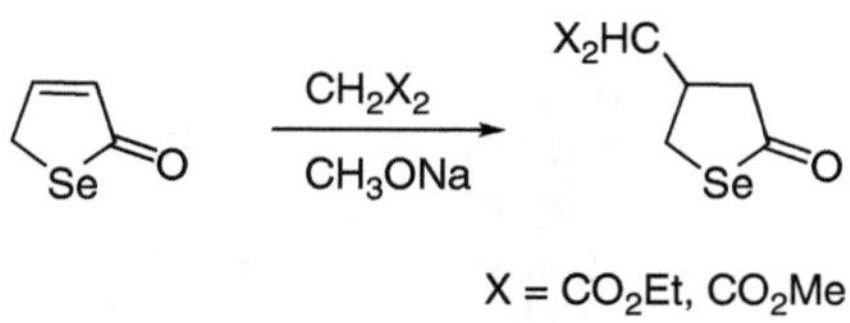

Scheme 10.6

10.3 Synthesis of Amino Selenophenes and their Transformation Studies

2- and 3-aminoselenophene are mostly known as unstable compounds but can be stabilized as salts or when acetylated or formylated. The methods of synthesis are generally similar to those used in the thiophene series.

10.3.1 Synthesis of 2- and 3- Aminoselenophenes

The first synthesis of the amino compound has been made from the nitro derivative. Nitration of selenophene in acetic acid gave a mixture of 2- and 3-nitroselenophene and 2,4-dinitroselenophene. Separation of the major 2-nitro derivative and reduction with tin in hydrochloric acid afforded the corresponding 2-aminoselenophene as an hexachlorostannate [17]. Treatment with acetic anhydride lead to the formation of the 2-acetamido derivative [18] (Scheme 10.7).

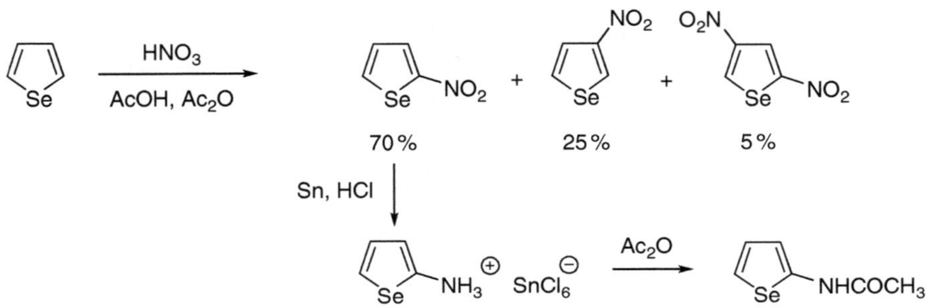

Scheme 10.7

As the access to selenophene is not easy, synthesis of 2-aminoselenophene derivatives appeared again with the extension of Gewald's thiophene synthesis to selenophene. Refluxing elemental selenium, ketones and cyanoacetate or malodinitrile in the presence of a base allowed the preparation of 2-aminoselenophenes substituted in the three position by a cyano or a carbethoxy group [19–22] (Scheme 10.8). The presence of these groups in that position helps to stabilize the amine.

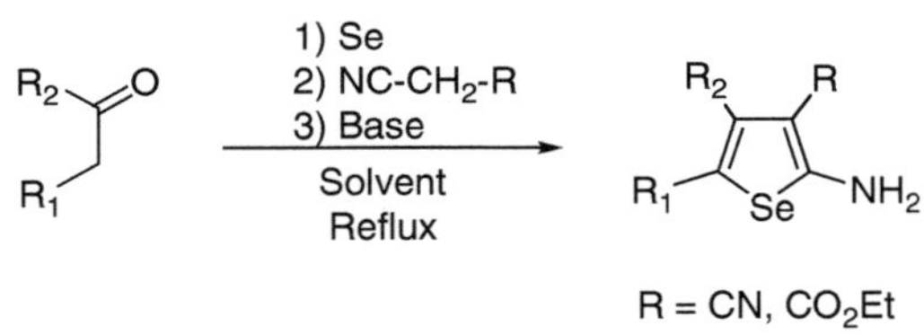

Scheme 10.8

The reaction could be run from ylidene-malodinitrile with selenium in the presence of triethylamine [23] and also under microwave and ultrasound conditions [21]. One example of synthesis of 2, 5-diaminoselenophene has been described. Reacting tetracyanoethylene with hydrogen selenide leads to 2,5-diamino 3,4-dicyanoselenophene [24].

10.3.2 Synthesis of 3-Aminoselenophenes

Different routes to 3-aminoselenophenes have been developed starting from the heterocycle or building the ring with simultaneous introduction of the amino function. Nitration of 2-cyano selenophene gave a mixture of the 5- and 4- isomers. The 4-nitro isomer was separated and hydrolysed to the acid which was decarboxylated by refluxing in quinoline in the presence of copper bronze. The three nitro derivative isolated was reduced with tin in hydrochloric acid. The hexachlorostannate obtained was reacted with formylacetate in basic media to give the formamidoselenophene [17] (Scheme 10.9).

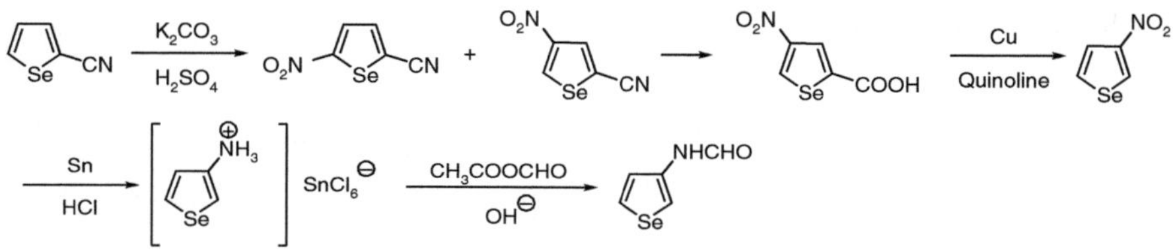

Scheme 10.9

Reaction of 3-bromo 2-formyl selenophene with sodium azide in DMSO afforded the 3-azidoderivative which was reduced to the amine using hydrogen sulfide (Scheme 10.10) [25]. Oxidation of the aldehyde followed by the same reduction gave the 3-amino 2-selenophene carboxylic acid (Scheme 10.10).

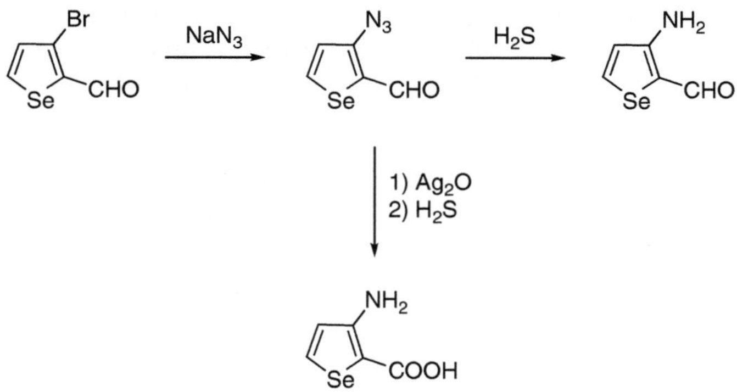

Scheme 10.10

Construction of the ring with introduction of the amino function in three position was first made by reacting chlorophenylmethylidene malonitrile with sodium selenide, generated in situ by reacting selenium with sodium borohydride, and ethyl bromoacetate [26] (Scheme 10.11).

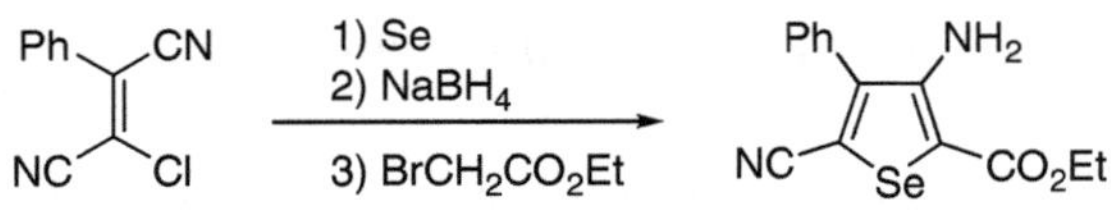

Scheme 10.11

Ketene dithioacetals have been for the synthesis of 3-aminothiophenes. The starting ketenes are easily prepared from carbon disulfide. Carbon diselenide is not commercially available and its preparation in greater amount is not easy due to the noxious odor of the compound. Taking advantage of the reactivity of the ketene dimethyl dithioacetal from malonitrile, reaction with sodium selenide followed by addition of activated halides and cyclisation allowed the synthesis of 3-aminoselenophenes [11, 27] (Scheme 10.12).

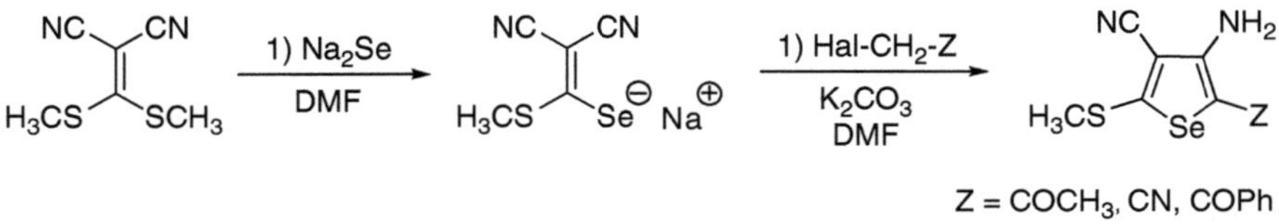

Scheme 10.12

The reaction was run in a "one-pot" procedure. Sodium selenide was prepared by reduction of black selenium with rongalite in basic media filtered and suspended in DMF. The ketene dithioacetal was then added followed later by the active halide. Cyclisation with potassium carbonate gave the 3-aminoselenophenes. The ability of substitution of the methylsulfanyl group was exploited to prepare in a "one-pot" procedure 5-*N*,*N*-dialkylamino 3-aminoselenophenes. The reaction is described in Scheme 10.13 [28].

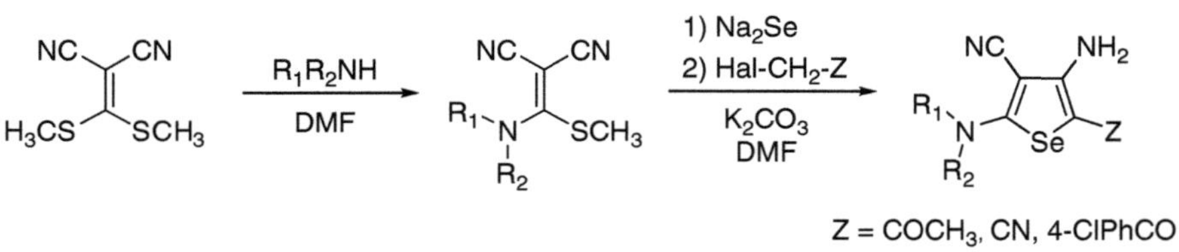

Scheme 10.13

β-Substituted β-chloro acroleins have been used for the synthesis of thiophenes, selenophenes and tellurophenes. Transformation of the aldehyde into nitrile led to the corresponding β-substituted β-chloro acrylonitriles starting material for the synthesis of 2-cyano and 2-nitro 3-aminoselenophenes [30, 31] (Scheme 10.14).

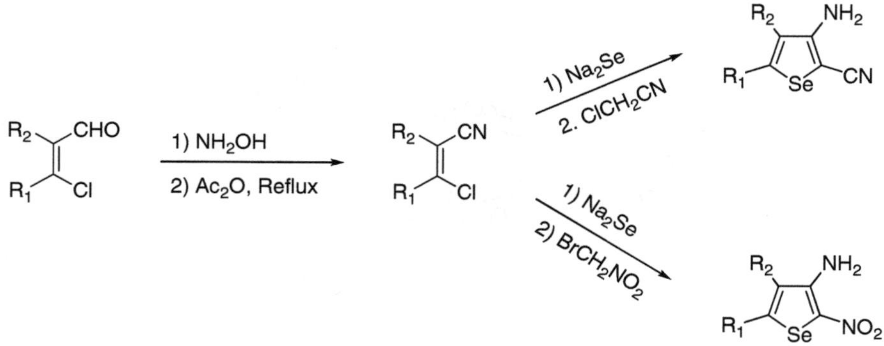

Scheme 10.14

Many 3-aminoselenophenes condensed to aromatic nitrogen heterocycles have been described in the literature. Starting material were 2-chloro 3-cyanopyridines [31], 2-chloro 3-cyanoquinolines [32], 3-cyano-2*H*-pyridoselones [33–35], or 3,3′-dicyano 2,2′-dipyridylselenides [36].

10.3.3 Reactions of Aminoselenophenes

10.3.3.1 Reactions of 2-Aminoselenophenes

2-Aminoselenophene hexachlorostannate, obtained by reduction of 2-nitroselenophene, was condensed with 1, 1, 3, 3-tetraethoxypropane in the presence of zinc chloride to afford selenolo[2,3 *b*]pyridine [18] (Scheme 10.15).

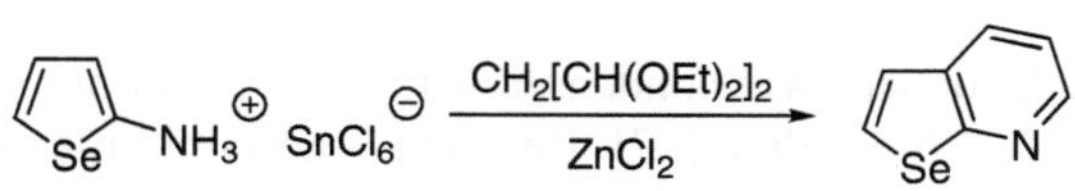

Scheme 10.15

2-amino-3-cyano selenophenes have been used as starting material for the construction of polycyclic heterocycles. Condensation with formamide, carbon disulfide or phenylisothiocyanate afforded the corresponding aminopyrimidine, pyrimidodithione and pyrimidoiminothione [23] (Scheme 10.16).

Transformation of the cyano group into a dihydroimidazole followed by cyclisation led to more condensed nitrogen containing heterocycles [23] (Scheme 10.17).

Friedländer's condensation applied to 2-amino 3-cyano selenophenes allowed the preparation of selenophene analogs of Tacrine®, an acetylcholinesterase inhibitor used in Alzheimer's cure [37] (Scheme 10.18).

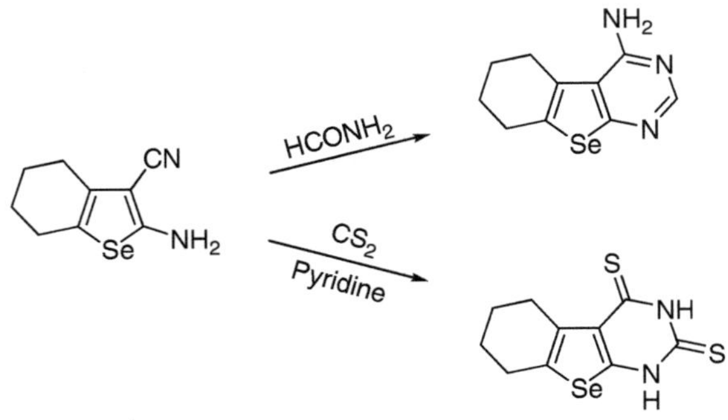

Scheme 10.16

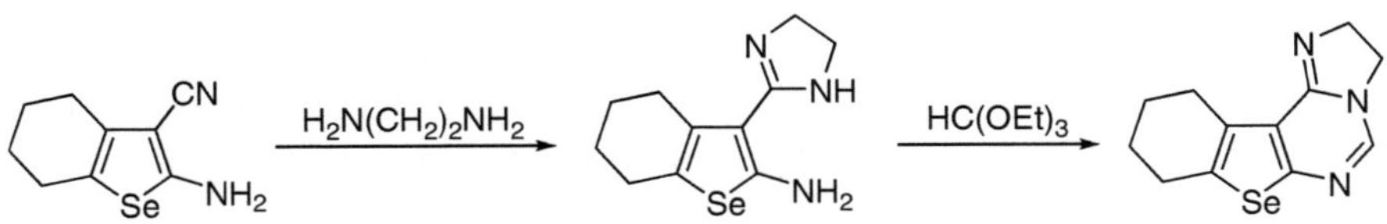

Scheme 10.17

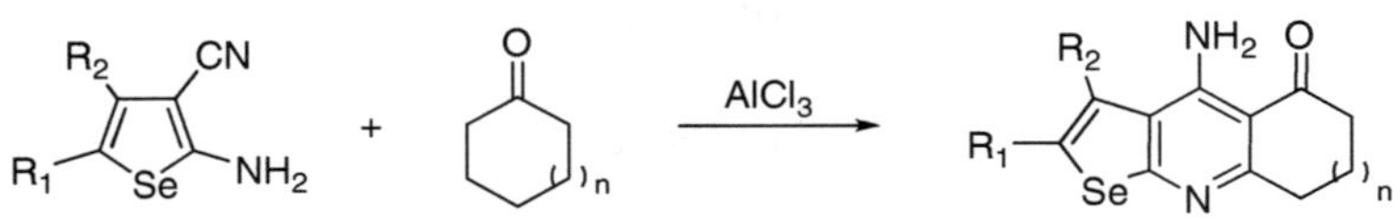

Scheme 10.18

10.3.3.2 Reactions of 3-Aminoselenophenes

3-Acetamidoselenophene was condensed with 1, 1, 3, 3-tetrathoxypropane to the 2-methylselenolo[2, 3-*b*]pyridine [17]. Reacting 3-acetamido(formamido)2-acetyl (formyl) selenophenes with ammonium formate in the presence of ammonia at 160°C allowed the synthesis of substituted seleno[3, 2-*d*]pyrimidines [17]. 3-Aminoselenophenes substituted in the two position with an electron-withdrawing group are interesting starting compounds for further synthesis especially by using the proximity of the two functions. Condensation of 3-aminoselenophene 2-carbaldehyde with pyruvic acid allowed the access to selenolo[2, 3-*b*]carboxylic acid [39] (Scheme 10.19).

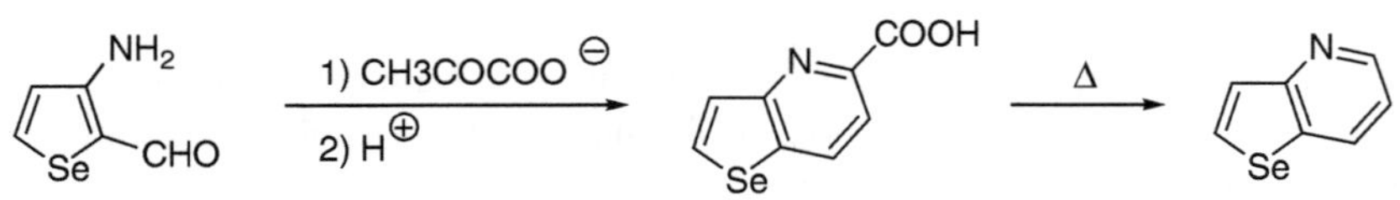

Scheme 10.19

Condensation of the amino-carbaldehyde with acetone in basic media gave the methyl substituted selenolo[3, 2-*b*]pyridine [17]. By diazotation and reaction with potassium selenocyanate, selenocyanation of 3-amino 2-selenophenecarbaldehyde and 3-amino 2-acetyl selenophene was achieved [38]. Treatment of the acetyl derivative with ammonia gave the oxoselenolo[3, 2-*b*]selenophene [38] (Scheme 10.20).

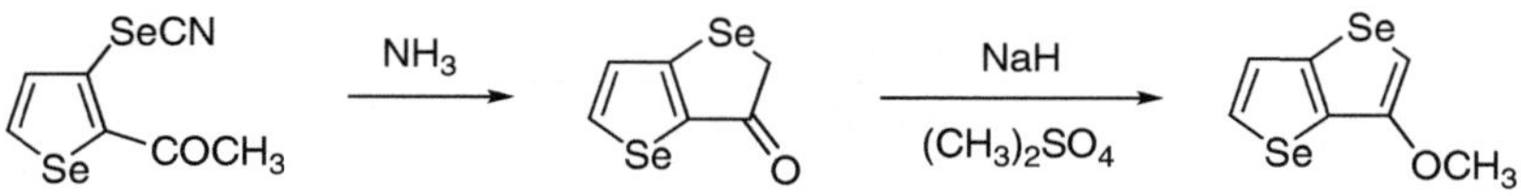

Scheme 10.20

Diazotation and reaction with sodium nitrite has also been used for introducing the azido function in the three position. In this way 3-azido 2-formyl, 2-acetyl and 2-nitro were prepared [39]. Some of these azido derivatives have been used for the synthesis of selenolo[3,2-*b*]pyrroles [39] or selenolo[2, 3-*b*]isoxazoles [40]. From 3-amino-2-cyanoselenophenes, condensation with different cyclanones in the presence of aluminum chloride in dichloroethane led to the other analogs of Tacrine® [29] (Scheme 10.21).

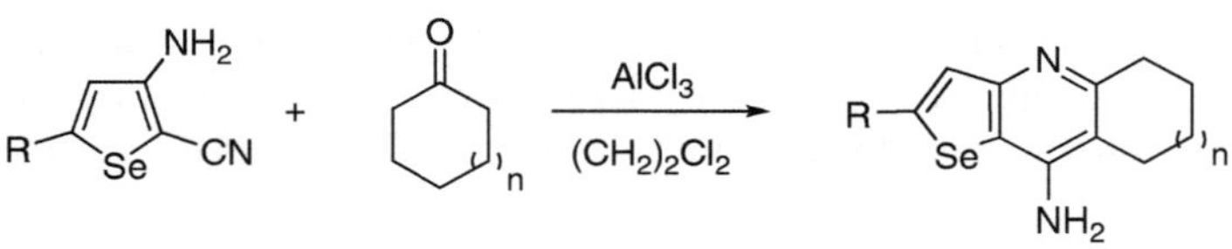

Scheme 10.21

Comparing classical heating and microwave irradiation showed that the reaction time could be brought from 12 h to 8–16 min depending on the solubility of the compounds. Yields could also be increased under microwave conditions. Condensation of 3-amino 2-cyanoselenophenes with formic acid in the presence of sulfuric acid under microwave irradiation afforded the corresponding selenolo[3, 2-*d*] pyrimidinones which could readily be transformed into the chloropyrimidines with phosphorus oxychloride [41] (Scheme 10.22).

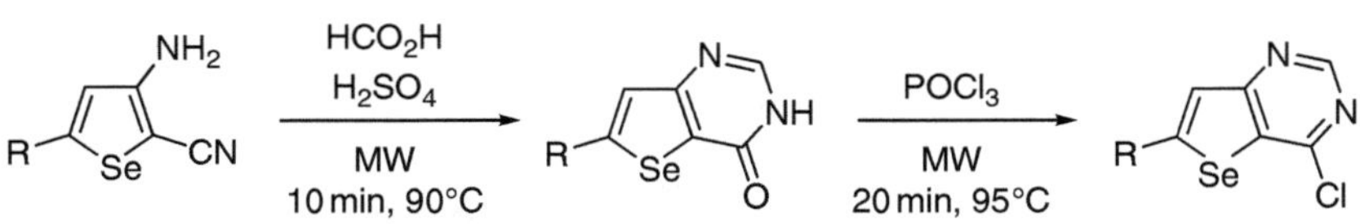

Scheme 10.22

The chloropyrimidines were used in either palladium catalysed Suzuki coupling reaction with boronic acids or functionalised by nucleophilic substitution with secondary amines or phenolates [41] (Scheme 10.23).

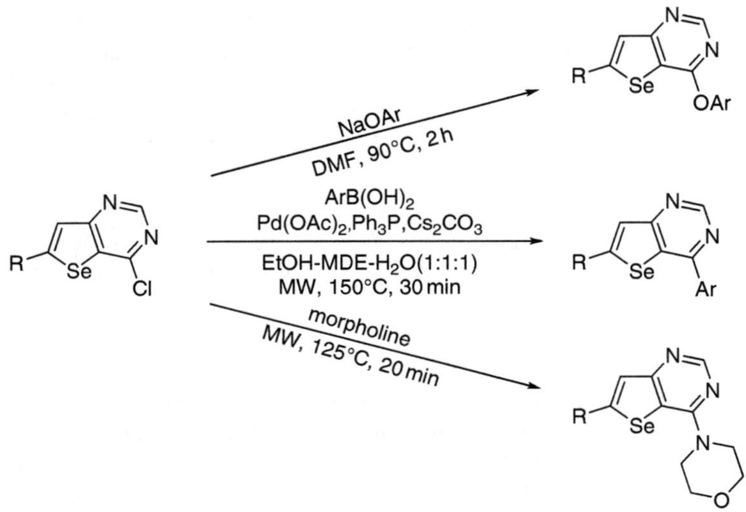

Scheme 10.23

Treatment of the amino-cyano-derivatives with sulfuric acid in the same conditions gave the 3-amino-2-carboxamides. Refluxing the latter with diethyloxalate gave the carboxylated selenolo[3, 2-*d*] pyrimidinones which could be transformed into the chloropyrimidines by the same way described earlier [41] (Scheme 10.24).

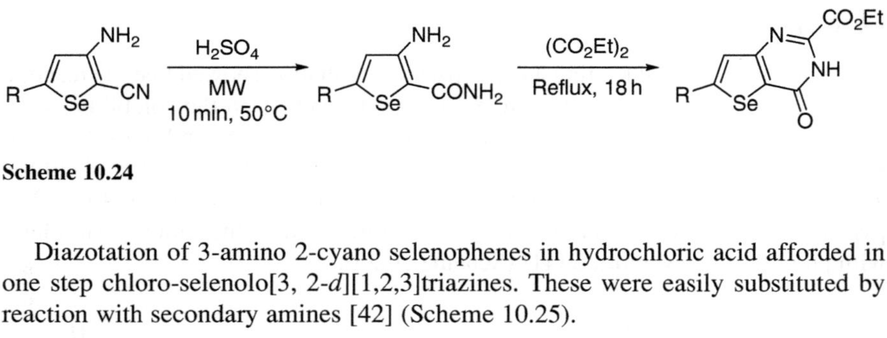

Scheme 10.24

Diazotation of 3-amino 2-cyano selenophenes in hydrochloric acid afforded in one step chloro-selenolo[3, 2-*d*][1,2,3]triazines. These were easily substituted by reaction with secondary amines [42] (Scheme 10.25).

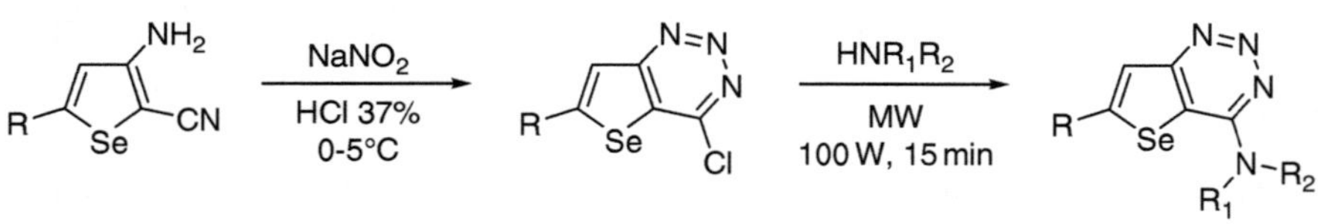

Scheme 10.25

10.4 Conclusion

Hydroxy and aminoselenophenes have not been very much studied over the years. Many synthetic possibilities still stay open starting from these derivatives. Possibly some of the success appearing in the analogous thiophene compounds will promote the research in this part of the organic chemistry of selenium and bring interesting results in different domains like medicinal chemistry or material science.

References

1. Klayman DL, Günther WHH (1973) Organic selenium compounds: their chemistry and biology. Wiley, New York
2. Schatz J (2002) Product class 11: selenophenes in science of synthesis. Thieme, Stuttgart
3. Sommen G (2005) Mini Rev Org Synth 2(4):375–388
4. Pelkey ET (2008) In: Katritzky AR, Ramsden CR, Scriven EFV, Taylor RJK (eds) Comprehensive heterocyclic chemistry III. Elsevier, Oxford
5. Morel J, Paulmier C, Semard C, Pastour P (1970) C R Acad Sci Ser C 270:825–829
6. Morel J, Paulmier C, Semard C, Pastour P (1971) Bull Soc Chim Fr 10:3547–3553
7. Cederlund B, Hörnfeldt AB (1976) Acta Chem Scand B30:101–108
8. Morel J, Paulmier C, Pastour P (1972) J Heteroc Chem 9:355
9. Morel J, Paulmier C, Semard C, Pastour P (1973) Bull Soc Chim Fr 7–8:2434–2441
10. Lantz R, Hörnfeldt AB (1972) Chem Script 2:9–15
11. Sommen G, Comel A, Kirsch G (2003) Synlett 6:855–857
12. Sommen G, Linden A, Heimgartner H (2007) Lett Org Chem 1:7–12
13. Balenkovic K, Cerar D, Filipovic L (1984) J Org Chem 81:1556–1561
14. Hinsberg O (1910) Freiburg i B Ber 43:901–906
15. Aqad E, Lakshmikhantam MV, Cava MP (2008) Org Lett 26:4283–4286
16. Henrio C, Morel J, Pastour P (1975) Ann Chim (Paris) 10:37–40
17. Ah-Kow G, Paulmier C, Pastour P (1976) Bull Soc Chim Fr 5–6:151–160
18. Outurquin F, Ah-Kow G, Paulmier C (1976) Bull Soc Chim Fr 5–6:883–888
19. Khirpak SM, Dobosh MA, Smolanka IV, Mikitchin AS (1973) Kh Geterosikl Soedin 3:326–328
20. Khirpak SM, Dobosh MA, Smolanka IV (1973) Patent S.U. 71–178225
21. Sibor J, Pazdera P (1996) Molecules 1:157–162
22. Aumann KM, Scammells PJ, White JM, Schiesser CH (2007) Org Biomol Chem 5:1276–1281
23. Abdel-Hafez SH (2005) Russ J Org Chem 41:396–401
24. Wudl F, Zellers CT, Nalewajek D (1980) J Org Chem 45:3211–3215
25. Gronowitz S, Westerlund C, Hornfeldt AB (1975) Acta Chem Scand B29:224–232
26. Gewald K, Hain U (1992) Monatsh Chem 123:455–459
27. Sommen G, Comel A, Kirsch G (2005) Phosphorus Sulfur Silicon Relat Elem 180:939–943
28. Thomae D, Perspicace E, Henryon D, Xu Z, Schneider S, Hesse S, Kirsch G, Seck P (2009) Tetrahedron 65:10453–10458
29. Thomae D, Kirsch G, Seck P (2008) Synthesis 10:1600–1606
30. Thomae D, Rodriguez Dominguez JC, Kirsch G, Seck P (2008) Tetrahedron 64:3232–3235
31. Aadil M, Kirsch G (1993) Phosphorus Sulfur Silicon Relat Elem 82:91–92
32. Abdel-Hafez SH, Hussein MA (2008) Arch Pharm (Weinheim, Germany) 341:240–246
33. Mortikov VY, Litvinov VP, Shestopalov AM, Sharanin YA, Apenova EE, Galegov GA, Abdulaev II, Asadullaev TB, Abdullaev FI (1991) Khim-Farmats Zh 25:41–44

34. Litvinov VP, Mortikov VY, Sharanin YA, Shestopalov AM, Zelinsky ND (1985) Synthesis 1:98–99
35. Abdel-Hafez SH, Ahmed RA, Abdel-Azim MA, Hassan KM (2007) J Chem Res 10:580–584
36. Abdel-Hafez SH, Abdel-Mohsen SA, El-Ossaily YA (2006) Phosphorus Sulfur Silicon Relat Elem 181:2297–2305
37. Seck P, Thomae D, Perspicace E, Hesse S, Kirsch G (2008) Book of abstracts, 54. Cold Spring Harbour Laboratory symposium on neurodegenerative diseases, CSHL, New York, 2008
38. Paulmier C (1979) C R Acad Sci Ser C 28:317–319
39. Gronowitz S, Westerlund C, Hornfeldt AB (1976) Acta Chem Scand B30:391–396
40. Gronowitz S, Westerlund C, Hornfeldt AB (1979) Chem Script 10:165–172
41. Hesse S, Chenet C, Thomae D, Kirsch G (2009) Synthesis 7:1204–1208
42. Perspicace E, Thomae D, Hamm G, Hesse S, Kirsch G, Seck P (2009) Synthesis 20:3472–3476

Chapter 11
Activation of Peroxides by Organoselenium Catalysts: A Synthetic and Biological Perspective

Eduardo E. Alberto and Antonio L. Braga

11.1 Introduction

The development of catalytic systems based on a sustainable and atom-efficient concept is presently a cutting edge research field in synthetic chemistry. In particular, a crescent interest in the use of hydrogen peroxide as the final oxidizing agent for oxidation reactions has been witnessed in the past years [1–3]. A number of advantages to the use of H_2O_2 over other oxidizing agents have led to an intensive search for the effective use of this reagent in oxidation reactions. Among others reasons, H_2O_2 is an environmentally safe reagent since its decomposition leads to the formation of water and oxygen; it is commercially available and inexpensive; easy to handle and the content of active oxygen is high. However, even though the oxidation of organic substrates employing H_2O_2 as the final oxidizing agent are in most cases thermodynamically favorable, their kinetic profile are unsatisfactory therefore making these reactions impractical. To overcome this limitation, the use of a suitable catalyst is necessary to activate hydrogen peroxide and promote the desired transformations with appropriate reaction rates. Frequently the activation of H_2O_2 is efficiently achieved by the use of catalytic amounts of certain metals such as vanadium, titanium, molybdenum, among others [4–6].

On the other hand, in the late 1970s organoselenium compounds appeared as an elegant alternative to the use of metals for catalytic activation of peroxides in oxidation reactions. Advantages such as their high reactivity and selectivity as well as their readily commercial availability make these compounds excellent catalysts for these synthetic transformations [7–13]. Inorganic selenium catalysts, especially selenium dioxide (SeO_2) is equally used as activator of peroxides in

E.E. Alberto
Chemistry Department, Federal University of Santa Maria, Santa Maria, RS, Brazil

A.L. Braga (✉)
Chemistry Department, Federal University of Santa Catarina, Florianópolis, SC, Brazil
e-mail: albraga@qmc.ufsc.br

J.D. Woollins and R.S. Laitinen (eds.), *Selenium and Tellurium Chemistry*,
DOI 10.1007/978-3-642-20699-3_11, © Springer-Verlag Berlin Heidelberg 2011

oxidation reactions, however, its application is beyond the scope of this manuscript [9, 14]. Two distinct categories of organoselenium compounds exhibit catalytic activity in oxidation reactions with peroxides: seleninic acids and selenoxides. In the presence of an excess of H_2O_2, diselenide **1** is readily converted to seleninic acid **2** (Scheme 11.1) [15, 16]. Further oxidation of **2** with H_2O_2 yields perseleninic acid **3** which is the active oxygen transfer agent [17]. Although **3** can be isolated and fully characterized, it smoothly decomposes to seleninic acid **2**. Therefore, for convenience it can be prepared in situ through the reaction of excess H_2O_2 with the pattern diselenide **1**.

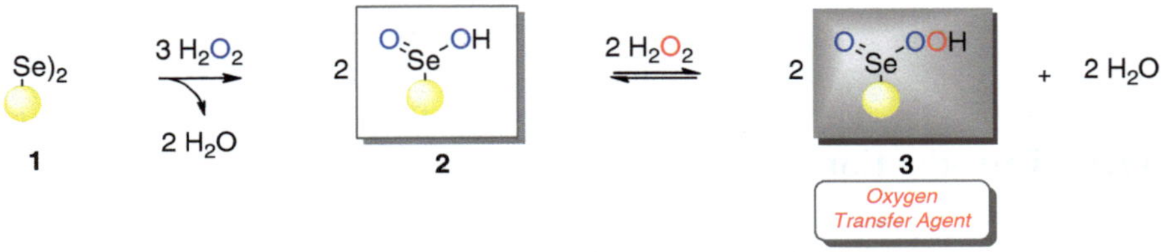

Scheme 11.1 Preparation of perseleninic acid **3** by reaction of diselenide **1** with H_2O_2

In parallel, selenoxide **5** can be prepared from the corresponding selenide **4** through a variety of methods (Scheme 11.2) [18–20]. This specie is quite stable when lacks β-hydrogens to selenium, which avoids selenoxide *syn* elimination [21, 22]. In an aqueous environment, selenoxide **5** is in equilibrium with dihydroxy selenane **6**. Reaction of **6** with peroxide produces hydroxy perhydroxy selenane **7**, a superior oxidizing agent than selenoxide **5**.

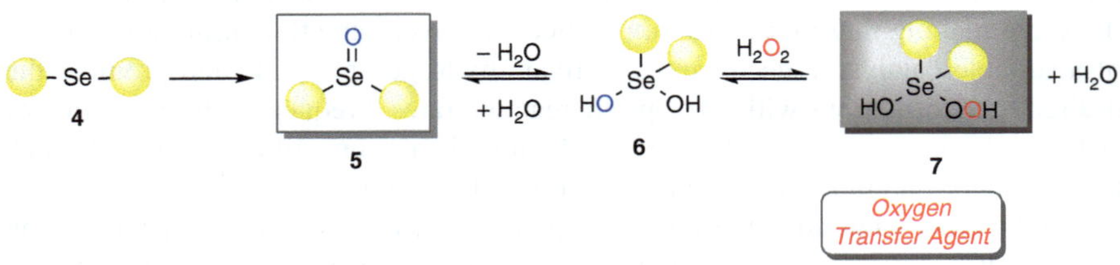

Scheme 11.2 Preparation of hydroxy perhydroxy selenane **7** by reaction of selenoxide **5** with H_2O_2

Both diselenide **1** and selenoxide **5** can be catalytically used as efficient activators of peroxides in oxidation reactions. They are converted in situ to the active oxygen transfer compounds, perseleninic acid **3** and hydroxy perhydroxy selenane **7**, respectively, in reactions with an excess of peroxide. Moreover, their catalytic performances and selectivities are strongly modulated by the nature of substituents attached to the selenium moiety, mainly due to the electronic demand that these substituents confer to selenium.

11.2 Baeyer-Villiger Reaction

Baeyer-Villiger oxidation is one of the reactions most commonly employed by organic chemists to convert carbonyl compounds to a variety of different products [23–26]. Due to the versatility of this synthetic tool, the search for new strategies to improve its efficiency has been the subject of renewed interest. Moreover, an especial emphasis has been devoted for the development of Baeyer-Villiger fashioned reactions in a "greener" perspective [27].

The activation of H_2O_2 by seleninic acids for use in the Baeyer-Villiger oxidation reaction was firstly described in the late 1970s, even though the seleninic acids were being used as stoichiometric reagents [28]. The utility of organoselenium compounds as catalysts for this important synthetic transformation promoted by H_2O_2 is illustrated in Scheme 11.3. Seleninic acid **2**, which is produced by oxidation of the corresponding diselenide, is converted in situ to perseleninic acid **3**. Subsequently, **3** reacts with a carbonyl compound **8** to produce the adduct **9**, which after migration of R^1 or R^2 leads to the formation of product **10** and regenerates seleninic acid **2** for the catalytic cycle.

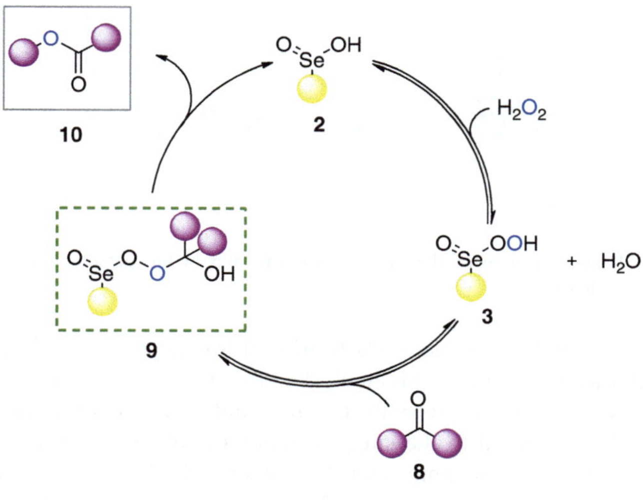

Scheme 11.3 Catalytic cycle in the seleninic acid/H_2O_2 approach to the Baeyer-Villiger reaction

The catalyst loading for an efficient transformation is usually in a range of 1–10 mol-% (of the corresponding seleninic acid) depending on the reaction conditions. The efficiency and the selectivity of the catalyst as well as the rate of the reaction have a close relationship with the nature of the substituents attached to the seleninic acid framework (Chart 11.1). Compared to SeO_2, alkyl or benzeneseleninic acids **11** and **12** respectively, seleninic acids with electron-withdrawing groups (e.g.: nitro **13** [17, 29, 30] or trifluoromethyl **14** [31]) are better catalysts for Baeyer-Villiger oxidations. Among them, catalyst **15** with two

CF_3 groups in the *meta* position is a superior catalyst in terms of activity and selectivity [31].

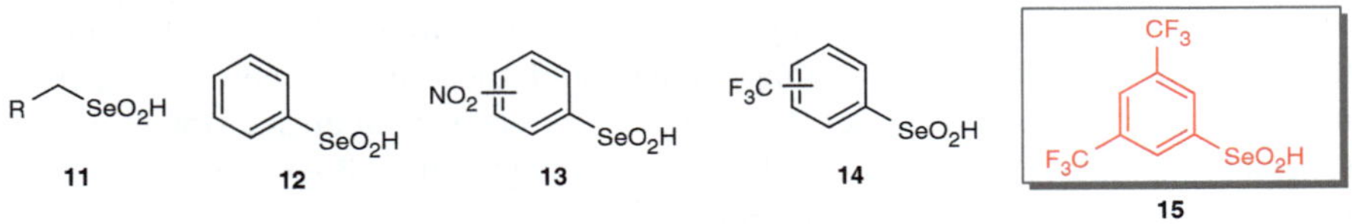

Chart 11.1 Seleninic acids used as catalysts in Baeyer-Villiger reactions

The enhanced performance of electron poor seleninic acids like **13** and **14** over benzeneseleninic acid **12** can be attributed to the influence of electron-withdrawing groups in adduct **9**. These groups can stabilize the negative charge which is developed when **9** collapses to seleninic acid **2** and the Baeyer-Villiger product **10** (Scheme 11.4). The electronic nature of the substituents attached to selenium can facilitate the elimination of seleninic acid **2** and consequently, modulate the reaction rate.

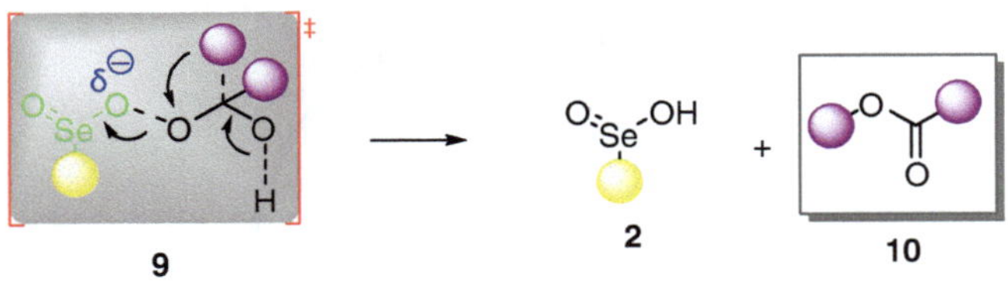

Scheme 11.4 Transition state of the rate-determining-step in the perseleninic acid catalyzed Baeyer-Villiger oxidation

However, although seleninic acids bearing nitro substituents **13** are powerful electron-withdrawing groups, and should be stronger acids than those carrying a CF_3 group **14**, the former precatalysts (diselenides) are poorer catalysts. This behavior can be attributed to the sluggish oxidation of the pattern diselenide **16** with H_2O_2 to produce seleninic acid **13** (Scheme 11.5) [31]. Nevertheless, in stoichiometric reactions employing preformed seleninic acids **13**, they activate H_2O_2 due to their strong acidic character, and are by far better catalysts [17, 29, 30].

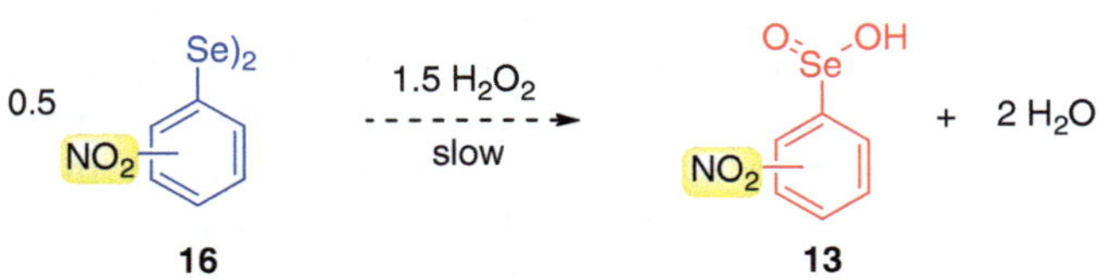

Scheme 11.5 Conversion of diselenides bearing nitro groups **16** to the corresponding seleninic acids **13**

Other structurally diverse catalysts designed to activate H_2O_2 for Baeyer-Villiger oxidation are depicted in Chart 11.2. Interestingly, diselenide **17** which is quite active when R = Tf and a poor catalyst when R = Me or H [32]. Chiral diselenide **18** showed a reasonable ability to promote the oxidation of cyclic ketones, however, only moderate enantiomeric ratios were obtained [33]. Cyclic selenenamide **19**, which is known as ebselen, displays catalytic activity in oxidation reactions promoted by peroxides [34]. Selenoxides **20**, likewise to seleninic acids, can conveniently be used as efficient H_2O_2 activators for Baeyer-Villiger reactions. Their reactivity is also modulated by the electronic nature of the substituents, benzyl 3,5-bis(trifluoromethyl) phenyl selenoxide (R = 3,5-CF_3) being the most active catalyst [35]. Seleninic acid immobilized on a solid support has also been designed. Although the availability of structurally modified resins does not allow the preparation of a wide range of catalysts, especially concerning electronic effects, polystyrene-bound benzeneseleninic acid **21** can be easily recycled and reused, which is a desirable feature [36, 37]. In the same way, catalyst **22** bearing perfluorooctyl substituents is suitable for oxidation reactions under monophasic, biphasic or triphasic conditions associated with perfluorinated solvents. In homogeneous reactions, catalyst **22** proved to be more active than 3,5-bis (trifluoromethyl)benzene-seleninic acid **15**. Despite of formation of emulsion was observed in biphasic reaction mixtures, under triphasic conditions, catalyst **22** could be recycled, with a slight decrease in the formation of products after repeated reaction cycles [38].

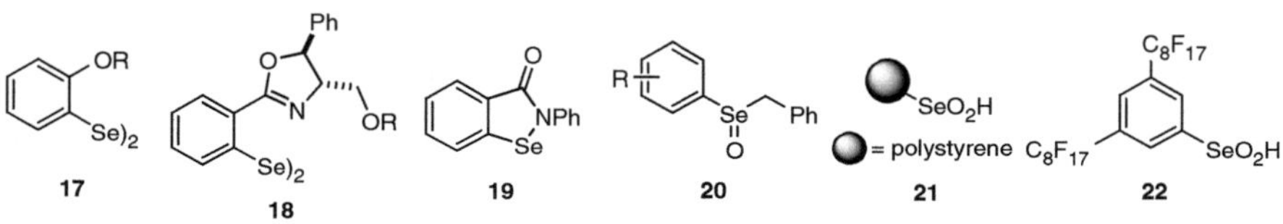

Chart 11.2 Representative organoselenium catalysts for Baeyer-Villiger reactions promoted by H_2O_2

The formation of products in the Baeyer-Villiger oxidation reaction promoted by H_2O_2 activated by organoselenium compounds is also modulated by the reaction conditions. Overall, good conversion rates and selectivities are achieved in THF under reflux [36, 37, 39] and particularly in halogenated solvents such as 1,1,1,3,3,3-hexafluoro-2-propanol, 2,2,2-trifluoroethanol or CH_2Cl_2 at room temperature [31]. Albeit halogenated solvents lead to better results, environmental and/or economic issues make their use somewhat prohibitive therefore, the development of new alternatives to their, especially the use of aqueous solutions, would be of interest [40–42].

In a synthetic perspective, the Baeyer-Villiger reaction is a useful tool to promote the oxidation of carbonyl containing substrates **8** (Table 11.1). The regiochemistry and, consequently, the selectivity in the formation of products **10a** or **10b** lies in the migratory aptitude of different groups (R^1 and R^2) in adduct **9**.

Depending on the reaction conditions, as well as the nature of the substrate, **10a** or **10b** may be isolated, or they may be hydrolyzed to the final products. Some representative examples of substrates, mostly ketones and aldehydes, which have been successfully oxidized with H_2O_2 catalyzed by seleninic acids **2** are listed in Table 11.1.

Table 11.1 Substrate scope and selectivity in the Baeyer-Villiger reaction promoted by the organoselenium/peroxide approach

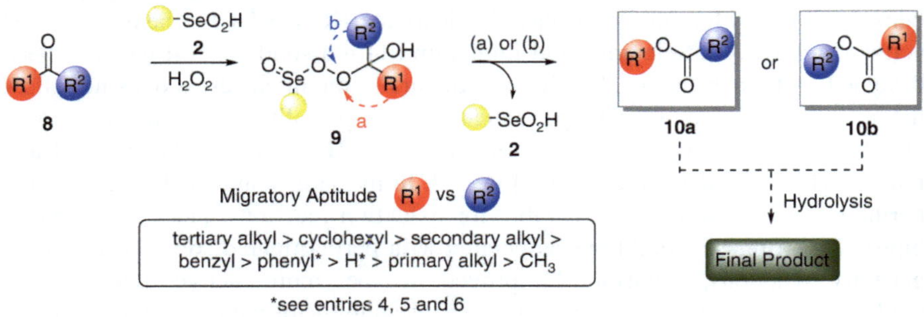

Entry	Substrate	Product(s)	Catalyst/reference
1	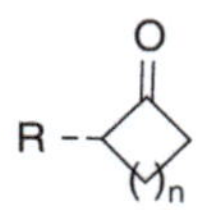 **23**	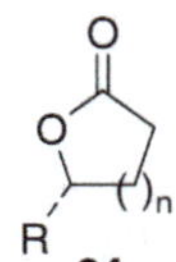 **24**	**15** [31], **17** [32], **18** [33], **20** [35], **21** [36], **22** [38]
2	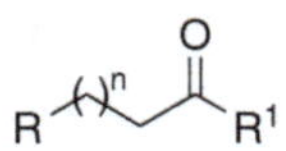 **25**	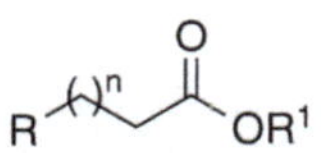 **26** + 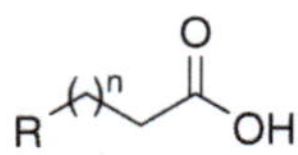 **27**	**15** [31]
3	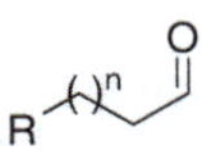 **28**	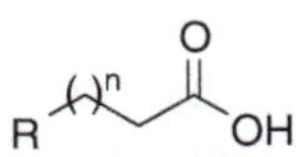 **29**	**12** [39], **15** [31], **21** [37], **22** [38]
4	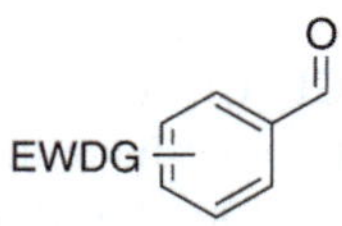 **30**	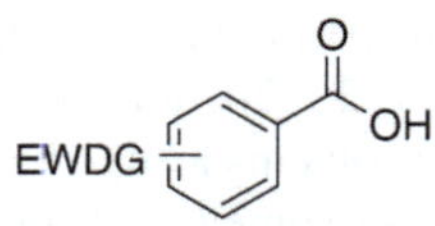 **31**	**12** [39], **15** [31], **19** [34], **21** [37], **22** [38]
5	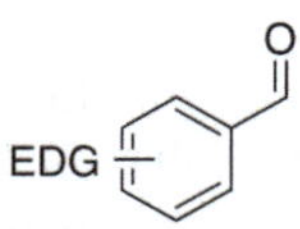 **32**	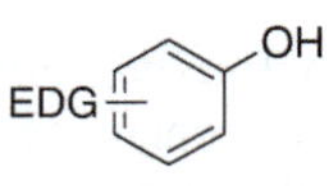 **33**	**13** [30], **15** [31], **20** [35], **22** [38]

(continued)

Table 11.1 (continued)

Entry	Substrate	Product(s)	Catalyst/reference
6	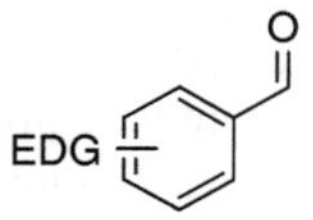 **32**	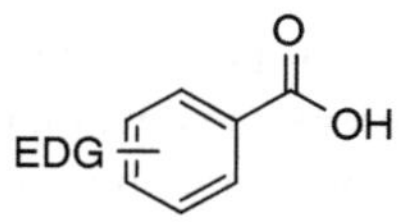 **34**	**12** [39], **21** [37], **19** [34]
7	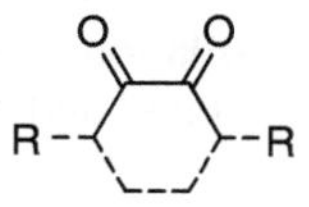 **35**	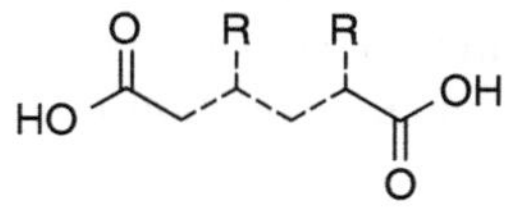 **36**	**15** [31]
8	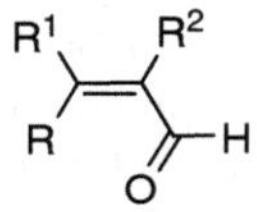 **37**	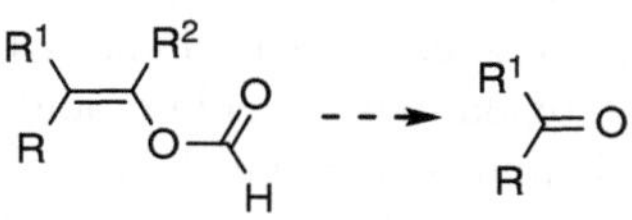 **38** **39**	**13** [29]

EWDG electron withdrawing group, *EDG* electron donating group

Cyclic ketones **23** are easily oxidized to the corresponding lactones **24** in near quantitative yields and selectivities $\geq$ 90% under mild reaction conditions (Entry 1). On the other hand, linear ketones **25** are more resistant towards Baeyer-Villiger oxidation, and formation of a mixture of ester **26** and carboxylic acid **27** can be observed depending on the reaction conditions due to the hydrolysis of ester **26** (Entry 2). Predominance in the migration of hydrogen rather than of the carbon chain is observed in the reaction of linear or electron poor aryl aldehydes **28** and **30**, respectively, leading to the almost exclusive conversion of these substrates to the corresponding carboxylic acids **29** and **31** (Entries 3 and 4). Interestingly, is the selectivity in the formation of products when electron rich aryl aldehydes **32** are subjected to different reaction conditions. While catalysts containing electron withdrawing groups attached to their structure in halogenated solvents at room temperature afford the corresponding formats, which are subsequently hydrolyzed to phenols **33**, carboxylic acids **34** are achieved using benzeneseleninic acids in THF under reflux (Entries 5 and 6). Moreover, when ebselen **19** is used as catalyst for oxidation of aldehydes bearing electron donating or withdrawing groups with *t*-BuOOH, the corresponding carboxylic acids are exclusively obtained (Entries 4 and 6) [34]. Reaction of 1,2-diones **35** smoothly produce anhydrides, which are converted to the dicarboxylic acids **36** as final products after hydrolysis (Entry 7). Noteworthy is the functional group tolerance of Baeyer-Villiger reaction in the oxidation of α-β carbonyl compounds **37**. These substrates are conveniently oxidized to vinyl formats **38**, which could be isolated or hydrolyzed to saturated ketones or aldehydes **39** (Entry 8).

11.3 Oxidation of Carbon-Carbon Double Bonds

Epoxidation reactions are of paramount interest in the organic synthesis scenario due to the common use of epoxides as building blocks for the preparation of more sophisticated products [43]. Therefore, the development of new synthetic protocols for the epoxidation of carbon-carbon double bonds is currently of great interest. In peroxide mediated oxidations, these valuable compounds are usually accessed by metal based catalysts [6]. On the other hand, the discovery of the ability that organoselenium catalysts exhibit to activate peroxides for oxidation reactions occurred simultaneously both for the oxidation of olefinic substrates and for the Baeyer-Villiger reaction.

Although in a first moment oxidation of alkenes was accomplished using stoichiometric amounts of organoselenium compounds [44], it rapidly evolved to the use of catalytic loadings of selenium based catalysts (1–10 mol-%) [45, 46]. Likewise in Baeyer-Villiger oxidations, seleninic acid **2** is the precursor of the active oxidizing agent, perseleninic acid **3**, which is generated by reaction of **2** with H_2O_2 (Scheme 11.6). Perseleninic acid **3** is able to transfer oxygen to the substrate **40** producing the product and regenerating **2** for the catalytic cycle. Commonly epoxide **41** is obtained as the major product, however, depending on the reaction conditions, the substrate and the catalyst employed, the corresponding *vic*-diol **42** can be isolated as the final product.

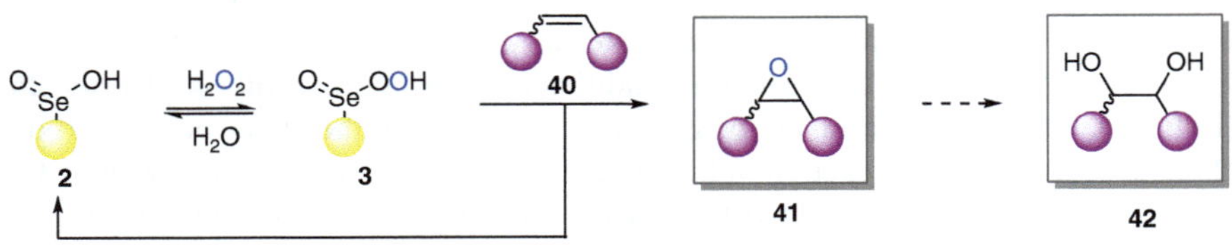

Scheme 11.6 Epoxidation promoted by H_2O_2 activated by seleninic acids

The ability that organoselenium compounds have to promote epoxidation reactions using hydrogen peroxide as oxidizing agent is strongly modulated by the nature of catalysts. Specifically, the electronic demand at selenium plays a crucial role modulating their activity. Although benzeneseleninic acid **12** is able to catalyze epoxidation reactions promoted by H_2O_2, insertion of electron withdrawing groups in the aryl moiety of arylseleninic acids significantly enhances the performance of the catalysts. For instance, nitro substituents, such as in seleninic acids **43a-b**, give better results, both in terms of yield and in the selectivity compared to **12** (Chart 11.3) [45]. Besides, among **43a-b**, compound **43b** assembled with two nitro groups, and consequently with more pronounced electron deficiency near to selenium, is a superior catalyst than **43a**. On the other hand, chiral perseleninic acid **44** is also an efficient catalyst, however, it is not able to transfer chirality to the epoxide obtained from the oxidation of prochiral olefins [46].

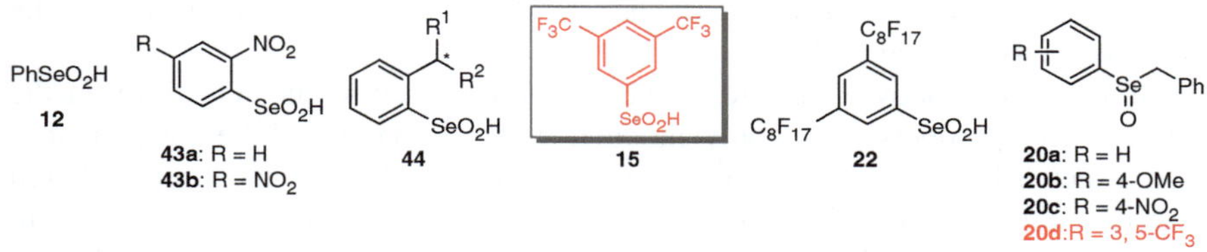

Chart 11.3 Representative organoselenium catalysts for epoxidation reactions

Albeit a variety of electron poor seleninic acids are excellent catalysts for the conversion of alkenes to epoxides mediated by peroxides, seleninic acid **15** containing two CF_3 groups attached in the *meta* position of the aryl moiety is a noteworthy catalyst. Compared to other catalysts, **15** exhibits enhanced performance in terms of efficiency and selectivity [47]. Similarly, seleninic acid bearing perfluoroalkyl substituents **22** fulfills the electronic requirements for improved catalytic performance towards epoxidation reaction of alkenes by H_2O_2. Furthermore, reactions in perfluorinated solvents allow the reuse of the reaction media (catalyst and solvent) without detectable loss of efficiency for at least ten consecutive reactions due to the immobilization of **22** in the fluorous phase [48]. Selenoxides **20a-d** (Chart 11.3) have also been used as organoselenium catalysts for epoxidation reactions with H_2O_2 [35]. Similarly to arylseleninic acids, the ability of selenoxides **20a-d** to catalyze epoxidation of olefins is strongly dependent on the electronic environment in the catalyst framework. Although all screened selenoxides exhibited some activity the best catalyst was benzyl 3,5-bis (trifluoromethyl) phenyl selenoxide **20d**, which is 13 times more active than benzyl phenyl selenoxide **20a**.

Mechanistically, the observed trend in the electronic demand of catalysts suggests that the rate limiting step in the reaction sequence should be the oxygen transfer from perseleninic acid **3** to the olefin **40**, rather than the formation of the active oxidizing agent, perseleninic acid **3**. The formation of epoxide **41** would occur via transition state **45** in which a negative charge is developed near to selenium (Scheme 11.7). This is consistent with the fact that catalysts bearing

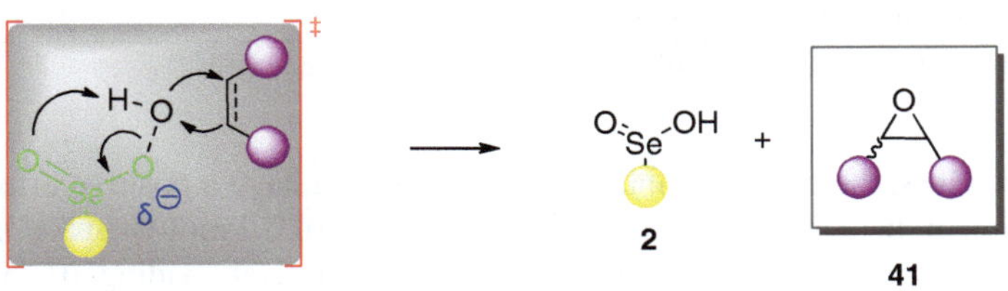

45

Scheme 11.7 Transition state for the rate-determining-step in the epoxidation of alkenes by H_2O_2 activated by perseleninic acids electron withdrawing groups increase the rate of the reaction, by stabilization of the negative charge in specie **45**.

As mentioned above, the rate of the reaction as well as the selectivity in the formation of products is not only modulated by the nature of the catalyst, it is also affected by the reaction conditions. The role of the solvent in epoxidation reactions can be accessed by the results of the oxidation of cyclohexene **46** with H_2O_2 catalyzed by **15** (Table 11.2, Entries 1–6). Among the tested solvents, 2,2,2-trifluoroethanol, which is miscible in H_2O_2 solution, is by far the best solvent, allowing the formation of **47** in 88% yield and 90% of selectivity (mmol epoxide/mmol of converted olefin). Quite recently, the successful use of glycerol-based solvents for epoxidation reactions catalyzed by diselenides was reported [49]. This solvent is a promising alternative to halogenated solvents due to the environmental issues associated with their use [40–42].

Table 11.2 Effect of different reaction conditions in the course of epoxidation of **46** by H_2O_2 catalyzed by **15**

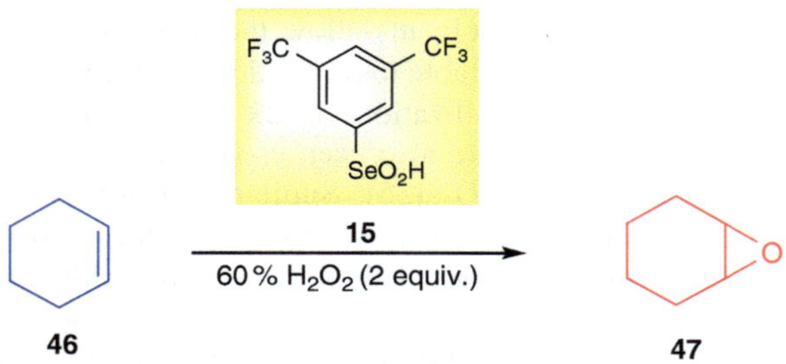

Entry	Solvent	**15** (mol-%)	Base[a]	GC yield (%)	Selectivity (%)[b]
1	2,2,2-Trifluorethanol	0.5	–	88	90
2	Dichloromethane	0.5	–	35	90
3	1,2-Dichloroethane	0.5	–	21	68
4	Nitromethane	0.5	–	19	69
5	Sulfolane	0.5	–	14	95
6	α,α,α-Trifluortoluene	0.5	–	7	49
7	2,2,2-Trifluorethanol	0.25	Pyrazole	75	96
8	2,2,2-Trifluorethanol	0.25	NaOAc	98	99
9	2,2,2-Trifluorethanol	0.25	Na_2HPO_4	82	91

[a]0.2 mol-% (related to olefin);[b]mmol epoxide/mmol of converted olefin

Nevertheless, the use of a catalytic amount of base have a positive effect on the epoxidation of **46** (Table 11.2, Entries 7–9). Addition of 0.2 mol-% of NaOAc to the reaction mixture allows a 50% reduction of the catalyst loading (0.25 mol-%) affording the desired product **47** in higher yield and selectivity compared to the reaction carried out in the absence of a base (Entries 1 and 8). This effect is attributed to the neutralization of the acidic solution of H_2O_2 (pH 2.5–5.0) [50]

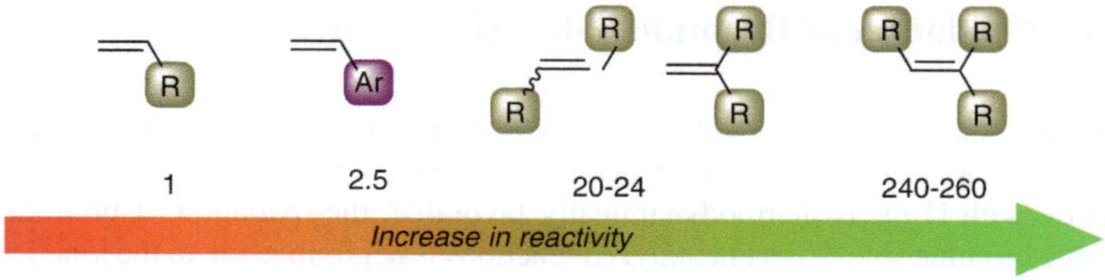

Fig. 11.1 Relative rate of reactivity of olefins towards epoxidation

and thus, avoiding the formation of *vic*-diols, common side products in epoxidation reactions, which are known to inhibit the catalytic activity [51]. The formation of these side products can also be conveniently reduced by the use of a drying agent such as $MgSO_4$ or by the use of more concentrated H_2O_2 solutions [46].

According to the available literature, the reactivity of different substrates towards epoxidation using the peroxide/organoselenium catalyst approach is depicted in Fig. 11.1. Substrates such as primary alkyl and aryl alkenes are more resistant towards epoxidation (relative reaction rates are 1 and 2.5, respectively). Conversely, secondary and tertiary olefins are much more reactive, with relative reaction rates in a range of 20–260 compared to the primary olefins. This trend in the substrate reactivity demonstrates that electron rich olefins are better substrates in this valuable synthetic transformation, and is in agreement with that results observed for epoxidations promoted by peracids [52]. Moreover, this evidence strengthens the assumption that the active oxidizing agent (perseleninic acid) is electrophilic and thus, activated by substituents with the ability to stabilize negative charges.

On the other hand, common byproducts found in the H_2O_2 promoting oxidation of carbon-carbon double bonds **48** are vic-diols **50** and **51** (Scheme 11.8). These valuable compounds can be prepared through ring opening of epoxides **49**, obtained by oxidation of olefins mediated by H_2O_2 activated by organoselenium catalysts [53]. In general, the use of aqueous solutions as well as longer reaction times, favor the formation of these products relative to the epoxides. The diasteroselectivity in the formation of products **50** and **51** is associated with the nature of the substrate [53].

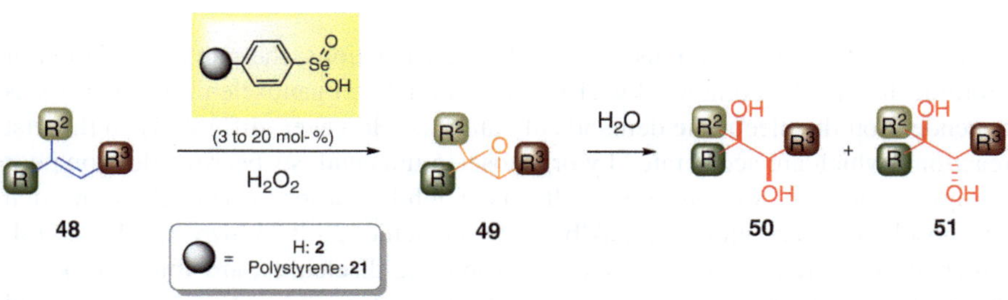

Scheme 11.8 Formation of *vic*-diols in oxidation of alkenes using the H_2O_2/seleninic acid method

11.4 Oxidation of Bromide Salts: Haloperoxidase Like

In nature, halide salts are oxidized by hydrogen peroxide due to the activity of haloperoxidase enzymes [54]. Specifically, in the case of bromide salts, although the reaction with H_2O_2 is thermodynamically favorable, the oxidation of bromide to positive bromine sources in uncatalyzed reactions is negligible due to the kinetics of this oxidation. In a number of organisms this problem is overcome by the use of the bromoperoxidase enzyme, a vanadium dependent enzyme, which is responsible for the biochemical conversion of bromide salts to oxidized bromide species [55, 56].

The wide applicability of brominated compounds in modern synthetic chemistry [57–59], has driven the development of new methodologies for their preparation, not only in an efficiently way, but also using environmentally sustainable protocols [60]. Taking these concepts into account and in connection with the well known ability of organoselenides to activate peroxides for a number of oxidation reactions, the use of organoselenium based catalysts has recently emerged as an elegant alternative to the oxidation of bromide salts employing H_2O_2 as the terminal oxidant. The reaction sequence for the oxidation of bromide salts by H_2O_2 promoted by organoselenium catalysts is depicted in Scheme 11.9. Seleninic acids **2**, cycle "A", or selenoxides **5**, cycle "B", when subjected to reaction with H_2O_2 are in equilibrium with the corresponding perseleninic acid **3** or hydroxy perhydroxy selenane **7**, respectively. These species, **3** and **7**, are the true oxidizing agents of bromide salts to "Br^+", which upon addition of a suitable substrate lead to the formation of brominated products.

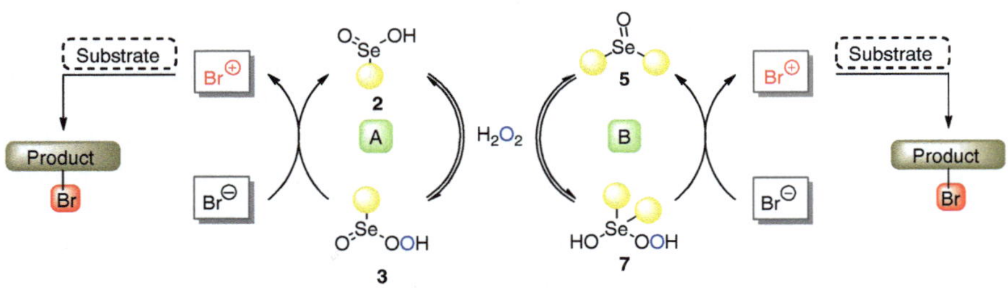

Scheme 11.9 Haloperoxidase like cycle of seleninic acids **2** and selenoxides **5**

Similarly to epoxidation and Baeyer-Villiger reactions, the rate of oxidation of bromide to "Br^+" promoted by H_2O_2 activated by organoselenium catalysts is dependent on the electronic demands of catalysts. However, differently to the first reactions, which are accelerated by organoselenium catalysts bearing electron poor groups, oxidation of bromide salts is slightly faster when electron rich organoselenides are employed. Albeit all selenenic acids shown in Chart 11.4, which are prepared in situ from the corresponding diselenide, are able to promote the desired reaction, benzeneseleninic acid **12** and 4-methoxyphenylseleninic acid **52** have the same activity and are 6-fold more efficient compared to 3,5-bis

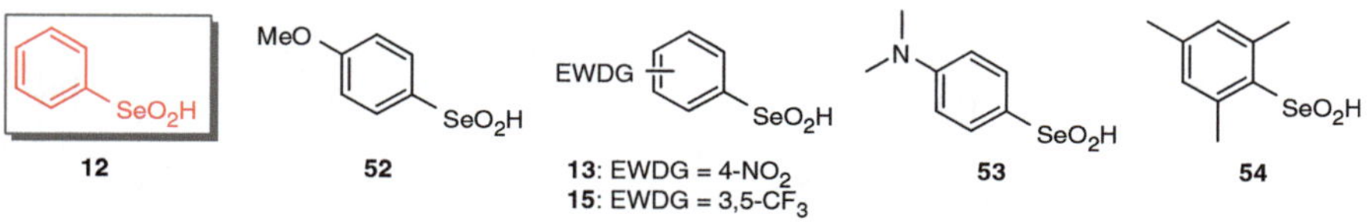

Chart 11.4 Haloperoxidase like organoselenium catalysts

(trifluoromethyl)phenylseleninic acid **15** which is by far the best catalyst in other oxidation reactions promoted by H_2O_2 [61]. Catalysts **12** and **52** are slightly better than electron poor 4-nitrophenylseleninic acid **13** and roughly 3-fold more active than 4-dimethylaminophenylseleninic acid **53** On the other hand, steric hindered seleninic acid **54** is also an efficient catalyst, which suggests that steric bulk groups have little influence on the reaction course. Additionally, identical results are obtained using either seleninic acids or employing those generated in situ by oxidation of diselenides with H_2O_2 [61].

Despite the contradictory results observed for **53**, which do not follow the trend of the electronic demand in aryl seleninic acids in the oxidation of bromide salts, its lower effectiveness may be reasoned as the contribution of the equilibrium between the resonance species **53** and **55** (Fig. 11.2). This equilibrium might be responsible for a slower formation of the corresponding perseleninic acid from **53**, impacting the reaction rate.

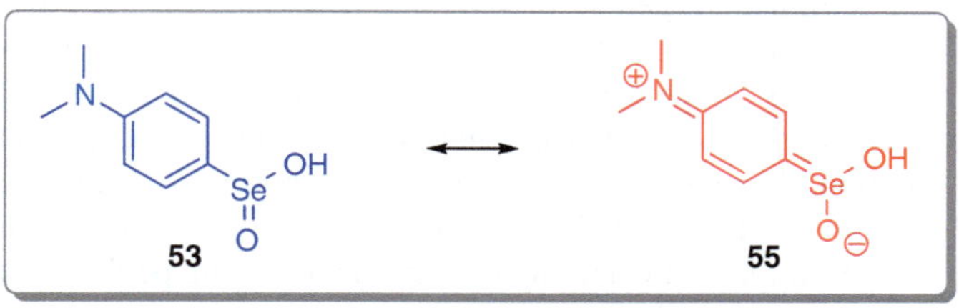

Fig. 11.2 Resonance equilibrium of seleninic acid **53**

The exact mechanism of oxidation of bromide salts to "Br^{+}" by H_2O_2 catalyzed by seleninic acids is not completely elucidated to date. Moreover, the nature of positive bromine species "Br^{+}" produced in the catalytic cycle remain unknown, and could be Br_2, Br_3^+ or even hypobromous acid (HOBr). Despite this lack of accurate information, the catalytic performance of seleninic acids depicted in Chart 11.4, suggests that in the rate limiting step there is little buildup of charge near to the aromatic ring of catalysts and that this reaction is, even if moderately, accelerated by electron rich catalysts. Thus, pathway "A" (Scheme 11.10) seems to be more plausible than pathway "B" in the reaction between a substituted perseleninic acid **3** and bromide. Addition of bromide to **3** following path "A" would produce **57**, which might itself act as a brominating agent or react with another equivalent of bromide to deliver Br_2 to the reaction media. In both

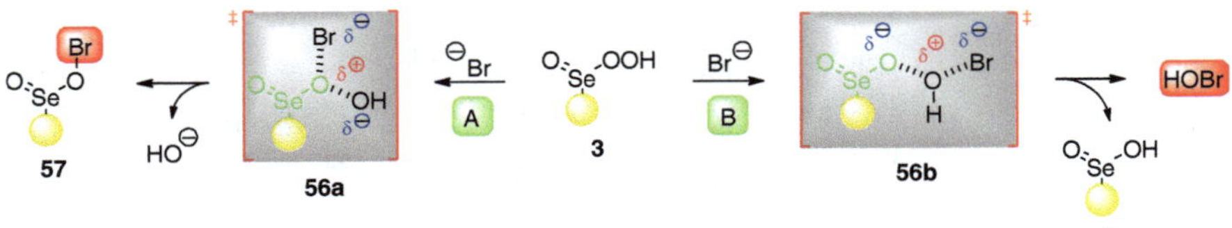

Scheme 11.10 Two possible reaction pathways in the haloperoxidase like activity of perseleninic acid **3**

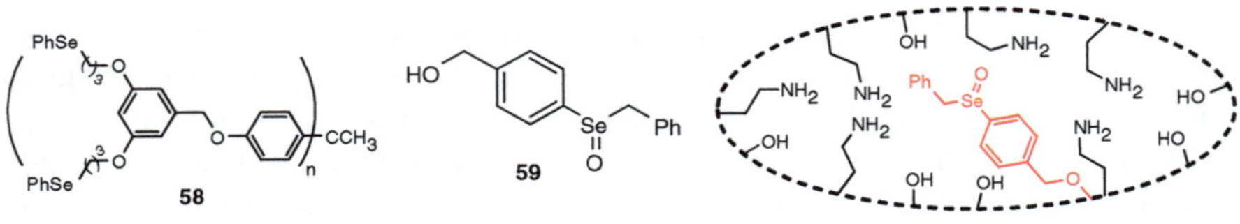

Chart 11.5 Selenoxides with haloperoxidase like activity

scenarios, seleninic acid is regenerated in the catalytic cycle. Formation of **57** would occur via transition state **56a**, in which a positive charge is developed near to selenium. Electron donating groups should be more effective stabilizing this species, and consequently increase the reaction rate. On the other hand, attack of bromide at the terminal oxygen of perseleninic acid, pathway "B", would afford HOBr via transition state **56b** where electron withdrawing groups would stabilize the incoming negative charge developed near to selenium [61].

Selenoxides, like those depicted in Chart 11.5, are also excellent activators of H_2O_2 for oxidation of bromide salts. Fréchet-type dendrimers [62, 63] terminating in $-O(CH_2)_3SePh$ groups **58**, which are converted in situ to the corresponding selenoxide, exhibit improved catalytic activity [64–66]. The performance of these catalysts is closely associated with the number of selenium groups attached to the dendrimer core (number of PhSe groups from 1 to 12, depending on the dendrimer core and *n*). Impressive turnover numbers (TON) of >60,000 mole of H_2O_2 consumed and TOF values of >2,000 mole of brominated product formed per mole of catalyst can be achieved using a dendrimer with 12 PhSe groups. Furthermore, the outstanding activity of catalysts with 6 or 12 PhSe groups exceeds the statistical contribution expected *per* seleno group in the dendrimer series [65]. This "dendrimer effect" experienced by higher generations of dendrimers is the result of autocatalysis, in which each PhSe group is converted to the selenoxide by the positive bromine specie generated in the background reaction. The selenoxide functionality is significantly more efficient in the activation of H_2O_2 than the corresponding selenide [66].

Selenoxide **59** exhibits a remarkably higher catalytic performance when covalently attached to a halide permeable xerogel, formed from 10/90 (mol/mol) of 3-aminopropyltriethoxysilane/tetramethoxysilane **60** than when used as a xerogel-free catalyst in solution [67]. The improved activity of **60** (the reaction is 23 times faster compared to that catalyzed by **59**) can, to a great extent, be ascribed to the active-site/surface cooperativity gained from the interactions of selenoxide with hydrogen bond donors and acceptors groups in the xerogel scaffold [68]. Moreover, since catalyst **60** is solid, it can be easily recovered by filtration after the reaction, allowing it to be recycled. On the other hand, although the catalyst **60** lacks β-hydrogens which avoids *syn*-elimination of selenoxide and it is not leached from the xerogel scaffold, its efficiency has a limited lifetime. A decrease in its catalytic activity can be detected after the production of 80 mol of brominated product per mol of **60**.

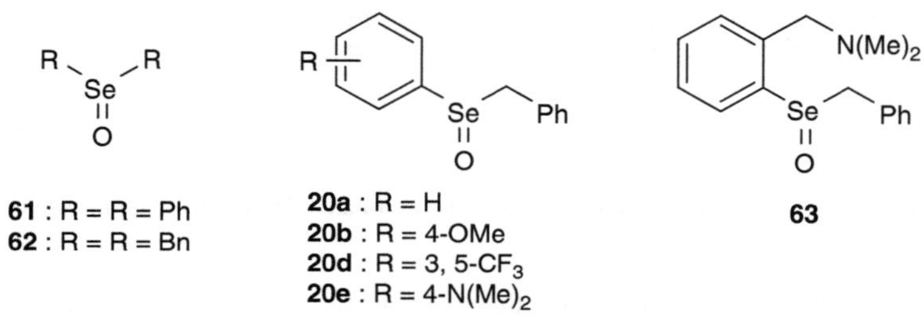

Chart 11.6 Selenoxides with different electronic demands for haloperoxidase like activity

Substituent effects, as in other synthetic transformations, modulate the activity of selenoxides towards oxidations promoted by H_2O_2. Particularly, in the oxidation of bromide salts, the selenoxides shown in Chart 11.6 follow the same trend observed for seleninic acids: the reaction velocity is enhanced by selenoxides bearing electron donating groups. Assigning diphenyl selenoxide **61** as a standard catalyst for comparison purposes, its activity is almost 7-fold lower than the electron richer dibenzyl selenoxide **62** and approximately 4-fold lower than benzyl phenyl selenoxide **20a**. The performances of 4-methoxyphenyl benzyl and 4-dimethylaminophenyl benzyl selenoxides **20b** and **20e** correspondingly, rise up to 6 and 18-fold, respectively, compared to that observed for **61**. Additionally, selenoxide **63**, bearing a chelating group, showed a remarkable activity, catalyzing the oxidation of bromide 28 times faster than that catalyzed by **61**. Besides, the activity of selenoxide **20d**, carrying a strong electron withdrawing group CF_3, is much lower than that exhibited by electron rich catalysts [69].

Although selenoxides behave similarly to seleninic acids catalyzing the oxidation of bromide salts to positive bromine species promoted by H_2O_2, the proposed reaction mechanism is, to some extent, different to that of seleninic acids (Scheme 11.11). Under typical reaction conditions, that is, using pH 6 buffered aqueous solutions, selenoxide **5** is in equilibrium with dihydroxy selenane **6**, which in the presence of H_2O_2 is in equilibrium with hydroxy perhydroxy selenane **7** [70]. Both species would be in equilibrium with hydroxyselenonium **64**, which can be formed by protonation of selenoxide **5**, or by elimination of one molecule of water from dihydroxy selenane **6**. In pH 6 solutions, the concentration of **64** at the above mentioned equilibrium should be higher than that of **5** and **6**. On the basis of the

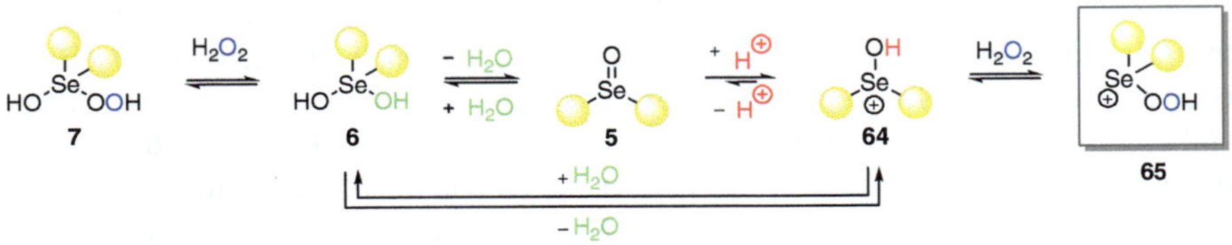

Scheme 11.11 Products of the reaction of selenoxide **5** with H_2O_2 in water solutions

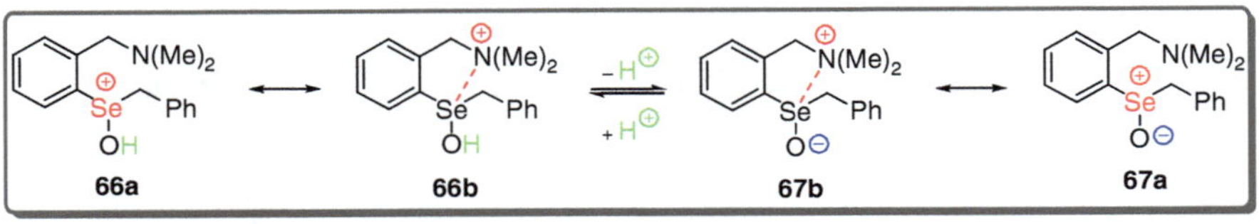

Scheme 11.12 Resonance contributors of **63** in its protonated/deprotonated equilibrium

influence of catalyst substituents on the reaction rate, the rate limiting step in the oxidation of bromide to a positive bromine specie may be the formation of the active oxidizing agent, possibly perhydroxyselenonium **65**, generated by reaction of hydroxyselenonium **64** with H_2O_2.

The influence of catalyst substituents on the kinetics of the reaction due to the rate of formation of perhydroxyselenonium **65** is illustrated in Scheme 11.12 using selenoxide **63** as a suitable example. In pH 6 solutions, hydroxyselenonium **64**, presumably the predominant specie in the equilibrium depicted in Scheme 11.11, is stabilized by electron donating and chelating groups. In the protonated/deprotonated equilibrium between **66a** and **67a**, the interaction of selenium with the electron lone par of the nitrogen would contribute to the stabilization of these species, by the resonance contributors **66b** and **67b** (Scheme 11.12). Moreover, this interaction should increase the basicity of the selenoxide oxygen. The increased basicity of oxygen in hydroxyselenonium **66a** would also account for its higher nucleophilicity, which support the faster formation of the active specie **65** by reaction of **64** with H_2O_2 in selenoxides bearing chelating groups. In the same way, electron donating groups would also account for the increase in the oxygen basicity and stability of the hydroxyselenonium species, even though in a lower extent compared to chelating groups [69].

Once that perhydroxyselenonium **65** is formed, two different pathways for the oxidation of bromide are possible (Scheme 11.13). In pathway "A", **65** reacts with bromide to produce **68**, which might itself act as a brominating agent or react with another equivalent of bromide to deliver Br_2 to the reaction medium. In both hypotheses, **64** is regenerated to restart the catalytic cycle with H_2O_2. On the other hand, in sequence "B" bromide would attack the terminal oxygen in **65** furnishing HOBr as positive bromine specie and **64** to resume the catalytic cycle.

The oxidation of bromide salts to "Br+" species promoted by H_2O_2 catalyzed by organoselenium compounds is usually carried out in pH 6 phosphate buffer or,

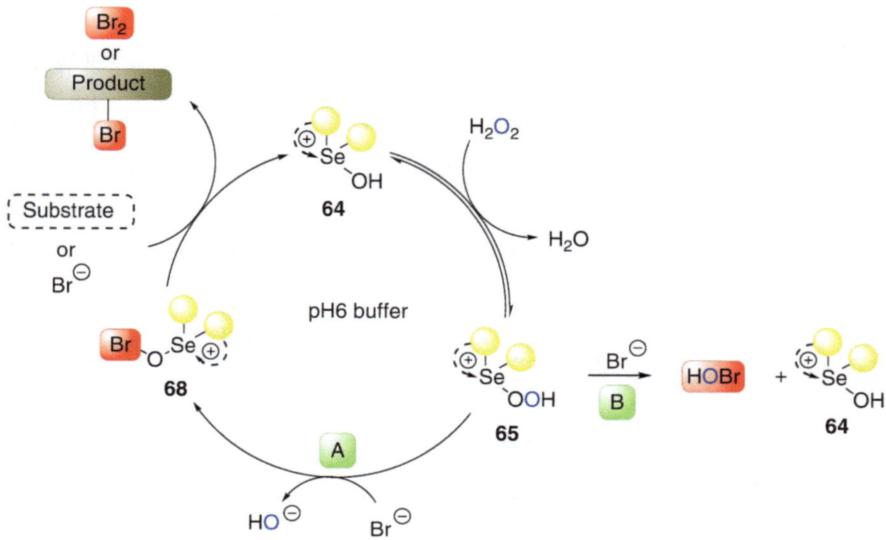

Scheme 11.13 Catalytic cycle of selenoxides as haloperoxidase mimics

$$Br^{\ominus} + H_2O_2 \longrightarrow Br^{\oplus} + 2\,HO^{\ominus} \qquad (1)$$

$$Br^{\oplus} + H_2O_2 \longrightarrow Br^{\ominus} + O_2 + 2\,H^{\oplus} \qquad (2)$$

$$2\,H_2O_2 \longrightarrow O_2 + 2\,H_2O \qquad (3)$$

Scheme 11.14 Disproportionation of H_2O_2 by bromide/bromine salts mixture

depending on the organic substrate solubility, in a 1:1 (v:v) mixture of buffer and an organic solvent, most commonly dichloromethane, diethyl ether or 1,4-dioxane. However, only a fraction of the "Br+" produced in the reaction is scavenged by the substrate (Scheme 11.14). It is well known that bromide is slowly oxidized by H_2O_2 to produce "Br^+" and two equivalents of hydroxide, (1). Conversely, "Br^+" reacts with H_2O_2 to produce molecular oxygen and two protons, (2). The sum of equations (1) and (2), consists in the disproportionation of H_2O_2 to oxygen and water (3) [71, 72].

In other words, the reaction of "Br^+" with a substrate is a competing reaction to the disproportionation of H_2O_2 by "Br^+" (Scheme 11.15). To increase the formation of the brominated organic products, an excess of 2–10 equivalents of H_2O_2 as well as the bromide source (NaBr), is generally required.

Synthetically, the oxidation of bromide to positive bromine species by H_2O_2 activated by organoselenium catalysts, allows access to a series of brominated organic products such as those depicted in Table 11.3.

Scheme 11.15 H_2O_2/substrate competitive scavenging reaction of positive bromide species

Table 11.3 Substrate scope in the haloperoxidase like reaction of organoselenides

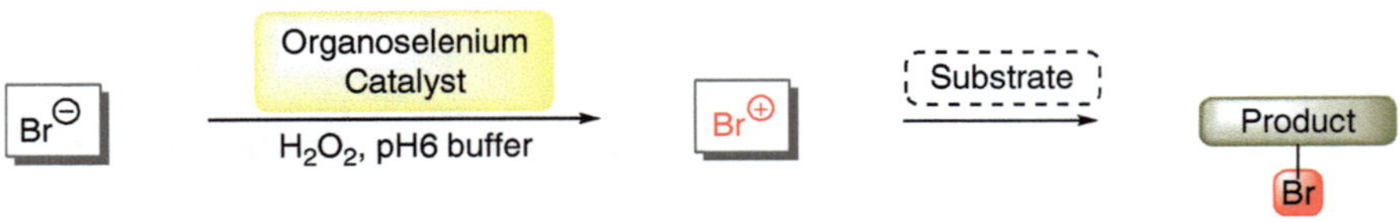

Entry	Substrate	Product(s)	Catalyst/ reference
1	 **46**	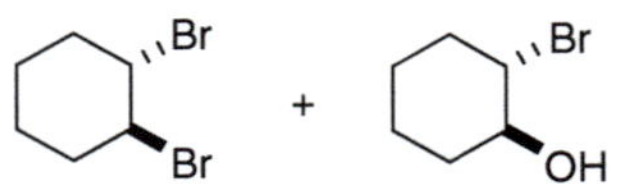 **69** + **70**	**58** [64–66]
2	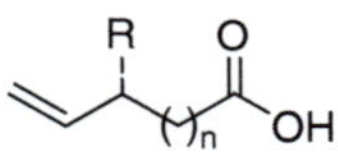 **71**	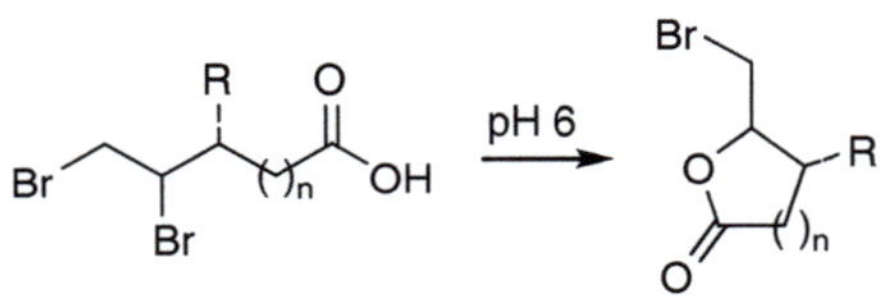 **72** → **73**	**12** [61], **60** [67], **63** [69]
3	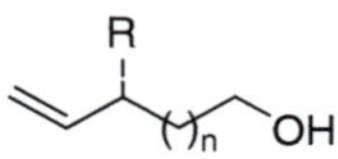 **74**	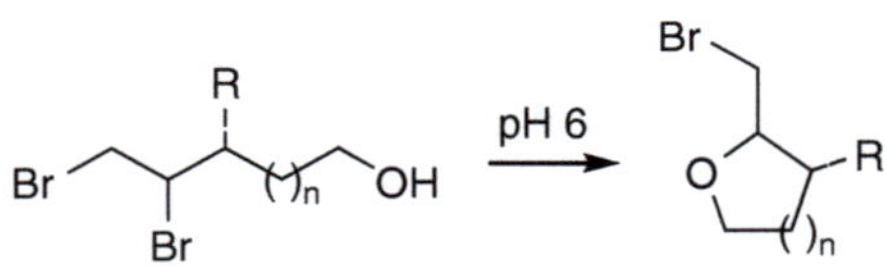 **75** → **76**	**12** [61]
4	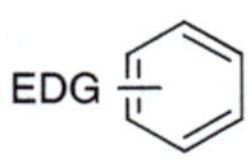 **77**	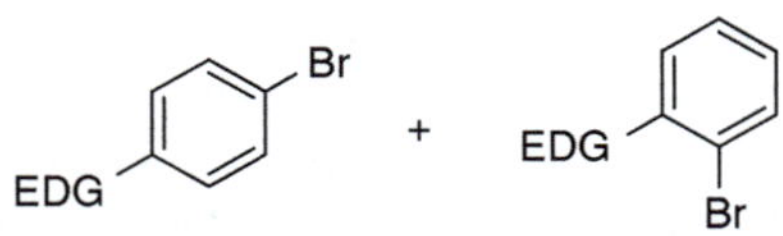 **78** + **79**	**12** [61], **60** [67], **63** [69]

EDG electron donating group

Alkenes and activated aryl compounds are suitable substrates to capture "Br+" formed in the oxidation reaction of bromide salts. Under two phase reaction conditions, CH_2Cl_2/aqueous pH 6 phosphate buffer, cyclohexene **46** is converted to *trans*-1,2-dibromocyclohexane **69** and *trans*-2-bromocyclohexanol **70** (Entry 1). Despite the faster formation of brominated products in catalyzed reactions compared to the uncatalyzed ones, the selectivity is somehow independent of the action of the catalyst, affording mixtures in a range of 1:1.2–1:1.9 of **69** and **70**, respectively. On the other hand, alkenoic acids **71** are selectively converted to the corresponding dibromo carboxylic acids **72**, which in pH 6 solutions, are smoothly converted to the bromolactones **73** as final products (Entry 2). Similarly, unsaturated alcohols **74** undergo reaction to afford dibromoalcohols **75** or cyclic bromoethers **76** (Entry 3). Aryl compounds bearing electron donating groups **77** are conveniently brominated, generally speaking, with a preferential regioselectivity toward the formation of *para*-aryl bromides **78** (Entry 4).

11.5 Oxidation of Thiols: Glutathione Peroxidase Like

In a synthetic perspective, the use of peroxides as terminal oxidants is highly desirable due to the advantages that they exhibit over other oxidizing agents. However, for living cells peroxides are extremely hazardous substances. These compounds are known as reactive oxygen species (ROS), which readily destroy key biological components and cause damage to cell membranes. Peroxides are produced in living cells during the metabolism of oxygen [73, 74], and several diseases, including neurodegenerative diseases, such as Alzheimer's and Parkinson's, as well as other physiological and inflammatory processes are linked to their activity [75].

Nature has solved this problem by the use of enzymes that are able to convert these harmful species into innocuous metabolites. As part of a complex and sophisticated detoxification system, selenoenzymes constitute important mammalian antioxidant defense protecting biomembranes and other cellular components from oxidative stress. One of these enzymes is glutathione peroxidase (GPx) which plays a pivotal role catalyzing the reduction of hazardous peroxides and their byproducts to water or alcohols [76, 77]. The most important amino acid in the active site of this enzyme is L-Selenocysteine **80** (Chart 11.7), which is responsible for reducing hydroperoxides at the expense of the tripeptide glutathione GSH **81** [78, 79].

The ability of GPx to promote the reduction of peroxides lies in the redox properties of the selenol moiety of L-Selenocysteine, the catalytic cycle of GPx is shown in Scheme 11.16. Initially, the selenol functionality in the enzyme, E-SeH

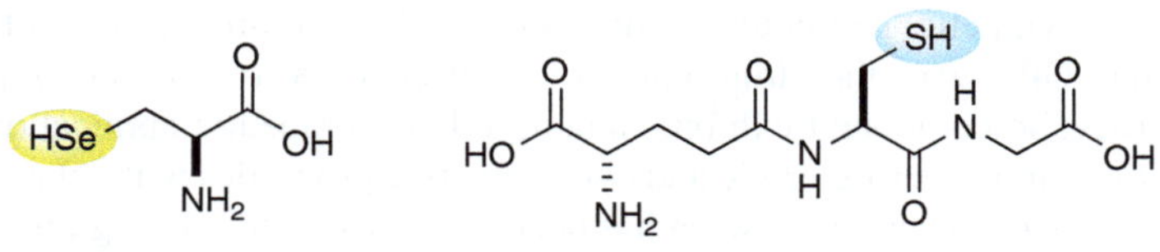

80 **81**

Chart 11.7 Selenocysteine **80** and Glutathione **81**

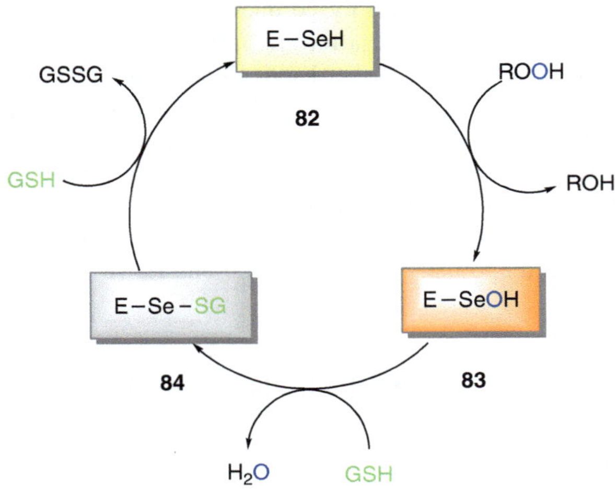

Scheme 11.16 Catalytic cycle of the enzyme Glutathione peroxidase

82, reacts with a peroxide to generate the corresponding alcohol or water and selenenic acid E-SeOH **83**. The latter then reacts with one equivalent of glutathione to produce water and the corresponding selenenyl sulfide E-Se-SG **84**. The last step is the reaction of glutathione with selenenyl sulfide **84** to produce the oxidized glutathione (GSSG) and regenerate the reduced enzyme **82** to resume the catalytic cycle [80, 81].

The total process consists of the reduction of one equivalent of a reactive oxygen specie at the expense of two equivalents of glutathione, producing two equivalents of water (when R = H) and oxidized glutathione (GSSG) Scheme 11.17.

Scheme 11.17 Global reaction of reduction of peroxides by GSH catalyzed by GPx

Since the discovery that selenium plays a pivotal role in GPx enzymes, the design and development of new synthetic chalcogen-based catalytic antioxidants has attracted considerable attention [82–89]. Small molecules of organoselenium compounds have emerged as excellent candidates to act as GPx mimics, due to their well-known ability to undergo a two-electron redox cycle between selenium (II) and (IV) species [90–93].

The first synthetic compound found to be able to mimic the activity of the enzyme GPx was the drug known as ebselen **19** (Chart 11.7) [94–99]. This heterocyclic compound exhibits, among other therapeutic applications, anti-inflammatory, anti-atherosclerotic and cytoprotective properties with relatively low toxicity [100–105]. Thereby, ebselen has been used as a standard drug for comparison purposes with other selenium compounds in terms of GPx like activity.

As a result of the great ability of ebselen to destroy peroxides at the expense of thiols, intensive research has been directed to the synthesis and evaluation of ebselen

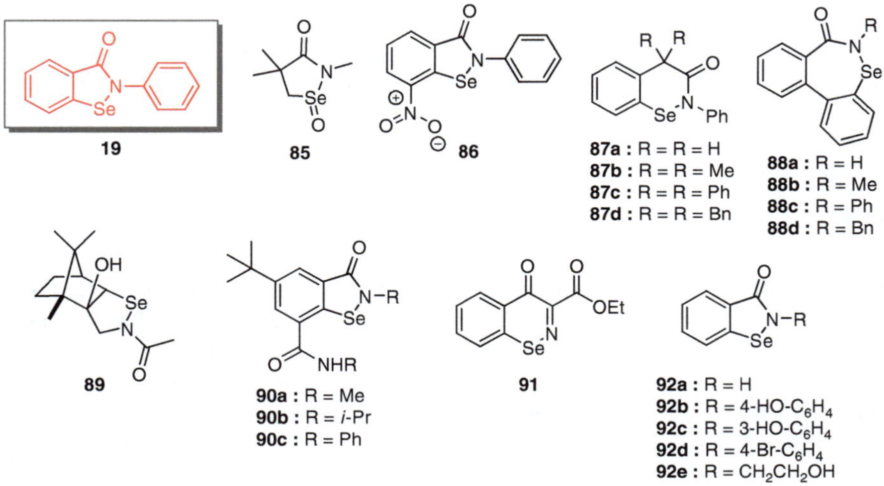

Chart 11.8 Ebselen derivatives with GPx like activity

derivatives with improved catalytic activity. A series of new selenenamides, including those depicted in Chart 11.8, has been successfully used for the oxidation of thiols by a variety of peroxides. Among them, compounds **85** [106], **86** [107], **87** [108], **88** [109], and **89** [110] are efficient catalysts to accomplish this transformation.

Substituent effects display a significant role modulating the activity of these catalysts. On account of this, the improved performance of compounds **90a-b**, which possess an *ortho*-coordinating amide group, and are more active than the parent ebselen, while **90c** is a poorer catalyst. The enhanced activity of these derivatives is attributed to the intramolecular nonbonding interactions between the oxygen of the amide moiety and selenium [111]. Compound **91**, a six-membered ring homologue of ebselen, is also able to promote the oxidation of thiols using H_2O_2, at an increased rate compared to that of ebselen **19** [112]. Ebselen analogues **92a-e**, containing different substituents attached to nitrogen are excellent catalysts for the oxidation of thiols using peroxides, the only exception being **92a**, which has a slower reaction rate [113].

On the other hand, although the catalytic effectiveness of ebselen and its derivatives in the oxidation of thiols by peroxides is well established, its mechanism of action has still not been completely elucidated [106–114]. Nonetheless, experimental evidences collected in isolated reactions with this compound as well as with potential intermediates in the catalytic cycle give some insight towards a better understanding of the exact mechanism (Scheme 11.18). The reaction of ebselen **19** with an excess of H_2O_2 produces exclusively the corresponding seleninic acid **93**, in which the selenium atom is stabilized by nonbonding interactions with the oxygen atom from the amide functionality. When **93** is reacted with an excess amount of thiol, selenenyl sulfide **94** is the product obtained (this reaction may occur via the formation of selenenic acid **96** as an intermediate). Additionally, selenenyl sulfide

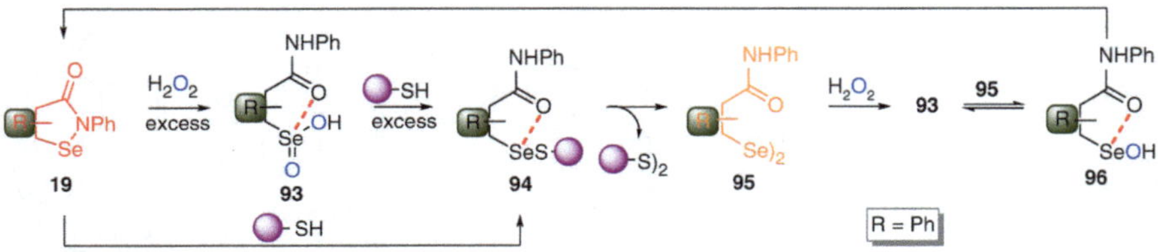

Scheme 11.18 Different reactions of ebselen **19** and its derivatives with H_2O_2 or thiol

94 is also obtained upon reaction of ebselen **19** with a thiol. This reaction is faster compared to that of ebselen **19** with H_2O_2. Selenenyl sulfide **94** gradually disproportionates to disulfide and diselenide **95** which is found as the sole product in the absence of H_2O_2. Besides, the reaction between H_2O_2 and diselenide **95** initially affords seleninic acid **93**, which in the equilibrium with the remaining diselenide **95** yields selenenic acid **96** [46, 115–117]. Selenenic acid **96** is rapidly converted to ebselen as the final product through the loss of one molecule of water, shifting the equilibrium of **95** and **93** to the formation of selenenic acid **96** and consequently, ebselen **19** [118].

According to these observations, a revised catalytic cycle has been proposed (Scheme 11.19) [118]. Firstly, ebselen reacts with a thiol to yield selenenyl sulfide **94**. Depending on the nature of the thiol, the formation of diselenide **95** from **94** is encumbered by thiol exchange reactions at selenium [119, 120]. Once that diselenide **95** is formed, it reacts with H_2O_2 to establish the seleninic acid **93**/ diselenide **95** equilibrium, which leads to the formation of selenenic acid **96**. In the last step two different pathways are possible: elimination of one molecule of water of **96** to regenerate ebselen **19**, or reaction of **96** with another equivalent of thiol to produce selenenyl sulfide **94**.

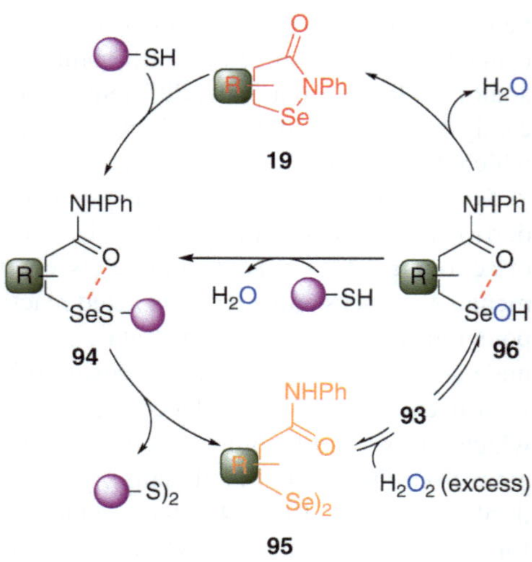

Scheme 11.19 Proposed catalytic cycle for the activity of ebselen as a GPx mimic

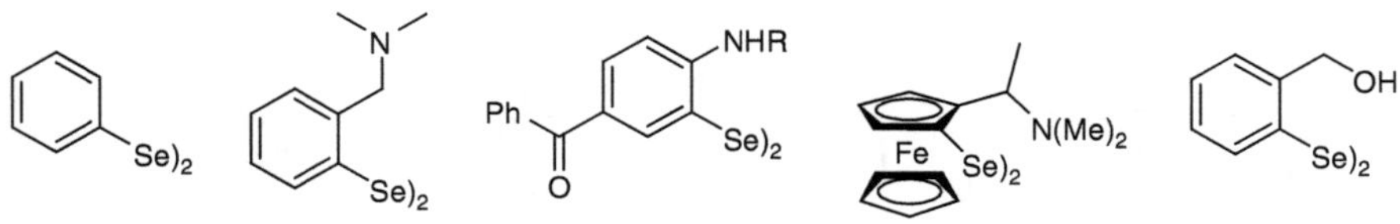

97 **98** **99a** : R = H **99b** : R = Me **100** **101**

Chart 11.9 Structurally diverse diselenides with GPx like activity

Considering that diselenides are active species in the catalytic cycle of ebselen and its derivatives for the oxidation of thiols by H_2O_2, it is reasonable to assume that diselenides themselves should be good catalysts. Not surprisingly, all diselenides shown in Chart 11.9 have some catalytic activity in this reaction. Diphenyl diselenide **97** for example is two times more efficient than the well established GPx mimic ebselen **19**. Nonetheless, similarly to other oxidation reactions promoted by peroxides activated by organoselenium catalysts, the performance of diselenides is also affected by substituent effects. In this regard, there is a notable increased effectiveness of diselenides bearing coordinating groups such as amines, alcohols or amides. For instance, the performance of diselenide **98** is approximately 5-fold higher than that of diphenyl diselenide **97** [121]. The same effect is observed for compounds **99** [122], **100** [123, 124] and **101** [125].

The addition of a methoxy group near to selenium in compound **102** (Chart 11.10) produces a positive effect on the efficiency of this catalyst, enhancing significantly its activity when compared to **98** [126]. Noteworthy is the considerable difference in the catalytic performance of amides **103a-b** and amines **104a-b** due to substituents attached to nitrogen. In amides the replacement of the hydrogen in the secondary amide **103b** by another methyl group in the tertiary amide **103a** leads to a 20-fold increase in the activity [127]. On the other hand, secondary amine **104b**, is approximately two times more effective catalyzing the oxidation of thiols with H_2O_2 than tertiary amine **104a** [128]. The effectiveness of amino ester derivatives bearing diselenides in promoting this oxidation, is closely related to the length of the carbon chain attached to selenium. Thus, while diselenides **105a-b** are poor catalysts, **105c** with longer carbon chain is more active than diphenyl diselenide [129].

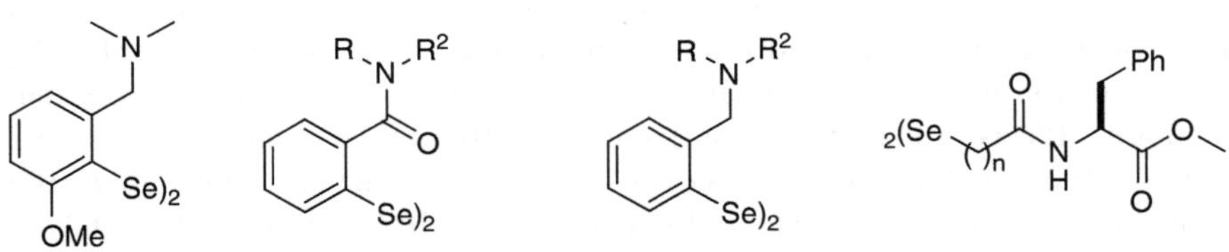

102 **103a** : R = R^2 = Me **103b** : R = Me, R^2 = H **104a** : R = R^2 = Me **104b** : R = Me, R^2 = H **105a** : n = 1 **105b** : n = 2 **105c** : n = 3

Chart 11.10 Diselenides bearing amine or amide functionality with GPx like activity

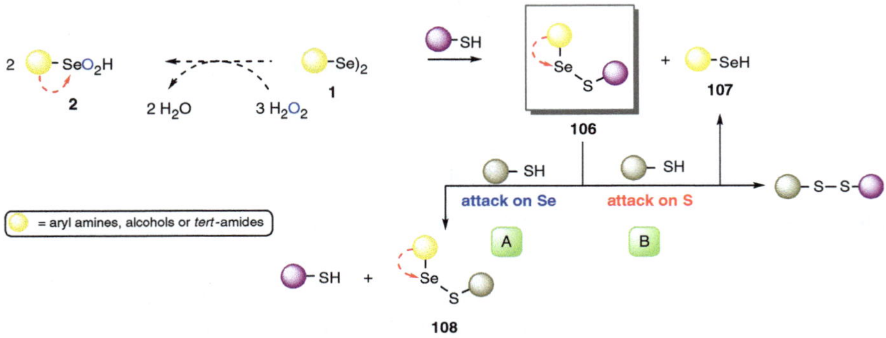

Scheme 11.20 Reaction of diselenide **1** in the presence of H_2O_2 and thiols; thiol exchange reaction of selenenyl sulfide **108**

Mechanistically, it is known that the reaction of diselenide **1** with a thiol generates selenenyl sulfide **106** and selenol **107** and that this reaction is faster than that of diselenide **1** with H_2O_2 to produce seleninic acid **2** (Scheme 11.20) [128, 130]. Despite of diselenides bearing chelating groups are considerable better catalysts for the oxidation of thiols using H_2O_2, depending on the strength of the nonbonding coordination between selenium and the heteroatom in selenenyl sulfide **106**, unproductive thiol exchange reactions may take place [131]. Experimental evidences as well as theoretical calculations suggest that strong nonbonding interactions in specie **106**, favor the attack of a second equivalent of thiol on selenium, pathway "A", which deplete the efficiency of these catalysts due to the undesired thiol exchange. Conversely, weak interactions in **106** would direct the attack of a second equivalent of thiol on sulfur, pathway "B", leading to the formation of disulfide and the active specie selenol **107**. Additionally, the nature of the thiol plays a crucial role in the degree of thiol exchange reactions. While with aryl thiols (e.g. thiophenol and thiobenzyl alcohol) the extent of thiol exchange is high, the use of glutathione as a cofactor drastically reduces this undesirable side reactions [119, 120, 132, 133].

Taking in account these evidences, a catalytic cycle for the oxidation of thiols by H_2O_2 activated by diselenides is depicted in Scheme 11.21. Selenol **107**, produced from the reaction of diselenide with thiol reacts with H_2O_2 to yield selenenic acid **109**, which can be further oxidized to seleninic acid if the concentration of H_2O_2 is high, or can react with one equivalent of thiol to afford selenenyl sulfide **106**. Similarly, **106** is produced from selenenic acid **109**, by reaction with an excess of thiol. The last step in the sequence is the reaction of the selenenyl sulfide **106** with thiol to give the disulfide and deliver selenol **107** to resume the catalytic cycle [128].

Even thought selenenamides such as ebselen **19** and its derivatives as well as diselenides are well known catalysts for the oxidation of thiols by H_2O_2, the GPx like ability of organoselenium compounds is not restricted to these two classes of compounds. Selenosubtilisin, an artificial selenoenzyme, and its derivatives display

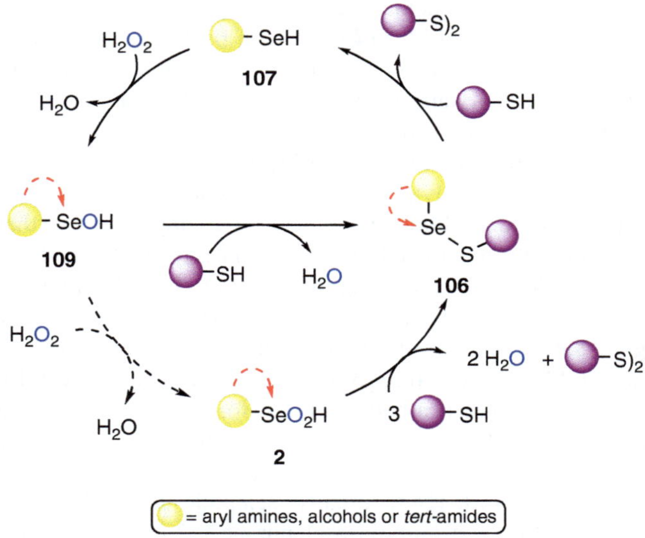

Scheme 11.21 GPx like cycle of diselenides

high efficiency mimicking the natural seleno enzyme GPx [134–136]. Moreover, all organoselenides some of which can be observed in Chart 11.11, possess some activity through a variety of mechanisms of action. Selenides **110** and **111** are procatalysts, the former reacts with thiol to produce the corresponding selenenyl

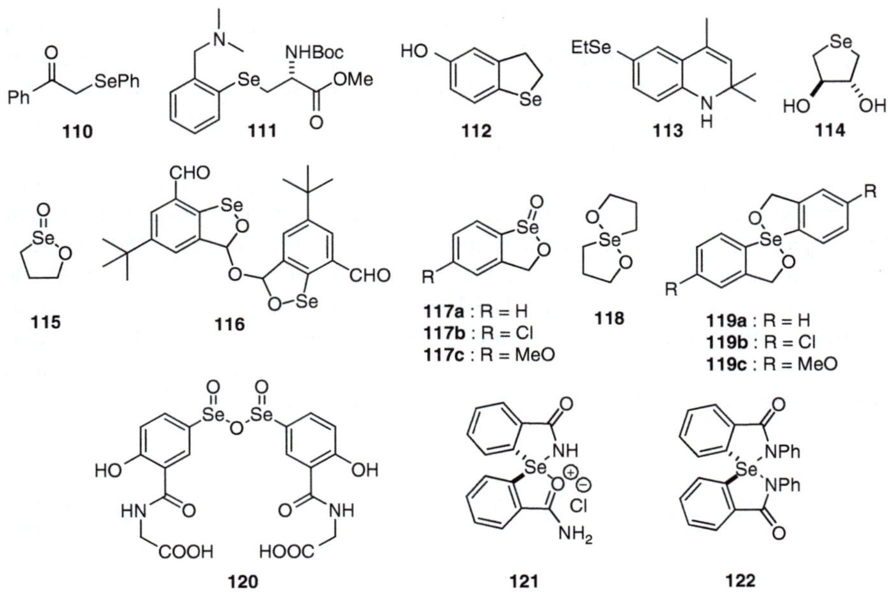

Chart 11.11 Representative selenides with GPx like activity

sulfide [137], and the latter generates selenenic acid on reaction with H_2O_2 [138]. On the other hand, selenides **112–114** react with peroxides to produce in situ the corresponding selenoxides which promote the oxidation of thiols to disulfides [139–141]. Notable, is the water solubility of selenide **114**, which is a desirable feature for further therapeutic investigations.

The cyclic seleninate ester **115** [132, 133] and selenenate ester **116** [142] are quite active species in the catalytic cycle of the activation of peroxides towards the oxidation of thiols. Besides, although aryl derivatives are less active than **115**, their effectiveness is strongly modulated by substituent effects. Also, should be noted that although aryl seleninate containing hydrogen or chloride in the *para* position **117a-b** are poorer catalysts, insertion of electron donating groups such as methoxy in **117c**, significantly enhances the performance of the compound, approaching that observed for **115** [143]. Similarly to catalyst **115**, the powerful GPx mimic spirodioxaselenanonane **118** is a better catalyst for the oxidation of thiols than its aryl congeners **119a-c** [144, 145]. Nonetheless, the performance of aryl spirodioxaselenanonanes **119** can be increased by attaching electron donating groups attached to their framework [143]. Finally, seleninic acid anhydride **120** [146], azaselenonium **121** [147] and spirodiazaselenurane **122** [148] are catalysts tailored to possess higher activity in the oxidation of thiols using peroxides.

11.6 Miscellaneous Oxidations

Although the most extensively explored synthetic transformations mentioned in this chapter are those discussed in the first sections, the full potential of organoselenium compounds as activators of peroxides towards oxidation reactions of organic substrates has yet to be realized. Because of the recognized general applicability of these catalysts associated with the use of peroxides as oxidizing agents, this approach is the method of choice for a variety of functional group interconversions.

The oxidation of alcohols is one of the synthetic transformations that have been the subject of renewed interest due to its widespread relevance [149, 150]. The use of organoselenium catalysts to accomplish this task was introduced in the late 1970s employing *t*-butyl hydroperoxide (TBHP) as the oxidizing agent [36, 151, 152]. In contrast to the reaction of diselenides with H_2O_2, which yields seleninic acid, seleninic acid anhydride **123** is the major product obtained when diselenide **1** reacts with TBHP (Scheme 11.22) [153–155]. Besides, when an excess of TBHP is used,

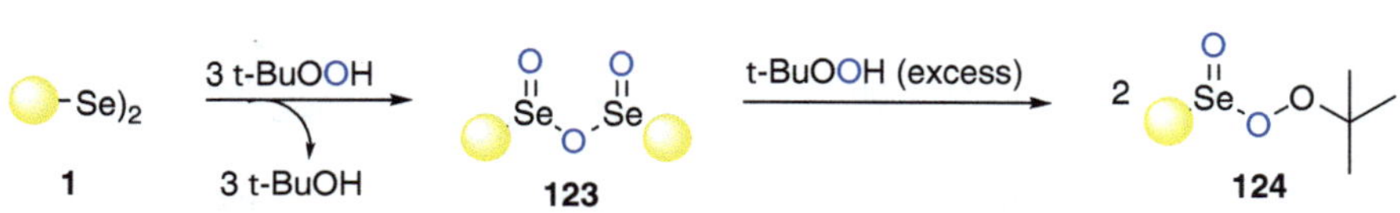

Scheme 11.22 Oxidation of diselenide **1** with TBHP

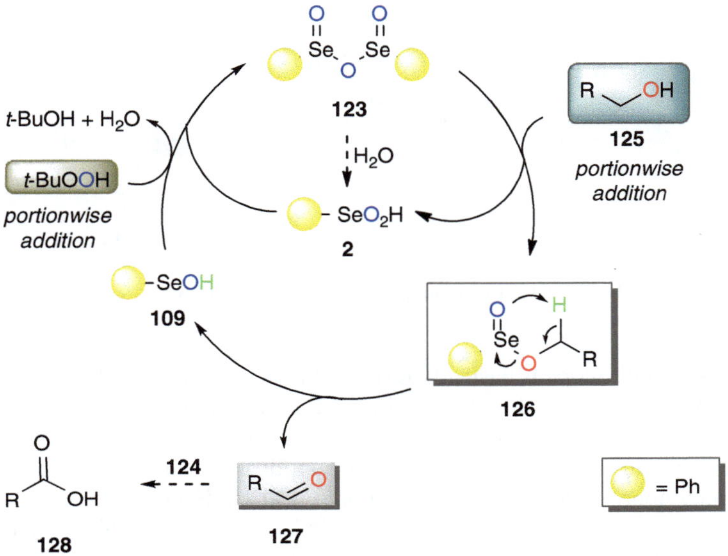

Scheme 11.23 Catalytic cycle for the oxidation of alcohols with TBHP and diselenide

formation of t-butyl perseleninic ester **124**, presumably a stronger oxidizing agent than **123**, may take place [156].

Seleninic acid anhydride **123** has found broad applicability as a stoichiometric reagent in the oxidation of alcohols [157, 158]. In the light of gaining a deeper understand regarding how catalytic amounts of seleninic acid anhydride **123** can affects the oxidation of alcohols, a revision of its mechanism has been recently proposed using diphenyl diselenide associated to TBHP as the oxidizing agent (Scheme 11.23) [159]. The revised mechanism considers the initial reaction of diselenide with TBHP to produce the corresponding anhydride **123**. After this reaction takes place, **123** reacts with the alcoholic substrate **125** to afford the adduct **126** and seleninic acid **2**. Subsequently, **126** undergoes rearrangement to yield the desired product **127** and selenenic acid **109**. Both selenenic and seleninic acids **109** and **2**, respectively, react with another equivalent of TBHP to regenerate the active anhydride **123** to restart the catalytic cycle.

Experimentally, best results are achieved using apolar solvents such as heptane or toluene under reflux in water free conditions. The presence of water reduces the activity of **123**, due to its hydrolysis to seleninic acid **2**, a reaction which competes with the addition of substrate **125** to anhydride **123**. Moreover, to avoid undesired overoxidation of products **127** to carboxylic acids **128** or the oxidation of solvents containing reactive carbon-hydrogen bonds, such as toluene, the concentration of TBHP should be kept low, to reduce the formation of **124** from **123**. To overcome these process drawbacks, water should be removed from the reaction medium throughout the process and the substrate and TBHP should be added in a portion wise fashion, to assure the completely consumption of seleninic acid anhydride

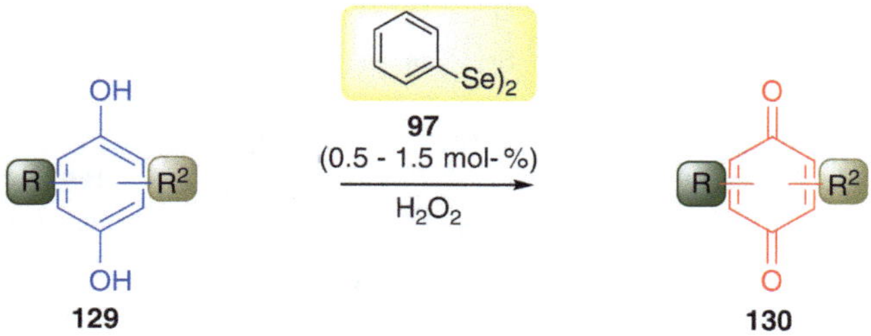

Scheme 11.24 Oxidation of 1,4-hydroquinones with H_2O_2 and $(PhSe)_2$

123. The substrate scope is basically restricted to activated alcohols such as benzyl and allyl alcohols, while alkyl alcohols are generally inert. On the other hand, the levels of selectivity are high, furnishing aldehydes as major products [36, 159].

Organoselenium catalysts also find application in the oxidation of 1,4-hydroquinones **129** to 1,4-benzoquinones **130** by H_2O_2 even on a kilogram scale (Scheme 11.24). Catalytic loadings of diphenyl diselenide **97** can effectively promote this transformation in the presence of a phase transfer catalyst, which enhances the rate of reaction in the biphasic system of aqueous H_2O_2 solution and substrate. Substituents such as alkyl, halogen or alkoxy in the substrate framework do not affect the course of the reaction [160].

Nitrogen containing compounds can also be oxidized employing the organoselenium/peroxide system. For instance, aldoximes **131** which are convenient aldehydes derivatives, are smoothly converted to esters **132** when treated with H_2O_2 and catalytic amounts of polystyrene bonded seleninic acid **21** [161] or 2-nitro-phenylseleninic acid **43a** [162] in alcoholic solutions (Scheme 11.25). Both aliphatic and aromatic aldoximes are suitable substrates for this reaction. On the other hand, esterification products are only obtained when primary and secondary alcohols are used, tertiary alcohols do not lead to the formation of products **132**.

In the same way, using catalytic loadings of ebselen **19** and hydrogen peroxide as the final oxidizing agent, N,N-dimethyl aryl hydrazones **133** are converted into aryl nitriles **134** and aryl ketazines **135** are oxidized to the corresponding carbonyl compound **136** (Scheme 11.26) [163].

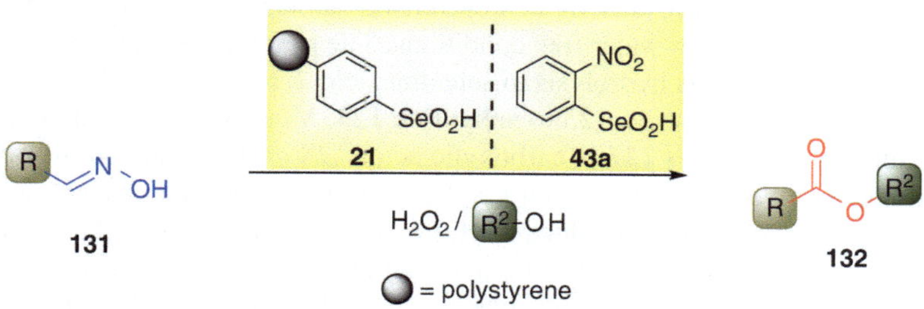

Scheme 11.25 Oxidation of aldoximes using catalytic amounts of seleninic acid and H_2O_2

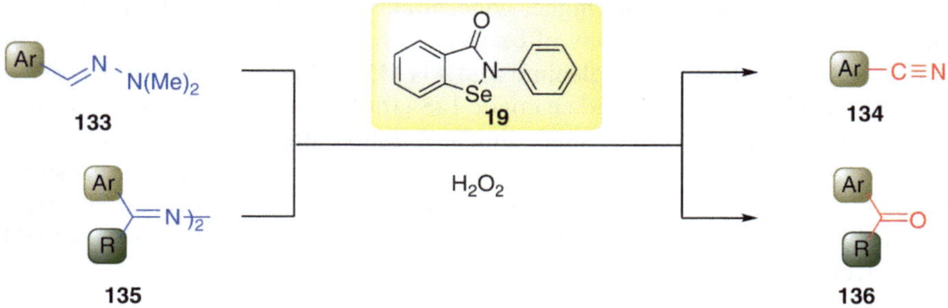

Scheme 11.26 Oxidation of hydrazones and ketazines with H_2O_2 catalyzed by ebselen

Recently, the preparation of nitroso compounds **138** from aromatic amines **137** through a reaction catalyzed by diphenyl diselenide **97** or benzeneseleninic acid **12** and H_2O_2 has been described (Scheme 11.27) [164]. This synthetic transformation allows the preparation in situ of **138** which is subsequently reacted with a diene **139** in a hetero-Diels-Alder reaction to afford a variety of products **140**. The oxidation of **137** with H_2O_2 and organoselenium catalysts **97** or **12** exhibits high selectivity, which is independent of the nature of the catalyst.

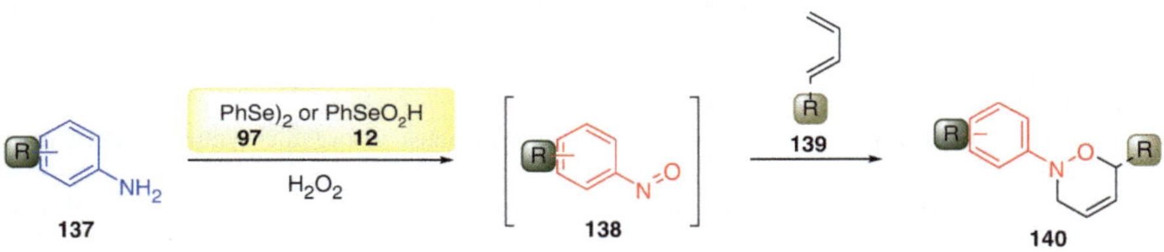

Scheme 11.27 Preparation of nitroso compounds by oxidation of aryl amines

Although the most studied oxidation reaction promoted by organoselenium catalysts and peroxides involving organosulfur compounds is the glutathione peroxidase like oxidation of thiols, organoselenium catalysts are also suitable activators of peroxides towards the oxidation of sulfides **141** to sulfoxides **142** or sulfones **143** (Scheme 11.28).

The selectivity of the reaction is predominantly influenced by the amount of peroxide employed. Lower concentrations of peroxide lead to sulfoxides **142** as

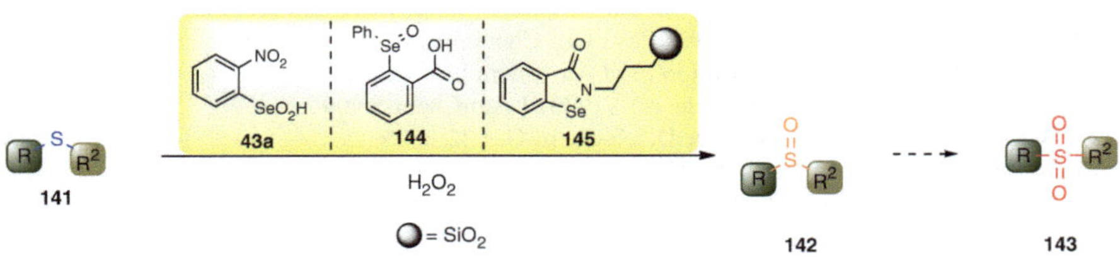

Scheme 11.28 Oxidation of sulfides to sulfoxides or sulfones

major products, whereas with higher concentrations of peroxide sulfones **143** are the preferred products. A variety of catalysts have been successfully employed in this valuable transformation: seleninic acid **43a** [45], selenoxide **144** [165], ebselen **19** [163] and its silica imprinted analogue **145** [166]. Worthwhile is the recyclability of catalyst **145**, since it is solid, simple filtration after the reaction allows its repeated reuse without detectable loss of efficiency over several runs. The substrate scope in the sulfide oxidation reaction is broad, allowing the preparation of simple and highly functionalized sulfoxides or sulfones in moderate to good levels of yield and selectivity.

References

1. Strukul G (ed) (1992) Catalytic oxidations with hydrogen peroxide as oxidant. Kluwer, Dordrecht
2. Jones CW (1999) Applications of hydrogen peroxide and derivatives. MPG Books Ltd, Cornwall
3. Adam W (2000) Peroxide chemistry mechanistic and preparative aspects of oxygen transfer. Wiley-VCH, Darmstadt
4. Sheldon RA, Kochi JK (1981) Metal-Catalyzed Oxidations of Organic Compounds: Mechanistic Principles and Synthetic Methodology Including Biochemical Processes. Academic Press, New York
5. Jorgensen KA (1989) Chem Rev 89:431
6. Lane BS, Burgess K (2003) Chem Rev 103:2457
7. Paulmier C (1986) Selenium reagents and intermediates in organic synthesis. Pergamon Press, Oxford
8. Back TG (1981) Organoselenium chemistry. A practical approach. Oxford University Press, Oxford
9. Mlochowski J, Brzaszcz M, Giurg M, Palus J, Wojtowicz H (2003) Eur J Org Chem 4329
10. Braga AL, Lüdke DS, Vargas F, Braga RC (2006) Synlett 1453
11. Braga AL, Lüdtke DS, Vargas F (2006) Curr Org Chem 10:1921
12. Freudendahl DM, Shahzad SA, Wirth T (2009) Eur J Org Chem 1649
13. Freudendahl DM, Santoro S, Shahzad SA, Santi C, Wirth T (2009) Angew Chem Int Ed 48:8409
14. Nicolaou KC, Petasis NA (1984) Selenium in natural products synthesis. CIS, Philadelphia
15. McCullough JD, Gould ES (1949) J Am Chem Soc 71:674
16. Sobkowiak A, Sawyer DT (1990) Inorg Chem 29:1248
17. Syper L, Mlochowski J (1987) Tetrahedron 43:207
18. Shimizu T, Enomoto M, Taka H, Kamigata N (1999) J Org Chem 64:8242
19. Khurana JM, Kandpal BM, Chauhan YK (2003) Phosphorus Sulfur Silicon Relat Elem 178:1369
20. Detty MR, Logan ME (2004) Adv Phys Org Chem 39:79
21. Reich HJ, Reich IL, Renga JM (1973) J Am Chem Soc 95:5813
22. Sharpless KB, Law RF, Teranishi AY (1973) J Am Chem Soc 95:6137
23. Baeyer A, Villiger V (1889) Ber Dtsch Chem Ges 32:3625
24. Krow GR (1993) Org React 43:251
25. Bolm C, Beckmann O, Luong O (1998) Metal-catalyzed Baeyer-Villiger reactions. Wiley–VCH, New York
26. Mihovilovic MD, Rudroff F, Groetzl B (2004) Curr Org Chem 8:1057
27. ten Brink GJ, Arends IWCE, Sheldon RA (2004) Chem Rev 104:4105

28. Grieco PA, Yokoyama Y, Gilman S, Ohfune Y (1977) J Chem Soc Chem Commun 870
29. Syper L (1987) Tetrahedron 43:2853
30. Syper L (1989) Synthesis 167
31. ten Brink GJ, Vis JM, Arends IWCE, Sheldon RA (2001) J Org Chem 66:2429
32. Ichikawa H, Usami Y, Arimoto M (2005) Tetrahedron Lett 46:8665
33. Miyake Y, Nishibayashi Y, Uemura S (2002) Bull Chem Soc Jpn 75:2233
34. Wójtowicz H, Brzaszcs M, Kloc K, Mlochowski J (2001) Tetrahedron 57:9743
35. Goodman MA, Detty MR (2006) Synlett 1100
36. Taylor RT, Flood LA (1983) J Org Chem 48:5160
37. Qian H, Shao LX, Huang X (2002) J Chem Res (S) 514
38. ten Brink GJ, Vis JM, Arends IWCE, Sheldon RA (2002) Tetrahedron 58:3977
39. Choi JK, Chang YK, Hong SY (1988) Tetrahedron Lett 29:1967
40. Anastas PT, Warner JC (1998) Green chemistry: theory and practice. Oxford University Press, New York
41. Minakata S, Komatsu M (2009) Chem Rev 109:711
42. Anastas P, Eghbali N (2010) Chem Soc Rev 39:301
43. Sienel G, Rieth R, Rowbottom KT (1987) Ullmann's encyclopedia of organic chemicals. Wiley-VCH, Weinheim
44. Grieco PA, Yokoyama Y, Gilman S, Mshizawa N (1977) J Org Chem 42:2034
45. Reich HJ, Chow F, Peake SL (1978) Synthesis 299
46. Hori T, Sharpless KB (1978) J Org Chem 43:1689
47. Brink GJ, Fernandes BCM, Vliet MCA, Arends IWCE, Sheldon RA (2001) J Chem Soc Perkin Trans 1:224
48. Betzemeier B, Lhermitte F, Knochel P (1999) Synlett 489
49. Marín HG, Toorn JC, Mayoral JA, García JI, Arends IWCE (2009) Green Chem 11:1605
50. Anelli PL, Banfi S, Montanari F, Quici V (1989) J Chem Soc Chem Commun 779
51. Al-Ajlouni AM, Espenson JH (1995) J Am Chem Soc 117:9243
52. Sienel G, Rieth R, Rowbottom KT (1999) Ullmann's encyclopedia of organic chemicals, vol 4. Wiley-VCH, Weinheim, p 1987
53. Santoro S, Santi C, Sabatini M, Testaferri L, Tiecco M (2008) Adv Synth Catal 350:2881
54. Neidleman SL, Geigert J (1986) Biohalogenation. Wiley, New York
55. Wever R, Kreen MBE (1990) Vanadium in biological systems. Kluwer Academic Publishers, Dordrecht
56. Butler A, Walker JV (1993) Chem Rev 93:1937
57. Kleemann A, Engel J (2001) Pharmaceutical substances. Thieme, New York
58. Dagani MJ, Barda HJ, Benya TJ, Sanders DC (2002) Ullmann's encyclopedia of industrial chemistry: bromine compounds. Wiley-VCH, Weinheim
59. Meijere A, Diederich F (2004) Metal-catalyzed cross-coupling reactions. Wiley-VCH, Weinheim
60. Podgorsek A, Zupan M, Iskra J (2009) Angew Chem Int Ed 48:8424
61. Drake MD, Bateman MA, Detty MR (2003) Organometallics 22:4158
62. Fréchet JMJ (1994) Science 263:1710
63. Grayson SM, Fréchet JMM (2001) Chem Rev 101:3819
64. Francavilla D, Bright FV, Detty MR (1999) Org Lett 1:1043
65. Francavilla D, Drake MD, Bright FV, Detty MR (2001) J Am Chem Soc 123:57
66. Drake MD, Bright FV, Detty MR (2003) J Am Chem Soc 125:12558
67. Bennett SM, Tang Y, McMaster D, Bright FV, Detty MR (2008) J Org Chem 73:6849
68. Notestein JM, Katz A (2006) Chem Eur J 12:3954
69. Goodman MA, Detty MR (2004) Organometallics 23:3016
70. Detty MR (1980) J Org Chem 45:274
71. Bray WC, Livingston RS (1923) J Am Chem Soc 45:251
72. Mohammad A, Liebhafsky HA (1934) J Am Chem Soc 56:680
73. Sies H (1985) Oxidative stress: introductory remarks. Academic, London

74. Sies H (1986) Angew Chem Int Ed 25:1058
75. Flohe L, Pryor WA (1982) Free radicals in biology. Academic, New York
76. Stadtman TC (1991) J Biol Chem 266:16257
77. Ursini F, Paoletti R (1994) Oxidative processes and antioxidants. Raven, New York
78. Flohé L (1989) Glutathione: chemical, biochemical, medical aspects. John Wiley & Sons, New York
79. Böck A (1994) Encyclopedia of inorganic chemistry: selenium proteins containing selenocysteine. John Wiley & Sons, Chichester
80. Wendel A, Pilz W, Ladenstein R, Sawatzki G, Weser U (1975) Biochim Biophys Acta 377:211
81. Epp O, Ladenstein R, Wendel A (1983) Eur J Biochem 133:51
82. Sies H (1058) Angew Chem Int Ed 1986:25
83. Mugesh G, Singh HB (2000) Chem Soc Rev 29:347
84. Mugesh G, du Mont W-W (2001) Chem Eur J 7:1365
85. Mugesh G, du Mont W-W, Sies H (2001) Chem Rev 101:2125
86. Jacob C, Giles GI, Giles NM, Sies H (2003) Angew Chem Int Ed 42:4742
87. Nogueira CW, Zeni G, Rocha JBT (2004) Chem Rev 104:6255
88. Alberto EE, Nascimento V, Braga AL (2010) J Braz Chem Soc 21:2032
89. Bhabak KP, Mugesh G (2010) Acc Chem Res 43:1408
90. Sharpless KB, Gordon KM, Lauer RF, Patrick DW, Singer SP, Young MW (1975) Chem Scr 8A:9
91. Reich HJ (1979) Acc Chem Res 12:22
92. Detty MR, Merkel PB, Powers SK (1988) J Am Chem Soc 110:5920
93. Detty MR (1991) Organometallics 10:702
94. Müller A, Cadenas E, Graf P, Sies H (1984) Biochem Pharmacol 33:3235
95. Wendel A, Fausel F, Safayhi H, Tiegs G, Otter R (1984) Biochem Pharmacol 33:3241
96. Pawham MJ, Kindt S (1984) Biochem Pharmacol 33:3247
97. Muller A, Gabriel H, Sies H (1985) Biochem Pharmacol 34:1185
98. Safayhi H, Tiegs G, Wendel A (1985) Biochem Pharmacol 34:2691
99. Wendel A, Tiegs G (1986) Biochem Pharmacol 35:2115
100. Ruwet A, Renson M (1969) Bull Soc Chim Belg 78:571
101. Weber R, Renson M (1979) Bull Soc Chim Fr 1124
102. Lambert C, Hilbert M, Christiaens L, Dereu N (1991) Synth Commun 21:85
103. Sies H (1993) Free Radical Biol Med 14:313
104. Fong MC, Schiesser CH (1997) J Org Chem 62:3103
105. Sies H, Masumoto H (1997) Adv Pharmacol 38:22229
106. Reich HJ, Jasperse CP (1987) J Am Chem Soc 109:5549
107. Parnham MJ, Biederman J, Bittner C, Dereu N, Leyck S, Wetzig H (1989) Ag Act 27:306
108. Jacquemin PV, Christianes LE, Renson MJ (1992) Tetrahedron Lett 33:3863
109. Mohsine A, Christiaens L (1996) Heterocycles 43:2567
110. Back TG, Dyck BP (1997) J Am Chem Soc 119:2079
111. Zade SS, Panda S, Tripathi SK, Singh HB, Wolmershäuser G (2004) Eur J Org Chem 3857
112. Messali M, Christiaens LE, Alshahateet SF, Kooli F (2007) Tetrahedron Lett 48:7448
113. Bhabak KP, Mugesh G (2007) Chem Eur J 13:4594
114. Glass RS, Farooqui F, Sabahi M, Ehler KW (1989) J Org Chem 54:1092
115. Reich HJ, Wollowitz S, Trend JE, Chow F, Wendelborn DF (1978) J Org Chem 43:1697
116. Gancarz RA, Kice JL (1981) Tetrahedron Lett 22:1661
117. Reich HJ, Jasperse CP (1988) J Org Chem 53:2389
118. Sarma BK, Mugesh G (2008) Chem Eur J 14:10603
119. Sarma BK, Mugesh G (2005) J Am Chem Soc 127:11477
120. Sarma BK, Mugesh G (2008) Org Biomol Chem 6:965
121. Wilson SR, Zucker PA, Huang RRC, Spector A (1989) J Am Chem Soc 111:5936
122. Galet V, Bernier JL, Hénichart JP, Lesieur D, Abadie C, Rochette L, Lindenbaum A, Chalas J, Faverie JFR, Pfeiffer B, Renard P (1994) J Med Chem 37:2903

123. Mugesh G, Panda A, Singh HB, Punekarb NS, Butcher RJ (1998) Chem Commun 2227
124. Mugesh G, Panda A, Singh HB, Punekar NS, Butcher RJ (2001) J Am Chem Soc 123:839
125. Tripathi SK, Patel U, Roy D, Sunoj RB, Singh HB, Wolmershäuser G, Butcher RJ (2005) J Org Chem 70:9237
126. Bhabak KP, Mugesh G (2008) Chem Eur J 14:8640
127. Bhabak KP, Mugesh G (2009) Chem Asian J 4:974
128. Bhabak KP, Mugesh G (2009) Chem Eur J 15:9846
129. Alberto EE, Soares LC, Sudati JH, Borges ACA, Rocha JBT, Braga AL (2009) Eur J Org Chem 4211
130. Iwaoka M, Tomoda S (1994) J Am Chem Soc 116:2557
131. For a recent review about selenium-heteroatom nonbonding interactions, see: Mukherjee AJ, Zade SS, Singh HB, Sunoj RB (2010) Chem Rev 110:4357
132. Back TG, Moussa Z (2002) J Am Chem Soc 124:12104
133. Back TG, Moussa Z (2003) J Am Chem Soc 125:13455
134. Wu ZP, Hilvert D (1992) J Am Chem Soc 112:5647
135. House KL, Dunlap RB, Odom JD, Wu ZP, Hilver D (1992) J Am Chem Soc 112:8573
136. Liu L, Mao SZ, Liu XM, Huang X, Xu JY, Liu JQ, Luo GM, Shen JC (2008) Biomacromol 9:363
137. Engman L, Andersson C, Morgenstern R, Cotgreave IA, Andersson CM, Hauberg A (1994) Tetrahedron 50:2929
138. Phadnis PP, Mugesh G (2005) Org Biomol Chem 3:2476
139. For 112: Kumar S, Johansson H, Engman L, Valgimigli L, Amorati R, Fumo MG, Pedulli GF (2007) J Org Chem 72:2583
140. For 113: Kumar S, Engman L, Valgimigli L, Amorati R, Fumo MG, Pedulli GF (2007) J Org Chem 72:6046
141. For 114: Kumakura F, Mishra B, Priyadarsini KI, Iwaoka M (2010) Eur J Org Chem 440
142. Zade SS, Singh HB, Butcher RJ (2004) Angew Chem Int Ed 43:4513
143. Press DJ, Mercier EA, Kuzma D, Back TG (2008) J Org Chem 73:4252
144. Back TG, Moussa Z, Parvez M (2004) Angew Chem Int Ed 43:1268
145. Back TG, Kuzma D, Parvez M (2005) J Org Chem 70:9230
146. Yu SC, Borchert A, Kuhn H, Ivanov I (2008) Chem Eur J 14:7066
147. Kuzma D, Parvez M, Back TG (2007) Org Biomol Chem 5:3213
148. Sarma BK, Manna D, Minoura M, Mugesh G (2010) J Am Chem Soc 132:5364
149. Ley SV (1999) Comprehensive organic synthesis. Pergamon, Oxford
150. Tojo G, Fernandez M (2006) Oxidation of alcohols to aldehydes and ketones. Springer, Berlin
151. Shimizu M, Kuwajima I (1979) Tetrahedron Lett 20:2801
152. Kuwajima I, Shimizu M, Urabe H (1982) J Org Chem 47:837
153. Barton DHR, Lester DJ, Ley SV (1978) J Chem Soc Chem Commun 276
154. Back TG (1978) J Chem Soc Chem Comm 278
155. Barton DHR, Cussans NJ, Ley SV (1978) J Chem Soc Chem Commun 393
156. Bloodworth AJ, Lapham DJ (1983) J Chem Soc Perkin Trans 1:471
157. Barton DHR, Brewster AG, Hui RAH, Lester DJ, Ley SV, Back TG (1978) J Chem Soc Chem Commun 952
158. Barton DHR, Brewster AG, Ley SV, Read CM, Rosenfeld MN (1981) J Chem Soc Perkin Trans 1:1473
159. van der Toorn JC, Kemperman G, Sheldon RA, Arends IWCE (2009) J Org Chem 74:3085
160. Pratt DV, Ruan F, Hopkins PB (1987) J Org Chem 52:5053
161. Shenga SR, Huang X (2002) J Chem Res (S) 491
162. Said SB, Skarzewski J, Mlochowski J (1992) Synth Commun 22:1851
163. Mlochowski J, Gurg M, Kubicz E, Said SB (1996) Synth Commun 26:291
164. Zhao D, Johansson M, Bäckvall JE (2007) Eur J Org Chem 4431
165. Drabowicz J, Lyzwa P, Luczak J, Mikolajczyk M, Laur P (1997) Phosphorus Sulfur Silicon Relat Elem 120&121:425
166. Wojtowicz H, Soroko G, Mlochowski J (2008) Synth Commun 38:2000

Chapter 12
Selenium and Human Health: Snapshots from the Frontiers of Selenium Biomedicine

Leopold Flohé

12.1 Some Historical Landmarks for Introduction

In 1957 selenium was reported to be an essential trace element for mammals [99] and 15 years later glutathione peroxidase became the first enzyme suggested to be a selenoprotein [92]. The verification of this hypothesis in 1973 [40, 93] and the concomitant discoveries of two bacterial selenoproteins [3, 109] opened up the flourishing field of selenium biochemistry, smoothened the political acceptance of selenium as an essential micronutrient for man and life stock [110] and proved to be instrumental to our present understanding of selenium's role in biology, as has recently been reviewed in detail [37]. Landmarks to be here recalled are (*i*) the identification of the typical selenium moiety of selenoproteins as a selenocysteyl residue [26, 44, 116] that is integrated into the peptide chain [50], (*ii*) the recognition of the "stop codon" TGA as selenocysteine codon [22, 124] and (*iii*) its "recoding from a distance" by secondary mRNA structures called "SECIS" for selenocysteine insertion sequence [13, 123], (*iv*) the identification of a selenocysteine-specific tRNA ($tRNA^{sec}$) [64, 65, 79] and (*v*) its loading mechanism involving selenophosphate-dependent transformation of seryl-t-RNA^{sec} into selenocysteyl-$tRNA^{sec}$ [41, 46, 66], (*vi*) the characterisation of major components of the selenocysteine incorporating machinery in bacteria [41, 42, 66, 67], archaea [91] and mammals [29, 35, 80, 108] and (*vii*) finally the analysis of the complete human selenoprotein genome comprising 25 selenoprotein genes encoding more than 25 distinct selenoproteins with most diversified functions [63]. By now, the human selenoproteom comprises three deiodinases (DI1-3), 7 glutathione peroxidases (GPx1-3, 6 and the three distinct expression forms of GPx4), selenophosphate synthetase 2 (SPS2), selenoprotein P (SePP), 3 thioredoxin reductases (TR1-3), the 15 kDa

L. Flohé (✉)
Department of Chemistry, Otto-von-Guerricke-Universität Madgeburg, Magdeburg, Germany
e-mail: l.flohe@t-online.de

J.D. Woollins and R.S. Laitinen (eds.), *Selenium and Tellurium Chemistry*,
DOI 10.1007/978-3-642-20699-3_12,

selenoprotein (SeP15), and the "letter selenoproteins" (SelH, I, K, M, N, O, R, S, T, V and W) for which an enzymatic function remains to be identified.

In parallel with the developments in biochemical selenology the novel trace element had justified the fondest hopes to improve human health, and a most surprising success was already reported in 1979 by an anonymous "Chinese Keshan Disease Research Group" [5]: A cardio-myopathy endemic in a large rural area of China, the Keshan disease, proved to be associated with severe selenium deficiency and could practically be eradicated by selenium supplementation (see Chap. 4). However, most of the other concepts promising health benefits from selenium have not, or not yet, been crowned by unambiguous clinical proof. In particular the epidemiology-based attempts to verify anti-carcinogenic and anti-atherogenic activities of selenium overwhelmingly failed in large-scale clinical trials. These disappointments are not easily explained. The trials might have suffered from inadequate study design and patient selection or might not have seriously enough considered the complexity of multi-factorial diseases. Alternatively, one may also conclude that the underlying concepts were simply too naive. The circumstance that the three first discovered seleno-enzymes were all peroxidases [40, 107, 111] had certainly favoured the misconception that selenium is just a "biological anti-oxidant". Accordingly, selenium supplements were expected to be wonder drugs for all kind of diseases that were presumed to be caused or aggravated by "oxidative stress". However, the glutathione peroxidases turned out to exert many functions beyond detoxifying peroxides [38] and, at the latest in 1990 when the first of the deiodinases was discovered to be a selenoprotein [10], the role of selenium in human physiology had to be revised.

The intention of this article therefore is to line out what can realistically be expected from selenium supplementation in view of the present knowledge in selenium biochemistry. This knowledge still being fragmentary, future revisions of the forecasts are predictable.

12.2 The Implications of Selenoprotein Biosynthesis

Biosynthesis of selenoproteins is in essence achieved by the ribosomal machinery for protein synthesis, but requires a particular set of auxiliary proteins to translate the stop codon UGA into selenocysteine. Together with the ribosome, the specific mRNA and $tRNA^{sec}$, these proteins make up the "selenosome", a huge (and steadily growing) supra-molecular complex that displays discrete differences between archaea, bacteria and higher animals. The pertinent state-of-the-art has been recently compiled in topical reviews [2, 32, 120] and shall here only be recalled as far as relevant to medical perspectives.

In a wood-carved way, the key events of mammalian selenoprotein synthesis may be summarized as follows: The selenocysteine to be incorporated into the protein is not used as such but has to be synthesized on the $t\text{-}RNA^{sec}$. This highly unusual and distinctly modified t-RNA [14, 21] is first loaded with serine, the latter is phosphorylated and the phophoseryl-t-RNA^{sec} is finally transformed into selenocysteyl-tRNA^{sec}, selenophosphate being the likely selenium donor substrate of a still poorly characterized selenocysteine synthase. The tRNA^{sec} is recognized by a specific elongation factor eEFsec that only becomes operative in chain elongation, if the t-RNA^{sec} is loaded with selenocysteine. An indispensable prerequisite for chain elongation is the recognition of the SECIS, the mRNA stem loop in the 3′-untranslated region that is typical of mammalian selenoprotein mRNAs and characterized by a quartet of unusual non-Watson-Crick base pairs in the stem and a conserved AAR motif in the apical loop. This particular structure is recognized by the SECIS-binding protein SBP2 and the ribosomal protein L30. Evidently SBP2 interacts with eEFsec in the ribosomal complex. If the tRNA is not loaded with selenocysteine or the SECIS is eliminated, UGA is read as nonsense codon and truncation of the growing peptide chain results. The basic assembly of ribosome, specific mRNA, tRNA^{sec}, eEFsec, L30 and SBP2 is amended by a realm of modulating factors and its function is complicated by competition with components of canonical protein synthesis, redox status of the cell and nuclear trafficking [2, 54, 102].

A first consequence of the mechanism of selenoprotein biosynthesis is that selenium, taken up in whatever form, has first to be metabolized to selenide to be used for selenophosphate biosynthesis. In case of meat-derived selenocysteine, this is achieved by a specific selenocysteine lyase, in case of plant- or yeast-derived selenomethionine by means of the trans-sulfuration pathway and with selenite in drinking water by the rather unspecific thioredoxin reductase or by non-enzymatic reaction with glutathione followed by reduction of the resulting seleno-di-glutathione (GS-Se-SG) by thioredoxin reductase (TR1) or glutathione reductase, while selenate has first to be reduced to selenite via the Se-isologue of 3-phosphoadenosine 5′-phosphosulphate [14, 73]. The conversion of these major selenium sources explains their qualitative nutritional equivalence, the bioavailability of selenite being lower than that of selenomethionine [70].

A second trivial, though often ignored, consequence is that the biosynthetic machinery, as a dynamic supra-molecular enzyme complex, has plausibly some plasticity but ultimately a genetically fixed maximum capacity. This implies that selenium is physiologically only used up to a certain dosage that is defined by optimum synthesis of selenoproteins. According to most recent studies this limit is reached in humans at a daily dosage of 35–70 μg per day, depending on whether GPx3 or selenoprotein P (SePP) is chosen as a biomarker [118]. The newly established requirements are, thus, in reasonable compliance with the Dietary Reference Intakes (DRI) of 55 μg per day for adults [70]. Selenide that can no longer be used for selenoprotein biosynthesis is rapidly methylated by S-adenosyl-methionine and exhaled as di-methyl-selenide or excreted as the tri-methyl-selenonium cation. Once the methylation capacity is exhausted, selenium compounds will display their full toxic potential. Notably, the toxicity of selenols is *inter alia* caused by autoxidation and redox cycling with concomitant formation of $^{\bullet}O_2^-$ and H_2O_2 and, thus, due to a *pro*-oxidant action of selenium [103]. For obvious reasons, the threshold of selenium toxicity in humans is not precisely known. Based on circumstantial evidence, the

"No-Observed-Adverse-Effect-Level" (NOAEL) is currently set at 800 μg per day [70].

Third, the machinery allows a differential regulation of selenoprotein expression, which results in the "hierarchy of selenoproteins", a phenomenon describing that the individual proteins respond differently to selenium restriction. SePP ranks lowest in this hierarchy in requiring the highest selenium levels for full expression [118]; a low position is also occupied by GPx1, GPx3, TR1 [106] and the selenoproteins W, H, and M [61], while GPx2 and GPx4 mark the upper end of the scale [16, 106]. How this differential expression of the selenoprotein genes is achieved is still under investigation. A decisive factor is mRNA stability. In selenium deficiency the mRNA of GPx1 is most readily degraded, while that of GPx4 is hardly affected [39, 106] and the GPx2 mRNA is increased [117]. The particular instability of GPx1 mRNA has been attributed to "nonsense-mediated decay", which describes the destruction of mRNAs with a premature stop codon upstream of an intron [54, 76], but this hypothesis has still to explain why the mRNAs of GPx2 and GPx4 are stable despite a similar location of the Sec-encoding UGA codon. Other factors implicated in affecting selenium-dependent expression of selenoproteins are SECIS stability, mRNA affinity to SBP2 and translational efficiency which, in turn depend on still poorly defined sequence characteristics of the individual mRNAs [2, 39, 54, 106]. The differential response of selenoproteins to the selenium status suggests that syndromes resulting from a low to moderate selenium deficiency are likely caused by a decline of selenoproteins ranking lowest in the hierarchy. In line with this postulate, an impaired hydroperoxide metabolism due to GPx1 deficiency is indeed one of the prevailing symptoms of selenium deficiency. However, even selenoproteins ranking high in the hierarchy may be affected by moderate selenium deficiency, since certain tissues critically depend on selenium transport by SePP which again ranks low in the hierarchy (see Chap. 3).

12.3 Selenium Distribution to Privileged Tissues

Already in the eighties selenium was shown not be evenly distributed in the body but preferentially delivered to endocrine organs, testis and brain and there incorporated into proteins distinct from those of, e.g., liver or blood [11]. How this targeted distribution is achieved, had for long remained enigmatic. A first answer was provided by knocking out the gene encoding SePP, the extracellular selenoprotein that, depending on species, contains 10 (man) to 17 (zebra fish) selenocysteine residues. Because of its extreme selenium content SePP had been suspected to be a storage or transport form of selenium [83]. The inverse genetics unequivocally verified this hypothesis: In mice without SePP selenium accumulated in the liver; plasma selenium dropped to about 30% in line with the knowledge that >60% of plasma selenium is represented by SePP; and kidney, brain and testis became severely selenium deficient [95].

The male SePP-deficient mice were infertile corroborating the essentiality of selenium for reproduction [90] (see Sect. 12.8). They also developed neurological symptoms revealing for the first time an essential role of selenium in brain function [90] (see below, Sect. 12.9).

These observations allow the conclusion that food-derived selenium is taken up by the liver and incorporated into SePP. SePP is then secreted into the circulation, taken up by peripheral tissues, degraded, and selenocysteine thus liberated used for *de novo* synthesis of selenoproteins, as described above. The uptake of SePP is mediated by specific receptors. The apolipoprotein E receptor-2 (ApoER2) has been identified as the endocytotic SePP receptor in brain [20], for Sertoli cells in testis [86] and in placenta [82], while Megalin appears to mediate the uptake in the kidney [24, 85] and also in brain [24] and placenta [82]. Evidently, the nature and density of these and possibly other receptors determine where SePP is endocytosed and selenium, thus, is accumulated. This basic scheme of privileged selenium supply, however, has likely to be amended, because SePP is not only synthesized in, and secreted from, liver. Extra-hepatic sites of SePP synthesis as observed in Leydig cells of testis and in the cerebellum [105] might function as local stores to assure selenium supply to organs that critically depend on selenium.

12.4 Selenium in Infectious Diseases

The protective role of selenium against viral infections became evident from the elegant studies on the aetiology of Keshan disease by Melinda Beck. The controversial discussion whether this clearly selenium-responsive cardio-myopathy was directly caused by selenium deficiency or viral infection could be settled by demonstrating that the disease results from a disturbed host/virus interaction followed by a genetic transformation of the virus itself [9]. All kind of infections cause an oxidative burst as part of the innate immune response by Toll-like-receptor-mediated activation of NADPH oxidases. The superoxide anion thus formed is dismutated to H_2O_2. H_2O_2 in turn enhances various self-protecting signaling cascades and is essentially controlled by GPx1 which ranks low in the hierarchy of selenoproteins [18]. A disturbed immune response may therefore be inferred in selenium deficiency. In line with this reasoning, selenium-deficient mice developed a more severe myocarditis when infected with a virulent strain of coxsackievirus B4 than did selenium-adequate ones. Moreover, when GPx1 knockout mice were infected with an amyocarditic strain, they developed myocarditis, while wild-type mice did not. Finally, it could be demonstrated that the virus in the knockout mice hat mutated into a virulent one, likely due to the mutagenic potential of the increased hydroperoxide tone [9]. Similar relationships of selenium deficiency and acquisition of virulence was later demonstrated for influenza virus and is now being investigated for other viral infection such as poliovirus and HIV [8].

Although corroborating epidemiological data are scarce, selenium deficiency should similarly aggravate bacterial infections, as they also, though by other

Toll-like receptors, trigger an analogous immune response. In support of this hypothesis, GPx1 knockout mice proved to be extremely sensitive to lipopolysaccharide (LPS), the prototype of the pathogen-associated pattern of compounds in gram-negative bacteria [58]. In this context it is also worth mentioning that intravenous application of selenite significantly improved survival in sepsis patients [4], although the high dosages applied (up to 1000 μg per day) leave open the question whether this therapeutic effect may be attributed to improved selenoprotein function or to a genuine pharmaco-dynamic efficacy of selenite.

12.5 Selenium and Cancer

The hope that selenium might prevent or even cure cancer was already born in the first half of the last century (for an interesting compilation of early investigations see [98]), corroborated by the epidemiological data of Shamberger and Frost suggesting a higher incidence of certain malignancies in selenium-deficient areras of the US [101], and seemingly proven by the study Nutritional Prevention of Cancer Trial (NPC), a large-scale placebo-controlled double-blind long-term prospective trial with 200 μg Se per day (as Se-enriched yeast) to demonstrate the protective effect of selenium against basal cell or squamous cell skin carcinoma in a population with a related history. The study failed to meet its primary end point but surprisingly revealed a significant reduction of prostate, lung and colorectal cancer and of overall cancer mortality [25]. Increasing the dosage to 400 μg Se per day in a subgroup proved to be contra-productive in respect to cancer mortality [89]. Unfortunately, the attempt to verify the efficacy of selenium in prostate cancer in the Selenium and Vitamin E Cancer Prevention Trial (SELECT) with a total of 35,533 men treated with either 200 μg Se (as L-seleno-methionine plus vitamin E) or placebo completely failed to support any chemo-preventive or therapeutic efficacy of selenium [72]. Moreover, the study was terminated, because the Se-supplemented group showed a tendency to develop diabetes (see Chap. 10). Collectively, therefore, the message of the prospective clinical trials simply is that there is no compelling evidence that conventional selenium supplements such as selenium enriched yeast or seleno-methionine at dosages far beyond the Daily Reference Intakes are of any value in cancer prevention.

Based on animal experiments, however, one may still speculate that special selenium compounds, in particular those prone to redox cycling, might prove to be useful as cytotoxic drugs in cancer therapy [103]. They might directly act as cytotoxic compounds or indirectly by oxidative activation of protective transcription factor systems such as the Nrf2/Keap1 system [57] or, like methylseleninic acid, by inactivating PKC-type kinases involved in proliferation and apoptosis [49]. However, the efficacy of such compounds still awaits clinical verification and proof of superiority to established cytotoxic therapies.

The clinical studies do by no means exclude that an adequate selenium supply contributes to cancer prevention. In fact, in the NPC trial the groups with the lowest, i.e. probably inadequate, selenium status at trial start appeared to benefit most from the supplementation and contributed essentially to the overall positive outcome of the trial, while the group with the highest selenium status showed an *increased* cancer incidence [33]. This observation, in line with the early epidemiological studies, strongly suggests that selenium deficiency is indeed a risk factor for carcinogenesis and that the preventive action of selenium is primarily mediated by selenoproteins. The role of selenoproteins in this context is, however, complex. The potential of the GPxs to reduce mutagenic peroxides may with confidence be considered to contribute to the prevention of carcinogenesis in the initiation phase. Moreover, GPx1 has also been shown to inhibit cell proliferation triggered by epidermal growth factor (EGF) [51]; similarly, GPx4 strongly inhibits cytokine-driven NF-κB activation and prostaglandin E_2-sustained tumor cell proliferation [53]; and GPx2 inhibits invasiveness and migration of tumor cells, which are decisive criteria of malignancy [7]. These effects of GPxs, which are clearly beneficial in the tumor initiation phase and, in part, at later stages of carcinogenesis, are however counteracted by other potentials of the GPxs themselves and by other selenoproteins. All GPxs, as far as investigated, inhibit apoptosis, which according to present thinking, favours tumor development. Moreover, GPx2 [7], TR1 [119] and SeP15 [56] clearly *enhance* the growth of tumor xenographs in nude mice, the classical model to test the efficacy of anti-tumor drugs. Nevertheless, circumstantial evidence, as well as the association of discrete genetic variations (polymorphisms) in selenoproteins with cancer, supports an overall protective role of the selenoproteins SePP, GPx1, GPx3, GPx4 and SeP15, as reviewed by [122] and [87].

The vexing scenario of selenoprotein action in respect to carcinogenesis has recently been reviewed [17, 52, 57, 104]. Collectively, compelling evidence accumulated that an optimum selenoprotein function contributes to the prevention of tumor initiation, in particular when the latter is driven by viral infection or other inflammatory processes, but there is no evidence whatsoever that any of the selenoproteins could efficiently counteract the growth of an established tumor.

12.6 Selenium and Cardiovascular Diseases

The potential benefit of selenium for the cardiovascular systems remains a controversial issue despite attractive hypotheses linking vascular integrity to selenoprotein functions. Irrespective of the assumed aetiology of cardiovascular diseases, be it infection, oxidized LDL, diabetes, endothelial dysfunction or high blood pressure, an oxidative stress may be inferred to contribute to atherogenesis and, accordingly, the “antioxidant” selenoproteins should counteract initiation and/or progression of the disease [104].

Clinical evidence supporting these ideas is still scarce. The early epidemiological studies suggesting an inverse correlation between selenium intake and cardiovascular diseases (compiled in [48]) could not be corroborated by any later study performed in the US, whereas some of the European studies did, as reviewed by

[87, 104]. [15] found a significant inverse correlation of baseline erythrocyte GPx1 and cardiovascular events in 636 German patients with suspected coronary heart disease, the highest quartile having just 30% of the risk compared to the lowest quartile. The data in line with the role of GPx1 in dampening endothelial dysfunction induced by homocysteine, which is known to cause oxidative challenge in vascular endothelium [115]. Another selenoprotein implicated in cardiovascular disease is the selenoprotein S (SelS). Its precise enzymatic function still remains to be elucidated, but it is discussed to prevent misfolding of proteins in the endoplasmic reticulum which would result in NF-κB-mediated inflammatory responses [45]. In a Finnish cohort study, two distinct variants of the SePP1 gene, rs8025174 and rs7178239, were found to be associated with increased risk of coronary heart disease or ischemic stroke, respectively [1].

In short, selenoproteins still deserves interest in the context of cardiovascular diseases and selenium deficiency might well be a risk factor for coronary heart disease.

12.7 Selenium in Muscle Function

Muscle pathology, the "white muscle disease", has been one of the first symptoms of selenium deficiency observed in life stock [48, 98]. Adequate selenium supply to assure appropriate muscle function proved also to be required in human patients depending on prolonged parenteral nutrition. The essentiality of selenoproteins in general to cardiac and skeletal muscle was unequivocally demonstrated by a myocyte-specific knockout of the gene encoding the t-RNA^{sec} (a systemic knockout is, of course, lethal), but the specific selenoproteins that are essential for muscle integrity have for long remained, and still are, largely obscure (for review see [69].

Interestingly, mutations in one of the *in silico* discovered selenoproteins [68] with unknown function, selenoprotein N (SelN), more recently proved to be the cause of a neuro-muscular disease known as the "congenital muscular dystrophy with spinal rigidity" [81]. SelN is a trans-membrane glycoprotein associated with the endoplasmic reticulum. Its essentiality for muscle integrity was confirmed by inverse genetics in the zebra fish [31], where it was also shown to physically and functionally interact with the ryanodine receptor calcium release channel [59]. However, SelN shares with other selenoproteins implicated in muscle function, such as SelK, SelT and SelW, that their function at the molecular level is not yet understood, redox regulation of Ca^{++} homeostasis and Ca^{++}-dependent signaling being hot topics of discussion [69].

12.8 Selenium and Male Fertility

The long known association of selenium deficiency with sperm dysfunction and male infertility in mammals [48] has found its preliminary explanation in an unexpected role of GPx4 in spermatogenesis. GPx4, originally discovered as a protein protecting biomembranes against lipid peroxidation, was found to be excessively expressed in spermatogenic cells, specifically in round spermatids [74], but almost undetectable in mature spermatozoa. Mass spectroscopic investigations, however, revealed that the GPx4 *protein* was still present in sperm; the active peroxidase had been transformed into an enzymatically inactive structural protein during the late phase of spermiogenesis [112]. It there, together with other cysteine-rich proteins, polymerizes to the keratin-like material that embeds the mitochondrial helix in the mid piece of spermatozoa. The structural and functional change ("moonlighting") of GPx4 must be triggered by a still poorly understood burst of hydroperoxide formation, that leads to oxidation, secretion and extracellular degradation of glutathione. In shortage of its natural substrate GSH, GPx4 undergoes a dead-end alternate substrate inhibition: The oxidized selenium moiety of the peroxidase selenylates surface-exposed cysteine residues of itself and of other proteins, thus forming a three-dimensional supra-molecular aggregate [77]. Lack of GPx4 in sperm results in structural instability of the sperm mid piece, disorganisation of the microtubules in the flagella, impaired directed mobility of spermatozoa (they swim in circles), loss of tail and, in consequence, abrogation of fertilisation capacity.

Philosophically, the process intrigues by demonstrating how a clearly "antioxidant" peroxidase can adopt an opposite role: making use of peroxides to oxidize itself and other proteins to create a biologically essential structure. The medical relevance of this moonlighting process has meanwhile been corroborated by significant inverse correlations of GPx protein content of sperm and fertilisation capacity [43, 55]. Deficiency of testicular GPx4 and infertility may well result from inadequate selenium intake in cows or sheep grazing on meadows extremely poor in selenium. In man, however, infertility is not commonly explained by insufficient selenium intake because of the high ranking of GPx4 in the hierarchy and the privileged Se transport to testis. More likely the decreased testicular GPx4 is caused by lack of gonadotropins [74], mutations in the GPx4 gene [75], or any defect in the machinery of selenoprotein synthesis or Se transport [36]. In fact, knockout of the SePP gene in mice resulted in the very same sperm pathology as seen in severe selenium deficiency [19]. The typical sperm defects were equally observed in mice having the mitochondrial GPx4 knocked out, revealing that it is the mitochondrial GPx4 which is pivotal to spermiogenesis and male fertility [94], whereas disruption of the nuclear expression form of GPx4 only leads to impaired chromatin condensation in sperm without affecting fertility [28].

12.9 Selenium and the Brain

The phenotype of SePP knockout mice reanimated research on selenium function in the central nervous system. These mice, apart from being growth-retarded, developed an atactic gait reminding of cerebellar injury and died from seizures after a few weeks [95]. As outlined above (Sect. 12.3), selenoprotein biosynthesis in brain depends on ApoER2-mediated Se supply by SePP and, accordingly, brain pathology in ApoER2 and SePP knockout mice is similar [20], and also a conditioned knockout of the $tRNA^{sec}$, intrigued by cerebellar symptoms and seizures [100]. Collectively, these results disclosed the essentiality of Se for brain function which had not been detected by alimentary selenium restriction. Since, the peculiar expression pattern of selenoproteins in brain [11], as well as sporadic reports on low plasma selenium in brain disorders, are increasingly reconsidered, and the search for molecular mechanisms of selenium-dependent brain functions enjoys a novel start, as reviewed by Zhang et al. [121] and Conrad and Schweizer [27]. Unsurprisingly, though, progress is slow. Still bound to the idea that selenium is just a biological antioxidant, most of the investigations focus on established or suspected antioxidant functions of the selenoproteins or, more fashionably, on their putative role in redox regulation. The selenoproteins currently most discussed in neuro-degenerative diseases are GPx1 and 4, TR1, SelH, SelM and SelW, but their precise function in brain is far from clear [121]. A few intriguing findings may be highlighted.

In line with an primarily antioxidant role of Se in brain, overexpression of GPx1 dampened toxicity of 6-hydroxydopamine, an established model for the development of Parkinson's disease [12]. Similarly, GPx1 knockout mice showed an exaggerated response to amyloid beta-protein which is implicated in the pathogenesis of Alzheimer's disease [30]. However, while SePP knockout mice die from seizures, GPx1 knockout mice are surprisingly more resistant against kainic acid–induced seizures.

The role of GPx4 in brain is becoming more transparent. When the GP4 was knocked out in the brain of mice (a systemic knockout is lethal), the mice became atactic and hyper-excitable, thus showing a phenotype very much resembling that of systemic SePP knockout and brain-specific knockout of the $tRNA^{sec}$ [27]. The underlying neuro-degeneration could tentatively be explained by the high efficiency of GPx4 to silence lipoxygenases. Conditioned GPx4 knockout cells were shown to indeed die from lipid peroxidation triggered by 12,15-lipoxygenase (12,15-LOX), an enzyme that generates lipid hydroperoxides of lipids firmly integrated into membranes, i.e. the preferential substrates of GPx4. Like other LOXs, also the 12,15-LOX requires traces of hydroperoxides to become active, which implies that this enzyme remains dormant in the presence of GPx4 but initiates an unbalanced oxidative membrane destruction in absence of GPx4. Moreover, the 12,15-LOX products activate AIF1, a pro-apoptotic factor that programs cells to die [27]. Physiologically, the GPx4/12,15-LOX couple is believed to orchestrate the remodelling of intracellular organelles and, by AIF1-mediated apoptosis, of entire tissues during differentiation or organogenesis, respectively. Accordingly, any disturbance of the delicate balance between these two enzymes must have fatal consequences, LOX-dependent programmed cell death being the outcome of GPx4 deficiency. This peculiar phenomenon may explain the neuro-degeneration seen in SePP and GPx4 knockout mice and likely also the essentiality of GPx4 in embryogenesis. which is not shared by GPx1-3. Interestingly, only the cytosolic form of GPx4 has this pivotal cyto-protective function, as selective disruption of the nuclear or the mitochondrial expression of GPx4, as well as knockout of GPx1, 2 and 3, yields viable off-springs without any obvious brain pathology.

Out of the 3 TRs, TR1 is evidently the most important one in brain development. Brain-specific disruption of TR1 expression (but not of the mitochondrial TR) resulted in hypoplasia of the cerebellum, again associated with ataxia [27]. The predominant histological abnormalities were ectopically localized Pukinje cells with aberrant dendritic arborisation and reduced density of disoriented Bergmann glia cells, the latter being known to be responsible for neuronal migration during development of the cerebellum. The molecular basis of the peculiar pathology of TR1 knockout may be suspected in any of the many roles of the TR1 substrate thioredoxin, be it a peroxiredoxin-mediated hydroperoxide reduction, nucleic acid synthesis or any of the regulatory roles of this pleiotropic redox regulator [71].

SelM, a microsomal selenoprotein, has more recently been implicated to regulate Ca^{++} homeostasis in neuronal cells [88]. This protein is characterized by a CxxU motif suggestive of a redox function and reportedly decreases cytosolic Ca^{++} upon H_2O_2 exposure and apoptosis, the precise molecular links still being unclear. Also SelT has been implicated in brain Ca^{++} homeostasis but reported to increase intracellular Ca^{++} in response to a cAMP-stimulating trophic neuropeptide [47].

12.10 Selenium and Endocrine Systems

12.10.1 Selenium in Thyroid Function

As mentioned, endocrine organs such as the thyroid gland, the pituitary and ovaries are most privileged in respect to selenium supply and retention. A prominent role of selenium is evident for the thyroid gland, since the entire metabolism of thyroid hormones not only depends on iodine, as is more or less precisely known for centuries, but also on selenium. The type1 and type 2 deiodinases (5′-deiodinases; DI1 and DI2), which metabolize the precursor thyroxin (3,5,3′,5′tetraiodothyronine) to the active thyroid hormone 3,5,3′-triiodothyronine (T3), as well as the type 3 deiodinase (DI3), which produces the inactive 3,3′,5′-triiodothyronine (reverse T3; rT3), turned out to be selenoproteins [10, 62]. Further, the iodination within the thyroid is exerted by thyroid peroxidase with aid of H_2O_2 which in turn has to be tightly regulated by GPx1, 3 and/or 4. Combined deficiencies of the two essential trace elements J and Se may, therefore, be expected to impair thyroid function more drastically than any isolated trace element deficiency. Severe deficiency of both trace elements has indeed been found associated with myxoedematic cretinism in central Africa [114] and is also implicated in the aetiology of the Kashin-Beck disease, an osteo-arthropathy prevailing in the selenium deficient belt of China. In both cases, though, disease manifestation requires additional aetiological factors, thiocyanate overload in the African cretinism and mycotoxins in Kashin Beck disease [113].

Surprisingly, however, an isolated deficiency of Se appears to rarely affect thyroid function, primarily because of the high ranking of the deiodinases in the hierarchy of selenoproteins. Even SePP knockout mice, which consistently show severe selenium deficiency in kidney and brain (see above, Sect. 12.3), have an essentially normal thyroid without any significant disturbance in thyroid or tissue levels of T4, T3 and the stimulatory hormone TSH [97]. Moreover, even GPx activity and GPx1 mRNA, which typically is degraded in Se deficiency (Sect. 12.2), were found unaltered in the thyroid gland of SePP knockout mice. This unexpected finding reveals that the thyroid gland can retain its selenium by a mechanism distinct from the SePP-dependent pathway discussed above (Sect. 12.3).

The clinical relevance of Se-dependent thyroid hormone synthesis became obvious from genetic defects in SBP2, a key player in selenoprotein synthesis (Sect. 12.2). The patients present with high T4 and rT3, low T3 and normal or elevated TSH [34]. The common clinical denominator in partial SBP2 deficiency appears to be retarded growth [34], while a nonsense mutation in the SBP2 gene was associated with delayed bone maturation, congenital myopathy, impaired mental development, and disturbed motor coordination [6]. Se supplementation failed to restore selenoprotein synthesis and thyroid function [96].

Collectively, the state of the art teaches that Se deficiency alone is hardly any aetiologically relevant factor in thyroid diseases. Se supplementation is, therefore, of little promise for improving thyroid function, so far proved to be not efficacious when the machinery of selenoprotein synthesis was defective, and was neither consistently beneficial in thyroid disorders such as autoimmune thyroiditis [60].

12.10.2 Selenium and Diabetes Mellitus

Diabetes mellitus has been widely discussed to be *inter alia* an "oxidative stress disease". Accordingly, affected patients were expected to benefit from an optimized peroxide metabolism by Se supplementation. However, the attempt to create a super-healthy mouse by systemic overexpression of GPx1 yielded a surprise: These mice became obese, were hyperglycemic and hyperinsulinemic, insulin-resistant and had elevated leptin levels. In short, they developed type-2 diabetes mellitus [78]. That this observation might also be relevant to human health care is suggested by two clinical studies: (*i*) In pregnant women, a positive correlation was observed between insulin resistance and GPx activity in red blood cells, which represents GPx1 [23], and (*ii*) the only effect of the supra-nutritional Se supplementation applied in the SELECT study was an increased risk of type-2 diabetes [72]. Taken together, the clinical observations imply that even an "optimal" GPx1 level might be risky.

The likely link between insulin function and selenium is H_2O_2, the substrate of GPx1 which is one of the selenoproteins that vary considerably with Se intake. H_2O_2 is formed during insulin signaling and presumed to enhance signal transduction by oxidative inactivation of counteracting protein phosphatases, here in particular

PTP1B. An attenuated phosphorylation of the insulin receptor and of the downstream transducer Akt, indicating reduced signaling activity, was indeed observed in the GPx1 overexpressing mice [78]. Moreover, maximizing GPx1 activity by Se supplementation in mice resulted in increased activity of PTP1B *via* prevention of (inhibitory) glutathionylation of PTP1B, which dampens insulin signaling [84]. Thus, GPx1 impairs insulin function by steeling the H_2O_2 required for appropriate signalling.

Since the inhibitory role of GPx1 in insulin signaling comes into play within the physiological range of activity and since diabetes is considered a risk factor for cardio-vascular disease, it is tempting to speculate that the diabetogenic potential of GPx1 might have contributed to the inconsistent outcome of Se supplementation studies in cardiovascular diseases (see Sect. 12.6). Moreover, a regulatory role of GPxs may be suspected not only for insulin signaling, since analogous redox regulation evidently operates in many signaling cascades [18, 38].

12.11 Conclusions and Perspectives

Collectively, emerging knowledge does no longer justify the view that Se, by its "antioxidant" role, just prevents the organism from getting rancid due to unspecific oxidative stress. We are confronted with a most diversified scenario of selenoproteins, most of which have been demonstrated or, by structural criteria, are suspected to be oxido-reductases. However, even the glutathione peroxidases, which are specialized for hydroperoxide reduction, have adopted roles beyond fighting oxidative stress. It may suffice to recall the action of GPx1 in modulating signaling cascades or the roles of GPx4 in spermiogenesis and regulation of apoptosis. Even more surprises are expected from the characterisation of the less characterized selenoproteins.

As to Se supplementation, we can proudly look back on having practically eradicated Keshan disease and defined the Se requirements for parenteral nutrition. Circumstantial evidence further allows the assumption that keeping the selenoproteins near optimum activity is advisable for prevention of a variety of diseases comprising infections, inflammatory disorders and cancer. A general need for Se supplementation can be assumed for extremely deficient rural areas with little or no food exchange, which means not for reasonably eating people in industrialized countries. If for some reason concerned of a deficiency, a daily supplement of 50 μg may be permitted, which even on top of the food-bound intake will hardly reach any toxicity limit. By experience, supplementation with Se dosages beyond the DRI is worthless for essentially healthy people and may even be hazardous. Selenium compounds that are expected to have therapeutic activities beyond assuring selenoprotein synthesis should not be mislabelled as supplements but considered as drugs to be tested for efficacy and safety according to internationally accepted standards.

Looking back on the progress achieved over the last for decades and forward to the white areas of selenium biomedicine, future surprises can be predicted.

The tasks ahead are defined: unravelling the biological role of all the selenoproteins just known by sequence, understanding the distinct tissue-specific functions of selenoproteins and the in part still enigmatic trafficking of selenium to sites of demand. As in the past decade, inverse genetics and molecular biology in general will be of outstanding importance for meeting these challenges. They will also help to pave the way towards individualized therapeutic options. For mice, the possibility to correct pathologies resulting from defective Se transport by over-supplementation has already been demonstrated [100]. Similarly, defects in the synthesis machinery or in selenoprotein genes themselves may be, or may not be, overcome by increased Se supply and, thus define individual Se requirements.

Another new horizon of applied selenium biomedicine is showing up in microbiology. Genome analyses of pathogenic bacteria and protozoa disclosed an abundance of selenoproteins, most of them not being functionally characterized. If of vital importance for the pathogens, these proteins might become an attractive scenario of molecular targets to design novel antibiotic drugs.

References

1. Alanne M, Kristiansson K, Auro K, Silander K, Kuulasmaa K, Peltonen L, Salomaa V, Perola M (2007) Hum Genet 122:355–365
2. Allmang C, Wurth L, Krol A (2009) Biochim Biophys Acta 1790:1415–1423
3. Andreesen JR, Ljungdahl LG (1973) J Bacteriol 116:867–873
4. Angstwurm MW, Engelmann L, Zimmermann T, Lehmann C, Spes CH, Abel P, Strauss R, Meier-Hellmann A, Insel R, Radke J, Schüttler J, Gärtner R R (2007) Crit Care Med 35:118–126
5. Anonymous (1979) Chin Med J Engl 92:471–476
6. Azevedo MF, Barra GB, Naves LA, Ribeiro Velasco LF, Godoy Garcia Castro P, de Castro LC, Amato AA, Miniard A, Driscoll D, Schomburg L, de Assis Rocha Neves F (2010) J Clin Endocrinol Metab 95:4066–4071
7. Banning A, Kipp A, Schmitmeier S, Lowinger M, Florian S, Krehl S, Thalmann S, Thierbach R, Steinberg P, Brigelius-Flohé R (2008) Cancer Res 68:9746–9753
8. Beck MA (2006) In: Hatfield DL, Berry MJ, Gladyshev VN (eds) Selenium its molecular biology and role in human health. Springer, New York, pp 287–298
9. Beck MA, Esworthy RS, Ho YS, Chu FF (1998) FASEB J 12:1143–1149
10. Behne D, Kyriakopoulos A, Meinhold H, Köhrle J (1990) Biochem Biophys Res Commun 173:1143–1149
11. Behne D, Hilmert H, Scheid S, Gessner H, Elger W (1988) Biochim Biophys Acta 966:12–21
12. Bensadoun JC, Mirochnitchenko O, Inouye M, Aebischer P, Zurn AD (1998) Eur J Neurosci 10:3231–3236
13. Berry MJ, Banu L, Chen YY, Mandel SJ, Kieffer JD, Harney JW, Larsen PR (1991) Nature 353:273–276
14. Birringer M, Pilawa S, Flohé L (2002) Nat Prod Rep 19:693–718
15. Blankenberg S, Rupprecht HJ, Bickel C, Torzewski M, Hafner G, Tiret L, Smieja M, Cambien F, Meyer J, Lackner KJ (2003) N Engl J Med 349:1605–1613
16. Brigelius-Flohé R (1999) Free Radic Biol Med 27:951–965
17. Brigelius-Flohé R, Kipp A (2009) Biochim Biophys Acta 1790:1555–1568
18. Brigelius-Flohé R, Flohé L (2011) Antioxid Redox Signal, DOI: 10.1089/ars.2010.3534
19. Burk FR, Olson GE, Hill KE (2006) In: Hatfield DL, Berry MJ, Gladyshev VN (eds) Selenium its molecular biology and role in human health. Springer, New York, pp 111–122

20. Burk RF, Hill KE, Olson GE, Weeber EJ, Motley AK, Winfrey VP, Austin LM (2007) J Neurosci 27:6207–6211
21. Carlson BA, Xu XM, Shrimali R, Sengupta A, Yoo MH, Irons R, Zhong N, Hatfield DL, Lee BJ, Lobanov AV, Gladyshev VN (2006) In: Hatfield DL, Berry MJ, Gladyshev VN (eds) Selenium its molecular biology and role in human health. Springer, New York, pp 29–37
22. Chambers I, Frampton J, Goldfarb P, Affara N, McBain W, Harrison PR (1986) EMBO J 5:1221–1227
23. Chen X, Scholl TO, Leskiw MJ, Donaldson MR, Stein TP (2003) J Clin Endocrinol Metab 88:5963–5968
24. Chiu-Ugalde J, Theilig F, Behrends T, Drebes J, Sieland C, Subbarayal P, Kohrle J, Hammes A, Schomburg L, Schweizer U (2010) Biochem J 431:103–111
25. Clark LC, Dalkin B, Krongrad A, Combs GF Jr, Turnbull BW, Slate EH, Witherington R, Herlong JH, Janosko E, Carpenter D, Borosso C, Falk S, Rounder J (1998) Br J Urol 81: 730–734
26. Cone JE, Del Rio RM, Davis JN, Stadtman TC (1976) Proc Natl Acad Sci USA 73:2659–2663
27. Conrad M, Schweizer U (2010) Antioxid Redox Signal 12:851–865
28. Conrad M, Moreno SG, Sinowatz F, Ursini F, Kolle S, Roveri A, Brielmeier M, Wurst W, Maiorino M, Bornkamm GW (2005) Mol Cell Biol 25:7637–7644
29. Copeland PR, Driscoll DM (1999) J Biol Chem 274:25447–25454
30. Crack PJ, Cimdins K, Ali U, Hertzog PJ, Iannello RC (2006) J Neural Transm 113:645–657
31. Deniziak M, Thisse C, Rederstorff M, Hindelang C, Thisse B, Lescure A (2007) Exp Cell Res 313:156–167
32. Donovan J, Copeland PR (2010) Antioxid Redox Signal 12:881–892
33. Duffield-Lillico AJ, Reid ME, Turnbull BW, Combs GF Jr, Slate EH, Fischbach LA, Marshall JR, Clark LC (2002) Cancer Epidemiol Biomark Prev 11:630–639
34. Dumitrescu AM, Di Cosmo C, Liao XH, Weiss RE, Refetoff S (2010) Antioxid Redox Signal 12:905–920
35. Fagegaltier D, Hubert N, Yamada K, Mizutani T, Carbon P, Krol A (2000) EMBO J 19: 4796–4805
36. Flohé L (2007) Biol Chem 388:987–995
37. Flohé L (2009) Biochim Biophys Acta 1790:1389–1403
38. Flohé L (2010) Meth Enzymol 473:1–39, 2010
39. Flohé L, Brigelius-Flohé R (2006) In: Hatfield DL, Berry MJ, Gladyshev VN (eds) Selenium its molecular biology and role in human health. Springer, New York, pp 161–172
40. Flohé L, Günzler WA, Schock HH (1973) FEBS Lett 32:132–134
41. Forchhammer K, Böck A (1991) J Biol Chem 266:6324–6328
42. Forchhammer K, Leinfelder W, Böck A (1989) Nature 342:453–456
43. Foresta C, Flohé L, Garolla A, Roveri A, Ursini F, Maiorino M (2002) Biol Reprod 67: 967–971
44. Forstrom JW, Zakowski JJ, Tappel AL (1978) Biochemistry 17:2639–2644
45. Gao Y, Hannan NR, Wanyonyi S, Konstantopolous N, Pagnon J, Feng HC, Jowett JB, Kim KH, Walder K, Collier GR (2006) Cytokine 33:246–251
46. Glass RS, Singh WP, Jung W, Veres Z, Scholz TD, Stadtman TC (1993) Biochemistry 32: 12555–12559
47. Grumolato L, Ghzili H, Montero-Hadjadje M, Gasman S, Lesage J, Tanguy Y, Galas L, Ait-Ali D, Leprince J, Guerineau NC, Elkahloun AG, Fournier A, Vieau D, Vaudry H, Anouar Y (2008) FASEB J 22:1756–1768
48. Gun S, Harr JR, Levander OR, Olson OE, Schroeder HJ, Allaway WH, Lakin HW, Boaz TD (1976) Selenium. National Academy Press, Washington, DC
49. Gundimeda U, Schiffman JE, Chhabra D, Wong J, Wu A, Gopalakrishna R (2008) J Biol Chem 283:34519–34531
50. Günzler WA, Steffens GJ, Grossmann A, Kim SM, Ötting F, Wendel A, Flohé L (1984) Hoppe Seyler's Z Physiol Chem 365:195–212

51. Handy DE, Lubos E, Yang Y, Galbraith JD, Kelly N, Zhang YY, Leopold JA, Loscalzo J (2009) J Biol Chem 284:11913–11921
52. Hatfield DL, Yoo MH, Carlson BA, Gladyshev VN (2009) Biochim Biophys Acta 1790: 1541–1545
53. Heirman I, Ginneberge D, Brigelius-Flohé R, Hendrickx N, Agostinis P, Brouckaert P, Rottiers P, Grooten J (2006) Free Radic Biol Med 40:285–294
54. Hoffmann PR, Berry MJ (2006) In: Hatfield DL, Berry MJ, Gladyshev VN (eds) Selenium its molecular biology and role in human health. Springer, New York, pp 73–82
55. Imai H, Suzuki K, Ishizaka K, Ichinose S, Oshima H, Okayasu I, Emoto K, Umeda M, Nakagawa Y (2001) Biol Reprod 64:674–683
56. Irons R, Tsuji PA, Carlson BA, Ouyang P, Yoo MH, Xu XM, Hatfield DL, Gladyshev VN, Davis CD (2010) Cancer Prev Res Phila 3:630–639
57. Jackson MI, Combs GF Jr (2008) Curr Opin Clin Nutr Metab Care 11:718–726
58. Jaeschke H, Ho YS, Fisher MA, Lawson JA, Farhood A (1999) Hepatology 29:443–450
59. Jurynec MJ, Xia R, Mackrill JJ, Gunther D, Crawford T, Flanigan KM, Abramson JJ, Howard MT, Grunwald DJ (2008) Proc Natl Acad Sci USA 105:12485–12490
60. Karanikas G, Schuetz M, Kontur S, Duan H, Kommata S, Schoen R, Antoni A, Kletter K, Dudczak R, Willheim M (2008) Thyroid 18:7–12
61. Kipp A, Banning A, van Schothorst EM, Meplan C, Schomburg L, Evelo C, Coort S, Gaj S, Keijer J, Hesketh J, Brigelius-Flohé R (2009) Mol Nutr Food Res 53:1561–1572
62. Köhrle J (2005) Thyroid 15:841–853
63. Kryukov GV, Castellano S, Novoselov SV, Lobanov AV, Zehtab O, Guigo R, Gladyshev VN (2003) Science 300:1439–1443
64. Lee BJ, Worland PJ, Davis JN, Stadtman TC, Hatfield DL (1989) J Biol Chem 264:9724–9727
65. Leinfelder W, Zehelein E, Mandrand-Berthelot MA, Böck A (1988) Nature 331:723–725
66. Leinfelder W, Forchhammer K, Veprek B, Zehelein F, Böck A (1990) Proc Natl Acad Sci USA 87:543–547
67. Leinfelder W, Forchhammer K, Zinoni F, Sawers G, Mandrand-Berthelot MA, Böck A A (1988) J Bacteriol 170:540–546
68. Lescure A, Gautheret D, Carbon P, Krol A (1999) J Biol Chem 274:38147–38154
69. Lescure A, Rederstorff M, Krol A, Guicheney P, Allamand V (2009) Biochim Biophys Acta 1790:1569–1574
70. Levander OA, Burk RF (2006) In: Hatfield DL, Berry MJ, Gladyshev VN (eds) Selenium its molecular biology and role in human health. Springer, New York, pp 399–410
71. Lillig CH, Holmgren A (2007) Antioxid Redox Signal 9:25–47
72. Lippman SM, Klein EA, Goodman PJ, Lucia MS, Thompson IM, Ford LG, Parnes HL, Minasian LM, Gaziano JM, Hartline JA, Parsons JK, Bearden JD 3rd, Crawford ED, Goodman GE, Claudio J, Winquist E, Cook ED, Karp DD, Walther P, Lieber MM, Kristal AR, Darke AK, Arnold KB, Ganz PA, Santella RM, Albanes D, Taylor PR, Probstfield JL, Jagpal TJ, Crowley JJ, Meyskens ML Jr, Baker LH, Coltman CA Jr (2009) JAMA 301:39–51
73. Lu J, Berndt C, Holmgren A (2009) Biochim Biophys Acta 1790:1513–1519
74. Maiorino M, Wissing JB, Brigelius-Flohé R, Calabrese F, Roveri A, Steinert P, Ursini F, Flohé L (1998) FASEB J 12:1359–1370
75. Maiorino M, Bosello V, Ursini F, Foresta C, Garolla A, Scapin M, Sztajer H, Flohé L (2003) Biol Reprod 68:1134–1141
76. Maquat LE (2001) Biofactors 14:37–42
77. Mauri P, Benazzi L, Flohé L, Maiorino M, Pietta PG, Pilawa S, Roveri A, Ursini F (2003) Biol Chem 384:575–588
78. McClung JP, Roneker CA, Mu W, Lisk DJ, Langlais P, Liu F, Lei XG (2004) Proc Natl Acad Sci USA 101:8852–8857
79. Mizutani T, Hitaka T (1988) FEBS Lett 232:243–248
80. Mizutani T, Tanabe K, Yamada K (1998) FEBS Lett 429:189–193

81. Moghadaszadeh B, Petit N, Jaillard C, Brockington M, Roy SQ, Merlini L, Romero N, Estournet B, Desguerre I, Chaigne D, Muntoni F, Topaloglu H, Guicheney P (2001) Nat Genet 29:17–18
82. Motley AK, Hill KE, Winfrey VP, Olson GE, Burk RF (2010) Presented at Selenium 2010, Kyoto, Abstract O-266
83. Motsenbocker MA, Tappel AL (1982) Biochim Biophys Acta 719:147–153
84. Mueller AS, Bosse AC, Most E, Klomann SD, Schneider S, Pallauf J (2009) J Nutr Biochem 20:235–247
85. Olson GE, Winfrey VP, Hill KE, Burk RF (2008) J Biol Chem 283:6854–6860
86. Olson GE, Winfrey VP, Nagdas SK, Hill KE, Burk RF (2007) J Biol Chem 282:12290–12297
87. Rayman MP (2009) Biochim Biophys Acta 1790:1533–1540
88. Reeves MA, Bellinger FP, Berry MJ (2010) Antioxid Redox Signal 12:809–818
89. Reid ME, Duffield-Lillico AJ, Slate E, Natarajan N, Turnbull B, Jacobs E, Combs GF Jr, Alberts DS, Clark LC, Marshall JR (2008) Nutr Cancer 60:155–163
90. Renko K, Werner M, Renner-Muller I, Cooper TG, Yeung CH, Hollenbach B, Scharpf M, Kohrle J, Schomburg L, Schweizer U (2008) Biochem J 409:741–749
91. Rother M, Wilting R, Commans S, Böck A (2000) J Mol Biol 299:351–358
92. Rotruck JT, Hoekstra WG, Pope AL, Ganther HE, Swanson A, Hafemann D (1972) Fed Proc 31:691
93. Rotruck JT, Pope AL, Ganther HE, Swanson AB, Hafeman DG, Hoekstra WG (1973) Science 179:588–590
94. Schneider M, Forster H, Boersma A, Seiler A, Wehnes H, Sinowatz F, Neumüller C, Deutsch MJ, Walch A, Hrabe de Angelis M, Wurst W, Ursini F, Roveri A, Maleszewski M, Maiorino M, Conrad M (2009) FASEB J 23:3233–3242
95. Schomburg L, Schweizer U, Holtmann B, Flohé L, Sendtner M, Köhrle J (2003) Biochem J 370:397–402
96. Schomburg L, Dumitrescu AM, Liao XH, Bin-Abbas B, Hoeflich J, Köhrle J, Refetoff S (2009) Thyroid 19:277–281
97. Schomburg L, Riese C, Michaelis M, Griebert E, Klein MO, Sapin R, Schweizer U, Kohrle J (2006) Endocrinology 147:1306–1313
98. Schrauzer GN (1998) Selen, 3rd edn. Barth, Heidelberg, Leipzig
99. Schwarz K, Foltz CM (1957) J Am Chem Soc 79:3292–3293
100. Schweizer U, Schomburg L (2006) In: Hatfield DL, Berry MJ, Gladyshev VN (eds) Selenium its molecular biology and role in human health. Springer, New York, pp 233–248
101. Shamberger RJ, Frost DV (1969) Can Med Assoc J 100:682
102. Small-Howard A, Morozova N, Stoytcheva Z, Forry EP, Mansell JB, Harney JW, Carlson BA, Xu XM, Hatfield DL, Berry MJ (2006) Mol Cell Biol 26:2337–2346
103. Spallholz JE (1994) Free Radic Biol Med 17:45–64
104. Steinbrenner H, Sies H (2009) Biochim Biophys Acta 1790:1478–1485
105. Steinert P, Bächner D, Flohé L (1998) Biol Chem 379:683–691
106. Sunde RA (2006) In: Hatfield DL, Berry MJ, Gladyshev VN (eds) Selenium its molecular biology and role in human health. Springer, New York, pp 149–160
107. Takahashi K, Cohen HJ (1986) Blood 68:640–645
108. Tujebajeva RM, Copeland PR, Xu XM, Carlson BA, Harney JW, Driscoll DM, Hatfield DL, Berry MJ (2000) EMBO Rep 1:158–163
109. Turner DC, Stadtman TC (1973) Arch Biochem Biophys 154:366–381
110. Ursini F, Maiorino M (2010) Antioxid Redox Signal 13:1617–1622
111. Ursini F, Maiorino M, Gregolin C (1985) Biochim Biophys Acta 839:62–70
112. Ursini F, Heim S, Kiess M, Maiorino M, Roveri A, Wissing J, Flohé L (1999) Science 285:1393–1396
113. Vanderpas J (2006) Annu Rev Nutr 26:293–322
114. Vanderpas JB, Contempre B, Duale NL, Goossens W, Bebe N, Thorpe R, Ntambue K, Dumont J, Thilly CH, Diplock AT (1990) Am J Clin Nutr 52:1087–1093

115. Weiss N, Heydrick SJ, Postea O, Keller C, Keaney JF Jr, Loscalzo J (2003) Clin Chem Lab Med 41:1455–1461
116. Wendel A, Kerner B, Graupe K (1978) Presented at the 25 Konferenzder Gesellschaft für Biologische Chemie Functions of Glutathione in Liver and Kidney, Schloss Reisensburg, July 8–11
117. Wingler K, Böcher M, Flohé L, Kollmus H, Brigelius-Flohé R (1999) Eur J Biochem 259: 149–157
118. Xia Y, Hill KE, Li P, Xu J, Zhou D, Motley AK, Wang L, Byrne DW, Burk RF (2010) Am J Clin Nutr 92:525–531
119. Yoo MH, Carlson BA, Tsuji P, Irons R, Gladyshev VN, Hatfield DL (2010) Meth Enzymol 474:255–275
120. Yoshizawa S, Böck A (2009) Biochim Biophys Acta 1790:1404–1414
121. Zhang S, Rocourt C, Cheng WH (2010) Mech Ageing Dev 131:253–260
122. Zhuo P, Diamond AM (2009) Biochim Biophys Acta 1790:1546–1554
123. Zinoni F, Heider J, Böck A (1990) Proc Natl Acad Sci USA 87:4660–4664
124. Zinoni F, Birkmann A, Stadtman TC, Böck A (1986) Proc Natl Acad Sci USA 83:4650–4654

Chapter 13
Metal Complexes Containing P-Se Ligands

Chen-Wei Liu and J. Derek Woollins

13.1 Introduction

In comparison to P-S systems the coordination of P-Se and P-Te ligands is not a large field. Many of the well known P-S ligands (e.g. dithiophosphoric acid derivatives) have important commercial applications but their direct heavier analogues are not known or rare. Here we provide an introduction to the literature of P-Se and P-Te coordination chemistry from the historical naked anions through to modern day derivatives.

13.2 Naked Anions

There are a number of examples of metal complexes containing naked anionic P-Se ligands and these have been reviewed [1, 2]. By analogy with reactions of P_4 using tris(2-diphenylphosphinoethyl)amine (np3) complexes of nickel and palladium with P_4X_3 (X = S or Se) have been reported and like P_4 apical co-ordination *via* phosphorus is observed. On co-ordination of P_4X_3, the tripodal counter-ligand flexes and the nitrogen atom is no longer bound in the product. Alternatively, a basal phosphorus of the P_4X_3 cage may be replaced. Reaction of P_4X_3 (X = S or Se) with [{RhC1(COD)}$_2$] (COD = 1,5-cyclooctadiene) in the presence of triphos[1,1,1-tris(diphenylphosphinomethyl) ethane] gives [Rh(triphos)(P_3X_3)]. These complexes can be regarded formally as consisting of a metal fragment which stabilises the otherwise unknown $[P_3X_3]^-$ anions. In the X-ray structure of the compound with X = S it is clear that the rhodium has

C.-W. Liu
Department of Chemistry, National Dong Hwa University, Taiwan, China

J.D. Woollins (✉)
School of Chemistry, University of St Andrews, Fife KY16 9ST, UK
e-mail: Jdw3@st-and.ac.uk

J.D. Woollins and R.S. Laitinen (eds.), *Selenium and Tellurium Chemistry*,
DOI 10.1007/978-3-642-20699-3_13,

replaced a basal phosphorus atom and is co-ordinated by three phosphorus atoms from the triphos and by one phosphorus and two sulfur atoms from the P_3S_3 unit

The (formally tri-anionic) cyclic fragments, P_2S, P_2Se and As_2Se, may also be stabilised by co-ordination Thus, reaction of $[Co(H_2O)_6]^{2+}$ with firstly triphos and then E_4X_3 (E = P, X = S or Se; E = As, X = Se) gives $[Co(triphos)(As_2X)]^+$. Several binuclear compounds containing bridging E_2X have been prepared in a fashion (Eq. 13.1) analogous to the preparation of the homoatomic P_3 compounds.

$$Co(BF_4)2.6H_2O + triphos + P_4X_3 \rightarrow [(triphos)Co(EX)[BF_4]] \quad (13.1)$$

Treatment of $[Co(triphos)(E_2S)]^+$ with $[Pt(C_2H_4)(PPh_3)_2]$ gives a rare example of a mixed cobalt-platinum species, $[(triphos)Co(E_2S)Pt(PPh_3)_2]^+$. The X-ray structure of the As_2S compound reveals that the $Pt(PPh_3)_2$ group has added across, and cleaved, a sulfur-arsenic bond but all three atoms of the As_2S ring remain co-ordinated to the cobalt

Kolis and others extended the above systems elegantly [3]. Thus they showed that $[WSe]^{2-}$ readily attacks P_4Se_4 glass to form the first metal phosphorus selenide complex (Eq. 13.2).

$$[PPh_4]_2[WSe_4] + P_4Se_4 \rightarrow [PPh_4]_2[Se = W(PSe_4)(PSe_2)] \quad (13.2)$$

They also found that many of the anionic clusters, particularly those whose salts are soluble in polar organic solvents, will react with metal carbonyls to form transition metal main group cage complexes. Most of these complexes are the result of oxidation of the metal center with concomitant reduction of the main group cage. For example, $[P_2Se_8]^{2-}$ reacts with iron pentacarbonyl to form a dimer containing a bridging $[PSe_5]^{3-}$ unit (Eq. 13.3) [3].

$$3Fe(CO)_5 + [P_2Se_8]^{2-} + PPh_4Br \rightarrow [PPh_4]_2[Fe_2(CO)_4(PSe_5)_2] \quad (13.3)$$

13.3 Phosphorodiselenoate and Phosphinodiselenoate Complexes

The synthesis, chemistry and structural behaviour of the complexes of phosphodithioate ligands is plentiful and widely discussed [4, 5], and there have been a few examples of the unsymmetrical phosphonodithioate complexes reported [6–9]. Symmetrical phosphorodiselenoate and phosphinodiselenoate ligands form metal complexes with alkali metals, transition and non-transition metals and have also been reported to be used in the formation of large cubic clusters.

We synthesised four examples of metal complexes directly from Woollins' Reagent **WR** [10] by reaction with *cis*-$PtCl_2(PR_3)_2$ (PR_3 = ½ dppe, PEt_3, PMe_2Ph and PPh_2Me) in THF to give *cis*-$Pt(Se_3PPh)(PR_3)_2$ in *ca* 70% yield, the first reported examples of molecules containing a $PtSe_2P$ ring (Eq. 13.4).

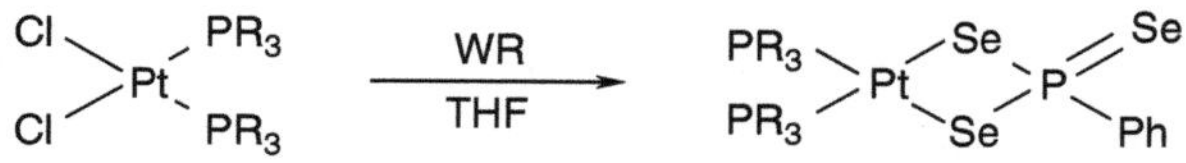

(13.4)

Kuchen and Knop [11, 12] reported the synthesis of the sodium salts of dialkylphosphinic acids, as well as the acids and ester derivatives, as long ago as 1964. They found that the sodium salt $Na[Et_2PSe_2]$ could be prepared by firstly reacting a P(III) chlorophosphine with selenium to form the P(V) chlorophosphine selenide then treating this with NaSeH (prepared from a stoichiometric amount of hydrogen selenide and sodium ethoxide in an ethanol solution, Eq. 13.5).

$$Et_2PCl + Se \rightarrow Et_2P(Se)Cl \rightarrow Na[Et_2PSe_2] + H_2Se \qquad (13.5)$$

$Na[Et_2PSe_2]$ was reported to react with hydrochloric acid to form the diethyl diselenophosphinic acid which can readily decomposed to form bis(diethylseleno-phosphoryl) selenide and the corresponding triselenide. This sodium salt was also reported to form Zn(II), Cd(II), Pb(II), Pd(II), Rh(III), Bi(III), In(III) and Tl(I) complexes but unfortunately very little experimental data was made available and no spectroscopic data was recorded [11, 12]. This work was further extended by Kuchen and Hertel to include the IR and electronic spectra for these compounds [13].

In 1991, we [14] reported the synthesis of a sodium complex of the diphenyl phosphinodiselenoate ligand, $Na_2[Ph_2PSe_2]_2{\cdot}THF{\cdot}5H_2O$ which exists as a polymeric structure, with a central polymeric core built up of $Na(H_2O)_6$ and Na $(H_2O)_3$(thf)(Se) units with additional hydrogen-bonded $[Ph_2PSe_2]^-$ anions.

A more in-depth study of alkali metal complexes of such phosphinodiselenoate and related selenophosphorus ligands was conducted by Davies and co-workers [15–17]. They reported the first crystallographic study of solvated lithium salts of a selenophosphinite anion, diselenophosphinoate anion and a triselenophosphonate dianion , as well as full structural characterisation of these compounds in solution using multinuclear and variable-temperature NMR spectroscopic studies [16]. Davies extended the synthetic procedure in Eq. 13.6 to the reaction of 3 equivalents of elemental selenium with a dilithiated primary phosphine, $CyPLi_2$ (Cy = c-C_6H_5). Two Se atoms inserted into the P-Li bonds and the third Se atom oxidised the phosphorus (III) to phosphorus (V) to yield the dilithium triselenophosphonate complex. Using an analogous synthetic route, Davies and co-workers were also able to prepare tellurium analogues [16].

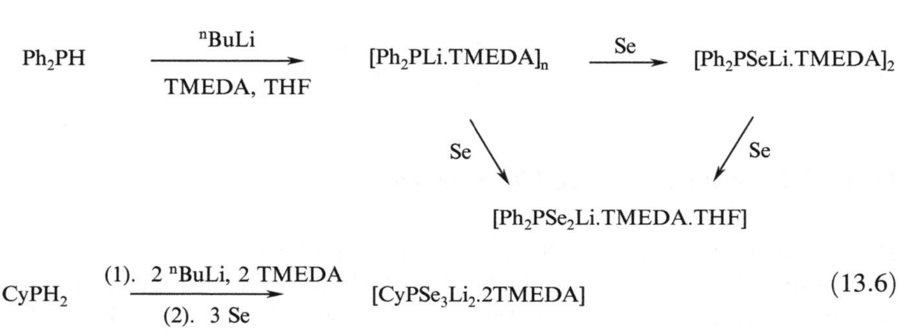

(13.6)

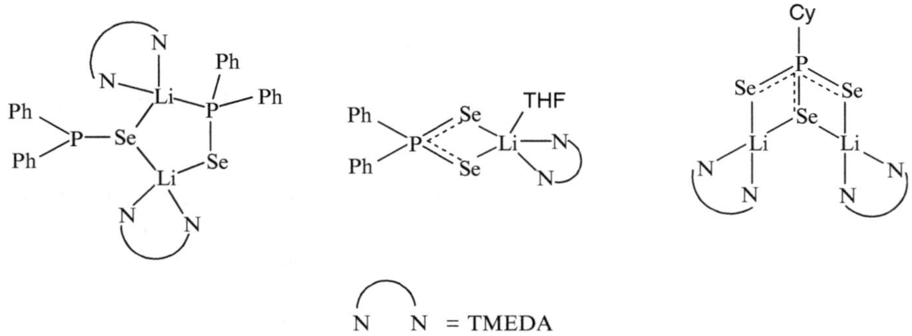

Fig. 13.1 The structures of selenophosphinite anion, diselenophosphinoate anion and a triselenophosphonate dianions prepared by Davies [16]

Davies et al also reported [17] the reaction of potassium diphenylphosphide with selenium which gives the potassium salt $[K(Se_2PPh_2)(THF)_2]_2$, this reacts further with $InCl_3$ to yield $[In(Se_2PPh_2)_3]$. Both of these compounds exist as linked dimers in the solid state, held together by Se···Se intermolecular van der Waals interactions (Fig. 13.1).

The coordination chemistry of phosphinodiselenoate ligands has also been extended to other metals, Gastaldi and Porta [18] reported the X-ray structure of bis(diphenyldiselenophosphinato) nickel(II) showing it to display a similar structure to known dithiophosphinato nickel(II) complexes. The nickel atom resides on a crystallographic inversion centre and is coordinated by four selenium atoms in a square planar geometry with symmetric $NiSe_2P$ rings.

Gelmini and Stephan reported the synthesis of a diphenyldiselenophosphinato complex of titanium [19] $Cp_2TiSe_2PPh_2$ from the reaction of Cp_2TiCl_2 in THF with 2 equivalents of $LiSePPh_2$ firstly forming the intermediate '$Cp_2TiSePPh_2$', which upon standing in THF/hexane undergoes complete conversion to $Cp_2TiSe_2PPh_2$ within 24 h (Eq. 13.7).

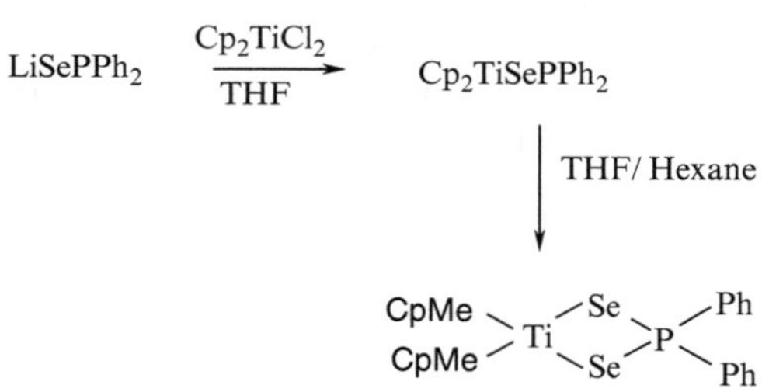

(13.7)

13.4 Diselenophosphate Complexes

The coordination chemistry of diselenophosphates [20] has been extensively studied by Liu and coworkers. Here, molecules which have not been included in the previous review [21] are reported. These cover diselenophosphates of Mn, Re, Fe,

Ru, Ag, Au, and Pb compounds. In addition, hetero-metallic clusters containing $[P(Se)(O^iPr)_2]^-$, the conjugated base of secondary phosphine selenide, as the bridging ligand with P-bonded to iron and Se-bonded to hetero-metals such as mercury cadmium, copper, and silver are described.

The first report of metal complexes containing diselenophosphate $[\{Se_2P(OR)_2\}]^-$ ligands (abbr. as dsep) was published by Zingaro et al. in 1968 [22]. More recent innovations for the use of dsep ligands in coordination chemistry lies in the synthesis of multi-metallic cluster compounds [21]. Surprisingly no structure of free dialkyl diselenophosphate ligands $[(RO)_2PSe_2]^-$ or their salts is known mainly because of their thermally unstable nature. In addition, they are highly susceptible to aerial oxidation and thus Ibers and his co-workers were able to achieve the structural characterization of two oxidized forms of the dsep ligand, $[(EtO)_2P(Se)Se]_2Se$ and $[(^iPrO)_2P(Se)Se]_2$ [23].

After tedious re-crystallizations, the isopropyl derivative of ammonium salt $(NH_4)[Se_2P(O^iPr)_2]$, **1**, was unequivocally corroborated by single crystal X-ray diffraction analysis in 2009 [24]. As expected the P atom is attached to two Se and two O atoms in a distorted tetrahedral geometry. The ammonium cation forms H-bonds with four selenium atoms from three, adjacent anionic $[Se_2P(O^iPr)_2]^-$ units to form a 1D supramolecular array along the *a* axis (Fig. 13.2).

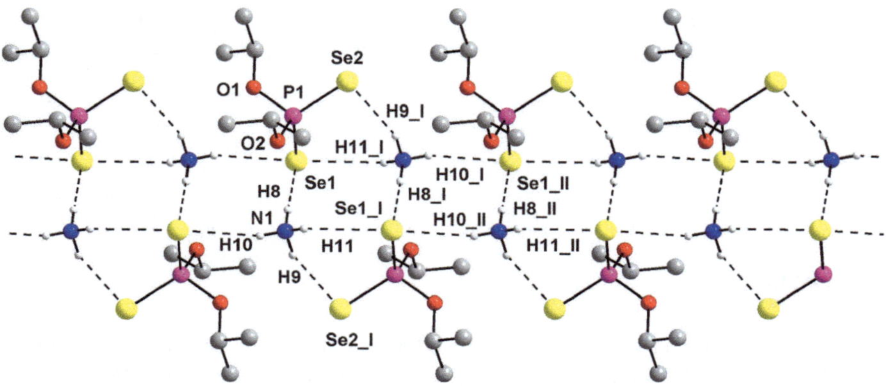

Fig. 13.2 H-bonded supramolecular array of cations and anions in **1** along the *a* axis

Mononuclear manganese complexes $Mn(CO)_4[Se_2P(OR)_2]$ (R = ^iPr, **2a**; Et, **2b**) are synthesized from either $[-Se(Se)P(O^iPr)_2]_2$ (**A**) or $[Se\{-Se(Se)P(OEt)_2\}_2]$ (**B**) with equimolar amount of $Mn(CO)_5Br$ in acetone at room temperature for 24h (Scheme 13.1) [25]. However the corresponding Re complexes $Re(CO)_4[Se_2P(OR)_2]$ (R = ^iPr, **3a**; Et,**3b**) are obtained by refluxing either $[-Se(Se)P(O^iPr)_2]_2$ or $[Se\{-Se(Se)P(OEt)_2\}_2]$ with $Re(CO)_5Br$ in a 1:2 M ratio in toluene for an hour. The higher temperature is required for the generation of **3a-b**, compared to that of **2a-b**, attributable to the higher stability of $Re(CO)_5Br$ over its Mn analogue. Dinuclear compounds $Mn_2(CO)_6[Se_2P(OR)_2]_2$ (R = ^iPr, **4a**; Et, **4b**) could be generated by refluxing mononuclear Mn complexes **2a-b** in acetonitrile for an hour. Both di-rhenium complexes $Re_2(CO)_6[Se_2P(OR)_2]_2$ (R = ^{i}Pr, **5a**; Et, **5b**) can also be obtained (Scheme 13.1) by refluxing the corresponding mononuclear rhenium compound (**3a-b**) in acetonitrile for 1 h.

Efforts have been made to yield the mononuclear complexes **2** and **3** directly by stirring or refluxing $M(CO)_5Br$ (M = Mn, Re) with dsep ligands in solution, but to no avail. Apparently the reduction of the ligands **A** and **B** that yields dsep ligand during the complexation process plays a crucial role in knocking out a CO from the starting metal ion complex. Most probably the reduction of the ligand takes place at the cost of oxidation of CO, which eventually was released from the coordination sphere of metal center.

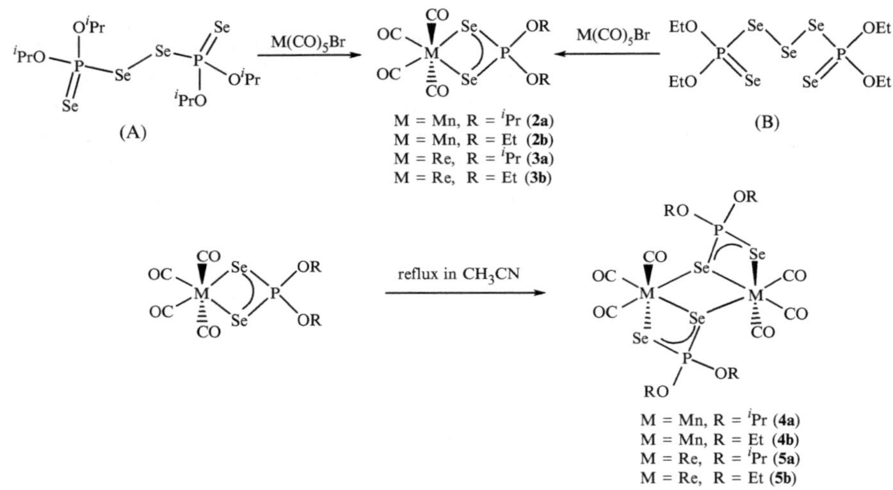

Scheme 13.1

It has been known from the literature that the oxidative cleavage of $[Cp^*Fe(CO)_2]_2$ with $(RO)_2P(S)SSP(S)(OR)_2$ (Cp^* = C_5H_5, C_5H_4Me, C_5Me_5; R = Et, ^{i}Pr) in cyclohexane at 50–60°C can produce complexes of the type, $Cp^*Fe(CO)_2[\eta^1\text{-}SP(S)(OR)_2]$, in 70–88% yield [26]. Utilizing the same methodology, its selenium analogue, CpFe $(CO)_2[\eta^1\text{-}SeP(Se)(O^iPr)_2]$, **6** (Fig. 13.3a), can be prepared in 86% yield even at ambient temperature (Scheme 13.2). These two reactions suggest that the cleavage

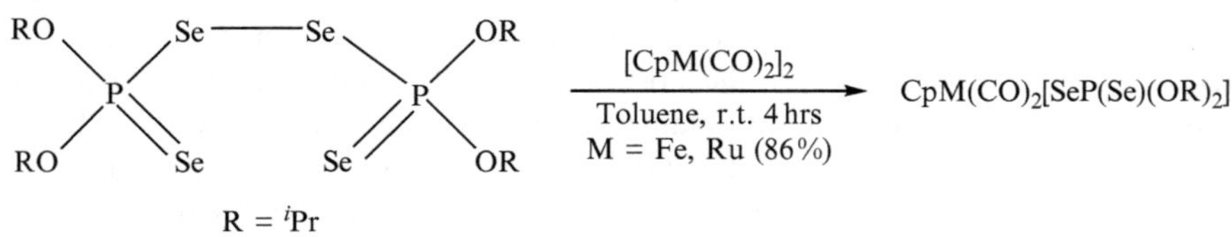

Scheme 13.2

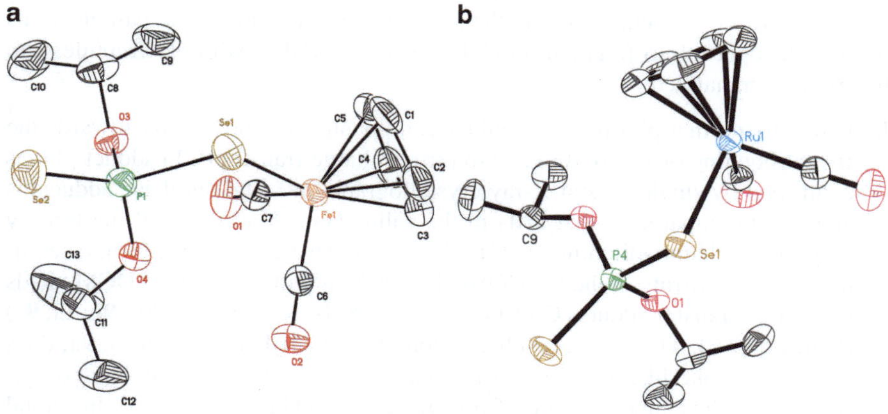

Fig. 13.3 (**a**) Thermal ellipsoid drawing of **6**, and (**b**) thermal ellipsoid drawing of **7**. H atoms were omitted for clarity

of Se-Se bond is easier than that of S-S bond [7]. The phosphor-1, 1-diselenoato compound of ruthenium, $CpRu(CO)_2[\eta^1\text{-}SeP(Se)(O^iPr)_2]$, **7** (Fig. 13.3b), can also be synthesized in a similar fashion (Liu CW Unpublished results).

When the reaction of $[CpFe(CO)_2]_2$ with dialkyl diselenophosphate was performed in refluxing toluene for 4 h, surprisingly, neutral selenophosphito-iron compounds, $CpFe(CO)_2P(Se)(OR)_2$ **8** (R = ^{i}Pr, Pr, Et), could be isolated [27, 28]. Presumably the selenophosphito moiety, a conjugated base of secondary phosphine selenide, is generated via the cleavage of the P-Se bond of dsep by iron carbonyl radicals. The structure of isopropyl derivative depicted in Fig. 13.4 clearly suggests that the Se center is relatively open to interact with various types of electrophilic reagents.

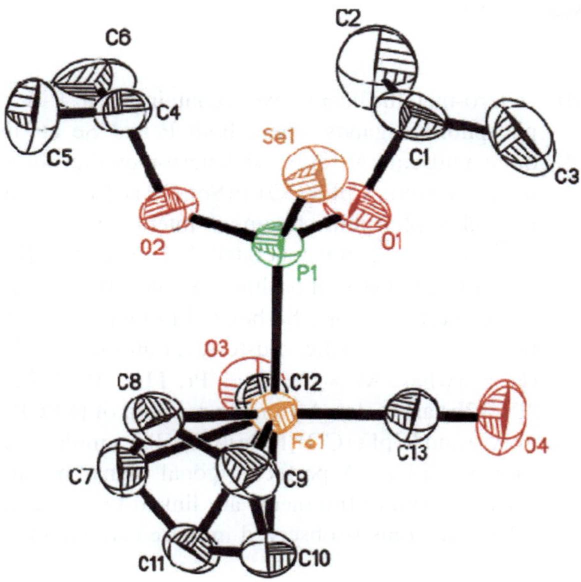

Fig. 13.4 The X-ray structure of complex **8a**. H atoms were omitted for clarity

Notable reactions benefited from the activation of selenium's lone-pair electrons of the selenophosphito fragment in $CpFe(CO)_2P(Se)(OR)_2$ with electrophiles can be briefly summarized:

I. It was found that phosphine selenides can donate electron density towards the σ* antibonding orbital of diiodine to form a charge-transfer (CT) adduct [29] as confirmed by single crystal X-ray crystallography as *d*(I–I) in these adducts is elongated compared to that in I_2 in the solid state. In addition all the tertiary phosphine selenide-diiodine 1:1 CT adducts known are spoke like [30]. Accordingly simple stirring of the $Cp(CO)_2FeP(Se)(OR)_2$ with I_2 at 0°C in DCM yields the charge-transfer adducts $Cp(CO)_2FeP(OR)_2SeI_2$ (R = ^{i}Pr, **9a**; ^{n}Pr, **9b**; Et, **9c**) (Scheme 13.3). The structural elucidation of $Cp(CO)_2FeP(O^iPr)_2SeI_2$, **9a**, does exhibit a spoke-like charge-transfer adduct of diiodine on the organoiron-substituted phosphine selenide, $[Cp(CO)_2FeP(Se)(O^iPr)_2]$ (Fig. 13.5). In crystal structure of **9a** the I–I distance is 2.974 Å, which indeed is longer than 2.772 Å in free I_2 [31]. The Se–I–I linkage is essentially linear (172.272(17)°) in **9a**.
II. Organoiron-substituted phosphine selenides $Cp(CO)_2FeP(Se)(OR)_2$ (R = ^{i}Pr (**8a**), ^{n}Pr (**8b**), Et (**8c**)) were capable of reacting as a nucleophile towards Me_3OBF_4 to formed Se-methylated products (**10a–c,** Scheme 13.3). The molecular structure of **10b** is characterized by X-ray diffraction (Fig. 13.5) [9].

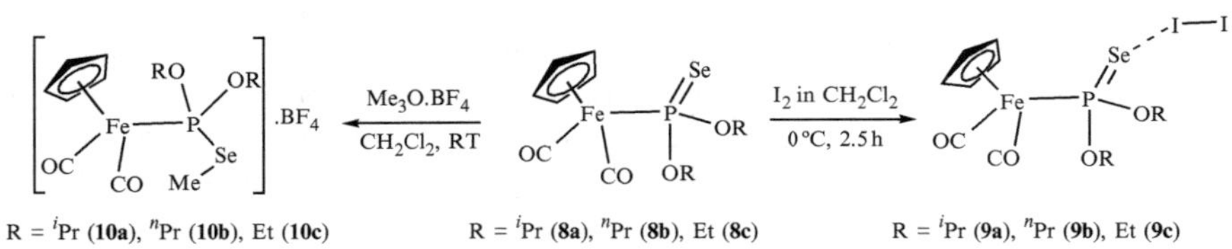

Scheme 13.3

III. Hetero-metallic complexes containing a P-Se fragment of organoselenophosphorus ligands where both P and Se atoms bridged metal centers are quite unusual. A series of hetero-metallic cluster compounds based on the metalloligand, $CpFe(CO)_2P(Se)(OR)_2$ [8], where the lone-pair electrons of the selenophosphito fragment interact with Lewis acids such as Cu^I, Ag^I, Cd^{II}, and Hg^{II}, are generated [32, 33]. The $[P(Se)(O^iPr)_2]^-$, the conjugate base of secondary phosphine selenide $[HP(Se)(O^iPr)_2]$, acts as a bridge with P-bonded to iron and Se-bonded to mercury (cadmium, copper, and silver) in these hetero-metallic clusters. Four clusters $[M\{CpFe(CO)_2P(Se)(OR)_2\}_3]$ (PF_6), (where M = Cu, R = ^{i}Pr, **11a,** ^{n}Pr, **11b**, and M = Ag, R = ^{i}Pr, **12a**, ^{n}Pr, **12b**) are isolated from the reaction of $[M(CH_3CN)_4(PF_6)]$ (where M = Cu or Ag) and $[CpFe(CO)_2P(Se)(OR)_2]$ in a molar ratio of 1:3 in acetonitrile at 0°C (Scheme 13.4). A perfect trigonal planar metal center in which three iron-selenophosphito fragments are linked to the central copper or silver atom via selenium atoms is observed in these cationic clusters. Besides, the reaction of

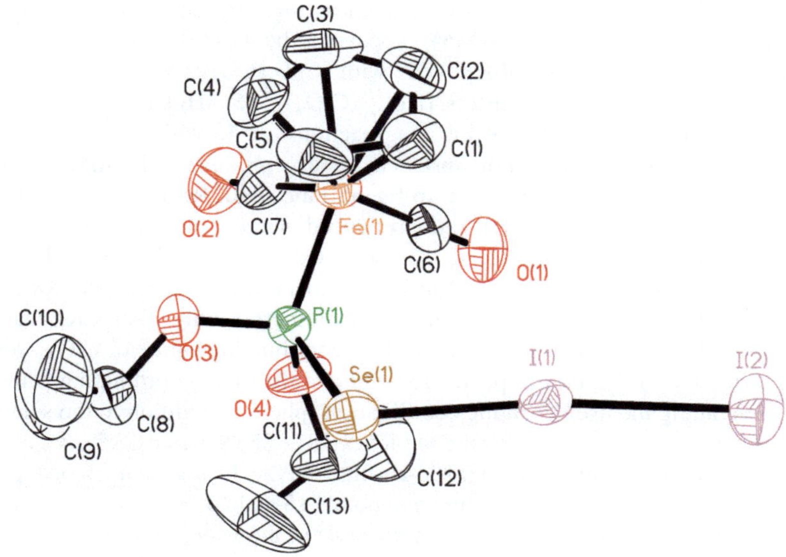

Fig. 13.5 The X-ray structure of complex **9a**. H atoms were omitted for clarity

$[CpFe(CO)_2P(Se)(O^iPr)_2]$ with cuprous halides in acetone produce two mixed-metal, $Cu_2^IFe_2^{II}$ clusters, $[Cu(\mu\text{-}X)\{CpFe(CO)_2P(Se)(O^iPr)_2\}]_2$ (X = Cl, **13**; Br, **14**) (Scheme 13.4). Each copper center in neutral heteronuclear clusters **13** and **14** is also trigonally coordinated to two halide ions and a selenium atom from selenophosphito-iron moiety. The structure can also be delineated as a dimeric unit which is generated by an inversion center and has a Cu_2X_2 parallelogram core.

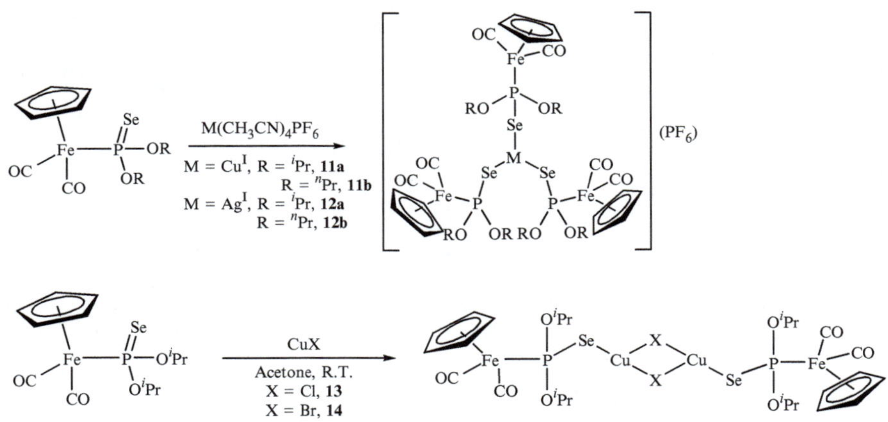

Scheme 13.4

The reactivity of group 12 elements such as mercury and cadmium with the metalloligand, $CpFe(CO)_2P(Se)(O^iPr)_2$, appears to be more diverse [27, 33]. Thus reactions of perchlorate salts of Hg^{II}/Cd^{II} with $[CpFe(CO)_2P(Se)(O^iPr)_2]$ (denoted as **L**), produced di-cationic clusters $[HgL_2](ClO_4)_2$, **15**, $[HgL_3](ClO_4)_2$, **16**, and $[CdL_3(H_2O)](ClO_4)_2$, **17**. However, the reactions of **L** with Hg^{II}/Cd^{II} halides yielded neutral complexes. For instances, HgI_2 produced $[Hg_3I_6L_2]$, **18**, or $[HgI_2L_2]$, **19** depending on the metal to ligand ratio used. Reaction of **L** with any of the Hg^{II}/Cd^{II} halide in a 1:1 ratio produced neutral clusters $[HgX(\mu\text{-}X)L]_2$ (M = Hg; X = Cl, **20**; Br, **21**; I, **22**. M = Cd; X = Cl, **23**; Br, **24**; I, **25**) (Scheme 13.5). Thus variation of M to L ratio made it possible for the formation of compounds with different metal/ligand stoichiometries only when ClO_4^- and I^- salts of Hg(II) were used [14]. Clearly the less-coordinating nature of ClO_4^- results in the formation of dicationic species **15** ~ **17** and the more coordinating nature of halides engaging themselves in the coordination sphere in addition to the selenium donor ligand, L, to produce neutral complexes **18** ~ **25** (Scheme 13.5).

Two-coordinated mercury center [Se-Hg-Se 166.95(7)°] and near trigonal planar geometry of mercury are observed in compounds **15** and **16**, respectively. In **17** the three-coordinated cadmium seems to be unstable as the water molecule is attached to Cd in the apical position besides three **L**. The Hg_3 cluster **18** has each terminal Hg in an extremely distorted trigonal geometry and slightly distorted tetrahedral for the central Hg atom. Whereas the coordination sphere of the terminal Hg atoms is occupied by two iodo groups and one **L** unit, the central Hg is surrounded by four

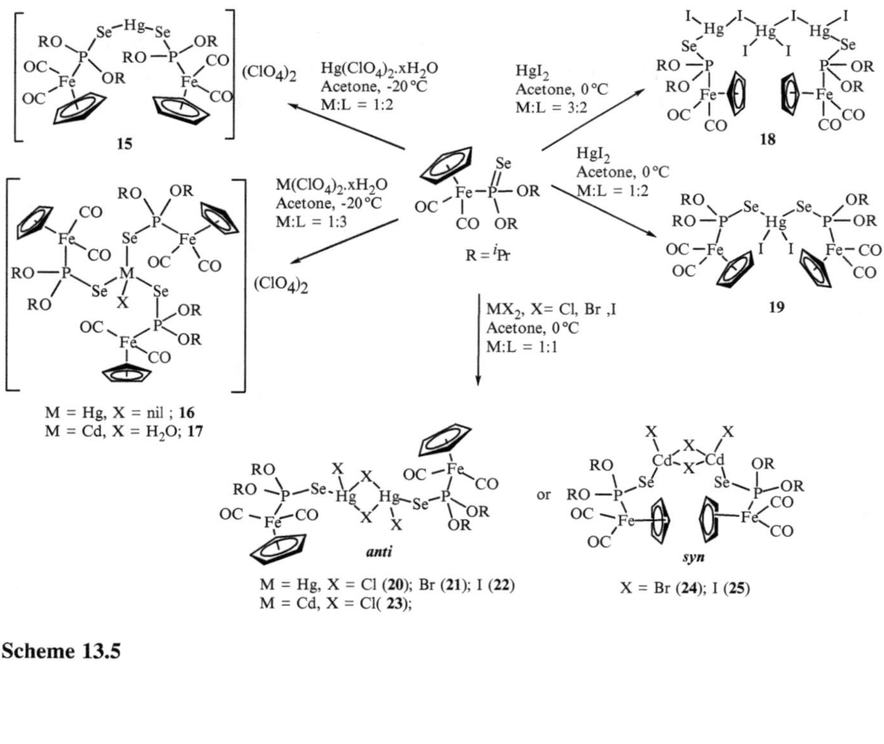

Scheme 13.5

iodo ligands. The Hg^{II} in **19** was connected to two iodo and two **L** units in distorted tetrahedral. A metalloligand, **L** (through its Se), and a terminal halogen were attached to each of the Hg^{II} of the Hg_2X_2 parallelogram core in **20** and **21**. The cadmium complex **23**, with a Cd_2Cl_2 parallelogram core, was iso-structural with the mercury complex **20**. Although bonding and connectivity in **23** was similar to those in **24** and **25**, the conformation of the bulky **L** ligands was unusually *syn* in **24** and **25** unlike *anti* orientation in **20**, **21** and **23**.

While the reaction of copper(I) salts with dsep ligands in various molar ratios tend to form cluster compounds [21] some polymeric species of silver diselenophosphate appear to crystallize quite easily. For example, the reaction of equal molar ratio of $Ag(CH_3CN)_4(PF_6)$ with $NH_4Se_2P(OEt)_2$ yields a polymer, $[AgSe_2P(OEt)_2]_n$ **26** (Fig. 13.6). The 4:3 M ratio reaction in THF always produces pentanuclear extended chain polymers, $[Ag_5\{Se_2P(OEt)_2\}]_n(PF_6)_n$ **27**. Its structural elucidation reveals that each repeating unit in the 1D cationic polymeric chain consists of five silver atoms in which four constitute a tetrahedron and the fifth silver atom, acting as a bridge, links two Ag4 tetrahedra via four Se atoms of the neighboring dsep ligands (Fig. 13.7) [27]. Subsequently a hydride-centered octanuclear silver cluster formulated as $Ag_8(H)[Se_2P(OEt)_2]_6^+$ **28** [34],

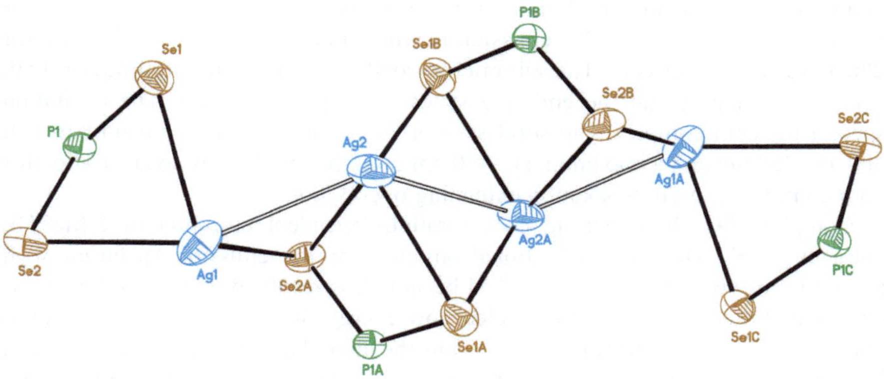

Fig. 13.6 A perspective view of **26** with ethoxy groups omitted for clarity

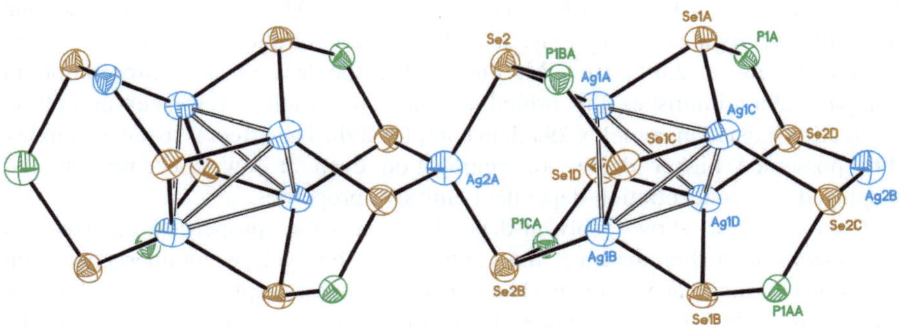

Fig. 13.7 A perspective view of **27** with ethoxy groups omitted for clarity

can be generated in high yield from the reaction of 1D chain with borohydrides. Apparently it is formed via anion template approaches and a variety of anion-encapsulated silver clusters can be anticipated to form via different shapes of anion [35].

Gold compounds containing dsep ligands were virtually unknown until this group reported the dimetallic Au(I) complexes, $[AuSe_2P(OR)_2]_2$ (R = ^{i}Pr, **29a**; Et, **29b**), obtained as yellow powders from the reaction of AuCl(tht) with stoichiometric amount of $NH_4[Se_2P(OR)_2]$ over a course of 4h in THF at −50°C under nitrogen atmosphere [16]. It took more than 12 years' effort to grow single crystals appropriate for X-ray diffraction. In solid state, both compounds are built into an one-dimensional chain based on the near co-linear alignment of dimetallacycle entities, $[AuSe_2P(OR)_2]_2$, with shorter intra- and slightly longer inter Au-Au distances. The gold centers in a $[AuSe_2P(OR)_2]_2$ basic unit are doubly bridged by two selenium atoms of dsep ligands, resulting in a puckered eight-membered ring with a short transannular Au-Au interaction (Fig. 13.8). Whereas Au—Au—Au angles of 171.4(3) and 174.1(3)° are revealed in **29a**, the packing in **29b** displays a strict linear Au—Au—Au chain.

Intriguingly both compounds **29a** and **29b** display photoluminescence and their photophysical data are summarized in Table 13.1. Both compounds exhibit weak orange emission in the solid state at room temperature and the emission color becomes intense at 77 K. The emission maxima appeared at 580 and 575 nm for **29a** and **29b**, respectively. The sub-microsecond to microsecond lifetimes for both complexes suggest that the emissive states are triplet in nature. The excitation spectra for both complexes in solid state at 77 K showed a broad band between 300 and 450 nm and a maximum at ~470 nm even though there is no absorption for both complexes in dilute solution extending over 370 nm.

Complex **29a** shows strong concentration-dependent emission in 2-MeTHF glass at 77 K. The clear vibrational structure of the emission spectrum with maximum of 466 nm at 3.3×10^{-5} M became less resolved with emission maximum red shifted to 540 nm with a clear low energy shoulder at 625 nm tailing to 700 nm at 3.3×10^{-3} M (Fig. 13.9a). Thus the aggregate structures may appear in high concentration glass due to the strong aurophilic interaction. On the other hand, complex **29b** displayed less concentration dependent emission properties compared to complex **29a** in 2-MeTHF glass at 77 K. The emission barely shifted from 620 nm at 2.0×10^{-5} M to 650 nm at 7.4×10^{-3} M, which indicates notable molecular aggregates already exist at 77 K glass for complex **29b** even at concentration as low as 2.0×10^{-5} M. Presumably the less linear conformation in complex **29a** diminishes the orbital overlap and results in weaker aurophilic interaction existed in complex **29a** than complex **29b.** The effect is to have complex **29b** possessed higher degree of aggregation even at low concentration and displayed less concentration-dependent emissive properties.

Complex **29a** shows solvent-dependent emission properties as well as thermochromism. For instance, the complex exhibits yellow luminescence with emission maximum at 565 nm in THF, whereas emission displays orange color with maximum at 595 nm in acetone. Interestingly, in dichloromethane glass, the complex **29a** (~10^{-2} M) shows vivid thermochromism with a color change from yellow (570 nm) to green (558 nm) upon increasing the temperature from 77 to 177 K (Fig. 13.9b).

Table 13.1 Photophysical data for complexes **29a** and **29b** at 77 K

	Excitation spectra		Emission spectra	
Compound	Medium	λ_{max}, nm	λ_{em}, nm	τ, μs
	Solid	470	580	0.17
		294, 311 (3.3×10^{-5} M)	464	0.15, 3.0
		341, 412 (5.4×10^{-4} M)	509, 540	1.1, 8.3
29a	Glass[a]	440 (3.3×10^{-3} M)	500, 540, 625	0.3, 8.6, 32.4
	Solid	388, 467	575	3.7, 12.5
		350 (2.0×10^{-5} M)	630	0.16, 15.3
		361, 390 (4.0×10^{-4} M)	650	8.4, 23.1
29b	Glass[a]	361, 418 (7.4×10^{-3} M)	650	9.1, 22.3

[a]Measured in 2-MeTHF

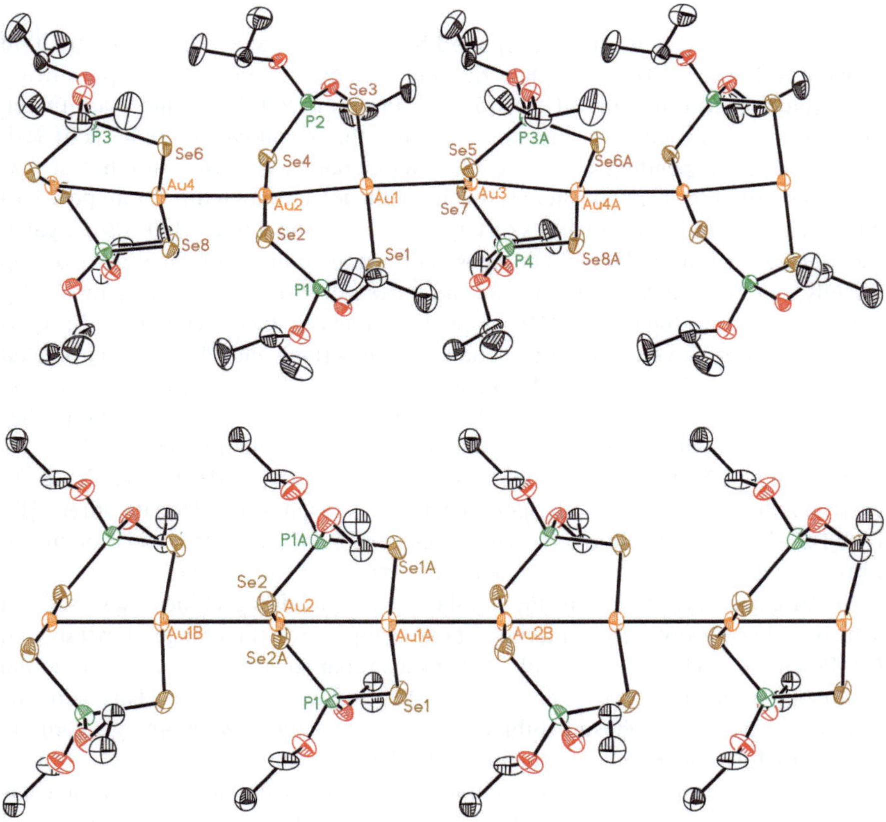

Fig. 13.8 Perspective views of **29a** (*top*) and **29b** (*bottom*)

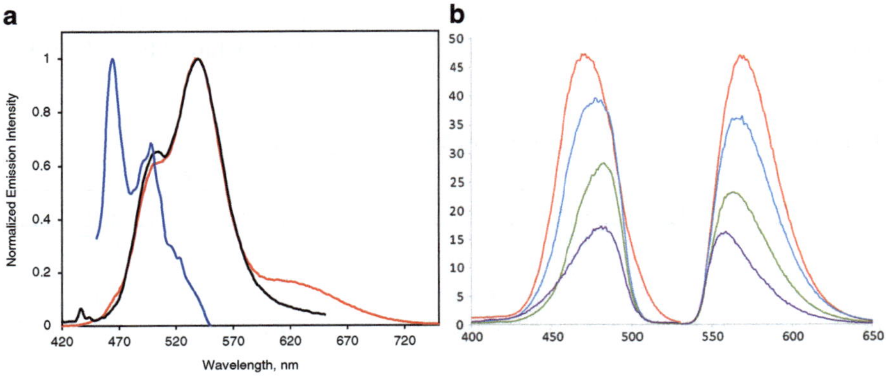

Fig. 13.9 (**a**) Normalized emission spectra of complex **29a** in 2-MeTHF glass at 77 K. *Blue curve* (3.3×10^{-5} M), *black curve* (5.1×10^{-4} M), and *red curve* (3.4×10^{-3} M). (**b**) Excitation (*left*) and emission (*right*) spectra of **29a** in 0.01 M CH_2Cl_2: *red*, 77 K; *blue*, 137 K; *green*, 157 K; *purple*, 177 K

$Pb\{Se_2P(OEt)_2\}_2$ **30** was first reported by Zingaro et. al. [22], but the structure was not well characterized in either the solid state or solution. Thus simple stirring of a stoichiometric mixture of $NH_4[Se_2P(O^iPr)_2]$ and $Pb(OAc)_2$ under a nitrogen atmosphere produced two compounds with the chemical formula $[Pb\{Se_2P(O^iPr)_2\}_2]_n$ **31** depending on the reaction temperature. Reaction in ether at 0°C produces **31α** (Fig. 13.10a) whereas that in methanol at room temperature produces **31β** (Fig. 13.10b). Two structures differed in the binding modes of the dsep ligand. Each repeating unit in **31α** was composed of a lead atom coordinated by two dsep ligands, one in a chelating mode and the other in a bridging-dangling mode. By contrast, the dsep ligands in **31β** adopted a bimetallic-biconnective (μ_1-S, μ_1-S) binding pattern. Several Se…Se secondary interactions and Pb…Se non-bonded interactions co-exist in both polymorphs. $[Pb\{Se_2P(O^iPr)_2\}_2]_n$ could be successfully utilized as a single source precursor (SSP) for growing lead selenide (PbSe) nano-structures with different morphologies via the solvothermal process [36].

We found [37] that the reaction of WR with NaOR (R = Me, Et and ^{i}Pr) in the corresponding alcohol gives the non-symmetric phosphonodiselenoate anions $[Ph(RO)PSe_2]^-$ as their sodium salts. The structure of a product of the oxidation of one of these salts gave an interesting cubane like structure (Fig. 13.11).

In an analogous manner to the similar sulfur containing anions these sodium salts have been shown to react to give stable complexes with a range of metals (Ni, Cd, Pb and Sn). The nickel complex adopts a similar square planar structure to the analogous sulfur complexes. The structure of a cadmium complex is also analogous to those displayed by phosphodithioate cadmium compounds, adopting a dimeric M_2L_4 structure containing an 8-membered $Cd_2P_2Se_4$ ring.

Two distinctly different lead complexes have been obtained, one which has an OEt substituent (Fig. 13.12) consists of dimeric pairs similar to those displayed for the lead phosphonodithioate structures, whilst the other structure containing

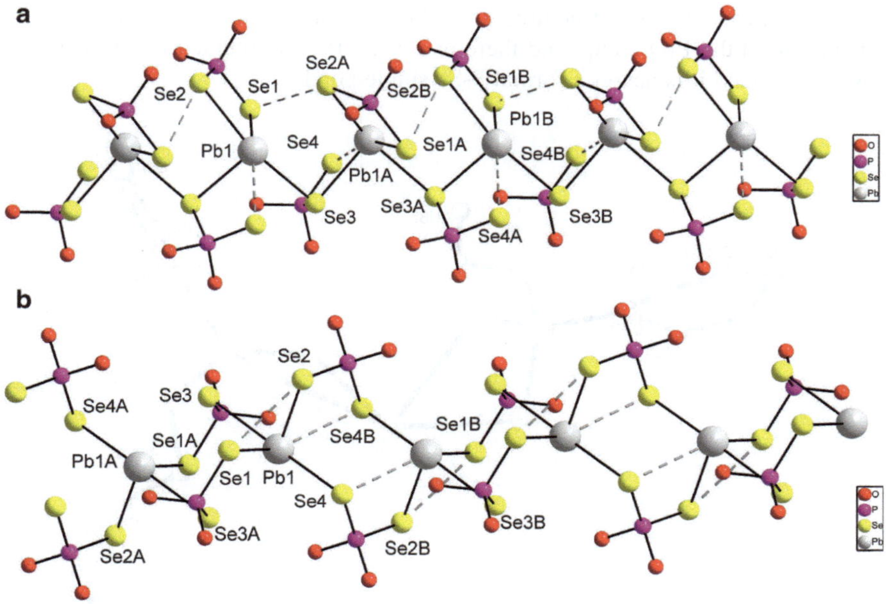

Fig. 13.10 Perspective views of (**a**) **31α** and (**b**) **31β** with isopropyl groups omitted for clarity. *Dotted lines* show the non-bonding interactions

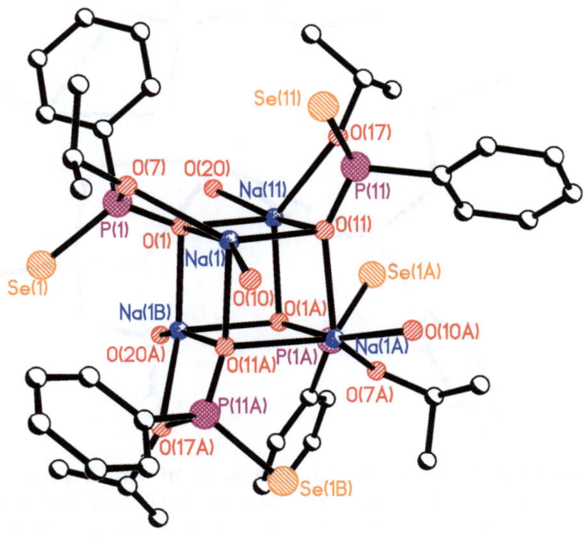

Fig. 13.11 The X-ray structure of [NaPh(RO)PSe_2].(iPrOH)

an OMe substituent adopts, what appears to be, a completely different structural motif - the complex displays a novel dimeric ML_4 structure built around a central 4-membered P_2Se_2 ring. (Fig. 13.13) Comparison of Fig. 13.12 and 13.13 enables us to recognise that the structures are related by a simple change in the P-Se separation, in the EtO compound there is a weak P. . .Se interaction whilst in the OMe compound it is nearer a formal P-Se single bond.

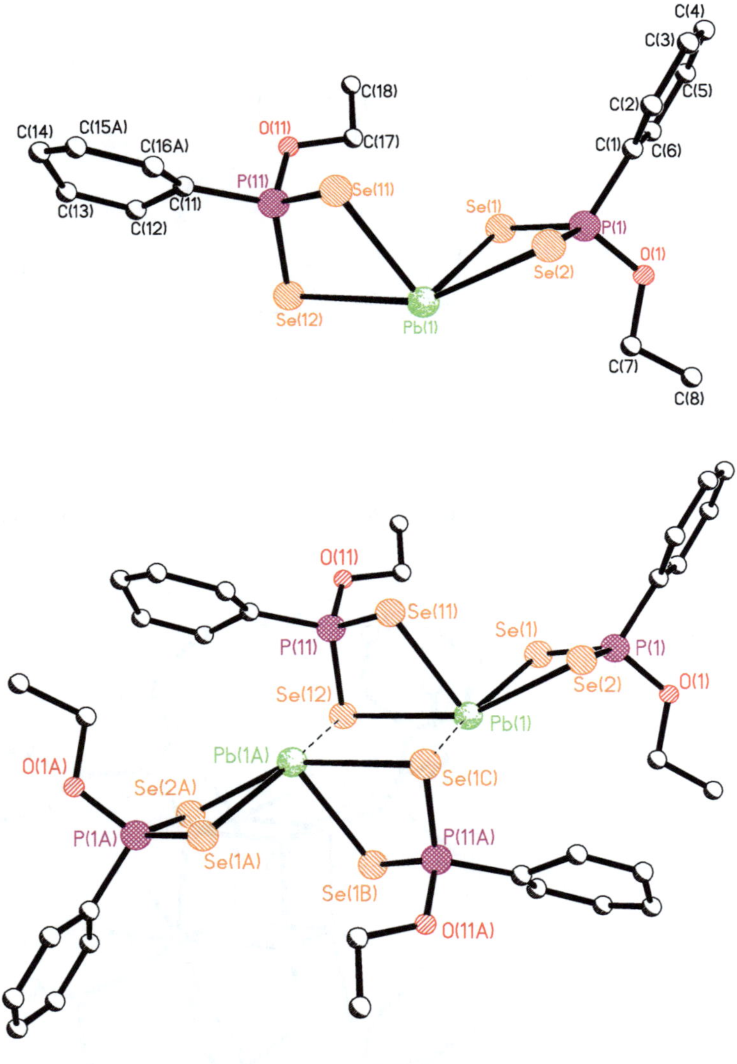

Fig. 13.12 The X-ray structure of $Pb(Se_2P(OEt)Ph)_2$. *Upper diagram*, single monomeric unit with a "vacant" site at Pb; *lower diagram*, pair of units forming dimeric structure. C(15) and C(16) are disordered and the figure only shows one orientation [C(15a) and C(16a)]

Rothenberger and co-workers have used **WR** and **LR** in a range of elegant syntheses to prepare new complexes and clusters which have been reviewed elsewhere [38]. but can be summarised here. Thus reactions [39] of Cu(I) thiolates

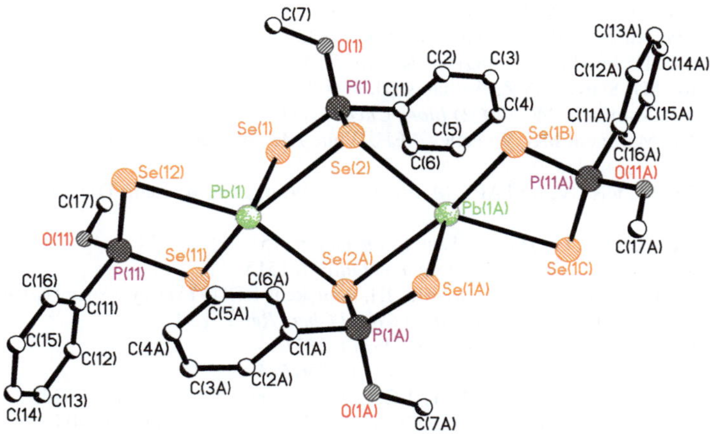

Fig. 13.13 The X-ray structure of $Pb(Se_2P(OMe)Ph)_2$

and **WR** gives copper complexes with unusual P/Se anions such as $[PhP(Se)StBu]^-$ and $[PhSe_2PPSePh]^{2-}$. Reactions [40] of **WR** with alkali-metal proceeds to give polymeric systems containing $[PhPSe_3]^{2-}$ or $[PhPSe_2SeSeSe_2PPh]^{2-}$.

13.5 Conclusion

It is clear that replacement of sulfur by selenium in P-E ligands has a profound effect on the chemistry. The diversity of new systems containing P-Se ligands is remarkable and likely to continue to expand in the next few years. We also anticipate that some of the new M-P-Se systems will find application in the synthesis of new materials. We look forward to the development of new and exciting P-Te chemistry.

References

1. Wood PT, Woollins JD (1986) *Transit Met Chem* 11:358–360
2. Drake GW, Kolis JW (1994) *Coord Chem Rev* 137:131–178
3. Zhao J, Pennington WT, Kolis JW (1992) *J Chem Soc, Chem Commun* 26
4. Haiduc I, Sowerby DB (1995) *Polyhedron* 14:3389
5. Haiduc I, Sowerby DB (1996) *Polyhedron* 15:2469
6. Gray IP, Milton HL, Slawin AMZ, Woollins JD (2003) *Dalton Trans* 3450
7. Gray IP, Slawin AMZ, Woollins JD (2004) *New J Chem* 28:1383
8. Gray IP, Slawin AMZ, Woollins JD (2004) *Dalton Trans* 2477
9. Gray IP, Slawin AMZ, Woollins JD (2004) *Z Anorg Allgm Chem* 630:1851
10. Jones R, Williams DJ, Woollins JD (1987) *Polyhedron* 6:539; Parkin IP, Pilkington MJ, Slawin AMZ, Williams DJ (1990) *Polyhedron* 7:987

11. Kuchen W, Knop B (1964) *Angew Chem Int Ed* 3:507
12. Kuchen W, Knop B (1965) *Angew Chem Int Ed* 4:244
13. Kuchen W, Hertel H (1969) *Angew Chem Int Ed* 8:89
14. Pilkington MJ, Slawin AMZ, Williams DJ, Woollins JD (1991) *Polyhedron* 10:2641
15. Davies RP, Martinelli MG (2002) *Inorg Chem* 41:348
16. Davies RP, Martinelli MG, Wheatley AEH, White AJP, Williams DJ (2003) *Eur J Inorg Chem* 3409
17. Davies RP, Francis CV, Jurd APS, Martinelli MG, White AJP, Williams DJ (2004) *Inorg Chem* 43:4802
18. Gastaldi L, Porta P (1977) *Cryst Struct Commun* 6:175
19. Gelmini L, Stephan DW (1987) *Organometallics* 6:1515
20. Leemput VD, Hummelink TW, Noordik JH, Beurskens PT (1975) *Cryst Struct Commun* 4:167
21. Lobana TS, Wang J-C, Liu CW (2007) *Coord Chem Rev* 251:91
22. Krishnan V, Zingaro RA (1969) *Inorg Chem* 8:2337
23. Bereau V, Ibers JA (2000) *Acta Crystallogr C* C56:584
24. Sarkar B, Fang C-S, You L-Y, Wang J-C, Liu CW (2009) *New J Chem* 33:626
25. Fang C-S, Huang Y-J, Sarkar B, Liu CW (2009) *J Organomet Chem* 694:404
26. Moran M, Cuadrado I (1985) *J Organomet Chem* 295:353
27. Liu CW, Chen J-M, Santra BK, Wen S-Y, Liaw B-J, Wang J-C (2006) *Inorg Chem* 45:8820
28. Chiou L-S, Fang C-S, Sarkar B, Liu L-K, Leong M, Liu CW (2009) *Organometallics* 28:4958
29. Lippolis V, Isaia F, Devillanova FA (eds) (2007) Handbook of chalcogen chemistry: new perspectives in sulfur, selenium, tellurium. RSC, Cambridge
30. Godfrey SM, Jackson SL, McAuliffe CA, Pritchard RG (1997) *J Chem Soc, Dalton Trans* 4499
31. Odagi K, Nakayama H, Ishii K (1990) *Bull Chem Soc Jpn* 63:3277
32. Santra BK, Chen C-L, Sarkar B, Liu CW (2008) *Dalton Trans* 2270
33. Sarkar B, Wen S-Y, Wang J-H, Chiou L-S, Santra BK, Liao P-K, Wang J-C, Liu CW (2009) *Inorg Chem* 48:5129
34. Liu CW, Chang H-W, Sarkar B, Saillard J-Y, Kahlal S, Wu Y-Y (2010) *Inorg Chem* 49:468
35. You H-J, Fang C-S, Lin J-L, Sun S-S, Liu CW (2010) *Inorg Chem* 49:7641. doi:10.1021/ic101390e
36. Chang W-S, Lin Y-F, Sarkar B, Chang Y-M, Liu L-K, Liu CW (2010) *Dalton Trans* 39: 2821–2830
37. Gray IP, Slawin AMZ, Woollins JD (2005) *Dalton Trans* 2188–2194
38. Shi W, Shafaei-Fallah M, Zhang L, Anson CE, Matern E, Rothenberger A (2007) *Chem Eur J* 13:598–603
39. Shi W, Shafaei-Fallah M, Rothenberger A (2007) *Dalton Trans* 4255–4257
40. Shi W, Shafaei-FallahM, Anson CE, Rothenberger A (2006) *Dalton Trans* 2979–2983

Index

J.D. Woollins and R.S. Laitinen (eds.), *Selenium and Tellurium Chemistry*,
DOI 10.1007/978-3-642-20699-3,

MIX
Papier aus verantwortungsvollen Quellen
Paper from responsible sources
FSC® C105338

If you have any concerns about our products,
you can contact us on
ProductSafety@springernature.com

In case Publisher is established outside the EU,
the EU authorized representative is:
Springer Nature Customer Service Center GmbH
Europaplatz 3, 69115 Heidelberg, Germany

Printed by Libri Plureos GmbH
in Hamburg, Germany